TD2 6√2932

WITHDRAWN

THE
OXFORD ENGINEERING SCIENCE SERIES
GENERAL EDITORS

L. C. WOODS, W. H. WITTRICK, A. L. CULLEN.

PLANAR OPTICAL WAVEGUIDES AND FIBRES

BY

H.-G. UNGER

CLARENDON PRESS · OXFORD
1977

Oxford University Press, Walton Street, Oxford OX2 6DP

OXFORD LONDON GLASGOW NEW YORK
TORONTO MELBOURNE WELLINGTON CAPE TOWN
IBADAN NAIROBI DAR ES SALAAM LUSAKA ADDIS ABABA
KUALA LUMPUR SINGAPORE JAKARTA HONG KONG TOKYO
DELHI BOMBAY CALCUTTA MADRAS KARACHI

ISBN 0 19 856133 4

(C) Oxford University Press 1977

All rights reserved. No part of this publication may be reproduced, stored in a retrieval system, or transmitted, in any form or by any means, electronic, mechanical, photocopying, recording, or otherwise, without the prior permission of Oxford University Press

Printed in Great Britain by Thomson Litho Ltd, East Kilbride, Scotland

PREFACE

The advent of the laser and recent advances in opto-electronics and electro-optics have opened up the infrared and visible part of the electromagnetic spectrum for communications and general data processing applications. Planar optical waveguides such as films and strips or strip-derived structures are needed in these applications to form distributed components and to connect components and subsystems. Glass fibres serve as the transmission medium in optical communications. This book analyses light propagation in optical waveguides with a view to these applications and treats their design and fabrication as well as their excitation by couplers, connectors, and transitions.

It starts with plane-wave absorption, scattering, and dispersion in the bulk of optical waveguide materials and their total reflection at interfaces. Film-guide modes are then derived from a ray picture and as field solutions, including graded-index films. Interface scattering and curvature loss are analysed for film modes, as well as their excitation by grating and prism couplers. As planar guides with transverse confinement, film-lens guides, strips, strip-loaded films, and rib and bulge guides are analyzed using a generalized ray concept. Their excitation by directional couplers follows from a general theory of coupled waveguides.

Step- and graded-index fibres are first treated by ray optical methods which are then complemented by more exact field solutions. Thus guided and leaky mode characteristics are obtained for multimode and single-mode operation. The dispersion analysis takes account of mode, material, and profile dispersion and arrives at near optimum profiles. For imperfect fibres, including those with micro-bending, the solution of coupled wave and coupled power equations and a power diffusion approximation allow one to specify fibre tolerances for certain signal degradations. Methods for fibre fabrication and cabling are presented which produce optical fibre cables with low loss and distortion.

PREFACE

The theory of wave propagation and waveguide design is presented in a sufficiently complete and detailed way to suit all practical purposes, but it is not in all respects rigorous. The sections on waveguide technology have more the character of an introduction to the many different principles and methods of waveguide fabrication.

For the inception of this book, I am indebted to Professor A. L. Cullen, University College, London, who suggested that I should write it and advised me in conceiving its style and contents. I also owe a great deal of assistance to my associates at the Institut fuer Hochfrequenztechnik of the Technische Universitaet Braunschweig, in particular to Dr.-Ing. K. Petermann who reviewed most of the theoretical sections and made extensive suggestions. Colleagues of mine to whom I am indebted for reviewing other parts of the manuscript are Dipl.-Ing. K. Behm, Dipl.-Ing. J.H. Hinken, Dr.-Ing. M.H. Kuhn, Dipl.-Ing. U. Ruetze, Dr.-Ing. U. Unrau, and Dr.-Ing. H.D. Friedrichs. Dr.-Ing. H.D. Friedrichs also helped substantially in editing and preparing the manuscript for the publisher. I am also thankful to Miss K. Hartfiel for typing the entire manuscript and to Mrs. B. Titze for preparing the illustrations.

November 1976.   H.-G.U.

CONTENTS

LIST OF PRINCIPAL NOTATION	ix
0. INTRODUCTION AND BRIEF HISTORY	1
References	6
1. UNIFORM PLANE WAVES	8
1.1. Plane-wave propagation and reflection	8
1.2. Plane-wave scattering and absorption	23
1.3. Plane-wave dispersion	34
References	56
2. DIELECTRIC FILMS	58
2.1. Film modes	62
2.2. Guided modes of the symmetrical dielectric slab	83
2.3. Field solution for guided modes	92
2.4. Guided-mode absorption	106
2.5. Guided-mode scattering	114
2.6. Grating couplers	138
2.7. Prism couplers	148
2.8. Curved slabs	157
2.9. Slabs and films with graded index	171
References	187
3. PLANAR GUIDES WITH TRANSVERSE CONFINEMENT	189
3.1. Film lenses and lens guides	189
3.2. Strip guides	197
3.3. Strip-loaded film guides	214
3.4. Rib guides	225
3.5. Bulge guides	238
3.6. Coupled waveguides and directional couplers	252
3.7. Planar waveguide technology	267
References	287
4. CLADDED-CORE FIBRES	290
4.1. Ray picture of core modes	292
4.2. Field solution for fibre modes	309
4.3. Guided modes for unlimited cladding	321
4.4. Weakly guiding fibres	336
4.5. Leaky modes in weakly guiding fibres	369
4.6. Weakly guiding fibre with jacket	388
4.7. Guided-mode attenuation	401
4.8. Single-mode fibres	411
4.9. Multimode fibres	430
References	443
5. GRADED-INDEX FIBRES	444
5.1. Ray analysis	445
5.2. Field solutions for modes in a parabolic index profile	465

5. continued

   5.3. Field solutions for multimode graded-index fibres 478
   5.4. Variational analysis and power-series analysis ... 487
   5.5. Single-mode operation 500
   5.6. Delay differences 507
   5.7. Impulse response 534
   References 550

6. FIBRE IMPERFECTIONS 552

   6.1. Coupled wave equations 553
   6.2. Random imperfections and single-mode loss 563
   6.3. Curvature coupling in fibres 568
   6.4. Microbending loss in single-mode fibres 575
   6.5. Signal distortion in single-mode transmission 586
   6.6. Steady-state power flow in multimode fibres .. 598
   6.7. Power transients in multimode fibres 610
   6.8. Power diffusion approximation 620
   6.9. Microbending in multimode fibres 638
   References 648

7. FIBRE FABRICATION AND CABLING 649

   7.1. Glasses for fibres 649
   7.2. Chemical vapour deposition for quartz-glass fibres 654
   7.3. Double crucible process for compound glass fibres 668
   7.4. Rod-in-tube method 676
   7.5. Polymer-clad fibres 678
   7.6. Single fibres or fibre bundles in cables 681
   7.7. Fibre coating and jacketing 682
   7.8. Fibre packaging 685
   7.9. Cable strength and strain relief 690
   7.10. Cable design 693
   References 696

8. FIBRE JUNCTIONS AND TRANSITIONS 699

   8.1. Analysis of fibre mode excitation 700
   8.2. Fibre connectors and splices 709
   8.3. Optical transitions 719
   8.4. Source to fibre transitions 727
   References 741

INDEX 742

LIST OF PRINCIPAL NOTATION

a	wave amplitude; loss constant; fibre, core, or profile radius
a_c	radius of core glass
$a_{E,H}$	asymmetry parameter
a_i	field factor
a_m	amplitude of forward-travelling wave
$(a)_m$	Pockhammer's symbol
A	mode amplitude
$A_i(z)$	amplitude distribution
$A(\omega)$	spectral component of wave form
$Ai(z)$	Airy function
b	strip width
$b(t)$	output wave form
b_m	amplitude of backward-travelling wave m
B	phase parameter; bandwidth
$\vec{B}$	magnetic induction
$B_d(\omega)$	spectral distribution of output waveform
B_m	constant of dielectric material
B_{rf}	phase parameter (rib guide)
B_{rms}	r.m.s. width of spectral distribution
$B_{y,z}$	autocorrelation distance of random distribution
$B(\theta)$	brightness
$B(\omega)$	spectral component of wave form
$Bi(z)$	Airy function
c	velocity of light in vacuum; ratio of cladding to core radius; nozzle radius
$c, c_{mn}(z)$	coupling coefficient (between modes m and n)
C	contour of cross-section
C_{mn}	coupling coefficient between modes m and n
C_s	characteristic quantity of a radiating source
d	distance; width; thickness (film, slab); loss factor; profile parameter
$d(x)$	coupling function

LIST OF PRINCIPAL NOTATION

d_b	bulge thickness
d_c	correlation distance
d_e	effective width
$d_{E,H}$	distance of apparent reflectors (E-, H-modes)
d_m	average coupling factor
d_{mn}	coupling factor
d_o	optimum value of profile parameter d; coupling factor
d	
d_r	rib thickness
D	directivity; diffusion constant
$\vec{D}$	displacement density
e	base of natural logarithm
$\vec{e}(\vec{r})$	transverse electric field distribution
$\exp(z) = e^z$	exponential function
E	energy
$\vec{E}$	electric field vector
E_d	dispersion energy
E_g	effective energy-gap
$E_{g,s}$	elastic modulus
$E_n(r,\phi)$	transverse electric field distribution of mode n
$E(\kappa)$	complete elliptic integral of the second kind
E_ν	energy of mode ν
f	frequency; focal length
$f(z)$	distribution function
f_c	carrier frequency
f_1	boundary displacement
$F(a,b,c,x)$	Gaussian hypergeometric series
$F(\Delta\beta_n)$	limited Fourier transform
$F_{\ell K}(r)$	orthogonal set of functions
F_x	field component (H_x, E_x)
g	damping factor; gain constant; gap width
$g(f)$	line form; line shape
$g(r/a)$	profile function
g_1	gain by stimulated emission

LIST OF PRINCIPAL NOTATION

h	Planck's constant
$\hbar$	Planck's constant divided by 2π
$\vec{h}(\vec{r})$	transverse magnetic field distribution
H	flexural rigidity
$\vec{H}$	magnetic field vector
$H_m(x)$	Hermite polynomial
$H_m^{(n)}(x)$	Hankel function
I	moment of inertia
$I(z)$	current distribution
$I_m(z)$	modified Bessel function
$\vec{\imath}\,l$	electric dipole moment
j	imaginary unit ($= \sqrt{-1}$)
J	density of injection current
$\vec{J}$	electric current density
$\hat{J}_l(z)$	combination of Bessel functions
$J_m(z)$	Bessel function
J_s	surface distribution of electric current density
k	wave number in free space
k_B	Boltzmann's constant
k_i	wave number in region i ($= n_i k$)
k_n	wave number in medium with n
$\vec{k}_n$	wave vector
$k_{r,\varphi,z}$	circular cylinder component of $\vec{k}$
k_t	transverse wave number
$k_{x,y,z}$	Cartesian component of $\vec{k}$
K	ion-exchange parameter
$K(\kappa)$	complete elliptic integral of the first kind
$\hat{K}_l(z)$	combination of Hankel functions
$K_m(z)$	modified Hankel function
l	integral number; number of periods; circumferential mode order; volume loading factor
l_c	coupling length; ion exchange length
$ln(z)$	natural logarithm ($= \log_e z$)
l_v	period length of fibre displacement

LIST OF PRINCIPAL NOTATION

L	distance; fibre length
L_c	coupling length
L_k	correlation distance
L_p	period length
$L_q^{(l)}$	generalized Laguerre polynomial
LP_{lp}	linearly polarized mode
m	integral number; mode order; compound mode number; electron rest mass
m^*	electron effective mass
M	total number of guided modes
$M(a,b,x)$	Kummer's confluent hypergeometric function
M_{cl}	total number of modes up to cladding loss α
M_j	molecular weight
M_l	total number of leaky modes
$M_n(z)$	time moments of impulse response
M_t	total number of modes
M_α	total number of modes up to leakage loss α
n	refractive index, $\mathrm{Re}(\hat{n})$
$\hat{n}$	complex refractive index
n''	absorption coefficient, $\mathrm{Im}(\hat{n})$
n'	group index
N	effective index; electron concentration; step number
N_1	oscillator concentration
N_A	Avogadro's number
N.A.	numerical aperture
O	opto-electronic coefficient
p	integral number; number of modes; radial mode-order; launching efficiency; photoelastic coefficient
p_o	coupling efficiency
p_{oo}	launching efficiency
P	power
$P(t,z)$	impulse response
$P(\xi)$	polynomial in ξ
$\overline{P_m}$	average power in modes of group m
P_n	power of mode n

LIST OF PRINCIPAL NOTATION

q	elementary charge
$q_n(t)$	nth component of pulse tail
r	radial coordinate
$\vec{r}$	space vector
r_e	effective mode radius
$r_{e,m}$	reflection coefficient
r_t	turning point
R	radius of curvature; retention length
s	beam width; distance
S	radiation intensity; power flow density; cross-section
$\mathcal{S}$	Poynting vector
$S(\psi)$	stationary form
t	time; transverse phase parameter; strip-layer thickness; coating thickness
t_0, t_1	pulse width
$t_d(z)$	average time delay
$t_{e,m}$	transmission coefficient
t_k	coherence time
T	absolute temperature; power transmission coefficient; pulse width
T_F	Fresnel coefficient
TE	<u>t</u>ransverse <u>e</u>lectric (mode)
TM	<u>t</u>ransverse <u>m</u>agnetic (mode)
u	transverse or radial phase parameter
$\vec{u}$	unit vector
u_c	u at cut-off
u_{lp}	pth zero of J_l
u_s	transverse phase parameter (strip)
$\underline{U}_j$	statistical mode vector
$U_j(x)$	statistical mode function
U_m	combination of cylinder functions
v	phase velocity; transverse or radial attenuation parameter; pulling speed
$\vec{v}$	velocity of electron

LIST OF PRINCIPAL NOTATION

v_c	speed of glass flow
v_g	group velocity
v_s	transverse attenuation parameter (strip)
V	generalized frequency parameter; film or fibre parameter; volume
V_b	bulge parameter
V_c	V at cut-off
V_{cM}	V_c for Mth mode
V_n	amplitude factor of mode n
V_{rf}	rib parameter
V_s	waveguide parameter
V_{st}	strip parameter
w	transverse or radial attenuation parameter; width of Gaussian beam; spot size; mode radius
W	normalized scattering loss
W_i	statistical mode amplitude
W_m	combination of cylinder functions
x	Cartesian coordinate
x_j	mole fraction
x_t	turning point
X_m	combination of cylinder functions
y	Cartesian coordinate; profile dispersion parameter
y_t	turning point
Y	wave admittance
Y_m	combination of cylinder functions
z	Cartesian coordinate; cylinder coordinate
$z_{E,H}$	beam shift
Z	wave impedance
$Z_m(z)$	combination of cylinder functions
α	attenuation constant; exponent of power-law profile
$\vec{\alpha}$	attenuation vector
α_{opt}	optimum value of α

LIST OF PRINCIPAL NOTATION

β	phase constant
$\vec{\beta}$	phase vector
β_p	phase constant of quasi-mode
β_T	isothermal compressibility
γ, γ_1	spatial frequency
γ_m	propagation constant of mode m
Γ_n	attenuation constant of statistical mode
δ	Dirac's delta function; relative difference of permittivities; relative gap width; relative bulge thickness
δ_{ki}	Kronecker delta (= 0 if $k \neq i$; = 1 if $k = i$)
Δ	increment; relative index difference
Δf	band with (1/e)
Δn	refractive-index difference
Δn_{max}	maximum index deviation
Δt	delay difference
Δx	offset between axes
Δz	spacing between fibre ends; differential length
$\Delta \alpha$	tilt angle between axes
$\Delta \beta$	difference of phase constants
$\Delta \tau$	delay difference per length
ε	permittivity
ε''	dielectric loss factor
ε_o	permittivity in vacuum
ε_l	Neumann number
$\varepsilon_q, \varepsilon'_q$	Fourier coefficients
ε_r	relative permittivity
ζ	angular spatial frequency
η	intrinsic wave impedance; quantum efficiency; launching efficiency
η_l	combination of cylinder functions

LIST OF PRINCIPAL NOTATION

ϑ	polar angle
ϑ_c	limiting angle
θ	angle of incidence; ray angle; propagation angle
θ_c	limiting angle
κ	modulus of elliptic integral
$\kappa(z)$	curvature
$\kappa_l(v)$	combination of cylinder functions
κ_{mn}	coupling coefficient between modes m and n
λ	wavelength
Λ, Λ_t	beat wavelength
μ	integral number; permeability
μ_0	permeability of vacuum
ν	integral number, mode index
$\nu(\beta)$	number of modes with phase constants smaller than β
ξ	angular spatial frequency; strain relief factor
$\xi_{1,2,3}$	roots of cubic polynomial
ρ	density of material; radius of curvature, radial variable
$\rho_{1,2}$	turning point
$\sigma, \sigma(z)$	r.m.s. pulse-width
σ_B	tensile strength
σ_{IM}	r.m.s. pulse-width from intermodal pulse broadening
σ_M	r.m.s. pulse-width from intramodal pulse broadening
σ_0	average tension
τ	group delay per unit fibre length
τ_g	group delay
φ	phase angle; cylinder coordinate
$\varphi(x)$	autocorrelation function

LIST OF PRINCIPAL NOTATION   xvii

ϕ	phase angle; phase shift
$\phi(\omega)$	power spectrum
χ	electric susceptibility; eigenvalue
ψ	phase retardation
$\psi(r)$	wave function
ω	angular frequency
ω_d	dispersion frequency
ω_m	resonance frequency
ω_p	plasma frequency
Ω	angular frequency
Ω_p	angular spatial frequency
$\underline{\underline{A}}$	square matrix
$\mathrm{Im}(z)$	imaginary part of z
$\mathrm{Re}(z)$	real part of z
z^*	conjugate complex value of z
$\langle \ldots \rangle$	average value
∇	nabla operator
∇^2	Laplacian
∇_t	transverse gradient
∇_t^2	transverse Laplacian

INTRODUCTION AND BRIEF HISTORY

Optical fibres confine and guide light by total reflection. For this purpose they consist of a centre region of transparent and optically dense material surrounded by an optically less dense material. Planar optical waveguides in the form of films and strips work usually on the same principle; they also consist of a transparent material that is deposited on, or surrounded by, a substrate with a lower refractive index. Light rays that are incident on the front face of such a structure at a small enough angle within the limits of total reflection of its sidewalls are trapped by the film, strip, or fibre and guided by it. This explanation of light guidance using geometric optics takes no account of the character of light as an electromagnetic wave; it was recognized even before the wave character of light became generally accepted. If the wave character is taken into account, it is found that of the continuum of light rays within the limit of total reflection of a particular structure, only a limited number with discrete angles are allowed as guided waves of the structure. In addition to total reflection these allowed rays satisfy the condition that after successive reflections at the sidewalls, they repeat in phase and thus interfere constructively when they superimpose upon themselves. Only when a ray obeys this phase condition will it fit into the structure and form a self-consistent field distribution. Otherwise it interferes destructively with itself and just dies out. The quantization of waves into a discrete set of guided modes was only discovered when Maxwell's equations were solved for particular guiding structures. For a round dielectric cylinder such solutions were first obtained by Hondros and Debye (1910). These authors were, however, more interested in the propagation of radio-frequency waves along dielectric wires, as were Zahn (1916) and Schriever (1920), who first investigated such wave propagation experimentally. Most of the theoretical and experimental work that followed well into the 1950s was likewise concerned with radio and microwave propagation along dielectric waveguides; it led to their application as transmission lines and in circuit components and antennas for these waves. During this time the quasi-optic character of dielectric waveguides received little attention except for one detailed theoretical study by Buchholz (1938).

The first dielectric waveguide in which discrete modes of propagation were studied at optical frequencies was the glass-cladded glass fibre developed originally for fibre optics imaging applications (Kapany 1967). The work on optical waveguides received much impetus from the advent of the laser in the early 1960s and the possibility of using coherent light in communications and other applications where radio- and microwaves had hitherto served so well. The first objective was to find a low-loss wide-band transmission medium for long-distance communication. The glass fibre guide as such a transmission line was first proposed by Kao and Hockham (1966) and also by Börner (1966). At first it seemed quite daring to make such a proposal in the light of the high absorption which even good optical glasses exhibited at the time, but Kapron, Keck, and Maurer (1970) succeeded in reducing the transmission loss of quartz-glass fibres below 20 dB km^{-1} initially and thereby opened up bright prospects for fibre guides in optical communications. Since glass fibres were first proposed as a transmission medium for optical signals, the literature on them has virtually exploded and cannot be surveyed in this brief history of optical waveguides. Review papers by Miller, Marcatili, and Li (1973) and by Clarricoats (1976) offer such a survey. Monographs by Marcuse (1972, 1974), by Kapany and Burke (1972), and by Arnaud (1976) together give a very complete account of fibre guide theory. Today, the wealth of literature as well as these monographs seem to provide a firm enough theoretical foundation to answer all questions that arise in communications applications of fibres. Much work has also been done on the fabrication of fibres and the development of fibre cables. Altogether the field of optical fibres and fibre cables has matured to a stage that justifies its exposition in a comprehensive form.

The invention of the laser and the development of coherent optics have created not only the need for a long-distance transmission medium but also for guiding structures with which to build optical components and connect them into optical circuits. Preferably these optical waveguides should allow the planar fabrication of components and their integration into planar optical circuits, a configuration which has proven so efficient and shown so many advantages in electronic circuits at all frequencies up into the microwave range. Planar fabrication and integration require planar optical waveguides in the form of films and strips. Together with round dielectric waveguides, wave propagation in the dielectric slab guide of unlimited transverse extent was well understood and it was likewise used in microwave engineering. Its transversely limited form, the rectangular dielectric waveguide, was first studied theoretically by Schlosser (1964) but the initial experiments were only of a model character and worked again with microwaves (Schlosser and Unger 1966). However, soon afterwards, film waveguides as well as other planar components and circuitry for applications in the infrared range were constructed by Anderson (1965). These and other early efforts in the development of planar optical waveguides and circuits received additional support and stimulus when Miller (1969) surveyed the advantages and possibilities of this technology and coined the term 'integrated optics' for it. Since then not only the theory of wave propagation, excitation, and coupling in optical films and strips has been advanced to encompass all aspects in the analysis and design of these guiding structures, but also precise and reliable technologies have been developed with which to fabricate planar optical waveguides and integrated optical circuits. Tamir (1975) reviews the progress in integrated optics in the introduction to a comprehensive book on this subject, to which a number of prominent specialists have contributed. Important earlier publications on integrated optics have been selected and edited in a reprint by Marcuse (1973).

In this book we attempt a pragmatic exposition of the theory of wave propagation in planar optical waveguides and fibres with a view to their applications in integrated optics and optical communications. This exposition of waveguide theory is complemented by fairly comprehensive yet relatively brief accounts of the design and fabrication of optical waveguides and the technologies which have been developed and applied for them. While a sufficiently complete and detailed theory of optical waveguides is presented to suit most practical purposes, the chapters and sections on technology can serve only as an introduction. They acquaint the reader with the principles and methods of waveguide fabrication but for actual work in this field refer him to the extensive and rapidly expanding literature. The technologies of planar optical waveguides and integrated circuits on the one hand and fibre fabrication and cabling on the other are founded on such a variety of different physical principles and use so many diverse methods that to treat them completely and in detail would go beyond the scope of a book on optical waveguides.

In this book the analysis of wave propagation in the different optical waveguides always starts with the simple geometric optics concept of rays and their total reflection at the guide boundaries. When, instead of a sidewall with an abruptly changing refractive index, the particular guide has a gradually decreasing index, then the laws of geometric optics that govern ray deflection in such a graded-index medium are used to trace rays. In the geometric optics approximation a light ray represents the uniform plane-wave solution of Maxwell's equations. As such it is applied to develop the ray picture of wave propagation in films and other optical waveguides. The ray concept is, however, also generalized to represent, in particular, film waves and modes of propagation in other multilayer structures. With this generalization a ray picture of wave propagation is then similarly developed for strip guides and such planar waveguides as derive from the basic strip structure.

When the phase condition for selfconsistent field distributions is introduced in the ray picture for any particular waveguide, it is quantized into a finite number of discrete modes of propagation. The ray picture, together with its quantization into modes, proves adequate to describe light propagation in optical waveguides that guide a large number of modes. As long as the waveguide is uniform and of perfect geometry, all propagation characteristics of practical relevance in integrated optics and optical communications may be obtained from it. A more rigorous solution of Maxwell's equations than the ray approximation provides is necessary when low-order modes are to be utilized in structures that guide only these few modes or that are even single-mode guides. Field solutions of Maxwell's equations are also desirable or even necessary when the field distribution of individual modes is required; for example, to analyse their excitation or their coupling to other modes in waveguide couplers or at waveguide imperfections.

For a particular waveguide or class of guiding structures we will therefore first develop the ray picture, complete with its quantization into guided modes. Subsequently we derive solutions of Maxwell's equations, but only to the degree of accuracy that is really needed for the analysis and the design of waveguides in integrated optics and optical communications. These field solutions enable us to analyse the excitation and coupling of modes and the effects of waveguide imperfections. The final sections and chapters on planar waveguides and fibres are devoted to their technologies and their respective incorporation into integrated circuits and fibre cables.

The ray picture of films as well as that of fibres starts with plane uniform waves, which are represented by rays. Many propagation characteristics of plane uniform waves in a given medium transfer directly and with little modification to the modes of propagation of the waveguide. We therefore devote the first chapter to uniform plane-wave propagation in the bulk of waveguide materials, their absorption, scattering, and dispersion, as well as the reflection of uniform plane-wave beams at plane interfaces between different waveguide materials.

REFERENCES

Anderson, D.B. (1965). Application of semiconductor technology to coherent optical transducers and spatial filters. In *Optical and Electro-Optical Information Processing* (ed. J. Tippett). MIT Press, Cambridge, Mass., 221-34.

Arnaud, J.A. (1976). *Beam and fibre optics*. Academic Press, New York.

Börner, M. (1966). Mehrstufiges Übertragungssystem für in Pulscodemodulation dargestellte Nachrichten. DBP 1 254 513, vol.21.

Buchholz, H. (1938). Die Quasioptik der Ultrakurzwellenweiter. *Elektrische Nachrichtentechnik* 15, 297-320.

Clarricoats, P.J.B. (1976). Theory of optical fibre waveguides - a review. *Progress in Optics* (ed. E. Wolf), vol.14. North-Holland, Amsterdam.

Hondros, D. and Debye, P. (1910). Elektromagnetische Wellen an dielektrischen Drähten. *Ann. Phys.* 32, 465-76.

Kao, K.C. and Hockham, G.A. (1966). Dielectric fibre surface waveguides for optical frequencies. *Proc. IEE* 113, 1151-58.

Kapany, N.S. (1967). *Fiber optics principles and applications*, Academic Press, New York.

——— and Burke, J.J. (1972). *Optical waveguides*, Academic Press, New York.

Kapron, F.P., Keck, D.B., and Maurer, R.D. (1970). Radiation losses in glass optical waveguides. *Appl. Phys. Lett.* 17, 423-5.

Marcuse, D. (1972). *Light transmission optics*, Van Nostrand Reinhold, Princeton, N.J.

——— (1973). *Integrated optics*. IEEE Press, New York.

——— (1974). *Theory of dielectric optical waveguides*, Academic Press, New York.

Miller, S.E. (1969). Integrated Optics: An introduction. *Bell Syst. tech. J.* 48, 2059-69.

———, Marcatili, E.A.J., and Li, T. (1973). Research toward optical fiber transmission systems. *Proc. IEEE* 61, 1703-51.

Schlosser, W. (1964). Der rechteckige dielektrische Draht. *Arch.elekt.Übertr.* 18, 403-410.

Schlosser, W. and Unger, H.G. (1966). Partially filled waveguides and surface waveguides of rectangular cross-section. In *Advances in Microwaves* 1, Academic Press, New York, 319-92.

Schriever, O. (1920). Elektromagnetische Wellen an dielektrischen Drähten. *Ann. Phys.* 63, 645-73.

Tamir, T. (1975). *Integrated optics*, Springer Verlag, Berlin.

Zahn, H. (1916). Über den Nachweis elektromagnetischer Wellen an dielektrischen Drähten. *Ann. Phys.* 49, 907-33.

1
UNIFORM PLANE WAVES

Light, radiated from any source and propagating through free space has locally the form of a uniform plane wave if the source and any scattering objects or boundaries are far enough removed. When penetrating a material of sufficient extent and uniformity, light also propagates as a uniform plane wave in the bulk of the material. Such uniform plane waves constitute the simplest wave solution of Maxwell's equations, and as such, serve well to examine wave propagation in free space and in the bulk of materials, and also to characterize these media with respect to wave propagation.

Furthermore, any source-free solution of Maxwell's equations in uniform regions may be represented as a superposition of uniform plane waves. Light waves propagating in optical guiding structures or radiating from these structures are such source-free solutions. For homogeneous regions inside or around these guiding structures, these light waves may be represented as a superposition of uniform plane waves. Uniform plane waves therefore form a basic tool in studying optical guiding structures. Many of the propagating characteristics of uniform plane waves in free space or in the bulk of materials hold similarly for propagation in optical guiding structures.

1.1. PLANE-WAVE PROPAGATION AND REFLECTION

Light waves which are guided by dielectric slabs or films consist of plane electromagnetic waves propagating inside these layers and confined to them by partial or total internal reflection at the boundaries of the layer. Plane-wave propagation in homogeneous dielectric media and reflection at plane boundaries between media of different optical characteristics form the basis for the analysis of wave guidance in dielectric films.

Film guides as well as most other optical waveguides usually consist of materials with characteristics nearly independent of light intensity. These materials are also non-magnetic, in particular at optical frequencies. Furthermore, and except for special applications, they behave isotropically or their anisotropic character is of no significance in optical wave guidance. In such materials the constituent equations

$$\vec{D} = n^2 \varepsilon_0 \vec{E} \qquad \vec{B} = \mu_0 \vec{H} \qquad (1.1)$$

relate the electric field vector $\vec{E}$ to the displacement density $\vec{D}$, and the magnetic field vector $\vec{H}$ to the magnetic induction $\vec{B}$. ε_0 and μ_0 denote the dielectric permittivity and magnetic permeability, respectively, in vacuum, and n the refractive index of the medium, with $\varepsilon_r = n^2$ as its relative permittivity.

With linear constituent equations such as (1.1) the superposition principle holds, and Maxwell's equations for source-free fields,

$$\nabla \times \vec{E} = -\partial \vec{B}/\partial t \qquad \nabla \times \vec{H} = \partial \vec{D}/\partial t, \qquad (1.2)$$

need to be solved only for time-harmonic electromagnetic fields, depending on time t as $\sin(\omega t + \varphi)$ with the angular frequency $\omega = 2\pi f$ and φ an arbitrary phase-angle independent of time. Solutions for any other time dependence may then be obtained by superimposing time-harmonic solutions of different frequencies.

For a time-harmonic field, Maxwell's equations reduce to

$$\nabla \times \vec{E} = -j\omega\mu_0 \vec{H} \qquad \nabla \times \vec{H} = j\omega\varepsilon_0 n^2 \vec{E}, \qquad (1.3)$$

where $\vec{E}$ and $\vec{H}$ now denote phasors of the electromagnetic field vectors. Any absorption in the medium may here be accounted for by a complex refractive index

$$\hat{n} = n - jn'' \qquad (1.4)$$

where n is the actual index of refraction and n'' the absorption coefficient of the material. Dielectric materials for optical waveguides are nearly transparent and hence have low absorption so that $n'' \ll n$.

To solve Maxwell's equations for Cartesian components of the field vectors we eliminate one or the other field vector by taking the curl of one equation and then substitute from the other equation. With the divergence of both field vectors being zero for any homogeneous region, the double curl operation reduces to the Laplacian operator ∇^2, and we obtain solutions either from the wave equation for $\vec{E}$

$$(\nabla^2 + k_n^2)\vec{E} = 0 \quad \text{with} \quad \vec{H} = (j/(\omega\mu_0))\nabla\times\vec{E} \qquad (1.5)$$

or from the wave equation for $\vec{H}$

$$(\nabla^2 + k_n^2)\vec{H} = 0 \quad \text{with} \quad \vec{E} = -(j/(\omega\varepsilon_0 n^2))\nabla\times\vec{H}. \qquad (1.6)$$

In both wave equations $k_n = \omega n(\mu_0\varepsilon_0)^{\frac{1}{2}}$ represents the wave number in the dielectric medium. To distinguish the wave number of a medium with refractive index n from the free space wave number

$$k = \omega(\mu_0\varepsilon_0)^{\frac{1}{2}},$$

it carries the subscript n and is related to k by

$$k_n = nk.$$

Equations (1.5) have the plane-wave solution

$$\vec{E} = \vec{E}_0 \exp(-j\vec{k}_n\cdot\vec{r}) \qquad \vec{H} = (1/(\omega\mu_0))(\vec{k}_n\times\vec{E}) \qquad (1.7)$$

while the plane-wave solution of equations (1.6) appears as

$$\vec{H} = \vec{H}_0 \exp(-j\vec{k}_n\cdot\vec{r}) \qquad \vec{E} = -(1/(\omega\varepsilon_0 n^2))(\vec{k}_n\times\vec{H}). \qquad (1.8)$$

$\vec{r}$ is the space vector to the field point and $\vec{k}_n$ the wave vector, which for the solution of the wave equations in (1.5) and (1.6) has a scalar product

UNIFORM PLANE WAVES

$$\vec{k}_n \cdot \vec{k}_n = k_n^2 \qquad (1.9)$$

equal to the wave number squared.

The wave vector is in general complex:

$$\vec{k}_n = \vec{\beta} - j\vec{\alpha} \qquad (1.10)$$

with $\vec{\beta}$ the phase vector and $\vec{\alpha}$ the attenuation vector. The plane wave propagates in the direction of $\vec{\beta}$ with a phase velocity $v = \omega/|\vec{\beta}|$ and plane phase fronts to which $\vec{\beta}$ is normal. In the direction of $\vec{\alpha}$ the plane wave decays exponentially according to the attenuation constant $\alpha = |\vec{\alpha}|$.

Any region without absorption has a real refractive index $\hat{n} = n$ and hence also a real wave number $k_n = k_n' = \beta$. The separation condition (1.9) then requires the scalar product $\vec{k}_n \cdot \vec{k}_n$ likewise to be real or

$$\vec{\alpha} \cdot \vec{\beta} = 0. \qquad (1.11)$$

To fill this requirement we must have either $\alpha = 0$ or $\vec{\alpha}$ must be perpendicular to $\vec{\beta}$. The case $\alpha = 0$ represents uniform plane waves propagating in the direction of $\vec{\beta}$ with no change in amplitude either in the direction of propagation or along the phase fronts normal to $\vec{\beta}$. The electric field vector of a uniform plane wave is at right-angles to the wave vector $\vec{k}_n = \vec{\beta}$ and the magnetic field vector. The Poynting vector $\vec{S} = \vec{E} \times \vec{H}^*$, with $\vec{H}^*$ as the complex conjugate of $\vec{H}$, is in the direction of $\vec{k}_n$ and is real, just as $\vec{k}_n$ is real. Directions of the field vectors as well as the Poynting vector and wave vector of a uniform plane wave are illustrated in Fig.1.1.

The other case of $\vec{\alpha}$ perpendicular to $\vec{\beta}$ represents a non-uniform plane wave which is transversely evanescent. It propagates in the direction of $\vec{\beta}$ but decays exponentially according to α in a direction transverse to its direction of propagation. Fig.1.2 shows the wave vector components and phase fronts as well as lines of constant amplitude of such a transversely evanescent wave.

A small amount of absorption in a region with absorp-

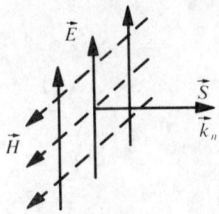

Fig.1.1. Fields, wave vector, and Poynting vector of a uniform plane wave.

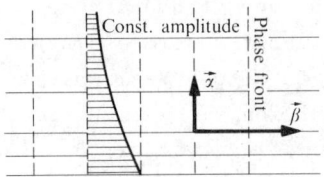

Fig.1.2. Wave vector, phase fronts, and planes of constant amplitude of a transversely evanescent plane wave.

tion coefficient n'' adds a small imaginary component $k_n'' = n''k$ to the wave number k_n, rendering it complex in the form

$$k_n = k_n' - jk_n''. \qquad (1.12)$$

The plane waves in Figs. 1.1 and 1.2 change little due to this small absorption. The uniform plane wave of Fig.1.1, in particular, maintains its uniform field distribution in transverse planes, but decays exponentially in its direction of propagation with an attenuation vector $\vec{\alpha}$ parallel to $\vec{\beta}$ and of magnitude

$$\alpha = k_n''.$$

The electric and magnetic field vectors of plane waves are at right-angles to each other. In the case of uniform plane waves they both lie in planes transverse to the direction of propagation. In the case of transverse evanescence the fields also have components in the direction of propagation.

Starting with an electric field from equation (1.7) which is perpendicular to $\vec{k}_n$, the magnetic field of a uniform plane wave will be perpendicular to $\vec{E}$ and $\vec{k}_n$, but for a transversely evanescent wave, will lie in the plane of Fig. 1.2 with a component also in the direction of $\vec{\beta}$. We therefore call such a wave a TE-wave because of its transverse electric character, or an H-wave because of its only field component in the propagation direction. Starting with a magnetic field from (1.8) perpendicular to $\vec{k}_n$ the electric field in the case of transverse evanescence has a component in the propagation direction. This case represents a TM- or E-wave. The classification of TE- (or H-) waves and TM- (or E-) waves proves useful not only for plane-wave propagation and reflection, as will become apparent presently, but also for waves in films and in guiding structures derived from films.

If a uniform plane wave travels towards a plane boundary between media of different refractive indices, part of it continues to propagate into the other medium and part of it reflects into the medium of incidence. Decomposing the incident wave into a polarization with the electric field parallel to the boundary and a polarization with the magnetic field parallel to it facilitates the analysis.

Fig.1.3 shows the wave component with its electric

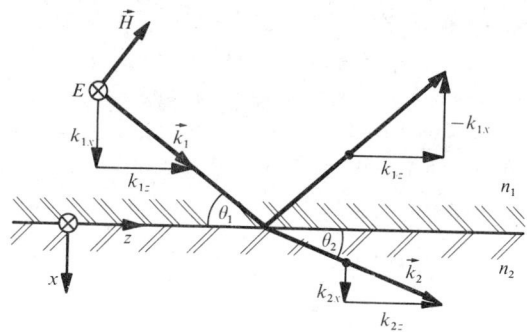

Fig.1.3. Wave vectors of incident reflected and refracted waves at a plane boundary between media with refractive indices $n_1 > n_2$.

field polarized parallel to the boundary directed into the plane of the figure. We choose Cartesian coordinates (x,y,z) with $x = 0$ as the plane boundary between region 1 with refractive index n_1 and region 2 with refractive index n_2; the plane of incidence is $y = 0$. We designate the wave number in the region with n_1 by

$$k_1 = n_1 k,$$

in the region with n_2 by

$$k_2 = n_2 k,$$

and also use the same designation for any wave vectors in the respective regions. The wave vector $\vec{k}_1$ of the incident wave has the components

$$k_{1x} = k_1 \sin\theta_1 \qquad k_{1y} = 0 \qquad k_{1z} = k_1 \cos\theta_1 \qquad (1.13)$$

with

$$\vec{k}_1 \cdot \vec{k}_1 \equiv k_{1x}^2 + k_{1z}^2 = k_1^2 \equiv \omega^2 n_1^2 \varepsilon_0 \mu_0, \qquad (1.14)$$

in accordance with the separation condition (1.9).

Reflected and transmitted waves follow from the boundary condition that all tangential field components are continuous across the boundary at $x = 0$. With the incident field independent of the y-coordinate, reflected and transmitted waves should likewise not change in the y-direction and have fields with the same z-dependence as the incident field. The reflected as well as the transmitted waves should therefore have the same z-component,

$$k_{2z} = k_{1z}, \qquad (1.15)$$

of their wave vector as the incident wave. The reflected wave moves in the same medium as the incident field; its wave vector obeys equation (1.14), leading to

$$k_{1x} = -k_1 \sin\theta_1, \quad k_{1y} = 0, \quad k_{1z} = k_1 \cos\theta_1 \qquad (1.16)$$

for the reflected wave. The reflected wave leaves the boundary with the same angle θ_1 as the incident wave arrives. The wave vector of the transmitted wave in region 2 must obey the separation condition

$$\vec{k}_2 \cdot \vec{k}_2 \equiv k_{2x}^2 + k_{2z}^2 = k_2^2 \equiv \omega^2 n_2^2 \varepsilon_0 \mu_0, \qquad (1.17)$$

and has the Cartesian components

$$k_{2x} = k_2 \sin\theta_2, \quad k_{2y} = 0, \quad k_{2z} = k_2 \cos\theta_2. \qquad (1.18)$$

From the condition (1.15) for identical z-dependence of all waves, we get

$$n_1 \cos\theta_1 = n_2 \cos\theta_2 \qquad (1.19)$$

which constitutes Snell's law of refraction. The angle of refraction, θ_2, stays larger than the angle of incidence as long as $n_2 > n_1$, otherwise θ_2 is smaller than θ_1 and diminishes to zero for

$$\cos\theta_{1c} = n_2/n_1. \qquad (1.20)$$

For an incident wave with $\theta_1 < \theta_{1c}$ in the case $n_2 < n_1$, total reflection occurs. Its consequences will be discussed presently.

Once the same z-dependence has been established by satisfying equations (1.15) and (1.19), the amplitudes of transmitted and refracted fields follow from matching tangential field components at the boundary. For the electric field polarized parallel to the boundary, matching the incident field components E_{yi} and $H_{zi} = (k_1/\omega\mu_0)E_{yi} \sin\theta_1$ and the reflected field components E_{yr} and $H_{zr} = -(k_1/\omega\mu_0)E_{yr} \sin\theta_1$ on one side of the boundary, to the refracted field components E_{y2} and $H_{z2} = (k_2/\omega\mu_0)E_{y2} \sin\theta_2$ on the other side of the boundary, leads to

$$E_{yi} + E_{yr} = E_{y2}$$

$$H_{zi} + H_{zr} = H_{z2}.$$

(1.21)

The reflection coefficient $r_e = E_{yr}/E_{yi}$ relates the reflected field amplitude to the incident field amplitude and the transmission coefficient $t_e = E_{y2}/E_{yi}$ the refracted field amplitude to the incident field amplitude. Both coefficients follow from (1.21) as

$$r_e = \frac{Z_2 - Z_1}{Z_2 + Z_1} \qquad t_e = \frac{2Z_2}{Z_2 + Z_1}$$

(1.22)

where $Z_1 = E_{yi}/H_{zi}$ and $Z_2 = E_{y2}/H_{z2}$ represent wave impedances normal to the boundary for the incident wave in region 1 and the refracted wave in region 2, respectively.

With the present polarization of $\vec{E}$ parallel to the boundary, these wave impedances are

$$Z_1 = \omega\mu_0/(k_1 \sin\theta_1) = \omega\mu_0/k_{1x},$$

$$Z_2 = \omega\mu_0/(k_2 \sin\theta_2) = \omega\mu_0/k_{2x}.$$

The reflection and transmission coefficients then follow as

$$r_e = \frac{k_{1x} - k_{2x}}{k_{1x} + k_{2x}}, \qquad t_e = \frac{2k_{1x}}{k_{1x} + k_{2x}}.$$

(1.23)

They depend only on those components of the wave vectors in both regions which are normal to the boundary.

For the incident wave polarized with its magnetic field parallel to the boundary its field components H_{yi} and $E_{zi} = -(k_1/(\omega\varepsilon_0 n_1^2))H_{yi}\sin\theta_1$, superimposed on H_{yr} and $E_{zr} = (k_1/(\omega\varepsilon_0 n_1^2))H_{yr}$ of the reflected wave, must be matched to H_{y2} and $E_{z2} = -(k_2/(\omega\varepsilon_0 n_2^2))H_{y2}\sin\theta_2$ of the refracted wave. From the corresponding equations

$$H_{yi} + H_{yr} = H_{y2}$$

$$E_{zi} + E_{zr} = E_{z2}$$

follow the reflection coefficient $r_m = H_{yr}/H_{yi}$ and transmission coefficient $t_m = H_{y2}/H_{yi}$ which relate the magnetic field amplitude $H = H_y$ of the reflected and refracted waves to the incident field amplitude. In contrast to equation (1.21) and to r_e and t_e, the field components H_y and E_z appear here instead of E_y and H_z. With the wave admittances normal to the boundary, $Y_1 = -H_{yi}/E_{zi}$ and $Y_2 = -H_{y2}/E_{z2}$, these reflection and transmission coefficients have the form

$$r_m = \frac{Y_2 - Y_1}{Y_1 + Y_2} \qquad t_m = \frac{2Y_2}{Y_1 + Y_2}$$

which constitute the dual to r_e and t_e in equation (1.22). Evaluating these wave admittances for incident and refracted field components results in

$$Y_1 = \omega\varepsilon_0 n_1^2/k_{1x} \qquad Y_2 = \omega\varepsilon_0 n_2^2/k_{2x}$$

and

$$r_m = \frac{n_2^2 k_{1x} - n_1^2 k_{2x}}{n_2^2 k_{1x} + n_1^2 k_{2x}} \qquad t_m = \frac{2n_2^2 k_{1x}}{n_2^2 k_{1x} + n_1^2 k_{2x}}, \qquad (1.24)$$

Therefore, with the magnetic field parallel to the boundary, reflection and transmission coefficients not only depend on the wave vector components normal to the boundary, but also on the index of refraction in both regions.

For loss-less media on both sides of the boundary, and as long as real positive values of the angle θ_2 follow from equation (1.19), all quantities in (1.23) and (1.24) likewise stay real and positive. Under these conditions reflection and transmission coefficients turn out to be real and of magnitude smaller than unity. This is the usual case of refraction and partial reflection, for which the power of the incident wave is split into reflected and refracted waves.

If, however, for $n_2 < n_1$ the complement to the angle of incidence decreases to

$$\theta_{1c} = \text{arc } \cos(n_2/n_1),$$

the refracted wave has $\theta_2 = 0$, and hence also $k_{2x} = 0$. In this limiting case the reflection coefficients increase to $r_e = r_m = 1$, indicating total reflection of the incident wave. Below this limiting angle, for $\theta_1 < \theta_{1c}$ we have $k_z = k_1 \cos \theta_1 > k_2$ and the separation condition (1.17) for the region with refractive index n_2 requires $k_{2x}^2 = k_2^2 - k_z^2 < 0$. Starting from $k_{2x} = 0$ for $\theta_1 = \theta_{1c}$, the x-component of the wave vector k_2 becomes imaginary for all $\theta_1 < \theta_{1c}$. With k_{1x} real and k_{2x} negative imaginary the reflection coefficients remain of unit amplitude as for $\theta_1 = \theta_{1c}$, but turn complex with phase angles

$$\phi_e = 2 \text{arc } \tan(jk_{2x}/k_{1x}) \tag{1.25}$$

$$\phi_m = 2 \text{arc } \tan[jn_1^2 k_{2x}/(n_2^2 k_{1x})]. \tag{1.26}$$

This total reflection occurs for all angles of incidence in the range $0 < \theta_1 < \theta_{1c}$. The reflected wave has the same amplitude as the incident wave but shifts in phase with respect to the incident wave according to ϕ_e and ϕ_m from equations (1.25) and (1.26). The superposition of incident and reflected waves of equal amplitudes in region 1 forms a standing wave in the x-direction with the transverse wave number k_{1x}. In region 2 the wave vector assumes the real z-component $k_{2x} = \beta_z = k_1 \cos \theta_1$ but an imaginary x-component $k_{2x} = -j\alpha_x$. Associated with total reflection, a non-uniform plane wave travels in the z-direction with plane phase fronts transverse to this direction, and its fields decay exponentially in the transverse direction away from the boundary with the attenuation constant α_x. This transversely evanescent wave has the real component of its Poynting vector in the z-direction only; therefore, power propagates only in the z-direction and no longer in the x-direction as well. Fig.1.4 shows the field distribution and phase fronts on both sides of the boundary in the case of total reflection.

For θ_1 just below the limiting angle θ_{1c} of total reflection, α_x remains small and the field in region 2 decays

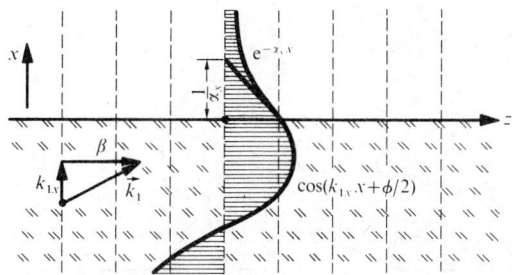

Fig.1.4. Field distribution and phase fronts on both sides of a boundary in case of total reflection.

only slowly away from the boundary. The phase shift ϕ of the reflected wave is also quite small for θ_1 near θ_{1c}. If, however, θ_1 approaches zero, we have $\alpha_x \simeq 2\pi(n_1^2 - n_2^2)^{1/2}/\lambda$ with λ the free-space wavelength. Under these conditions the field in region 2 decays much more rapidly in the x-direction, depending on the refractive index difference and λ. The phase shift of the reflected wave nearly approaches a phase reversal of 180°. All these characteristics of total reflection appear to be significant in optical waveguides consisting of different dielectric media.

A uniform plane wave has constant field amplitudes and phase angles transverse to its direction of propagation; it extends to infinity in all transverse directions with no change of these quantities. Actual light beams are always limited in the transverse direction, and we need to examine what changes such a transversely limited light beam experiences in total reflection. As a simple example of a plane wave with transversely changing amplitudes, we superimpose two uniform plane waves of equal amplitude but with slightly different angles of propagation $\theta_1 = \theta-\Delta\theta/2$ and $\theta_2 = \theta+\Delta\theta/2$. Fig.1.5 illustrates the superposition of these two waves. In the direction θ both waves have the same wave vector component $k_{11} = k_{12} = k_1 \cos(\Delta\theta/2)$, while perpendicular to this direction their wave vector components oppose each other, so that the superposition forms a standing wave in the transverse direction with the planes of zero field

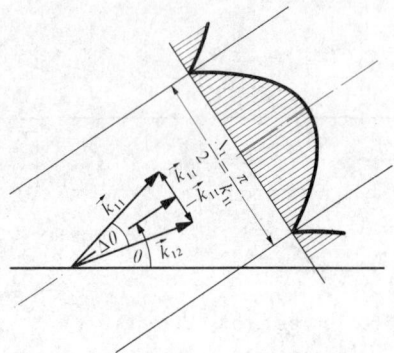

Fig.1.5. Superposition of two uniform plane waves differing by $\Delta\theta$ in propagation direction.

amplitude spaced by $\Lambda_t/2 = \pi/k_{1t} = \pi/[k_1 \sin(\Delta\theta/2)]$. In the transverse direction of the wave vector component k_{1t} the superposition of both waves has constant phase but sinusoidal amplitude distribution.

Any other amplitude distribution, and in particular a transversely limited light beam in the direction θ, may be formed by superimposing an infinite number of uniform plane waves ranged around θ in their propagation directions with different amplitudes and phase angles. Transversely limited light beams thus consist of bundles of rays each representing a plane uniform wave.

We return to the simple superposition of only two uniform plane waves and examine its total reflection at the boundary $x = 0$ to a region of lower refractive index in Fig.1.6. We assume a line of zero field amplitude meets the boundary at $z = 0$. On this line both incident waves of the superposition have opposite phases so that, for example, $E_{i1} = -E_{i2}$ at $x = 0$ and $z = 0$. Each wave is totally reflected but suffers a phase shift ϕ depending on the angle of incidence and the wave number in both regions, according to equations (1.25) and (1.26). With $E_{i1} = -E_{i2}$ at $x = 0$ and $z = 0$ the electric fields of the reflected waves depend on x and z as

UNIFORM PLANE WAVES

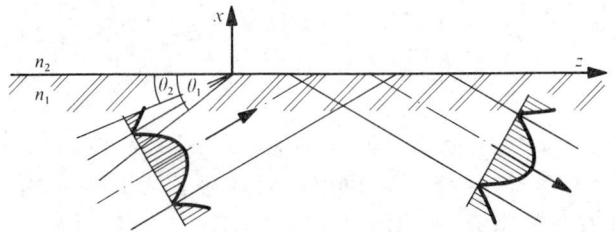

Fig.1.6. Beam shift in total reflection of a superposition of two uniform plane waves.

$$E_{r1} = E_{i1} \exp\{j(k_1 \sin\theta_1 x - k_1 \cos\theta_1 z + \phi_1)\}$$

$$E_{r2} = -E_{i1} \exp\{j(k_1 \sin\theta_2 x - k_1 \cos\theta_2 z + \phi_2)\}.$$

Their lines of zero field with $E_{r1} = -E_{r2}$ follow from

$$k_1(\sin\theta_1 x - \sin\theta_2 x - \cos\theta_1 z + \cos\theta_2 z) = \phi_2 - \phi_1 + m\pi \text{ with } m = 0,1,2,\ldots$$

They start at the boundary $x = 0$ at z-values according to

$$z = (\phi_1 - \phi_2 + m\pi)/[k_1(\cos\theta_1 - \cos\theta_2)].$$

For the zero field line of the incident superposition at $x = z = 0$ we have $m = 0$. This particular line shifts to

$$z = (\phi_1 - \phi_2)/[k_1(\cos\theta_1 - \cos\theta_2)]$$

in the reflected superposition. Designating the wave vector components parallel to the boundary by $k_1 \cos\theta = \beta$ and their difference by $k_1(\cos\theta_1 - \cos\theta_2) = \Delta\beta$ we observe the following shift of zero field lines in total reflection

$$z = \Delta\phi/\Delta\beta.$$

To generalize this result for any light beam formed by a

bundle of nearly parallel light rays we go to the limit of $\Delta\theta \to 0$ and obtain

$$z = d\phi/d\beta \qquad (1.27)$$

for the shift of a light beam in total reflection. Evaluating this expression for H-waves with ϕ_e from equation (1.25) the beam shift becomes

$$z_H = 2\cot\theta/\alpha_x , \qquad (1.28)$$

while for E-waves with ϕ_m from equation (1.26) the beam shift is

$$z_E = \frac{2n_2^2 \cot\theta}{n_1^2 \cos^2\theta - n_2^2 \sin^2\theta} \cdot \frac{1}{\alpha_x} . \qquad (1.29)$$

The transverse attenuation constant α_x of the evanescent fields beyond the boundary depends on θ as $\alpha_x = (k_1^2 \cos^2\theta - k_2^2)^{\frac{1}{2}}$.

The phenomenon is known as the Goos-Haenchen shift. At the limit of total reflection the transverse attenuation constant α_x goes to zero and the beam shift grows beyond all bounds, indicating that a beam under these circumstances travels for longer and longer distances along the boundary before it returns into the region of incidence.

Due to the Goos-Haenchen shift the reflected beam appears as a reflection not from the actual boundary but from a reflector which is somewhat recessed into the medium of lower refractive index. As shown in Fig.1.7 the apparent

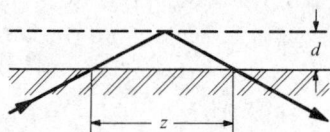

Fig.1.7. Apparent reflector at distance d beyond boundary of total reflection accounting for the Goos-Haenchen shift.

reflector surface is a distance $d = z\tan\theta/2$ from the actual boundary. For H-waves this distance is

$$d_H = 1/\alpha_x, \quad (1.30)$$

and corresponds to the plane in which the evanescent fields have decayed to $1/e$ of their magnitude at the boundary. For E-waves the apparent reflector is at a distance

$$d_E = \frac{n_2^2}{n_1^2 \cos^2\theta - n_2^2 \sin^2\theta} \cdot \frac{1}{\alpha_x} \quad (1.31)$$

Near the limit of total reflection we have $d_E = n_1^2/(n_2^2 \alpha_x)$ and hence $d_E > d_H$, while for grazing incidence with $\theta \to 0$ this relation changes to $d_E = n_2^2/(n_1^2 \alpha_x)$ which renders $d_E < d_H$.

1.2. PLANE-WAVE SCATTERING AND ABSORPTION

Optical waveguides consist mostly of solid materials. In the form of liquid core fibres and in waveguide dye lasers they also contain liquids. Light waves travelling through such solids and liquids suffer loss from scattering and absorption. In the bulk of these materials part of these losses, scattering as well as absorption, are caused by the random molecular structure and part by imperfections and impurities.

We consider scattering losses first. Associated with the random molecular structure are microscopic local variations of the refractive index. Imperfections of the material and impurities also cause index variations. As long as the latter index variations also have a random microscopic nature they may be included with those of the random molecular structure, and scattering as well as scattering losses from both types of index variations may be analysed independent of the particular form of optical waveguide. If, however, an index deviation extends over macroscopic dimensions its effect on light wave propagation will depend on the waveguide structure. Such scattering can therefore be analysed only for particular waveguide forms.

To be more specific only index deviations, which are small and extend over regions small in their dimensions compared to the light wavelength, will be treated here. We consider the relative permittivity $\varepsilon_r = n^2$ to have a random distribution throughout the material with variance $\langle (n^2 - \overline{n^2})^2 \rangle$ and assume the correlation distance d_c to be small compared to the light wavelength $\lambda/(\overline{n^2})^{\frac{1}{2}}$ in the material. The refractive index remains nearly constant within a cube of volume d_c^3, and deviates typically by $(n^2 - \overline{n^2})^{\frac{1}{2}} \ll (\overline{n^2})^{\frac{1}{2}}$ from its r.m.s. value $(\overline{n^2})^{\frac{1}{2}}$.

The field scattered by such a cube may be calculated by introducing an effective current source for the index deviation. For a source-free region with varying index n the second of Maxwell's equations appears as

$$\nabla \times \vec{H} = j\omega\varepsilon_0 n^2 \vec{E}$$

which may be written with a constant r.m.s. index $(\overline{n^2})^{\frac{1}{2}}$ in the form

$$\nabla \times \vec{H} = j\omega\varepsilon_0 \overline{n^2} \vec{E} + \vec{J}.$$

The fictitious electric current density

$$\vec{J} = j\omega\varepsilon_0 (n^2 - \overline{n^2}) \vec{E}$$

takes account of the index variation and, within a cube d_c^3, acts as an oscillating electric dipole of moment

$$\vec{I}l = j\omega\varepsilon_0 (n^2 - \overline{n^2}) d_c^3 \vec{E} .$$

The primary plane-wave propagating in the uniform medium with refractive index $(\overline{n^2})^{\frac{1}{2}}$ excites this dipole with its electric field $\vec{E}$. The dipole has an axially symmetric radiation characteristic with respect to the polarization of the primary electric field. The radiation intensity has its maximum in the plane to which $\vec{E}$ is perpendicular. It is given by

$$S = S_{max} \sin^2 \vartheta \qquad (1.32)$$

depending on the polar angle ϑ between $\vec{E}$ and the respective direction of radiation. In planes containing the electric field vector the dipole radiates as shown in Fig.1.8. The

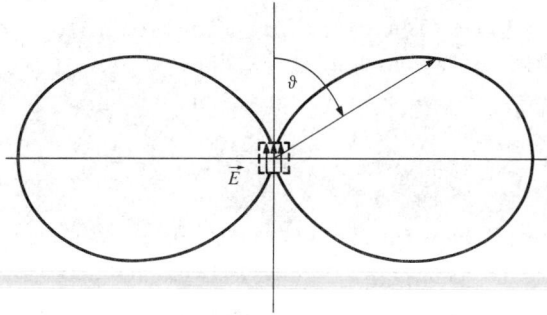

Fig.1.8. Radiation characteristic of Rayleigh scattering in planes containing the electric field vector.

total power radiated by the dipole,

$$dP = \frac{2\pi}{3} \left(\frac{\mu_0}{\varepsilon_0}\right)^{\frac{1}{2}} \frac{|Il|^2}{\lambda^2} (\overline{n^2})^{\frac{1}{2}} , \qquad (1.33)$$

subtracts from the power of the primary wave and scatters in all directions as shown in Fig.1.8. The infinitesimal portion scattered in the direction of the primary wave and superimposing on it, may be neglected here.

Neighbouring cubes of size d_c^3 as well as all other volume elements, being further than the correlation distance apart, have no correlation in their index deviation and scatter independently of each other. The total power therefore follows from integrating over the average of equation (1.33).

A <u>uniform</u> plane wave with power flow density $S = |E|^2 (\overline{n^2})^{\frac{1}{2}} (\varepsilon_0/\mu_0)^{\frac{1}{2}}$ traversing a layer of thickness d_c perpendicular to its direction of propagation thus loses power per unit area of this layer according to

$$\frac{\langle dP \rangle}{d_c^3} = \frac{8\pi^3}{3\lambda^4} \langle (n^2 - \overline{n^2})^2 \rangle \, d_c^3 S .$$

This power is subtracted from the power flow density S of the primary wave at the same rate:

$$-\frac{dS}{dz} = \frac{8\pi^3}{3\lambda^4} \langle (n^2 - \overline{n^2})^2 \rangle \, d_c^3 \, S \, . \tag{1.34}$$

Because of this scattering loss, the power decays exponentially in the direction of propagation according to $S = S_0 \exp(-2\alpha z)$. The attenuation constant α of this exponential decay follows from equation (1.34):

$$\alpha = \frac{4\pi^3}{3\lambda^4} \langle (n^2 - \overline{n^2})^2 \rangle d_c^3 \, . \tag{1.35}$$

The fourth-power dependence of loss on wavelength λ represents the law of Rayleigh scattering and always occurs when scattering objects have dimensions small compared to the wavelength.

To evaluate this Rayleigh scattering any further requires specification of the state and composition of the material. In pure liquids, thermally driven fluctuations in the number of molecules within a region which is substantially smaller than the wavelength cause the index of refraction to vary locally. From thermo-dynamical considerations the product of the variance of relative permittivity and the correlation cube is (Pinnow, Rich, Ostermayer, and DiDomenico 1973)

$$\langle (n^2 - \overline{n^2})^2 \rangle \, d_c^3 = (\overline{n^2})^4 p^2 \beta_T \, k_B T, \tag{1.36}$$

where p is the photoelastic coefficient of the liquid, and β_T its isothermal compressibility. k_B is Boltzmann's constant and T the absolute temperature. To understand the relation (1.36), note that $k_B T$ represents the driving force for the density fluctuations and β_T measures the extent to which this force compresses and changes the density of the liquid. According to the factor $(\overline{n^2})^4 p^2$ this density fluctuation converts to variations of the relative permittivity.

The parameters on the right-hand side of equation (1.36) are known for liquids of not too complicated a composition. From an evaluation of equations (1.36) and (1.35) for CCl_4,

Fig.1.9 shows the intrinsic scattering loss at room temperature as a function of wavelength.

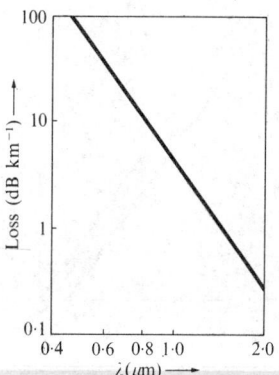

Fig.1.9. Intrinsic scattering loss of liquid CCl_4 at room temperature.

Of particular interest for optical waveguides is Rayleigh scattering of solid materials in the amorphous state such as glasses and certain plastics. Single-component glasses such as fused silica if heated beyond the softening temperature come into thermodynamic equilibrium and then exhibit a microscopic index variation due to thermal fluctuations according to the same formula (1.36) as for pure liquids. On cooling, these index variations freeze in at the softening point. At any ambient temperature below the softening point the index variations remain nearly constant and their variance follows from equation (1.36) with the softening temperature substituted for T. In particular, from the high softening temperature of fused silica (1940 K) one would infer much higher scattering losses in glasses than in liquids. However, glasses usually have much less compressibility then liquids. This effect not only compensates for the higher effective temperature but actually leads to lower scattering losses in some glasses as compared to liquids. Fig.1.10 shows the scattering loss of high quality fused silica from measured data (Pinnow *et al.* 1973).

If the glass consists not of one but of several material components the index of refraction varies locally not only

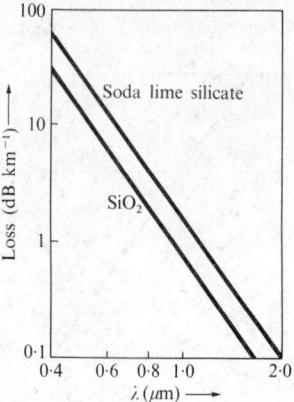

Fig.10. Intrinsic scattering loss of high quality fused silica and soda lime silicate glasses (Pinnow et al. 1973).

due to density fluctuations frozen in at the softening point, but also to a statistically random distribution of the polarizable components. This local variation in composition of multicomponent glasses causes additional scattering and losses. The product of the variance of permittivity and the correlation cube which determines the additional scattering loss follows from (Pinnow et al. 1973)

$$\langle (n^2 - \overline{n^2})^2 \rangle \, d_c^3 = \frac{4n^2}{\rho N_A} \sum_{j=1}^{m} \left[\left(\frac{\partial n}{\partial x_j} \right)_{\rho, T, x_i \neq x_j} + \left(\frac{\partial n}{\partial \rho} \right)_{T, x_i} \left(\frac{\partial \rho}{\partial x_j} \right)_{T, x_i \neq x_j} \right]^2 M_j x_j \quad (1.37)$$

where n and ρ are refractive index and density, respectively of the material, and M_j and x_j the molecular weight and mole fraction of its jth component. N_A is Avogadro's number. The partial derivatives of n may be evaluated according to Gladstone and Dale (1863), whereas Huggins and Sun (1943) have provided formulas for evaluating the partial derivatives of ρ. Both derivatives appear together with the molecular weight and the mole fraction of the respective component in equation (1.37).

The scattering loss from compositional variations together with the scattering loss due to density fluctuations yields the total scattering loss of chemically pure multicomponent glasses. Additional contributions to scattering result from imperfections and impurities. The scattering loss of a multicomponent glass, soda lime silicate, also appears in Fig.1.10. As compared to fused silica, such multicomponent glasses have lower softening temperatures and hence less scattering from density fluctuations. However, the additional scattering from compositional variations raises the scattering loss in Fig.1.10 above that of fused silica.

The other mechanism of power loss for light waves travelling through bulk material is absorption. Such absorption occurs when light photons interact with electrons or with vibrational states of the material constituents. During this interaction the material absorbs photons which excite electrons or certain vibrational states. Free electrons in metals interact very strongly with light at any wavelength, causing high absorption, so that light cannot penetrate metals. For bound electrons to absorb light the photons must have sufficient energy, $E = hf$, to excite these electrons (h represents Planck's constant). In optically transparent media bound electrons have so high an excitation energy that only photons of ultraviolet light have sufficient energy to interact with them and be absorbed. In simple silicate glasses, for example, electronic states of the oxygen ions determine the ultraviolet interband absorption edge.

Glasses normally contain additional elements, mostly as impurities, with electronic states of lower excitation energy. A number of metallic ions occur as such foreign elements in glasses. More easily excited than the intrinsic states of the material, they shift the ultraviolet absorption edge to a lower wavelength and cause additional absorption bands in the visible and infrared range of the light spectrum.

At even lower photon energies in the infrared the light begins to interact with vibrational states of the intrinsic material. Several infrared absorption bands appear in glasses due to this interaction so that beyond 5 µm wavelength these glasses become completely opaque.

This brief survey shows that transparent materials normally have low absorption only in the visible and near-infrared range of the light spectrum; to maintain this low absorption, they must be of sufficient purity. Examining the ultraviolet absorption edge of chemically pure glasses more closely, it is found to depend on the excitation energy of their O^{2-}-ions. In fused silica these ions are very tightly bound, leading to high ultraviolet transparency with the absorption edge at a short ultraviolet wavelength. The tail of this absorption edge, however, reaches to longer wavelengths and may even be noticed in the visible range when extremely low absorption is required. The random molecular structure of the material, due to thermally driven fluctuations frozen in below the softening point, produces varying local electric fields on the same microscopic scale as the density fluctuations. These local electric fields induce broadening of excitonic levels with energies near and somewhat below the band edge. The field-induced broadening of these levels draws the tail of the absorption edge out into the visible range with low yet noticeable absorption decaying exponentially with photon energy E according to

$$\alpha \sim \exp\left(\frac{E-E_g}{\Delta E}\right) \qquad (1.38)$$

E_g represents the effective energy gap of the material and ΔE is a characteristic quantity for such a material. Fitting experimental data on fused silica and soda lime silicate of high chemical purity to the exponential absorption tail of equation (1.38) results in the intrinsic absorption curves of Fig.1.11 (Pinnow *et al.* 1973). From the slope of these curves fused silica has a larger ΔE-value (0·5 eV) than soda lime silicate with a ΔE-value of only 0·3 eV. Together with the larger effective energy gap for fused silica of E_g = 13·4 eV, this material exhibits an intrinsic absorption loss, which is lower and rises more slowly with increasing frequency than for most other glasses. The wider effective energy gap of fused silica is due to its tightly bound

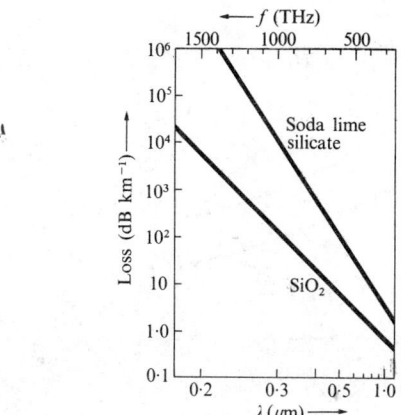

Fig.1.11. Intrinsic absorption loss of fused silica and soda lime silicate versus free space wavelength and frequency (Pinnow et al. 1973).

O^{2-}-ions, as opposed to some of the O^{2-}-ions in soda lime silicate, as well as in other silicate glasses, which have only dangling bonds. By comparing the ΔE-values for a number of liquids at different temperatures T and glasses of different softening temperatures T_s, it appears that ΔE is directly proportional to T or T_s, respectively. The local electric fields associated with density fluctuations of the random molecular structure therefore turn out to be responsible for this broadening of the absorption edge tails.

Rayleigh scattering loss and intrinsic absorption loss of chemically pure materials add together and form the total intrinsic loss of the respective material. As such they represent the lower limit of loss to be expected in these materials under perfect conditions. Fig.1.12 shows the intrinsic loss in fused silica and soda lime silicate in that part of the visible and near-infrared spectrum, which is of greatest interest for optical communications. The most promising luminescent diodes and semiconductor lasers as well as several of the most efficient solid state lasers have their emission lines in the spectral range represented by Fig.1.12. Fused silica appears as the most promising material with low scattering loss and lowest intrinsic ab-

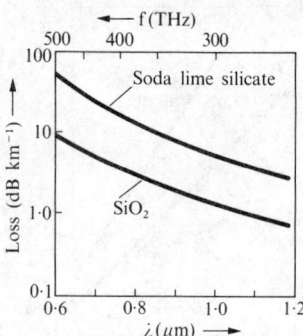

Fig.1.12. Total intrinsic loss in fused silica and soda lime silicate versus wavelength and frequency.

sorption, adding up to the lowest total intrinsic loss. At wavelengths of GaAlAs-lasers between $\lambda = 0\cdot 8$ and $0\cdot 9$ μm, its intrinsic loss may be 2 dB/km or even lower, while for the emission wavelength of Nd-YAG-lasers and Nd-Ultraphosphate-lasers near $\lambda = 1\cdot 06$ m the intrinsic loss even decreases below 1 dB/km.

Glasses maintain their low intrinsic loss in the visible and near infrared only when they are chemically pure and homogeneous. Foreign elements in the glass cause additional absorption and some scattering; phase separations in multicomponent glasses add even more to the scattering loss. More important for absorption are ions of the metals Cu, Ti, V, Cr, Mn, Fe, Co, and Ni. They have electronic transitions with energies low enough to be excited by photons in the visible or near infrared. Also the hydroxyl-ion OH^- shows a vibrational absorption near $2\cdot 8$ μm wavelength with second and third overtones near $0\cdot 95$ μm and $0\cdot 72$ μm, respectively.

For low concentrations of these impurities, the attenuation constant due to the absorption of any particular impurity-ion in a specific glass at a given frequency is directly proportional to the concentration of the impurity. It also depends on the embedding glass matrix, its past history from the glass-making process, and on temperature.

With so many factors influencing the impurity absorption, Fig.1.13 gives only an overall view of what may be typically expected as absorption from the most important impurity-ions. The attenuation constant is plotted against frequency for the different concentrations listed in Fig.1.13. The con-

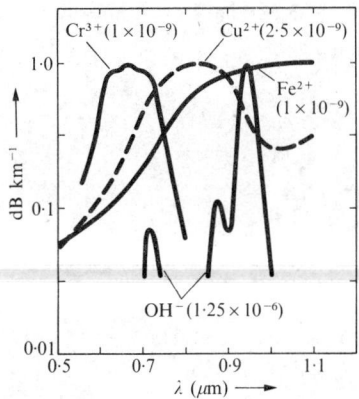

Fig.1.13. Absorption due to impurities.

centrations have been chosen for the peak loss of the strongest absorption band in the wavelength range of Fig.1.13. to be 1 dB/km. Judging only by its concentration and absorption loss between 0·8 and 0·9 μm wavelength as well as near λ = 1·06 μm, the iron-ion Fe^{2+} appears as the most critical impurity. In producing high-purity glasses, however, the effort, involved to reduce foreign elements to a certain low concentration differs considerably from element to element. In fused silica, for example, it turns out to be quite difficult to keep hydroxyl-ions at the not very low concentration for which Fig.1.13 shows absorption loss.

Scattering and absorption losses have been treated here with special emphasis on glasses, and in particular for such potentially low-loss glasses as fused silica and soda lime silicate. Although this treatment is complete in that it considers all important effects leading to loss of light when propagating through bulk materials, quantitative results may be quite different in other materials. Usually other materials for optical waveguides show much higher loss than high-quality glasses. In various applications of these waveguides in optical communication systems higher bulk material losses are tolerated when other characteristics of the particular material are of greater significance.

1.3. PLANE-WAVE DISPERSION

Uniform plane waves in homogeneous media propagate with the phase velocity

$$v = c/n = 1/\{n(\mu_0\varepsilon_0)^{\frac{1}{2}}\}$$

Their group velocity is

$$v_g = c/(n + \omega dn/d\omega). \qquad (1.39)$$

Their phase delay per unit length in the direction of propagation therefore is $\tau_p = 1/v = n/c$, while energy centred at ω in a monochromatic wave suffers the group delay

$$\tau_g = 1/v_g = (n + \omega \frac{dn}{d\omega})/c \qquad (1.40)$$

per unit length, again in the direction of propagation. This group delay per unit length is also called specific group delay.

If the refractive index n does not change with frequency, phase and group velocity will likewise remain constant and independent of frequency. All spectral components of a signal in a uniform plane wave will then propagate with one and the same velocity and will superimpose at any point along its path with the same dependence in time. Under these conditions signals suffer no distortion.

If, however, the medium shows dispersion, meaning that the refractive index changes with frequency, different spectral components of a signal travel with different velocities and shift in time with respect to each other along the path of propagation. Their superposition results in signals which are distorted compared to their original at the input.

Actually all solids and liquids, including those which are used in optical waveguides, show more or less dispersion, which is intimately related to the absorption of the material. The spectral range in which these media are transparent lies between the range of absorption due to vibrational resonances and the range of electronic absorption. In this range of transparency the electronic polarizability of the medium dominates in determining its index of refraction. Its dispersion is therefore mainly due to electronic excitation and is related to electronic absorption. A single such resonance at the angular frequency ω_1 contributes

$$\chi = \frac{B_1}{\omega_1^2 - \omega^2 + 2jg\omega} \tag{1.41}$$

to the complex electric susceptibility. Equation (1.41) is obtained by regarding the electronic transitions as classic electric dipole oscillators with a concentration per cubic space N_1. The constant $B_1 = N_1 q^2/(\varepsilon_0 m)$ is directly proportional to N_1, with q and m, the electron charge and mass respectively. The damping factor g takes account of radiation and other dissipative mechanisms in the oscillator.

Sufficiently below resonance, the real part of the electric susceptibility is approximated by

$$\text{Re}(\chi) = n^2 - 1 = \frac{B_1}{\omega_1^2 - \omega^2} \tag{1.42}$$

from (1.41). Electronic transitions in solids and liquids have not one but a large number of different transition frequencies. For interband transitions they practically form a continuum of transition frequencies. Nevertheless, the important interband transitions may be approximated by individual oscillators, and the contributions of these oscillators to the electric susceptibility may be found from (1.41). Its real part then appears as the following sum of contributions from electric dipole oscillators with discrete resonance frequencies ω_m:

$$n^2 - 1 = \sum_m \frac{B_m}{\omega_m^2 - \omega^2} . \tag{1.43}$$

However, this approximation only holds for frequencies ω sufficiently small compared to all significant resonance frequencies ω_m. To reduce it to an even simpler expression, we extract the term which represents the first strong resonance. Let this resonance have B_1 and ω_1 with $\omega_1 < \omega_m$. By itself this resonance would contribute with $B_1/(\omega_1^2 - \omega^2)$ to $n^2 - 1$. The remaining weaker contributions may now be accounted for to first order in ω^2/ω_m^2 by introducing an effective resonance frequency

$$\omega_0 = \omega_1 \left(\sum B_m \frac{\omega_1^2}{\omega_m^2} \right) \Big/ \left(\sum B_m \frac{\omega_1^4}{\omega_m^4} \right) \tag{1.44}$$

and the constant

$$B_0 = \omega_0^2 \sum (B_m/\omega_m^2) \tag{1.45}$$

of an equivalent oscillator. This results in the following simple dispersion formula for solids and liquids in their frequency range of transparency:

$$n^2 - 1 = B_0/(\omega_0^2 - \omega^2). \tag{1.46}$$

ω_1 represents the absorption edge of the material and ω_0 according to equation (1.44) is always somewhat larger than ω_1, typically one finds $\omega_0 \simeq 1 \cdot 5\, \omega_1$. By factorizing B_0 into $B_0 = \omega_0\, \omega_d$ the characteristic frequency ω_d is found, which also has a certain physical significance and is called the dispersion frequency. Higher ω_d-values lead to more dispersion. Crystalline solids of the same chemistry and structure have the same value for the dispersion frequency. In terms of the characteristic frequencies ω_0 and ω_d, the approximate dispersion formula appears as

$$n^2 - 1 = \omega_0\, \omega_d/(\omega_0^2 - \omega^2). \tag{1.47}$$

Experimental verification of the dispersion characteristic may be obtained when $1/(n^2 - 1)$ for a particular material is plotted versus ω^2 in its spectral range of transparency.

Such plots usually display the linear form of equation
(1.46) with deviations from a straight line only near the
lower and upper limits of the transparent range. Fig.1.14
shows $1/(n^2 - 1)$ versus $1/\lambda^2$ for NaF plotted from measured
data (Wemple and Di Domenico 1971). The straight line

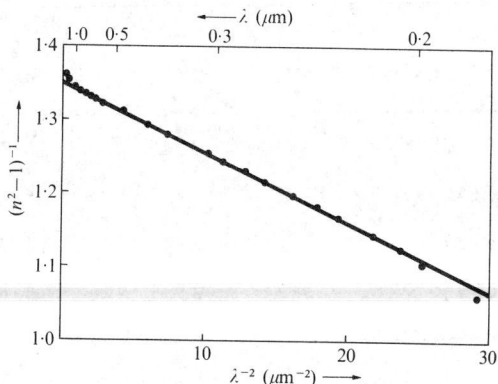

Fig.1.14. Measured points of refractive index factor $(n^2-1)^{-1}$ versus λ^{-2} for NaF (Wemple and Di Domenico 1971) compared to the straight line from equation (1.46) which is matched to the measured points.

fitting the measured points over most of the transparent
range corresponds to $\hbar\omega_0$ = 15 eV and $\hbar\omega_d$ = 11·3 eV, with
$\hbar = h/(2\pi)$ where h is Planck's constant. Near the long-
wavelength limit of Fig.1.14, the typical deviation from
the straight line to larger values is noticed, which is due
to negative contributions to $(n^2 - 1)$ from lattice vibrations. Near the short-wavelength limit of Fig.1.14 the
typical deviation from the straight line is to lower values
of $(n^2 - 1)^{-1}$ due to the proximity of the band edge or to
excitonic absorption.

For transparent glasses of good optical quality, the
measured characteristic of $(n^2 - 1)^{-1}$ versus ω^2 matches the
straight line of equation (1.47) even better and yields
values for the oscillator and dispersion energies $E_0 = \hbar\omega_0$
and $E_d = \hbar\omega_d$, respectively, which are listed in Table 1.1.
Of all the glasses appearing in this table fused quartz
(SiO_2) has the highest oscillator energy $E_0 = \hbar\omega_0$, corresponding to the highest ultraviolet absorption edge, and re-

TABLE 1.1.
Refractive index and dispersion parameters for a representative selection of glasses from the Schott Catalogue (Di Domenico 1972)

Material	n at $\lambda = 0.588 \mu m$	$E_0 = \hbar \omega_0$ (eV)	$\omega_0/2\pi$ (THz)	$E_d = \hbar \omega_d$ (eV)	$\omega_d/2\pi$ (THz)
SiO_2	1·4577	13·4	3240	14·7	3554
K9	1·5141	12·2	2950	15·3	3699
F2	1·6199	9·4	2273	14·5	3506
BaSF12	1·6697	9·74	2355	16·6	4013

flecting its potentially low absorption loss. In addition SiO_2 has a moderate value of the dispersion energy $E_d = \hbar \omega_d$, indicating that combined with the high oscillator energy it will have the lowest dispersion of all the glasses in Table 1.1.

To consider the effects of material dispersion on signal propagation in uniform plane waves we turn back to equation (1.40) for the group velocity of these waves in homogeneous regions. With

$$\frac{dn}{d\omega} = \frac{\omega_0 \, \omega_d \, \omega}{n(\omega_0^2 - \omega^2)^2} \tag{1.48}$$

from equation (1.47), the group velocity follows as

$$v_g = c / \left[n + \frac{\omega_0 \, \omega_d \, \omega^2}{n(\omega_0^2 - \omega^2)^2} \right]. \tag{1.49}$$

If the plane wave is monochromatic, consisting only of one coherent oscillation, but then carries a signal in the form of amplitude or phase modulation, its spectral components spread over a finite frequency band; besides the frequency of the monochromatic carrier it has sidebands below and above the carrier frequency. In a dispersive medium these spectral components travel with different velocities and, hence suffer different delays, with signal distortion as a consequence.

We consider a signal carried by a uniform plane wave which propagates in a dispersive medium of permittivity $\varepsilon = n^2 \varepsilon_0$ and permeability μ. The ratio of its electric field and magnetic field phasors forms the wave impedance of the medium,

$$E/H = (\mu/\varepsilon)^{\frac{1}{2}} = \eta. \qquad (1.50)$$

From the power flow density S of this wave,

$$S = \vec{E} \times \vec{H}^* = |E|^2/\eta^*, \qquad (1.51)$$

we define the wave amplitude

$$a = E/\sqrt{\eta}. \qquad (1.52)$$

The amplitude squared represents the wave's power flow density in the case of a real wave impedance:

$$S = a^2 \qquad \text{for } \eta \text{ real.} \qquad (1.53)$$

We now let this input carrier have an angular frequency ω_c and its amplitude be a complex function $a(t)$ of time. Allowing $a(t)$ to be complex takes into account both amplitude and phase modulation of the carrier. The expression

$$E = \eta^{\frac{1}{2}} a(t) e^{j \omega_c t} \qquad (1.54)$$

is then a complex representation of its input electric field, with its real time dependence following as

$$E(t) = 2^{\frac{1}{2}} \operatorname{Re}(E). \qquad (1.55)$$

The spectral components of the input wave form follow from the Fourier transform

$$A(\omega) = \int_{-\infty}^{\infty} a(t) \exp\{j(\omega_c - \omega)t\} dt \qquad (1.56)$$

assuming, of course, that $a(t)$ differs from zero only for a finite interval in time, so that the integral in equation (1.56) also remains finite. The actual input wave form represents the superposition of all spectral components contained in its Fourier transform and as such appears as the inverse transform or Fourier integral

$$a(t)\, e^{j\omega_c t} = \frac{1}{2\pi} \int_{-\infty}^{\infty} A(\omega) \exp j\omega t\, d\omega \,. \qquad (1.57)$$

While travelling the distance L each spectral component will suffer loss according to the attenuation constant α and phase delay according to the phase constant β, and emerge as

$$B(\omega) = A(\omega) \exp-(\alpha+j\beta)L \,. \qquad (1.58)$$

The superposition of all these components in form of the Fourier integral

$$b(t) = \frac{1}{2\pi} \int_{-\infty}^{\infty} A(\omega) \exp\{j\omega t - (\alpha+j\beta)L\} d\omega. \qquad (1.59)$$

represents the output wave form.

If the attenuation constant α and refractive index n are independent of frequency, then $\beta = \omega/v = \omega/v_g$ and

$$b(t) = \frac{e^{-\alpha L}}{2\pi} \int_{-\infty}^{\infty} A(\omega) \exp\{j\omega(t - L/v_g)\} d\omega. \qquad (1.60)$$

This expression contains the Fourier integral of the input wave form equation (1.57) with the time variable t replaced by $t - L/v$:

$$b(t) = \exp^{-\alpha L} a(t-L/v) \exp\{j\omega_c(t-L/v)\}. \qquad (1.61)$$

The output wave form thus appears as an undistorted replica of the temporal input, delayed in time by $t = L/v$, and attenuated according to $\exp(-\alpha L)$. For signal propagation without distortion in uniform plane waves the attenuation constant α as well as the refractive index n must be independent of frequency over the spectral range of the input light signal.

For wave propagation in a dispersive medium we assume the input wave envelope $a(t)$ to vary only slowly in time while modulating the oscillating carrier, which varies quite rapidly compared to the modulating signal. Its spectral components will then extend only over a narrow frequency range centred at ω_c. In transparent media at frequencies well below the absorption edge, any frequency dependence of the small attenuation constant α may be neglected over such narrow frequency bands. For the frequency dependence of the phase constant β, it suffices to introduce its Taylor expansion around the centre frequency ω_c, and include all terms up to second order:

$$\beta = \beta_c + \left.\frac{\partial \beta}{\partial \omega}\right|_{\omega_c} (\omega - \omega_c) + \frac{1}{2} \left.\frac{\partial^2 \beta}{\partial \omega^2}\right|_{\omega_c} (\omega - \omega_c)^2 . \quad (1.62)$$

Under these conditions the output wave form has the Fourier integral

$$b(t) = \frac{e^{-\alpha L}}{2\pi} \int_{-\infty}^{\infty} B_d(\omega) \exp\{j\omega(t - L/v)\} d\omega \quad (1.63)$$

with

$$B_d(\omega) = A(\omega) \exp\{-j(L/2)\beta''(\omega-\omega_c)^2\} \exp\{-j(L/v_g)(\omega-\omega_c)\} \quad (1.64)$$

where $\beta'' = \partial^2 \beta / \partial \omega^2 = -v'_g / v_g^2$ and $v'_g = dv_g / d\omega$. According to equation (1.63) the output wave form has a carrier reduced in amplitude by $\exp(-\alpha L)$ and retarded in phase by $\beta L = \omega L / v$. Its spectral distribution $B_d(\omega)$ is modified from the input distribution by two exponentials. The exponent of the last exponential term changes linearly with the frequency difference from the carrier frequency; due to this exponential the envelope is delayed in time by L/v_g. The first exponential has an exponent which depends on the frequency difference squared; it distorts the envelope due to the dispersion $\partial v_g / \partial \omega = -v_g^2 \beta''$ in group velocity.

The nature of the envelope distortion becomes apparent

when we consider a specific wave form. To obtain simple but representative relations, let the input wave form have a real $a(t)$ in the form of the Gaussian time function

$$a(t) = (2/t_0 \pi^{\frac{1}{2}})^{\frac{1}{2}} \exp(-2t^2/t_0^2) . \qquad (1.65)$$

The factor $(2/t_0 \pi^{\frac{1}{2}})^{\frac{1}{2}}$ has been added to normalize the Gaussian distribution:

$$\int_{-\infty}^{+\infty} a^2(t) \, dt = 1. \qquad (1.66)$$

The uniform plane wave then carries unit energy per unit cross-sectional area in this Gaussian pulse. At $t = \pm t_0/2$ the intensity a^2 drops to $1/e$ times its maximum value at $t = 0$. The time constant t_0 is therefore a convenient measure for the input pulse width and henceforth will be called the $1/e$-width of the pulse or just the pulse width.

The Gaussian input wave form has the Fourier spectrum

$$A(\omega) = \sqrt{t_0 \pi} \, \exp\{-\frac{(\omega_c-\omega)^2 t_0^2}{8}\}, \qquad (1.67)$$

which is of the same functional form as its amplitude function. Its power spectrum $A^2(\omega)$ has an $1/e$-width

$$\Delta f = \frac{\Delta \omega}{2\pi} = \frac{2}{\pi t_0} , \qquad (1.68)$$

which is even narrower than $1/t_0$ and demonstrates the unique feature that the product of pulse duration and bandwidth is minimal for the Gaussian distribution.

To determine the output wave form, the input spectrum $A(\omega)$ according to (1.67) is substituted into equation (1.64) and the Fourier integral of equation (1.63) evaluated under the assumption that $\Delta\omega$ from equation (1.68) is much smaller than the carrier frequency ω_c. The result may be written as

$$b(t) = \left(\frac{2}{t_0 \pi^{\frac{1}{2}}}\right)^{\frac{1}{2}} \frac{\exp(-\alpha L)}{[1+(4\beta''L/t_0^2)^2]^{\frac{1}{4}}} \exp\left\{-\frac{2(t-L/v_g)^2}{t_0^2[1+(4\beta''L/t_0^2)^2]}\right\}$$

$$\exp\{j[\omega_c t - \varphi(t)]\}. \quad (1.69)$$

Apart from an insignificant phase modulation according to

$$\tan\varphi(t) = \frac{\int_{-\infty}^{\infty} \exp(-\omega^2 t_0^2/8)\cos[\omega(t-L/v_g) - \beta''L\omega^2/2]d\omega}{\int_{-\infty}^{\infty} \exp(-\omega^2 t_0^2/8)\sin[\omega(t-L/v_g) - \beta''L\omega^2/2]d\omega}, \quad (1.70)$$

the output wave form again exhibits a Gaussian distribution. Its duration has, however, been extended to a pulse width of

$$t_1 = [t_0^2 + (4\beta''L/t_0)^2]^{\frac{1}{2}}, \quad (1.71)$$

while its peak amplitude has been reduced by $\exp(-\alpha L)[1+(4\beta''L/t_0^2)^2]^{-\frac{1}{4}}$. The exponential factor of this amplitude reduction is, of course, due to the attenuation αL of the medium, while the remaining factor takes account of the pulse spreading in time and thereby diminishing in peak amplitude. Compared to the exponential attenuation factor $\exp(-\alpha L)$ this remaining factor is under normal conditions much closer to unity and therefore need not be considered any further. The main effect of dispersion is in pulse distortion, first of all in the form of pulse widening. According to equation (1.71) the pulse widening increases when the medium shows more dispersion and consequently has a larger β''. The relative pulse widening is also more pronounced when the input pulse has shorter duration t_0 and spreads over a wider frequency spectrum, with its spectral width $\Delta f \sim 1/t_0$. With its spectral components spreading over a broader band, they will, of course, shift more and more in time with respect to each other the larger the dispersion β''.

In the range of normal dispersion, and when the refractive index depends on frequency as in equation (1.47), the dispersion β'' depends on frequency and on ω_0 and ω_d as

$$\beta'' = \frac{2\omega\omega_d\omega_0^3}{nc(\omega_0^2-\omega^2)^3} [1 + \frac{\omega^2}{\omega_0^2}] \ . \tag{1.72}$$

When ω approaches the characteristic frequency ω_0 corresponding to the effective energy gap of the material, the dispersion becomes stronger and stronger. In addition β'' also increases with the dispersion frequency ω_d.

Signals which are carried by uniform plane waves at optical frequencies through dispersive media will suffer additional distortion when the plane wave is not exactly coherent in time as well as in space. Ordinary light sources emit photons from many independent and uncorrelated processes, such as radiative recombinations of electrons and holes in semiconductors. Their light shows hardly any coherence, either in time or space. Nominally coherent light sources such as laser oscillators generate light by stimulated emission of radiation. A photon radiated by stimulated emission has a fixed phase relation to the stimulating photon and actually adds to it in phase. Thus certain modes of oscillation of the respective laser cavity are excited. The field distribution of the radiation from some of these modes resembles at least locally that of a plane uniform wave. Under ideal circumstances these waves are to a high degree coherent in time and space with a frequency spectrum consisting only of one very narrow line at the resonance frequency of the respective laser mode of oscillation. Under these ideal conditions the spectral line broadens only due to spontaneous emission of photons into the same mode of oscillation. Other influences, such as statistical fluctuations in time of the laser cavity parameters, lead to additional fluctuations in phase and amplitude of the nominally coherent radiation; they widen the spectral line even further. Strictly speaking, therefore, even the best lasers are sources of only partially coherent light, with the degree of coherence depending, among other factors, on the quality of the laser cavity for the particular mode of oscillation and the stability of its parameters in time.

To examine the effects of this partial coherence on

UNIFORM PLANE WAVES

signal propagation through dispersive media, we assume a plane uniform wave perfectly coherent in space transverse to its direction of propagation but only partially coherent in time. We write the time dependence of its field as

$$E(t) = \eta^{\frac{1}{2}} a(t) \exp(j\omega_c t) \qquad (1.73)$$

where ω_c designates the nominal carrier frequency and the complex amplitude $a(t)$ fluctuates randomly in time. The autocorrelation function of the random process is given by

$$\varphi(\tau) = \int_{-\infty}^{\infty} a(t)\, a^*(t-\tau)\, dt / (\int_{-\infty}^{\infty} |a(t)|^2\, dt) \qquad (1.74)$$

where $a^*(t)$ designates the complex conjugate of $a(t)$. To analyse the effects of material dispersion on such a partially coherent wave, we need to know its frequency spectrum. The spectral distribution of its intensity follows from the Fourier transform

$$S(\omega) = \frac{1}{\eta} \left| \int_{-\infty}^{\infty} E(t) \exp(-j\omega t) dt \right|^2 = \left| \int_{-\infty}^{\infty} a(t) \exp\{j(\omega_c - \omega)t\} dt \right|^2. \qquad (1.75)$$

Considering that for any stationary random process

$$\int_{-\infty}^{\infty} a(t) \exp(-j\Omega t) dt = \int_{-\infty}^{\infty} a(t-\tau) \exp(-j\Omega(t-\tau)) d\tau,$$

we obtain

$$S(\omega) = \int_{-\infty}^{\infty}\int_{-\infty}^{\infty} a(t) \exp\{j(\omega_c-\omega)t\} a^*(t-\tau) \exp\{-j(\omega_c-\omega)t\}$$
$$\exp\{j(\omega_c-\omega)\tau\} dt\, d\tau, \qquad (1.76)$$

which together with equation (1.74) leads to

$$S(\omega) = \int_{-\infty}^{\infty} |a(t)|^2 dt \cdot \phi(\omega-\omega_c), \qquad (1.77)$$

where $\phi(\omega)$ is defined as

$$\phi(\omega) = \int_{-\infty}^{\infty} \varphi(\tau) \exp(-j\omega\tau) d\tau. \qquad (1.78)$$

$\phi(\omega)$ represents the power spectrum of the random time function $a(t)$ normalized with respect to the energy $\int_{-\infty}^{\infty}|a(t)|^2 \, dt$ of the wave per unit transverse area. According to equation (1.78) this power spectrum is the Fourier transform of its autocorrelation function.

Turning back now to the question of how material dispersion affects signals carried by partially coherent waves, we assume a wave with the Gaussian autocorrelation

$$\varphi(\tau) = \exp\text{-}(\tau/t_k)^2 \qquad (1.79)$$

of its randomly fluctuating amplitude. The time constant t_k in this autocorrelation function represents a coherence time. Between instants of time separated by t_k the probability is only 1/e that the wave will maintain its coherence at a fixed point in space. The Gaussian distribution in equation (1.79) has been chosen for the autocorrelation function because its Fourier transform is of the same functional form, resulting in a power spectrum of the random amplitude fluctuations given by

$$\phi(\omega) = \sqrt{\pi} \, t_k \, \exp\text{-}(\omega t_k/2)^2 \, . \qquad (1.80)$$

This translates into a spectral distribution of the partially coherent wave which follows from

$$S(\omega)/\int_{-\infty}^{\infty} |a(t)|^2 \, dt = t_k \sqrt{\pi} \, \exp\{-(\omega-\omega_c)^2 \, t_k^2/4\}. \qquad (1.81)$$

The spectral intensity has its maximum at the nominal carrier frequency $f_c = \omega_c/(2\pi)$ and falls to 1/e of its maximum value at frequencies which differ by $1/(\pi \, t_k)$ from the centre frequency f_c. The 1/e-bandwidth of this partially coherent oscillation with Gaussian autocorrelation of coherence time t_k therefore appears as

$$B = 2/(\pi \, t_k). \qquad (1.82)$$

Assuming a Gaussian distribution for the autocorrelation function, we gain not only the advantage that the spectral distribution is Gaussian as well, but that it also allows a unified presentation of the effects of material dispersion on signal transmission, when the signals also modulate the amplitude or intensity of the light waves in the form of Gaussian pulses. Furthermore, even if partially coherent sources such as lasers or incoherent sources such as luminescent diodes do not exactly have a Gaussian autocorrelation function, the power spectrum very often resembles a Gaussian distribution quite closely. This indicates that the autocorrelation function also has a near-Gaussian distribution.

If a partially coherent source with a frequency spectrum of finite width, such as the Gaussian distribution in Fig.1.5(a), is modulated in intensity, each of its spectral components may be thought to carry this modulation. The modulation creates sidebands and each of the spectral components will be widened by the addition of these sidebands.

For an analytical representation of the power spectrum for the partially coherent wave with modulation, we consider first the unmodulated wave. Its power spectrum $\phi(f)$ in terms of the frequency f may be written as the convolution of $\phi(f)$ with Dirac's delta function:

$$\phi(f-f_c) = \int_{-\infty}^{\infty} \phi(f_1-f_c)\, \delta(f_1-f)\, df_1 \qquad (1.83)$$

which with

$$\int_{-\infty}^{\infty} \delta(f_1-f)\, df_1 = 1 \qquad (1.84)$$

and

$$\delta(f_1-f) = 0 \qquad \text{for} \qquad f_1 \neq f \qquad (1.85)$$

appears as an identity. Dirac's delta function in the convolution integral of equation (1.83) represents the infinitely narrow line of the spectral component at $f = f_1$, which forms part of the power spectrum of the partially coherent oscillation. If this oscillation is now modulated, and each of its

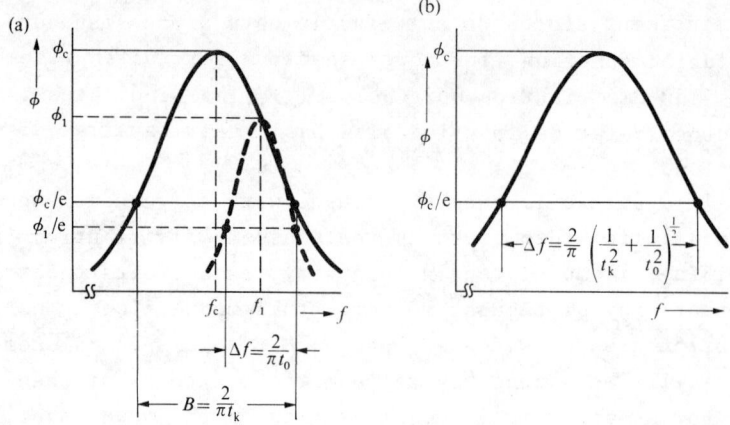

Fig.1.15. (a) Gaussian power spectrum of a partially coherent wave with Gaussian autocorrelation of coherence time t_k and widening of a spectral component f_1 due to Gaussian intensity modulation with pulse width t_0. (b) Gaussian power spectrum of the partially coherent oscillation widened by Gaussian intensity modulation.

spectral components carries this modulation likewise, the modulation will create sidebands for each of the spectral components. Their infinitely narrow lines $\delta(f_1-f)$ will be widened by the addition of these sidebands so that $\delta(f_1-f)$ in equation (1.83) must be replaced by a normalized line form $g(f_1-f)$ which describes the power spectrum of a coherent carrier of frequency f_1 with the respective modulation. The line form $g(f_1-f)$ must be normalized in the same fashion as is $\delta(f_1-f)$ by equation (1.84), requiring

$$\int_{-\infty}^{\infty} g(f_1-f) \, df_1 = 1. \qquad (1.86)$$

UNIFORM PLANE WAVES   49

Using this line form for the modulation, the power spectrum $\phi(f-f_c)$ of the partially coherent oscillation without modulation changes to

$$\phi_m(f-f_c) = \int_{-\infty}^{\infty} \phi(f_1-f_c)\, g(f_1-f)\, df_1 \qquad (1.87)$$

when it carries modulation with spectral distribution $g(f_1-f)$.

Returning to the example of Gaussian autocorrelation with coherence time t_k for the partially coherent oscillation, the unmodulated power spectrum is given by equation (1.81). Modulating the intensity of this oscillation by a Gaussian pulse of $1/e$-width t_0, we have from equation (1.67) that

$$g(f_1-f) = \pi t_0 \exp\{-\pi^2 (f_1-f)^2 t_0^2\}. \qquad (1.88)$$

A Gaussian pulse of a partially coherent carrier with Gaussian autocorrelation therefore has the power spectrum

$$\phi_m(f) = t_k t_0\, \pi^{3/2} \int_{-\infty}^{\infty} \exp[-\pi^2(f_1-f_c)^2 t_k^2 - \pi^2 (f_1-f)^2 t_0^2]\, df_1. \qquad (1.89)$$

To evaluate the convolution integral in this equation we transform to $\nu = f_1 - f_c$ as the new variable of integration and obtain

$$\phi_m(f) = t_k t_0\, \pi^{3/2} \exp\left\{-\pi^2 \frac{(f_c-f)^2}{t_k^{-2} + t_0^{-2}}\right\}$$

$$\int_{-\infty}^{\infty} \exp\left\{-\pi^2 [\nu + \frac{(f_c-f)t_0^2}{t_k^2 + t_0^2}]^2 (t_k^2 + t_0^2)\right\} d\nu. \quad (1.90)$$

The definite integral of this expression has the value $\pi^{-1/2}(t_k^2+t_0^2)^{-1/2}$, and equation (1.90) thus reduces to

$$\phi_m(f) = \frac{\pi}{(t_k^{-2} + t_0^{-2})^{\frac{1}{2}}} \exp\left\{-\pi^2 \frac{(f_c-f)^2}{t_k^{-2} + t_0^{-2}}\right\}. \qquad (1.91)$$

The resulting power spectrum of a partially coherent oscillation with Gaussian intensity modulation has an $1/e$-bandwidth

$$\Delta f = \frac{2}{\pi} \sqrt{(t_k^{-2} + t_0^{-2})}. \qquad (1.92)$$

As shown in Fig.1.15(b), it extends over a range in which the modulation bandwidth $2/(\pi\, t_0)$ of the Gaussian pulse with 1/e-duration t_0 adds quadratically to the spectral width $2/(\pi\, t_k)$ of the partially coherent oscillation with Gaussian autocorrelation of coherence time t_k.

We consider now a signal carried by a uniform plane wave which is partially coherent in time and has a Gaussian autocorrelation with coherence time t_k. The signal may be considered to modulate each of the spectral components of the partially coherent oscillation in the same way and then to travel in each of these components independently. While the wave propagates through the dispersive medium each of its spectral components suffers a time delay per unit length in the direction of propagation according to

$$\tau = \frac{1}{v_g} = \frac{d\beta}{d\omega}. \qquad (1.93)$$

Since for the dispersive medium this specific group delay changes with frequency; each of the spectral components takes a different time to propagate a certain distance. The signals which are carried by different spectral components of the partially coherent wave therefore arrive at different times. Superposition of these dispersed signals results in a distorted signal. If the frequency spectrum of the partially coherent wave is not too wide, the delay difference per unit frequency and unit length along the path of propagation is approximately $d\tau/df$, which with the specific group delay τ from equation (1.93) is given by

$$\frac{d\tau}{df} = 2\pi\beta''. \qquad (1.94)$$

For spectral components which are Δf apart and $\Delta f/2$ above and below a centre frequency f_c, the total delay difference over a length of propagation L is

$$\Delta t = L \frac{d\tau}{df} \Delta f = 2\pi L\, \beta''\, \Delta f. \qquad (1.95)$$

UNIFORM PLANE WAVES

Referring now to the example of a partially coherent wave with Gaussian autocorrelation of coherence time t_k, the 1/e-bandwidth is $\Delta f = 2/(\pi t_k)$, and the corresponding delay difference

$$\Delta t = 4L\ \beta''/t_k. \tag{1.96}$$

If a Gaussian pulse modulates the input wave with an 1/e-width of its intensity t_0, the output pulse has a Gaussian shape as well, but it is widened, not only according to equation (1.71) due to the bandwidth of the modulation, but also according to equation (1.96) due to partial coherence of the carrier. Both effects add quadratically in the resulting power spectrum of equation (1.91) and lead to the total 1/e-bandwidth in equation (1.92). In pulse widening these effects therefore also add quadratically, leading to an 1/e-width of the Gaussian pulse at the output:

$$t_1 = \left[t_0^2 + (4\beta''L)^2 \left(\frac{1}{t_0^2} + \frac{1}{t_k^2}\right)\right]^{\frac{1}{2}}. \tag{1.97}$$

This expression, as well as the previous equations (1.71) and (1.96), shows that signal distortion in dispersive media is governed by the specific group delay $\tau = d\beta/d\omega$ and its dependence on frequency $d\tau/df = 2\pi\ d^2\beta/d\omega^2$.

In order to obtain a convenient representation of these quantities in a dimensionless form, the specific group delay is multiplied by the light velocity $c = (\varepsilon_0\ \mu_0)^{-1/2}$ in free space:

$$c\tau = \frac{d\beta}{dk}. \tag{1.98}$$

With $\beta = nk$ for transparent media this product is designated by n',

$$n' = \frac{d(nk)}{dk} = c\tau, \tag{1.99}$$

and called the group index of the material. The group index is a convenient measure for the group delay of a material in its spectral range of transparency. Its relation to the group velocity, $n' = c/v_g$, leads from equation (1.49) to a representation of n' in terms of the refractive index n and the characteristic frequencies ω_0 and ω_d of the material:

$$n' = n + \frac{\omega_0 \omega_d \omega^2}{n(\omega_0^2 - \omega^2)^2} \; . \tag{1.100}$$

The spectral range of transparency lies substantially below both ω_0 and ω_d. The group index is therefore only slightly larger than the index of refraction. However, the difference between n' and n increases with increasing frequency. Equation (1.100) has been evaluated for some of the glasses listed in Table 1.1 using the characteristic frequencies from the same table. The resulting group index is plotted in Fig.1.16 as a function of frequency and free space wavelength.

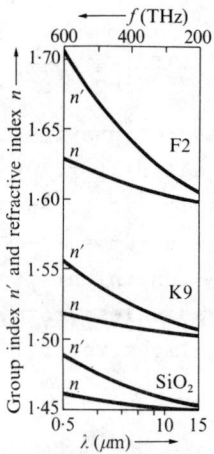

Fig.1.16. Refractive index n and group index n' from an evaluation of equations (1.47) and (1.100) using characteristic frequencies ω_0 and ω_d as listed in Table 1.1.

The second quantity of interest for signal distortion due to material dispersion is the derivative of specific group delay with respect to frequency

$$\frac{d\tau}{df} = 2\pi \frac{d^2\beta}{d\omega^2} \; . \tag{1.101}$$

UNIFORM PLANE WAVES

For a dimensionless presentation we multiply this derivative by the frequency f and the velocity of light c in free space. The resulting quantity may be expressed with the group index in the form

$$k \frac{dn'}{dk} = k \frac{d^2(nk)}{dk^2} = cf \frac{d\tau}{df},$$

and is called the dispersion coefficient. Using equations (1.46), (1.47), and (1.72), it may also be written in terms of the refractive index n and the characteristic frequencies ω_0 and ω_d of the material

$$k \frac{dn'}{dk} = 2 \frac{(n^2-1)^3}{n} \frac{\omega^2}{\omega_d^2} \left(1 + \frac{\omega^2}{\omega_0^2}\right). \qquad (1.102)$$

For several of the glasses listed in Table 1.1 the dispersion coefficient according to equation (1.102) has been evaluated using values for the refractive index and characteristic frequencies that also appear in the same table. Fig.1.17 shows these dispersion coefficients plotted against frequency f and free space wavelength λ. The curves demonstrate the predominantly square law dependence of dispersion

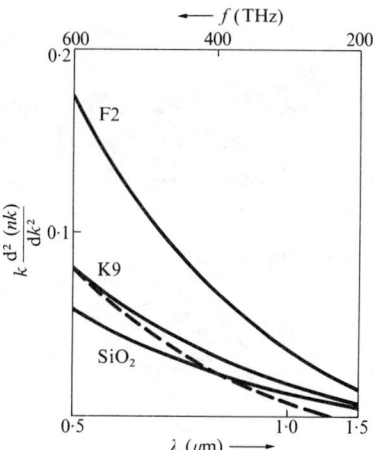

Fig.1.17. Dispersion coefficients $k d^2(kn)/dk^2$ from an evaluation of equation (1.102) using characteristic frequencies ω_0 and ω_d as listed in Table 1.1. Broken line for SiO_2 from a more accurate equation (Payne and Gambling 1975).

coefficient on frequency, caused mainly by the factor ω^2 in front of the brackets in equation (1.102). Of all the glasses listed in Table 1.1 fused silica has the smallest dispersion coefficient over the full spectral range of Fig.1.17. The effective energy gap is widest for SiO_2 of all the glasses in Table 1.1, and therefore the characteristic frequency ω_0 the highest, leading to the low dispersion coefficient, despite the intermediate value of ω_d for SiO_2 in Table 1.1.

The broken line in Fig.1.17 represents the dispersion coefficient for optical quality fused silica from a more accurate equation for the real part $(n^2 - 1)$ of the susceptibility (Payne and Gambling 1975). Instead of lumping all contributions to $(n^2 - 1)$ into just one resonance term as in equation (1.47), this more accurate representation uses three such terms in the form (Malitson 1965)

$$n^2 - 1 = \sum_{m=1}^{3} \frac{B_m}{\omega_m^2 - \omega^2}$$

to approximate measured results at 60 wavelengths from 0·21 to 3·71 μm with a least-squares fit. The values for the ω_m and B_m which resulted from the least-squares fit are

$\omega_1 = 2\cdot75370 \times 10^{16}$ s^{-1} $B_1 = 5\cdot27895 \times 10^{32}$ s^{-2}

$\omega_2 = 1\cdot62047 \times 10^{16}$ s^{-1} $B_2 = 1\cdot07122 \times 10^{32}$ s^{-2}

$\omega_3 = 1\cdot90342 \times 10^{14}$ s^{-1} $B_3 = 3\cdot25156 \times 10^{28}$ s^{-2}.

They correspond to resonances at

$$\lambda_1 = 0\cdot06840 \text{ μm},$$

$$\lambda_2 = 0\cdot11624 \text{ μm},$$

$$\lambda_3 = 9\cdot8961 \text{ μm}.$$

The first two of these resonances occur in the ultraviolet range and account for electronic interaction. The third of these resonances, however, is in the infrared range and accounts for the interaction with molecular vibrations.

The most significant property of this more accurate dispersion characteristic of fused silica is the zero crossing of its dispersion coefficient at a wavelength $\lambda = 1{\cdot}28$ μm. At or near this wavelength fused silica of good optical quality shows little or no material dispersion.

For convenient evaluation of the pulse widening of signals which are carried in the form of Gaussian pulses by partially coherent waves, we now introduce the dispersion coefficient into equation (1.97) and obtain

$$t_1 = \left[t_0^2 + \left(\frac{4L}{\omega c} k \frac{dn'}{dk} \right)^2 \left(\frac{1}{t_0^2} + \frac{1}{t_k^2} \right) \right]^{\frac{1}{2}}. \quad (1.103)$$

The 1/e-width t_1 of the output pulse from this expression is plotted in Fig.1.18 as a function of the 1/e-width t_0 of the input pulse for different values of

$$\beta''L = \frac{Lk}{\omega c} \frac{dn'}{dk} \quad (1.104)$$

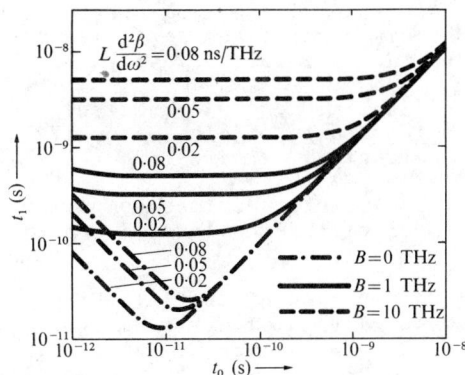

Fig.1.18. Output pulse width t_1 versus input pulse width t_0 for Gaussian pulses of partially coherent light with coherence time t_k or spectral width $B = 2/(\pi t_k)$ and for different dispersion factors $L d^2\beta/d\omega^2$ (Miyazaki 1974).

and the 1/e-bandwidth $\Delta f = 2/(\pi t_k)$ of the partially coherent carrier. As a function of t_0 the output pulse has a minimum width at

$$t_0 = 2(\beta''L)^{\frac{1}{2}} = 2\left(\frac{Lk}{\omega c}\frac{dn'}{dk}\right)^{\frac{1}{2}}, \qquad (1.105)$$

where the material dispersion widens the output pulse to

$$t_1 = [8\beta''L + (4\beta''L/t_k)^2]^{\frac{1}{2}} = \left[\frac{8L}{\omega c}k\frac{dn'}{dk} + \left(\frac{4L}{\omega c}\frac{k}{t_k}\frac{dn'}{dk}\right)^2\right]^{\frac{1}{2}}. \quad (1.106)$$

Input pulses which are narrower than t_0 from equation (1.105) have a wider output pulse because their modulation spectrum spreads over a wider range of frequencies. Input pulses which are wider than t_0 from equation (1.105) also have output pulses wider than the minimum from equation (1.106), but only because the input pulse is so wide to start with, and not so much from material dispersion.

REFERENCES

Di Domenico, J.M. (1972). Material dispersion in optical fibre waveguides. *Appl. Optics* 11, 652-4.

Gladstone, J.H. and Dale, T.P. (1863). Researches on the refraction, dispersion, and sensitiveness of liquids. *Phil. Trans. R. Soc.* 153, 317-43.

Huggins, M.L. and Sun, K.-H. (1943). Calculation of density and optical constants of glass from its composition in weight percentage. *J. Am. Ceram. Soc.* 26, 4-11.

Malitson, I.H. (1965). Interspecimen comparison of the refractive index of fused silica. *J. opt. Soc. Amer.* 55, 1205-09.

Miyazaki, Y. (1974). Pulses of partially coherent optical waves in dispersive dielectric waveguides. *Archiv. Elektron. & Übertragungstech.* 28, 160-72.

Payne, D.N. and Gambling, W.A. (1975). Zero material dispersion in optical fibres. *Electr. Lett.* 11, 176-8.

Pinnow, D.A., Rich, T.C., Ostermayer, J.F.W., and Di Domenico, J.M. (1973). Fundamental optical attenuation limits in the liquid and glassy state with application to fiber optical waveguide materials. *Appl. Phys. Lett.* 22, 527-9.

Wemple, S.H. and Di Domenico, J.M. (1971). Behaviour of the electronic dielectric constant in covalent ionic materials. *Phys. Rev.* B3, 1338-51.

2
DIELECTRIC FILMS

Optical film waveguides consist of a thin dielectric film of low optical absorption and an index of refraction n_1 deposited on a transparent substrate preferably of likewise low absorption and with an index of refraction n_2 which is somewhat smaller than n_1. The region above the film usually remains free space or air with refractive index $n_0 = 1$. Symmetrical films constitute the special case in which the regions on both sides of the film have the same refractive index $n_2 < n_1$, which, in case of a film suspended in free space or air reduces to $n_2 = 1$.

The symmetrical dielectric slab or film represents the simplest form of optical waveguide. It may therefore serve as a model to gain understanding of wave propagation in optical waveguides. Beyond this, it also occurs as a cross-sectional element of other more complicated guiding structures. The asymmetric dielectric film on a substrate, although rather more complicated, also appears to be quite a useful model for more general types of optical waveguides. More important, however, this dielectric film is actually used for guiding light and occurs in a number of optical components. In its simple and unique form and because of its universal usefulness throughout optical circuits it forms the basic element of integrated optics.

To survey the different modes of propagation of light waves in or near an asymmetric film guide we place an electromagnetic line source on the front face of the film, running parallel to it as in Fig.2.1. We assume this line source to have no y-dependence in the coordinates of Fig.2.1. For an actual experiment a slot in an opaque screen could simulate this line source and a plane uniform wave incident parallel to the z-axis could excite this slot. The slot or line source radiates into all directions from $\theta_1 = -\pi/2$ to $\theta_1 = \pi/2$ with respect to the z-axis. Decomposing its radiation into uniform plane waves, each wave may be represented by a ray whose angle θ_1 with respect to the plane of the film

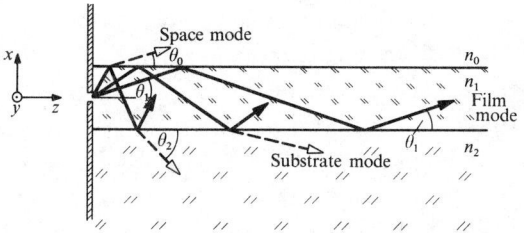

Fig.2.1. Rays of propagation for guided modes and radiation modes in a dielectric film guide.

corresponds to the direction of propagation of the respective plane-wave. As long as θ_1 is larger than

$$\theta_{oc} = \arccos(n_0/n_1)$$

the ray will be partially reflected upon striking the boundary between film and space above with a refracted ray radiating into space in the direction θ_0 with

$$n_0 \cos\theta_0 = n_1 \cos\theta_1 .$$

The reflected ray strikes the lower film boundary to the substrate at the angle $-\theta_1$, and for $n_2 > n_0$ suffers less reflection than at the upper boundary. The refracted ray radiates into the substrate with θ_2 where

$$n_2 \cos\theta_2 = n_1 \cos\theta_1 .$$

The process of partial reflection and refraction at the upper and lower film boundaries continues to repeat successively with more and more of the initial power in the uniform plane-wave represented by the ray leaking out into upper space and into the substrate. Because of the continuous radiation of this mode of propagation into space above as well as into the substrate below, the waves represented by rays with $\theta_{oc} < |\theta_1| < \pi/2$ are called radiation modes, or more specifically space radiation modes, (briefly, space modes) since part of their power radiates into the space above the film. There are actually solutions of Maxwell's equations for the film guide consisting of plane waves corresponding to $\theta_{oc} < |\theta_1| < \pi/2$ which decay exponentially in the z-direction and account for the leakage of power out into space and substrate. These solutions are called leaky waves.

Rays in Fig.2.1 with angles of propagation in the range

$$\theta_{2c} < |\theta_1| < \theta_{oc} \qquad (2.1)$$

with

$$\theta_{2c} = \text{arc } \cos(n_2/n_1) \qquad (2.2)$$

undergo total reflection at the upper boundary between film and space but only partial reflection at the film substrate boundary; refracted waves radiate into the substrate alone. Waves represented by rays with angles in the range of equation (2.1) also belong to the group of radiation modes. More specifically they are called substrate radiation modes, or briefly substrate modes, since all of their power radiates into the substrate. For the range of θ_1-values in equation (2.1), plane-wave solutions of Maxwell's equation exist which also decay exponentially in the z-direction and account directly for the leakage of power into the substrate.

Rays which radiate from the line source in Fig.2.1 at angles

$$\theta_1 < \theta_{2c} \qquad (2.3)$$

with respect to the z-axis experience total reflection, not only at the upper film boundary to free space, but also at the lower film boundary to the substrate. Any power carried by such rays remains confined to the film and is guided by it with the associated fields beyond the film boundaries decaying exponentially in the transverse direction. Waves represented by rays with angles θ_1 small enough according to equation (2.3) are guided by the film and are therefore called guided film modes, or briefly film modes.

As in equation (2.2) the limiting angle $\theta_1 = \theta_{2c}$ of rays inside the film for guided modes follows equivalently from

$$\sin \theta_{2c} = (n_1^2 - n_2^2)^{1/2}/n_1.$$

Rays incident from vacuum on the front face at an angle θ with respect to the z-axis are refracted upon entering the film to an angle θ_1 according to

$$\sin \theta = n_1 \sin \theta_1 .$$

If $\theta < \theta_c$ with

$$\sin \theta_c = (n_1^2 - n_2^2)^{1/2} , \qquad (2.4)$$

these rays fall into the range of guided modes; the film accepts them by total internal reflection at both boundaries. In optics the sine of the maximum angle of incidence for rays that a particular aperture still accepts is called the numerical aperture. This expression has also been adopted for optical waveguides so that equation (2.4) represents the numerical aperture of the film guide.

Of the three types of modes distinguished by different ranges of their angle θ_1 in Fig.2.1, film modes are of greatest practical significance because they are the only modes which are completely bound to the film and guided by it. Substrate modes radiate through all the substrate while space modes radiate both into the substrate and into space. While film modes are utilized for light guidance, substrate and space modes usually represent undesired and parasitic radiation. The following sections will therefore concentrate on the characteristics of film modes. Substrate and space modes must, however, also be considered because we cannot completely suppress them when exciting film modes as, for example, in Fig.2.1. Furthermore, when film modes travel through any film perturbations, they couple to the radiation modes and lose power to them or suffer from interference effects with radiation modes.

2.1. FILM MODES

Most of the propagation characteristics of film modes are conveniently derived by considering the plane uniform wave in Fig. 2.2 which travels inside the film on a zig-zag path and experiences total internal reflection when incident on the boundaries at an angle $\theta_1 < \theta_{2c}$. Fig.2.2 shows the

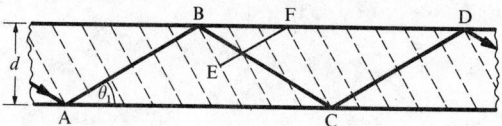

Fig.2.2. Zig-zag path for rays of a guided film mode.

typical ray path ABCD which represents this wave. Also indicated by dotted lines are phase fronts belonging to the uniform plane wave while travelling from A to B and from C to D, respectively. After two successive reflections, for example at B and C, the wave must repeat in phase in order to yield a self-consistent field distribution, which then forms a film mode. When not satisfying this phase condition the wave would interfere destructively with itself and just die out. The phase front containing B and E must change by the same phase angle when moving directly from E to F into the phase front through F and C as when undergoing the two reflections at B and C and moving from B to C in between. Any phase differences between these two paths may only amount to $\pm 2m\pi$ with m an integer number. Introducing ϕ_{10} and ϕ_{12} as phase shifts associated with the total internal reflection at B and C, respectively, the phase condition for the ray between B and C in Fig.2.2 is

$$- k_1 (\overline{BC} - \overline{EF}) + \phi_{10} + \phi_{12} = -2m\pi \qquad (2.5)$$

where for the longer distance $\overline{BC}$ between B and C

$$\overline{BC} = d/\sin \theta_1 ,$$

DIELECTRIC FILMS 63

and for the shorter distance $\overline{EF}$ between E and F

$$\overline{EF} = \overline{BF} \cos \theta_1$$

with

$$\overline{BF} = d/\tan \theta_1 - d \tan \theta_1 .$$

Substituting these expressions for the various distances into equation (2.5) leads to

$$-2k_1 d \sin \theta_1 + \phi_{10} + \phi_{12} = -2m\pi . \qquad (2.6)$$

This equation constitutes the characteristic or eigenvalue equation of film modes. Values for the angle θ_1 which solve this equation belong to film modes. A uniform plane wave which travels with any such angle on a zig-zag path fits into the film and forms a film mode.

The phase angles ϕ_{10} and ϕ_{12} take on different values depending on the polarization of the plane uniform wave. Film modes therefore must be distinguished with respect to the polarization of the plane uniform wave from which they originate. If the magnetic field is parallel to the plane of incidence, the film modes are designated as H-modes or TE-modes, because only the magnetic part of their total field has a component in the direction of propagation; also these film modes have a transverse electric (TE) field. To evaluate the characteristic equation for these H-modes the phase angle ϕ_e from equation (1.25) must be substituted for ϕ_{10} and ϕ_{12} in equation (2.6), leading to

$$\arctan \frac{jk_{0x}}{k_{1x}} + \arctan \frac{jk_{2x}}{k_{1x}} = d k_1 \sin \theta_1 - m\pi . \qquad (2.7)$$

For a representation of this eigenvalue equation in terms of dimensionless quantities, we take the product of the transverse component k_{1x} of the wave vector inside the film and the film thickness d and denote it by

$$u = k_{1x} d = d k_1 \sin \theta_1 . \qquad (2.8)$$

Similarly we take the transverse attenuation constants jk_{0x} and jk_{2x} of the evanescent fields outside the film times the film thickness and denote them by

$$v = jk_{2x} d \qquad w = jk_{0x} d. \qquad (2.9)$$

With the wave vector components in equation (2.7) thus normalized, the eigenvalue equation appears as

$$\arc\tan v/u + \arc\tan w/u = u - m\pi. \qquad (2.10)$$

Taking the tangent of this equation reduces it to the even simpler form

$$\tan u = \frac{u(v + w)}{u^2 - vw}. \qquad (2.11)$$

This equation, however, no longer shows the term $m\pi$ in which the integer m denotes the number of wavelengths by which the ray path from B to C in Fig.2.2 including the two reflections at B and C is optically longer than the direct ray from E to F.

Before treating the solutions of the eigenvalue equation (2.11) for film modes of the H- or TE-type, we will also obtain the corresponding eigenvalue equation for film modes which derive from plane uniform waves with their electric field vector parallel to the plane of incidence in Fig.2.2. These film modes have a total field of which only the electric field vector has a component in the direction of propagation while the magnetic field has only the transverse y-component. They are therefore designated as E-waves or TM-waves, respectively. The eigenvalue equation for these E- or TM-waves is obtained when expressions corresponding to ϕ_m from equation (1.26) are substituted for the two phase angles ϕ_{10} and ϕ_{12} of total reflection in equation (2.6). When the transverse components of wave vectors are again normalized as in equations (2.8) and (2.9) with respect to film thickness and the tangent of equation (2.6) is taken, the eigenvalue equation for film modes of the E- or TM-type follows as

$$\tan u = \frac{n_1^2 \, u(n_0^2 v + n_2^2 w)}{n_0^2 n_2^2 u^2 - n_1^4 vw} \,. \tag{2.12}$$

The parameters u, v, w representing the transverse phase and attenuation constants inside the film and in the substrate below as well as in the space above the film relate to the longitudinal component

$$k_z = \beta$$

of the wave vector and to the wave numbers k_0, k_1, k_2 of the three regions as follows:

$$u/d = k_{1x} = (k_1^2 - \beta^2)^{1/2}$$
$$v/d = jk_{2x} = (\beta^2 - k_2^2)^{1/2} \tag{2.13}$$
$$w/d = jk_{0x} = (\beta^2 - k_0^2)^{1/2} \,.$$

The longitudinal wave vector component β describes the z-dependence of the fields. If real values for β are obtained from the solution of the eigenvalue equation, they are phase constants for the propagation of film modes. One of the two eigenvalue equations, together with the three equations (2.13), forms a system of four equations for the four unknown quantities u, v, w, and the phase constant β. Each of the two eigenvalue equations together with equation (2.13) may have one solution for any of the numbers m in equation (2.6). Depending on this number m and on the polarization of fields for which the eigenvalue equation is taken, the film modes corresponding to these solutions are designated as H_m- and E_m-waves, respectively. The system of equations contains the transcendental eigenvalue equation. Therefore, in general only numerical methods will lead to solutions.

For one such method we note that with ϕ_{10} and ϕ_{12} written in terms of θ_1 for H-waves

$$\tan(\phi_{10}/2) = (n_1^2 \cos^2\theta_1 - n_0^2)^{1/2} / (n_1 \sin\theta_1)$$

$$\tan(\phi_{12}/2) = (n_1^2 \cos^2\theta_1 - n_2^2)^{1/2} / (n_1 \sin\theta_1),$$

and for E-waves

$$\tan(\phi_{10}/2) = n_1^2 (n_1^2 \cos^2\theta_1 - n_0^2)^{1/2} / (n_0^2 n_1 \sin\theta_1)$$

$$\tan(\phi_{12}/2) = n_1^2 (n_1^2 \cos^2\theta_1 - n_2^2)^{1/2} / (n_2^2 n_1 \sin\theta_1),$$

the eigenvalue equation (2.6) contains transcendental functions only of θ_1. For given n_0, n_1, n_2, and m we may therefore assign values to θ_1 in the range

$$\cos\theta_1 > n_2/n_1$$

of guided modes and compute kd from equation (2.6) and β/k from

$$\beta/k = n_1 \cos\theta_1. \qquad (2.14)$$

From this parameter evaluation we may then plot β/k as a function of kd.

For solutions of equations (2.11) and (2.12) which reveal the functional dependencies in general form, we consider films with refractive indices n_1 only slightly larger than the refractive index n_2 of the substrate. Both indices n_1 and n_2 are, however, assumed to be distinctly larger than the refractive index n_0 of the space above the film. This case deserves special interest because in many practical configurations of optical film guides, the three indices of refraction relate to each other in the specified manner. The space above the film is usually air and has $n_0 = 1$, while materials for film and substrate have indices in the range $n = 1 \cdot 4$ to $n = 2$ and therefore are definitely larger than n_0. In addition substrate and film very often consist of materials which are both transparent in the spectral range of operation and similar in many other respects except that the film index n_1 must be larger than the substrate index n_2 for total internal reflection to be possible. But only a small index difference will suffice, and for many practical applications is even desirable. This is the so-called case of a weakly guiding film in which the limiting angle θ_{2c}, given by

$$\sin \theta_{2c} = (n_1^2 - n_2^2)^{1/2}/n_1,$$

of total internal reflection at the boundary between film and substrate is quite small and film modes are only formed by rays with angles θ_1 even smaller than θ_{2c}. Optical film guides made from glass substrates with a glass film of somewhat higher index of refraction belong to this class of weakly guiding films. Usually the glass film has an index of refraction which is only of the order of one per cent larger than the refractive index of the substrate. For film modes of weakly guiding films values of θ_1, which are always below θ_{2c}, will remain quite small. The range of the transverse phase parameter u is then, according to equation (2.8), restricted to $u \ll k_1 d$ and will also remain quite small. Under these conditions the phase constant β according to equation (2.14) differs only little from $k_1 = n_1 k$ as well as from $k_2 = n_2 k$, since n_1 is only slightly larger than n_2. Therefore the transverse attenuation parameter v will also remain small, with $v \ll k_1 d$. However, the phase constant β exceeds the wave number $k_0 = k$ of the space above by a significant amount, so that the transverse attenuation parameter w will also be large compared to u and v. The inequality

$$w \gg u, v \qquad (2.15)$$

reflects the large step in refractive index between film and space. The range of θ_1, being limited to small values $\theta_1 < \theta_{2c}$ due to the weakly guiding character, always remains far below the limiting angle θ_{0c} for total internal reflection at the boundary between film and space because

$$\sin\theta_{0c} = (n_1^2 - n_0^2)^{1/2}/n_1.$$

In this range of θ_1 with respect to θ_{0c} the exponentially evanescent fields in the space above decay very rapidly according to the large transverse attenuation constant $\alpha_x = w/d$.

With the inequalities (2.15) of weakly guiding films, their eigenvalue equations (2.11) for H_m-waves and (2.12) for E_m-waves may be approximated by one and the same equation:

$$v = -u \cot u. \qquad (2.16)$$

Both eigenvalue equations reducing to only this one equation will, of course, let two solutions for each m, one from each of the two original equations, degenerate into just one solution of the approximate eigenvalue equation. Therefore in the approximation for weakly guiding films, E_m-waves and H_m-waves turn out to be degenerate with respect to their eigenvalue β. To each solution for β from (2.11) and (2.12) belongs one E_m- and one H_m-wave.

The approximate form (2.16) of the eigenvalue equation for weakly guiding films contains neither the parameter w nor any of the refractive indices explicitly. Its solutions therefore allow a universal representation of propagation characteristics of film modes for any kind of weakly guiding film. In order to obtain this solution graphically, we plot v according to equation (2.16) as a function of u as in Fig.2.3. Only the first quadrant of the (v,u)-plane is of interest because film modes have only positive real values of u and v. Besides equation (2.16) as plotted in Fig.2.3, u and v must also satisfy equation (2.13). The sum of their squares, which from equation (2.13) is $u^2 + v^2 = (k_1^2 - k_2^2)d^2$, forms a circle in the uv-plane of Fig.2.3 with radius

$$V = 2\pi (n_1^2 - n_2^2)^{1/2} d/\lambda. \qquad (2.17)$$

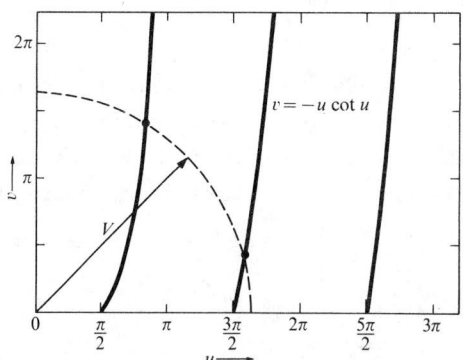

Fig.2.3. Graphical solution of the eigenvalue equation for guided modes of weakly guiding films.

Points at which this circle intersects branches of the curve $v = -u \cot u$ are solutions of the eigenvalue problem. Any such point represents two film waves.

From these cross-over points in Fig.2.3 the transverse wave parameters u and v are obtained as a function of V, the radius of the circle in Fig.2.3. V according to equation (2.17) contains the relevant film parameters; they are the refractive indices n_1 and n_2 and film thickness d as well as the frequency f expressed by its free-space wavelength $\lambda = c/f$. For the weakly guiding film the solutions of the eigenvalue problem in the form of the transverse wave parameters u and v depend only on the parameter V. V in turn combines film guide characteristics and frequency; it is therefore referred to as the generalized frequency parameter for the film, or just as the film parameter.

The transverse wave parameters u and v describe the transverse field distribution of the particular film mode, u as the transverse phase parameter in the film, and v as the transverse attenuation parameter in the substrate. Of even greater interest than the transverse wave parameters is the phase constant β in the direction of propagation of the film mode. The quantity

$$B = \frac{\beta^2 - n_2^2 k^2}{(n_1^2 - n_2^2)k^2} \qquad (2.18)$$

contains this phase constant and allows the dependence of β on the film parameter V to be universally represented, just as u and v may universally be represented as functions of V only. This representation follows from equations (2.13) and (2.17), according to which

$$B = v^2/V^2 . \qquad (2.19)$$

Moreover for the weakly guiding film with $n_1 \simeq n_2$, the expression

$$B \simeq \frac{\beta - n_2 k}{(n_1 - n_2)k} \qquad (2.20)$$

approximates the phase parameter quite well. For the weakly guiding film therefore, the phase parameter B is the difference between the phase constant of the respective film mode and the wave number of the substrate material divided by the difference of wave numbers of film and substrate materials.

Fig.2.4 shows for the different film modes the phase parameter B as a function of the film parameter V. No film

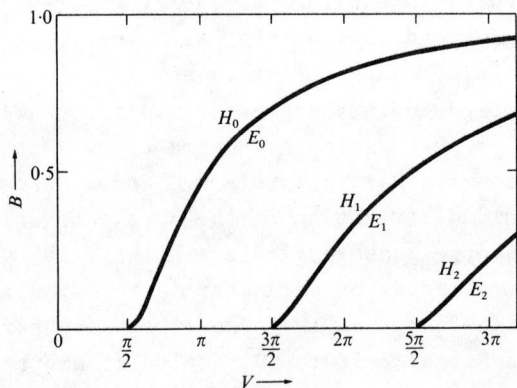

Fig.2.4. Phase parameter of guided modes as a function of the film parameter in weakly guiding films.

modes exist in the weakly guiding film as long as $V < \pi/2$. Only when the light wavelength is short enough, or its frequency high enough, and when the film is thick enough and the refractive indices differ sufficiently for the film parameter to become $V > \pi/2$, will the circle in Fig.2.3 first cross only the lowest branch of the curve $v = -u \cot u$. For this solution we have $m = 0$ in equation (2.6); it therefore represents the H_0- and E_0-modes of the film. At the cut-off, for $V = \pi/2$, the phase parameter starts at $B = 0$. The phase constant β at cut-off therefore equals the wave number $n_2 k$ of the substrate. The film modes at cut-off propagate as uniform plane waves in the substrate with the phase constant of the substrate. The transverse attenuation parameter $v = V(B)^{\frac{1}{2}}$ is likewise zero at cut-off, while the transverse phase parameter is $u = \pi/2$ just as V is $\pi/2$. At cut-off therefore, where the film modes begin to exist, their fields extend infinitely wide into the substrate, with no dependence on the transverse coordinates for $v = 0$. This field distribution into the substrate at cut-off corresponds to the limit of total reflection at the substrate boundary of the film. Inside the film at cut-off the transverse distribution of the electric field E_y for the H_0-mode, and of the magnetic field H_y for the E_0-mode forms a quarter of a standing sine-wave with its zero at the film boundary to the upper space and its maximum at the substrate boundary of the film.

As the film parameter V increases from its cut-off point at $V = \pi/2$, the cross-over points of the lowest-order solution move up on the lowest branch of the curve $v = -u \cot u$. In this process v becomes larger and larger and eventually approaches V while u goes asymptotically to $u = \pi$. At the same time the phase parameter B approaches unity, indicating that away from cut-off, the film modes propagate more and more with the wave number $n_1 k$ of the film. With the transverse attenuation parameter growing larger and larger as V increases, the fields penetrate less and less into the substrate; they concentrate more and more into the film, where with u approaching π, they distribute in the form of half a standing sine-wave, with the field equalling zero or being near zero at the film boundaries. Fig.2.5(a) shows the transverse distribution of the electric field E_y for the H_0-mode and the magnetic field H_y for the E_0-mode at an intermediate value of $V > \pi/2$. Because the transverse attenuation parameter w for the space above the

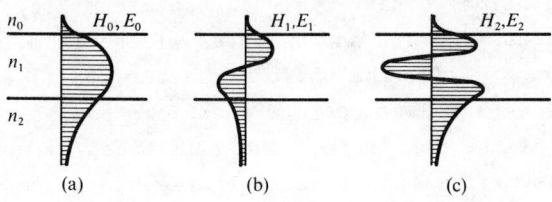

Fig.2.5. Transverse distribution of the electric field component E_y for H-modes and the magnetic field component H_y for E-modes, respectively (a) H_0- and E_0-modes, (b) H_1- and E_1-modes, (c) H_2- and E_2-modes.

film assumes very large values for weakly guiding films, the fields decay correspondingly faster into the upper space, than they do into the substrate.

With increasing film parameter V, whenever it crosses one of the values $(2m + 1)\pi/2$, a new pair of film modes comes into existence. Here m still has the same meaning as in equation (2.6). The new film modes with cut-off at

$$V_m = (2m + 1)\pi/2$$

are therefore the H_m- and E_m-modes. At cut-off their transverse attenuation parameters v into the substrate start again at zero so that their fields extend infinitely wide into the substrate with no transverse dependence. Their phase parameters β also start from zero at cut-off, so that all film modes at cut-off propagate with the wave number $n_2 k$ of the substrate as phase constant β. The transverse phase parameter u inside the film has values

$$u_{mc} = V_m = (2m + 1)\pi/2 \tag{2.21}$$

at cut-off. The transverse field distribution inside the film of E_y (H_y) for H_m-(E_m)-modes at cut-off therefore forms standing waves with field zeros at the boundary to upper space, maximum values at the substrate boundary, and $2m + 1$ quarter waves inbetween.

As the film parameter V moves up and away from the respective cut-off value V_m, the transverse attenuation parameter v increases from zero and eventually approaches V. The field then penetrates less into the substrate and eventually concentrates entirely within the film. The phase parameter B approaches unity in the process, so that far above cut-off all film modes propagate with the wave number $n_1 k$ as phase constant. The transverse phase parameter u within the film moves from u_{mc} according to equation (2.21) to

$$u_{m\infty} = (m + 1)\pi \qquad (2.22)$$

as V increases from its respective cut-off value V_m to values far above cut-off. Far above cut-off, therefore, the transverse fields E_y of H_m-waves, and H_y of E_m-waves distribute with $(m + 1)$ half sine-waves across the film and have near-zero fields at both film boundaries. Figs. 2.5(b) and (c) show these transverse field distributions for H_1- and E_1-waves as well as for H_2- and E_2-waves, again at intermediate values of V somewhat above the respective cut-off value V_m. The order m of the film mode shows up in these transverse field distributions as the number of transverse field zeros across the film. In equation (2.5) this order m also stands for the integer multiples of 2π by which the direct ray is shorter in optical length than the indirect ray via the two reflections at the boundaries. The angle θ_1 for the direction of propagation of the plane uniform waves, which make up the film modes starts at cut-off with the limiting angle θ_{2c} for total internal reflection at the substrate boundary, and goes to zero far above cut-off. The uniform elementary waves then propagate more and more as in a homogeneous space uniformly filled with the film material. For any film guide at a given light wavelength, only a limited number of film modes exist as solution of equations (2.13) and (2.16). If $V = (2m + 1)\pi/2$ the film guides $M = 2m + 2$ modes, including all H_m- and E_m-modes up to the order $m = V/\pi - 1/2$. In terms of the film parameter the total number of guided modes is hence given by the largest integer smaller than

$$M = 2V/\pi + 1$$

for $V > \pi/2$.

The exact eigenvalue equations (2.11) and (2.12) for film modes are only approximated by equation (2.16) for weakly guiding films with $(n_1 - n_2) \ll n_1$. Therefore all solutions of equation (2.16) together with equations (2.13) represent only approximations to the exact solutions. As one particular consequence of this approximation, any H_m-mode appears to be degenerate with the E_m-mode of the same order m. To estimate the quality of this approximation and the possible deviations of the exact solutions from these approximations, we will now obtain a more accurate solution of equations (2.11) and (2.12) by perturbation. We base this perturbation on the solution of equation (2.16), designating the corresponding values of u and v by u_0 and v_0, respectively, as the zero-order approximations to the exact solution. For a first-order perturbation we let

$$u = u_0 + \Delta u \quad \text{and} \quad v = v_0 + \Delta v.$$

From equation (2.13) we have to first order

$$u_0^2 + v_0^2 + 2(u_0 \Delta u + v_0 \Delta v) = V^2,$$

which requires

$$u_0 \Delta u = -v_0 \Delta v,$$

since $u_0^2 + v_0^2 = V^2$. The phase parameter $B = v^2/V^2$ to first order is

$$B = (v_0^2 + 2v_0 \Delta v)/V^2. \tag{2.23}$$

The first-order perturbation Δv of the transverse decay parameter v depends on whether we consider an H_m-wave with its eigenvalue equation (2.11) or an E_m-wave with its eigenvalue equation (2.12). From equation (2.11) follows the first-order perturbation

$$\Delta v = \frac{u_0^4}{w v_0 (u_0^2 + v_0^3)} \qquad (2.24)$$

while equation (2.12) yields the first-order perturbation

$$\Delta v = \frac{n_0^2 u_0^4}{n_1^2 w v_0 (u_0^2 + v_0^3)} \qquad (2.25)$$

which differs from equation (2.24) only by the ratio of relative film permittivities n_0^2/n_1^2 on the right-hand side. Both expressions (2.24) and (2.25) give perturbations of v and therefore hold only as long as $\Delta v \ll v_0$. They fail near cut-off where v_0 itself starts from zero. The transverse attenuation parameter w for the upper space is large compared to u_0 and v_0 and, in accordance with the first-order perturbation, may be approximated from equation (2.13) by

$$w = \left(\frac{n_1^2 - n_0^2}{n_1^2 - n_2^2}\right)^{\frac{1}{2}} V. \qquad (2.26)$$

For a convenient representation we normalize $n = n_1/n_0$ as well as $\Delta n = (n_1 - n_2)/n_0$, and approximate $(n_1 + n_2)/n_0$ by $2n$. If the space above the film is vacuum or air we have $n_0 = 1$, so that n and Δn are the refractive index and index difference of film and substrate respectively. Equation (2.26) now appears as

$$w = \left(\frac{n^2 - 1}{2n \, \Delta n}\right)^{\frac{1}{2}} V.$$

Introducing this expression into equation (2.24) and substituting for Δv from equation (2.24) into equation (2.23) leads to the following approximation for the phase parameter of H_m-waves:

$$B = \frac{v_0^2}{V^2} + \frac{2 u_0^4}{V^3 (u_0^2 + v_0^3)} \left(\frac{2n \, \Delta n}{n^2 - 1}\right)^{\frac{1}{2}} \qquad (2.27)$$

Substituting into equation (2.25) leads to the corresponding approximation of B for E_m-waves:

$$B = \frac{v_0^2}{V} + \frac{2u_0^4}{V^3(u_0^2 + v_0^3)n^2}\left(\frac{2n\,\Delta n}{n^2-1}\right)^{\frac{1}{2}}. \qquad (2.28)$$

Both expressions are accurate to first order in $(\Delta n)^{1/2}$, that is to first order in the numerical aperture according to equation (2.4). The difference between both approximations appears in the factor $1/n^2$ which multiplies the perturbation for E_m-waves only, and not the perturbation for H_m-waves.

All quantities which compose these perturbations in equations (2.24) and (2.25) as well as in equations (2.27) and (2.28) always remain positive. Therefore the perturbations are always positive too. Due to the factor $1/n^2$, the perturbation for any particular E_m-wave is always smaller than for the H_m-wave of the same order. This difference breaks the degeneracy. Given a specific film guide at a certain light wavelength, the H_m-wave with its larger transverse attenuation parameter v extends its fields less into the substrate and is more tightly bound to the film than the E_m-wave of same order m. Also the phase parameter for the H_m-wave is larger than for the E_m-wave of same order, making the H_m-wave propagate more nearly with the wave number of the film material than the corresponding E_m-wave. In Fig.2.4 the designation H_m above any of the phase parameter curves and E_m below these curves is meant to indicate this difference.

On closer inspection, H_m- and E_m-waves also differ in their cut-off. To obtain an exact expression for the film parameter V at cut-off, instead of the approximate equation (2.21) for weakly guiding films, we note that as long as $n_2 > n_0$, cut-off occurs at $\theta_1 = \theta_{2c}$, the limiting angle for total reflection at the boundary between film and substrate. For this limit the fields extend infinitely wide into the substrate, and we have $v = 0$, so that from equation (2.13)

$$\beta = n_2 k, \quad u = (n_1^2 - n_2^2)^{1/2}kd, \quad w = (n_2^2 - n_0^2)^{1/2}kd \quad (2.29)$$

DIELECTRIC FILMS

and
$$V_m = u_{mc},$$

as in equation (2.21). Introducing these cut-off relations into the exact eigenvalue equation (2.11) and solving for V_m yields the exact cut-off for H_m-waves:

$$V_m^{(H)} = \arctan \left(\frac{n_2^2 - n_0^2}{n_1^2 - n_2^2}\right)^{1/2} + m\pi . \qquad (2.30)$$

The corresponding cut-off value for the film parameter in the case of E_m-waves is obtained when we substitute from equation (2.29) into the exact eigenvalue equation (2.12) for these modes:

$$V_m^{(E)} = \arctan \frac{n_1^2}{n_0^2} \left(\frac{n_2^2 - n_0^2}{n_1^2 - n_2^2}\right)^{1/2} + m\pi . \qquad (2.31)$$

Due to the factor $n_1^2/n_0^2 > 1$ by which the argument of the inverse tangent in equation (2.31) differs from that in equation (2.30), we have for any particular order m,

$$V_m^{(H)} < V_m^{(E)} .$$

With increasing V therefore, the first film mode to come into existence at $V_0^{(H)}$ from equation (2.30) is the H_0-mode; from equation (2.31) between $V_0^{(H)}$ and $V_0^{(E)}$ only this one mode exists. The film guide is therefore mono-mode or single-mode in the range

$$V_0^{(H)} < V < V_0^{(E)} .$$

In the terminology of waveguides, the mode with lowest cut-off is called the fundamental mode. In the film guide the H_0-mode is the fundamental mode. The terms mono-mode or single-mode are, however, usually not used in this strict sense in optical waveguides. In weakly guiding structures, such as our present weakly guiding film, the two lowest-order modes of orthogonal polarization differ so little in cut-off that an optical waveguide is still termed mono-mode or single-mode even when these two orthogonal polarizations of the lowest transverse order are both above cut-off.

Starting with the exact cut-off parameters from equation (2.30) and equation (2.31) the transverse phase and attenuation parameters as well as the phase parameter B just above cut-off may approximately be determined by perturbation. We let

$$V = V_m + \Delta V$$

with

$$\Delta V \ll 1$$

and, because we remain near cut-off,

$$v \ll V_m.$$

A first-order perturbation based on the cut-off solution for equation (2.11) then results in the near cut-off approximation for H-modes:

$$v = \left[V_m^{(H)} + \left(\frac{n_1^2 - n_2^2}{n_2^2 - n_0^2} \right)^{\frac{1}{2}} \right] \Delta V ; \qquad (2.32)$$

while similarly from equation (2.12), the first-order perturbation for E-modes is

$$v = \left[V_m^{(E)} + \frac{n_1^2 n_0^2 (n_1^2 - n_0^2)}{n_0^4 (n_1^2 - n_2^2) + n_1^4 (n_2^2 - n_0^2)} \left(\frac{n_1^2 - n_2^2}{n_2^2 - n_0^2} \right)^{\frac{1}{2}} \right] \Delta V \cdot \frac{n_2^2}{n_1^2} . \qquad (2.33)$$

These near-cut-off approximations close the gap between the exact cut-off values from equations (2.30) and (2.31) respectively, and the approximations for weakly guiding films, which with their first-order perturbations equations (2.24) and (2.25) fail near cut-off. They also provide means for analysing film mode characteristics near cut-off for large index-differences between film and substrate.

Far above cut-off, when

$$u \ll v \quad \text{as well as} \quad u \ll w, \tag{2.34}$$

we obtain another approximation by noting that

$$\lim_{V \to \infty} u = u_{m\infty} = (m + 1)\pi,$$

not only for weakly guiding films but for any index-difference between film and substrate. This limiting value for u occurs when V and with it v and w grow larger and larger. The right-hand sides of both eigenvalue equations (2.11) and (2.12) then go to zero from negative values so that the $u_{m\infty}$ follow as the zero crossings of the tangent function. Far above cut-off, we therefore express u as

$$u = u_{m\infty}/(1 + \nu) \tag{2.35}$$

with

$$\nu \ll 1.$$

Substituting u from equation (2.35) into the eigenvalue equation (2.11) for H-modes, and solving for ν under the conditions (2.34), we obtain the following approximation for ν:

$$\nu = \frac{[V^2 - u_{m\infty}^2]^{\frac{1}{2}} + [V^2(n_1^2-n_0^2)/(n_1^2-n_2^2) - u_{m\infty}^2]^{\frac{1}{2}}}{\{[V^2 - u_{m\infty}^2][V^2(n_1^2-n_0^2)/(n_1^2-n_2^2) - u_{m\infty}^2]\}^{\frac{1}{2}}}. \tag{2.36}$$

From the eigenvalue equation (2.12) for E-modes we find under the same conditions the approximation

$$\nu = \frac{n_0^2[V^2 - u_{m\infty}^2]^{\frac{1}{2}} + n_2^2[V^2(n_1^2-n_0^2)/(n_1^2-n_2^2) - u_{m\infty}^2]^{\frac{1}{2}}}{n_1^2\{[V^2 - u_{m\infty}^2][V^2(n_1^2-n_0^2)/(n_1^2-n_2^2) - u_{m\infty}^2]\}^{\frac{1}{2}}}. \tag{2.37}$$

For $\nu \ll 1$ it should not make much difference when the perturbation factor $1/(1 + \nu)$ in equation (2.35) is replaced by $(1 - \nu)$. Comparing the approximation with the exact solutions shows, however, that more accurate results obtain, in particular not so far from cut-off, when equation

(2.35) is used (Marcuse 1974). Altogether this far-from-cut-off approximation yields remarkably accurate results over substantial ranges even not so far from cut-off.

Fig.2.6 compares the phase parameter B as a function of the film parameter V from exact solutions of the eigen-

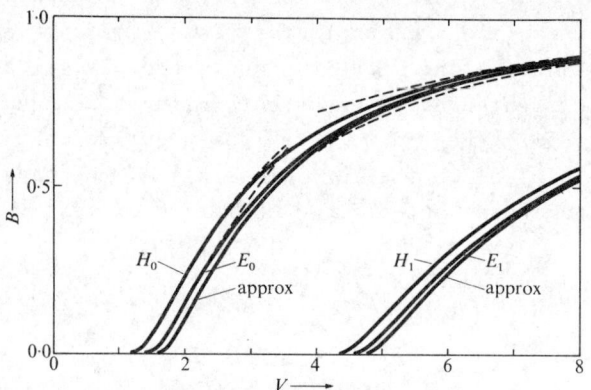

Fig.2.6. Phase parameter of guided modes in a film with $n_1 = 1\cdot53$ on a substrate with $n_2 = 1\cdot47$.

value equations with the approximation for weakly guiding films and with the near-cut-off and far-from-cut-off approximations. For the exact solutions and the latter two approximations the following refractive index values were chosen:

$$n_1 = 1\cdot53 \quad n_2 = 1\cdot47 \quad n_0 = 1.$$

They correspond to a polymerized organosilicon film of vinyltrimethylsilane on a pyrex substrate with the upper space empty. Film guides made from these materials have shown good characteristics, in particular low loss, for the film modes (Ulrich 1974). With their numerical aperture as large as $(n_1^2 - n_2^2)^{\frac{1}{2}} = 0\cdot425$ they are not really of weakly guiding nature. Nevertheless, quite accurate results are obtained from the weakly guiding film approximations.

For film guides which do not satisfy the weakly guiding film approximations, we need to solve the exact eigenvalue equations (2.11) and (2.12) numerically in order to obtain propagation characteristics of guided film modes. In the case of H-modes and their eigenvalue equation the numerical solution can be represented in universal form by expressing the transverse attenuation parameter w in terms of v and u by way of equation (2.13):

$$w^2 = v^2 + a_H (u^2 + v^2). \qquad (2.38)$$

The parameter

$$a_H = (n_2^2 - n_0^2)/(n_1^2 - n_2^2) \qquad (2.39)$$

appearing in (2.38) measures the asymmetry of the structure by being proportional to the difference in refractive index $(n_2 - n_0)$ between substrate and cover. For a symmetric structure with $n_2 = n_0$, we have $a_H = 0$, while the weakly guiding film with its pronounced asymmetry has very large values for a_H (Kogelnik and Ramaswamy 1974).

Substituting for w from equation (2.38) into equation (2.11) leads to a form of the eigenvalue equation for H-modes which, according to

$$\tan u = \frac{u\{v + [v^2 + a_H (u^2 + v^2)]^{\frac{1}{2}}\}}{u^2 - v[v^2 + a_H (u^2 + v^2)]^{\frac{1}{2}}}, \qquad (2.40)$$

contains only the transverse wave parameters u and v, and in addition the asymmetry parameter a_H. The solution of equation (2.40) together with equation (2.13) may now again be plotted in the form of the phase parameter $B = v^2/V^2$ as a function of $V = (u^2 + v^2)^{\frac{1}{2}}$ with a_H as a parameter. Fig.2.7 shows this diagram for a number of the lower-order H_m-modes. The curves for $a_H \to \infty$ represent the weakly guiding film approximation. We note from Fig.2.7 that the exact solutions deviate only slightly from the weakly guiding film approximation for a large range of values of the asymmetry parameter down to $a_H = 10$. The film guide for which Fig.2.6 compares the exact solution with different approximations has an asym-

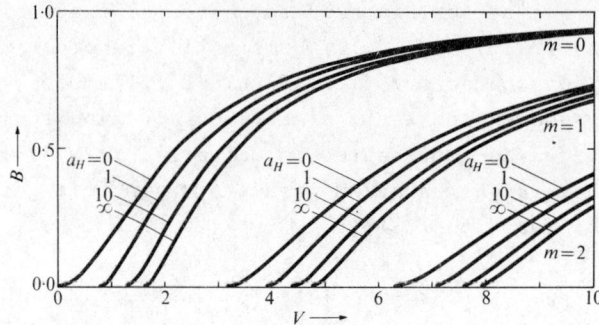

Fig.2.7. Phase parameter of low-order H-modes in films of different asymmetry (Kogelnik and Ramaswamy 1974).

metry parameter $a_H = 6\cdot 45$.

The universal representation for the phase parameter of H_m-modes may under weakly guiding conditions also be applied to E_m-modes by considering equation (2.38) for $n_1 \simeq n_2$ and hence large values of a_H. We then have

$$w^2 \simeq a_H (u^2 + v^2), \qquad (2.41)$$

and with this approximation the eigenvalue equation for H-modes reduces to

$$\tan u \simeq \frac{u \left[v + a_H^{\frac{1}{2}} (u^2 + v^2)^{\frac{1}{2}}\right]}{u^2 - a_H^{\frac{1}{2}} v(u^2 + v^2)^{\frac{1}{2}}} \qquad (2.42)$$

while equation (2.12) for E-modes appears approximately as

$$\tan u = \frac{u \left[v + a_E^{\frac{1}{2}} (u^2 + v^2)^{\frac{1}{2}}\right]}{u^2 - a_E^{\frac{1}{2}} v(u^2 + v^2)^{\frac{1}{2}}} \qquad (2.43)$$

with

$$a_E = (n_1/n_0)^4 a_H. \qquad (2.44)$$

Equation (2.43) has the same form as equation (2.42). Its solutions may therefore also be read from Fig.2.7 if the new asymmetry parameter a_E for E-modes according to equation (2.44) is used in its evaluation. The solution of equation

(2.43) as represented with $a_H = a_E$ in Fig.2.7 holds only for $n_1 \sim n_2$ and should only be applied for E-modes as long as

$$a_H \gg B \; ,$$

which is the prerequisite for the approximation (2.41). Nevertheless this solution is more accurate than the crude approximation of Fig.2.4, which appears in Fig.2.7 as the curves for $a_H \to \infty$.

The curves in Fig.2.7 when evaluated for E_m-modes reveal more details about their propagation characteristics. Because $a_E > a_H$ they show, for example, that for a given V-value the parameters B as well as v are always smaller for E_m-modes than for the H_m-modes of corresponding order m.

For symmetric structures with $a_H = 0$, the lowest-order H-mode has a phase parameter curve in Fig.2.7 starting at $V = 0$. This mode therefore appears to have zero cut-off. For E-modes the curves in Fig.2.7 hold for $a_H \gg B$ but definitely not for $a_H = 0$. Such symmetric structures and propagation characteristics of their guided modes have so much practical significance that we devote the next section to them.

2.2. GUIDED MODES OF THE SYMMETRICAL DIELECTRIC SLAB

When a dielectric film or slab is embedded on both sides in the same medium with a refractive index n_2 smaller than the refractive index n_1 of the film, the structure is called a symmetrical dielectric slab, or symmetrical film guide, or just slab guide. As the simplest form of optical waveguide, the slab guide may serve as a model for other more complicated guiding structures. But the slab guide itself is also used directly as an optical waveguide, or as an element in optical circuits.

The field in and around a slab guide again consists of radiation modes and guided modes. With the same medium on both sides of the slab the limiting angle θ_c of total reflection occurs at the same angle θ_{2c}, given by

$$\cos \theta_{2c} = n_2/n_1 \; ,$$

at both boundaries and radiation modes arise from uniform plane waves propagating inside the slab in directions $\theta_1 > \theta_{2c}$.

The guided modes of the slab guide arise from uniform plane waves inside the slab which are totally reflected at both its boundaries, with the limiting angle θ_{2c} now being equal on both sides, as well as the phase angle ϕ of total reflection. We again have H- and E-modes of the slab, and their eigenvalue equations follow from equation (2.11) and equation (2.12) with $n_0 = n_2$ as well as $w = v$. Replacing $\tan u$ in equation (2.11) by $\tan(u/2)$ according to

$$\tan u = \frac{2\tan(u/2)}{1 - \tan^2(u/2)}$$

the eigenvalue equation for H-modes of the slab appears as

$$\frac{\tan(u/2)}{1 - \tan^2(u/2)} = \frac{uv}{u^2 - v^2} .$$

This is a quadratic equation in $\tan(u/2)$ which has the following two solutions:

$$u \tan(u/2) = v \qquad (2.45)$$

$$-u \cot(u/2) = v . \qquad (2.46)$$

We now have two different eigenvalue equations for H-modes of the slab guide. Applying the same procedure to equation (2.12), it splits into the two following eigenvalue equations for E-modes of the slab:

$$u \tan(u/2) = v \frac{n_1^2}{n_2^2} \qquad (2.47)$$

$$-u \cot(u/2) = v \frac{n_1^2}{n_2^2} . \qquad (2.48)$$

Solutions of equations (2.45) and (2.47) correspond to solutions of the initial eigenvalue equation (2.6) when m is an even integer, while solutions of equations (2.46) and (2.48) correspond to odd integers m. Accordingly the transverse fields of H-modes from equation (2.45) and E-modes from equation (2.47) have an even or symmetrical distribution about the centre-plane of the slab, as Fig.2.8(a) shows for the H_0- or E_0-mode respectively. The H-modes from equation

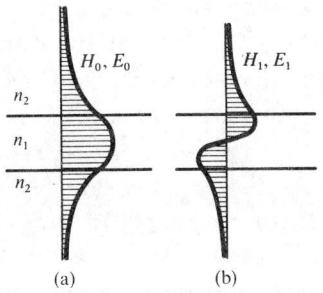

Fig.2.8. Transverse distribution of the electric field component E_y for H-modes and the magnetic field component H_y for E-modes, respectively, in symmetrical dielectric slabs. (a) H_0- and E_0-mode. (b) H_1- and E_1-mode.

(2.46) and the E-modes from equation (2.48), on the other hand, have an odd or asymmetrical distribution of their transverse fields about the centre-plane. Fig.2.8(b) shows the transverse field distribution of H_1- or E_1-modes as an example of such asymmetrical fields.

Modes with even and odd field distributions are to be expected for the slab guide because it has a symmetrical structure with respect to its centre plane. Any field distribution which is guided by the symmetrical slab therefore decomposes into even modes with a symmetrical field distribution and odd modes with an asymmetrical field distribution. In summary then, equation (2.45) belongs to even H-modes, equation (2.46) to odd H-modes, equation (2.47) to even E-modes, and equation (2.48) to odd E-modes.

For the solution of the eigenvalue equations of the symmetrical slab we note that equation (2.46) corresponds to the approximate form (2.16) of a weakly guiding film of thickness $d/2$. If in equation (2.46) we replace u by $2u$ and v by $2v$, we obtain equation (2.16). We can therefore read from Figs. 2.3 and 2.4 the transverse phase and attenuation parameters u and v as well as the longitudinal phase parameter B for odd H-modes of a symmetrical slab of thickness $2d$.

The weakly guiding film approximation regards the upper film boundary to free space as a perfect reflector because of its relatively large step in refractive index. More specifically, this boundary behaves as a perfectly conducting wall for plane uniform waves, which are incident at small angles $\theta_1 < \theta_{2c}$ and polarized perpendicular to the plane of incidence. Such a conducting wall forces the tangential electric field to be zero. The odd H-modes of the symmetrical slab satisfy this boundary condition in the centre plane of the slab. Therefore one half of the symmetrical slab guides these modes in the same way as the weakly guiding film and, with respect to these modes, is really equivalent to it.

A similar correspondence occurs for odd E-modes. Their plane uniform waves are polarized parallel to the plane of incidence and, when incident at small angles $\theta_1 < \theta_{2c}$ from the weakly guiding film on to the upper boundary to free space, see this boundary as a perfect magnetic wall. On the other hand, the odd E-modes of the symmetric slab satisfy the boundary condition of zero tangential magnetic field at a perfect magnetic wall in the centre plane. The odd E-modes of the symmetric slab therefore have similar field distribution and propagation characteristics as the E-modes in a weakly guiding film which consists just of one half of the symmetric slab. The similarity turns out to be an equivalence when the symmetric slab itself is only weakly guiding. In this case of $n_1 \simeq n_2$, the eigenvalue equation (2.41) is approximated by equation (2.39) and any of the odd E_m-modes of the symmetric slab becomes degenerate with the odd H_m-mode of the same odd order m. Under general circumstances, however, the odd E_m-modes are not degenerate with their H_m-mode partners and they also differ somewhat from the $E_{(m-1)/2}$-modes of the weakly guiding film of half the slab thickness.

A more fundamental difference between the asymmetric

film guide and the symmetric slab guide becomes apparent when the even H- and E-modes of the symmetric slab are examined as solutions of equations (2.45) and (2.47), respectively. These modes have no such correspondence to modes of the asymmetric film as have the odd modes to the modes of the weakly guiding film. They rather represent new solutions with partly different characteristics. We consider first equation (2.45) and even H_m-modes as its solutions. This eigenvalue equation does not explicitly contain any of the refractive indices. The characteristic parameters u, v, and B are therefore unique functions of the film or slab parameter V independent of the refractive indices. To obtain these functions we adopt the same procedure as in Fig. 2.3 for solving the characteristic equation, and plot $v = u \tan(u/2)$ as well as the circle of radius V, as shown in Fig.2.9. For completeness the positive branches of

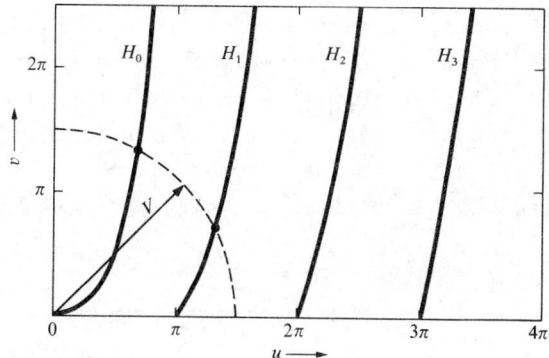

Fig.2.9. Graphical solution of the eigenvalue equation for H-modes of the symmetrical slab.

$v = -u \cot(u/2)$ for odd H-modes also appear in Fig.2.9. In contrast to the asymmetric film guide we now also find a solution, where the circle crosses the lowest branch of $v = u \tan(u/2)$, at arbitrarily small values of the film parameter, including $V = 0$. This lowest-order H-mode of the symmetric slab has no cut-off. The slab guides such a mode at any wavelength. For small values of V, however, the transverse attenuation parameter v remains small according to $v \simeq V^2/2$, and the fields extend far into the outer region on both sides of the slab. With a slab thickness d, they drop to 1/e of their magnitude at the slab surface only at a distance

$$x = d/v \simeq 2d/V^2$$

away from the slab surface. For the field extension to reduce to the order of the slab thickness, the film parameter must increase to the order of unity. Only for V-values of this order or larger will the field of the lowest H-mode be sufficiently concentrated in and near the slab to be well enough guided. Even though this mode theoretically has no cut-off, this practical aspect requires large enough V-values, and this more or less appears equivalent to a cut-off condition. For the remaining range of larger V-values the lowest-order H-mode behaves the same as the higher-order H-modes and the modes of the asymmetric film. While the field confines itself more and more to the slab and distributes transversely as half of a sine-wave, the phase constant changes gradually from the wave number $n_2 k$ of the outer region to the wave number $n_1 k$ of the slab material. Fig.2.10 illustrates the dispersion characteristic

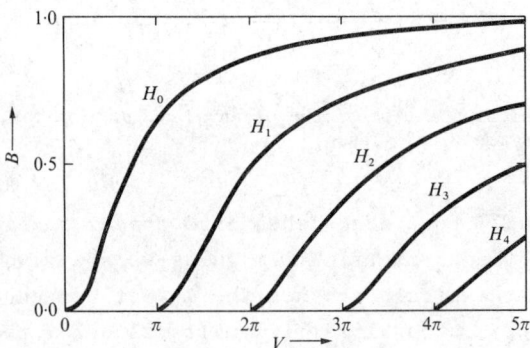

Fig.2.10. Phase parameter as a function of film parameter for H-modes of the symmetrical slab.

of a number of low-order even and odd H-modes of the symmetrical slab by plotting the phase parameter B as a function of the film parameter V.

The lowest-order E-mode of the symmetrical slab shows similar behaviour to the H-mode. For the weakly guiding slab with $n_1 \simeq n_2$, equation (2.45) approximates the eigenvalue equation (2.47) of even E-modes, and these modes become degenerate with even H-modes of the same order. Therefore, for the weakly guiding slab solutions for all E-modes may also be read from Fig.2.9, while Fig.2.10 also gives the phase parameter B for all E-modes.

A better approximation for E-mode parameters is found when we let

$$n_2 = n_1 - \Delta n$$

in equations (2.47) and (2.48) and consider

$$\Delta n \ll n_1 ,$$

a perturbation that shifts the solutions u_0, v_0 of equations (2.45) and (2.46) by Δu and Δv respectively to

$$u = u_0 + \Delta u$$

and

$$v = v_0 + \Delta v.$$

For the first-order perturbation of the transverse attenuation parameter v we find from equation (2.47), as well as from equation (2.48), that

$$\Delta v = - \frac{4 v_0 u_0^2}{(2 + v_0) V^2} \frac{\Delta n}{n_1} . \qquad (2.49)$$

This negative perturbation of v lets the fields of E-modes extend further into the outer medium than the fields of H-modes of the same order, just as in case of the asymmetric slab. In contrast to the corresponding approximations (2.24) and (2.25) the present perturbation is to first order in Δn, and not in $(\Delta n)^{1/2}$.

The change in v according to equation (2.49) leads to the new approximation for the phase parameter,

$$B = \frac{v_0^2}{V^2} \{1 - \frac{8n_0^2}{(2 + v_0)V^2} \frac{\Delta n}{n_1}\}. \qquad (2.50)$$

In this approximation E-modes of the symmetrical slab are no longer degenerate with corresponding H-modes, but, in accordance with their fields extending further into the outer medium, propagate with a smaller phase constant.

Of all guided modes of the symmetric slab, the H_0- and E_0-modes have theoretically no cut-off and exist under any conditions. The H_0-fields in particular are always more confined to the slab, and this mode has the larger phase constant. It therefore is considered to be the fundamental mode of the slab.

For guided modes of the asymmetric film which are far from cut-off, approximate solutions of their eigenvalue equations were found in the form of equations (2.36) and (2.37). These far-from-cut-off approximations hold just as well for the symmetrical slab, and yield equally good results by letting $n_0 = n_2$. For the E_0- and H_0-modes, which actually have zero cut-off, far from cut-off means that V must be large enough for their transverse phase parameter u to differ only little from its asymptotic value $u_{0\infty} = \pi$.

Near-cut-off approximations such as equations (2.32) and (2.33) for the guided modes of the asymmetric film may also be found for the symmetric slab. Due to the quite different cut-off characteristics of the symmetric slab modes, their near-cut-off approximations also have a different form. Starting with film parameter values

$$V = V_m + \Delta V$$

close to their cut-off-values

$$V_m = m\pi,$$

we find as first-order perturbation solutions of equations
(2.45) and (2.46) for H_m-modes with $m \neq 0$

$$v = m\pi \, \Delta V/2, \qquad (2.51)$$

while equations (2.47) and (2.48) yield the near-cut-off
approximations for E_m-modes with $m \neq 0$

$$v = \frac{1}{2} (n_2/n_1)^2 \, m\pi \, \Delta V.$$

An even better near-cut-off approximation for the H_0-mode is

$$v = (V^2 + 1)^{\frac{1}{2}} - 1 \qquad (2.52)$$

and for the E_0-mode,

$$v = \frac{n_1^2}{n_2^2} \left[\left(\frac{n_2^4}{n_1^4} V^2 + 1 \right)^{\frac{1}{2}} - 1 \right].$$

Fig.2.11 demonstrates the quality of these approximations by comparing the phase parameter B as a function of V from the exact solutions with the various approximate results. The refractive index ratio in Fig.2.11 is

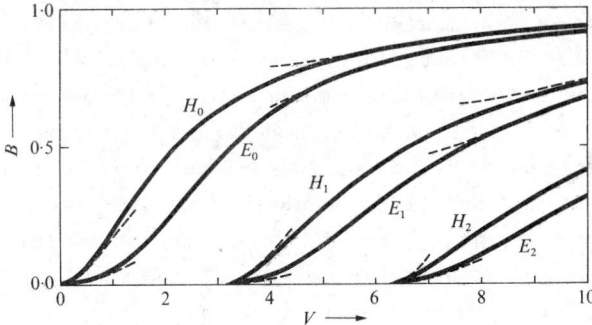

Fig.2.11. Phase parameter of low-order modes in a symmetrical film with $n_1/n_2 = 1\cdot 5$. Broken lines represent near-cut-off and far-above-cut-off approximations.

$n_1/n_2 = 1 \cdot 5$ and refers, for example, to a glass film suspended in free space.

2.3. FIELD SOLUTIONS FOR GUIDED MODES

The previous two sections have considered the guided modes of films and slabs as originating from plane uniform waves which experience total reflection at both boundaries and travel on a zig-zag path inside the film or slab. When satisfying the phase condition that after two successive reflections the plane uniform wave repeats in phase, it interferes constructively with itself, and propagates without attenuation along the guide. Such a plane uniform wave fits into the film or slab and forms a guided mode of propagation. This ray-optics method of mode analysis starts with plane uniform waves as solutions of Maxwell's equations, then satisfies the boundary conditions of the structure by considering the plane-wave reflection at the boundary, and eventually leads to a self-consistent field distribution for film and slab modes by obeying the phase condition.

The somewhat indirect approach of the ray-optics method yields a complete solution for all guided modes; all pertinent propagation characteristics may be derived from it. In addition it gives much insight into the mechanism of propagation and guidance of such modes, once the phenomenon of total reflection is understood. This information justifies the ray-optics method.

For more complicated structures the ray-optics approach can become rather laborious. For guides with three parallel boundaries instead of the two boundaries of a film, or for even more such parallel boundaries, it would be impractical to trace rays of uniform plane-waves, calculate their reflection, and find the phase condition for self-consistent solutions. Here it is much easier to solve Maxwell's equations directly subject to the boundary conditions of the particular planar structure.

We will derive these field solutions directly from Maxwell's equations for the guided modes of simple films and slabs. These solutions will not give much more information, than we could obtain from the previous ray-optics approach.

DIELECTRIC FILMS

The method can, however, be readily generalized to apply to multilayer structures with more than the three layers of film and slab guides.

In the Cartesian coordinates (x,y,z) of Fig.2.12 the layered structures, for which guided modes are to be found, are homogeneous in the y- and z-directions. We therefore

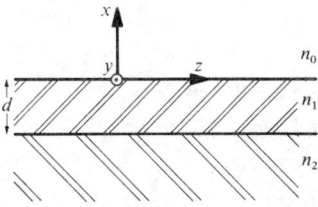

Fig.2.12. Film guide and coordinate system for field solution.

look only for waves with fields independent of y and hence have $\partial/\partial y = 0$, and which propagate in the z-direction with a propagation constant $j\beta$. Their fields therefore depend on z as $\exp(-j\beta z)$. Under these conditions Maxwell's equations in their source-free form (equation (1.3)) have the following Cartesian components. From $\nabla \times \vec{E} = -j\omega\mu_0 \vec{H}$ we obtain

$$j\beta E_y = -j\omega\mu_0 H_x \tag{2.53}$$

$$-j\beta E_x - \frac{\partial E_z}{\partial x} = -j\omega\mu_0 H_y \tag{2.54}$$

$$\frac{\partial E_y}{\partial x} = -j\omega\mu_0 H_z, \tag{2.55}$$

and from $\nabla \times \vec{H} = j\omega n^2 \epsilon_0 \vec{E}$ we obtain

$$j\beta H_y = j\omega n^2 \epsilon_0 E_x \tag{2.56}$$

$$-j\beta H_x - \frac{\partial H_z}{\partial x} = j\omega n^2 \epsilon_0 E_y \tag{2.57}$$

$$\frac{\partial H_y}{\partial x} = j\omega n^2 \epsilon_0 E_z . \tag{2.58}$$

Equations (2.53), (2.55), and (2.57) form a system that contains only the field components E_y, H_x, and H_z, while equations (2.54), (2.56), and (2.58) contain only H_y, E_x, and E_z. Both systems, of three equations each, are independent of each other, and so are their solutions. The general solution of equations (2.53) to (2.58) therefore splits into H-waves with H_z as the only longitudinal field component and E_y and H_x as transverse field components, and into E-waves with E_z as the only longitudinal field and the transverse field components H_y and E_x.

Eliminating H_x and H_z from the system of equations (2.53), (2.55), and (2.57) for H-waves leads to the following scalar wave equation for E_y:

$$\frac{\partial^2 E_y}{\partial x^2} + (n^2 k^2 - \beta^2) E_y = 0 \qquad (2.59)$$

with $k = \omega(\mu_0 \varepsilon_0)^{1/2}$ as the wave number of free space and n as the refractive index of the particular medium. The magnetic field components of H-waves are related to E_y by

$$H_x = \frac{-\beta}{\omega \mu_0} E_y, \qquad H_z = \frac{j}{\omega \mu_0} \frac{\partial E_y}{\partial x} . \qquad (2.60)$$

Similarly eliminating E_x and E_z from the system of equations (2.54), (2.56), and (2.58) for E-waves, leads to

$$\frac{\partial^2 H_y}{\partial x^2} + (n^2 k^2 - \beta^2) H_y = 0, \qquad (2.61)$$

and the electric field components are here related to H_y by

$$E_x = \frac{\beta}{n^2 \omega \varepsilon_0} H_y, \qquad E_z = \frac{-j}{n^2 \omega \varepsilon_0} \frac{\partial H_y}{\partial x} . \qquad (2.62)$$

Both wave equations (2.59) and (2.61) have the same form and identical coefficients. Their general solutions are formed by

$$E_y = A_n \cos k_{nx} x + B_n \sin k_{nx} x \quad \text{for } H\text{-waves},$$
$$H_y = C_n \cos k_{nx} x + D_n \sin k_{nx} x \quad \text{for } E\text{-waves},$$
$$(2.63)$$

where

$$k_{nx} = (n^2 k^2 - \beta^2)^{\frac{1}{2}}, \qquad (2.64)$$

from the coefficient of the wave equations, represent the wave number in the x-direction, and for given k and β still depends on the refractive index n. The factor $\exp(-j\beta z)$ for propagation in the positive z-direction is common to all field expressions and is not included in equation (2.63) nor in all other field expressions that follow. The factors A_n, B_n, C_n, D_n remain to be determined by boundary and excitation conditions.

To satisfy boundary conditions we consider first the outer regions of the layered guiding structures, which in the x-direction extend to infinity. Field components of any guided mode must decay exponentially in these regions, which requires $nk < \beta$ in order that $k_{nx} = -j\alpha_n$ remains purely imaginary in these regions, and the fields decay as $\exp(-\alpha_n |x|)$.

For explicit field expressions we must now consider a specific structure. As a sufficiently general example, we take the dielectric film with n_1 in Fig.2.12 on a substrate with n_2 and the upper space with n_0. For the exponential decay in the upper space we specify the factors in equation (2.63) so that

$$E_y = A \exp(-\alpha_0 x) \quad \text{for } H\text{-waves}$$
$$H_y = C \exp(-\alpha_0 x) \quad \text{for } E\text{-waves}$$
$$\text{and } x \geq 0. \qquad (2.65)$$

The factors A and C are regarded as excitation coefficients to be determined by the nature and strength of a source which excites these waves.

Inside the film, for $0 > x > -d$, we choose the factors A_1 to D_1 of the general solution (2.63) so that the field components E_y and H_y, respectively, are continuous at the boundary $x = 0$ to the upper space:

$$E_y = A\cos k_{1x} x + B\sin k_{1x} x \text{ for } H\text{-waves}$$
$$\text{and } 0 \geq x \geq -d. \quad (2.66)$$
$$H_y = C\cos k_{1x} x + D\sin k_{1x} x \text{ for } E\text{-waves}$$

This leaves the factors B and D still undetermined.

The fields in the substrate must decay exponentially in the negative x-direction, which requires $n_2 k < \beta$, in order that $k_{2x} = -j\alpha_2$ from equation (2.64) is purely imaginary. We then specify the factors A_2 to D_2 of the general solution so that its terms combine to a decaying exponential and the field components match E_y and H_y from equation (2.66) at the substrate boundary:

$$E_y = (A\cos u - B\sin u)\exp\{\alpha_2(x+d)\} \text{ for } H\text{-waves}$$
$$\text{and } x \leq -d.$$
$$H_y = (C\cos u - D\sin u)\exp\{\alpha_2(x+d)\} \text{ for } E\text{-waves} \quad (2.67)$$

Here we have abbreviated $u = k_{1x} d$ by introducing the transverse phase parameter u from equation (2.8). The expressions (2.65), (2.66), and (2.67) satisfy the boundary conditions at $x = 0$ and $x = -d$ for E_y and H_y, but not for any other field components which are also tangential to the boundary. In the case of H-waves, H_z is the other tangential component. With E_y given by equations (2.65), (2.66), and (2.67), H_z follows from equation (2.60) as

$$H_z = -\frac{j\alpha_0}{\omega\mu_0} A \exp(-\alpha_0 x) \quad \text{for } x \geq 0$$

$$H_z = -\frac{jk_{1x}}{\omega\mu_0}(A\sin k_{1x} x - B\cos k_{1x} x) \quad \text{for } 0 \geq x \geq -d \quad (2.68)$$

$$H_z = \frac{j\alpha_2}{\omega\mu_0}(A\cos u - B\sin u)\exp\{\alpha_2(x+d)\} \text{ for } x \leq -d.$$

Requiring H_z to go continuously through the boundaries leads to

$$wA + uB = 0 \quad (2.69)$$

for continuity at $x = 0$, and

$$(u \sin u - v \cos u)A + (u \cos u + v \sin u)B = 0 \quad (2.70)$$

for continuity at $x = -d$. Here we have introduced the transverse attenuation parameters $w = \alpha_0 d$ and $v = \alpha_2 d$ from equation (2.9). Equations (2.69) and (2.70) form a system of linear equations for the factors A and B. The system is homogeneous, so that for non-trivial solutions to exist, the coefficient determinant must vanish, that is

$$w(u \cos u + v \sin u) - u(u \sin u - v \cos u) = 0. \quad (2.71)$$

This is the characteristic equation for H-modes which, when solved for $\tan u$, has the previous form (2.11). For parameter values u, v, w which solve the characteristic equation, B is related to the excitation coefficient A by

$$B = -\frac{w}{u} A. \quad (2.72)$$

In the case of E-waves we have E_z, besides H_y, as the other component which is tangential to the boundaries. With equation (2.62) and H_y given by equations (2.65), (2.66), and (2.67), we obtain

$$E_z = (j\alpha_0/(n_0^2 \omega \varepsilon_0))C \exp(-\alpha_0 x) \quad \text{for } x \geq 0 \quad (2.73)$$

$$E_z = (jk_{1x}/(n_1^2 \omega \varepsilon_0))(C \sin k_{1x} x - D \cos k_{1x} x) \text{ for } 0 \geq x \geq -d$$

$$E_z = (-j\alpha_2/(n_2^2 \omega \varepsilon_0))(C \cos u - D \sin u)\exp\{\alpha_2(x+d)\} \text{ for } x \leq -d.$$

Requiring E_z to go continuously through the boundaries leads to

$$(w/n_0^2) C + (u/n_1^2) D = 0 \quad (2.74)$$

for continuity at $x = 0$, and

$$[(u/n_1^2)\sin u - (v/n_2^2)\cos u]C + [(u/n_1^2)\cos u + (v/n_2^2)\sin u]D = 0 \quad (2.75)$$

for continuity at $x = -d$. For non-trivial solutions of this homogeneous and linear system of equations for C and D the coefficient determinant must vanish, that is

$$\frac{w}{n_0^2}\left(\frac{u}{n_1^2}\cos u + \frac{v}{n_2^2}\sin u\right) - \frac{u}{n_1^2}\left(\frac{u}{n_1^2}\sin u - \frac{v}{n_2^2}\cos u\right) = 0. \quad (2.76)$$

This is the same characteristic equation as equation (2.12) from the ray-optics approach for E-modes. For values of u, v, w which solve this equation, the field coefficient D follows from the excitation coefficient C according to equation (2.74), as

$$D = -\frac{n_1^2}{n_0^2}\frac{w}{u} C. \quad (2.77)$$

The field solution of Maxwell's equations for the guided modes provides us directly with explicit expressions for all the field components. For a specific film guide at a given wavelength, once the characteristic equation has been solved for a particular mode, all the parameters u, v, and w are known. The expressions for the field components may then be evaluated in terms of just the excitation coefficients A and C.

For the weakly guiding film with n_1 only slightly larger than n_2, but a pronounced difference $n_1 > n_0$, the field expressions reduce approximately to particularly simple forms. For such films we have $w \gg u$ and also from equations (2.72) and (2.77) $B \gg A$ as well as $D \gg C$. The factors B and D, rather than A and C, should then be regarded as excitation coefficients, and the terms with A and C neglected altogether. We thus obtain for the transverse electric field of H-waves in the weakly guiding film

$$E_y \simeq \begin{cases} 0 & \text{for } x \geq 0 \\ B \sin(ux/d) & \text{for } 0 \geq x \geq -d \\ -B \sin u \, \exp\{v(1 + x/d)\} & \text{for } x \leq -d \end{cases} \quad (2.78)$$

and likewise for the transverse magnetic field of E-waves

$$H_y \simeq \begin{cases} 0 & \text{for } x \geq 0 \\ D\,\sin(ux/d) & \text{for } 0 \geq x \geq -d \\ -D\,\sin u\,\exp\{v(1+x/d)\} & \text{for } x \leq -d. \end{cases} \quad (2.79)$$

These simple expressions give the actual distribution for transverse fields of modes in weakly guiding films, as shown, for example, in Fig.2.5, to a rather good approximation.

Simple expressions are also found for the field distribution of modes in symmetric slabs. Because of the symmetry, the x-coordinate is shifted according to Fig.2.13 with

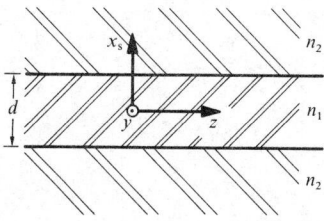

Fig.2.13. Symmetric dielectric slab with adequate coordinate system.

$x_s = x + d/2$ so that the plane $x_s = 0$ coincides with the centre plane of the slab. As a function of x_s, even H-modes have the transverse electric field

$$E_y = \begin{cases} A_e\,\cos(u/2)\exp\{-v(x_s-d/2)/d\} & \text{for } x_s \geq d/2 \\ A_e\,\cos(ux_s/d) & \text{for } -d/2 \leq x_s \leq d/2 \\ A_e\,\cos(u/2)\exp\{v(x_s+d/2)/d\} & \text{for } x_s \leq -d/2 \end{cases} \quad (2.80)$$

and odd H-modes have

$$E_y = \begin{cases} A_o \sin(u/2)\exp\{-v(x_s-d/2)/d\} & \text{for } x_s \geq d/2 \\ A_o \sin(ux_s/d) & \text{for } -d/2 \leq x_s \leq d/2 \\ -A_o \sin(u/2)\exp\{v(x_s+d/2)/d\} & \text{for } x_s \leq -d/2. \end{cases} \quad (2.81)$$

Likewise the transverse magnetic field of even E-modes appears as

$$H_y = \begin{cases} C_e \cos(u/2)\exp\{-v(x_s-d/2)/d\} & \text{for } x_s \geq d/2 \\ C_e \cos(ux_s/d) & \text{for } -d/2 \leq x_s \leq d/2 \\ C_e \cos(u/2)\exp\{v(x_s+d/2)/d\} & \text{for } x_s \leq -d/2, \end{cases} \quad (2.82)$$

while odd E-modes have

$$H_y = \begin{cases} C_o \sin(u/2)\exp\{-v(x_s-d/2)/d\} & \text{for } x_s \geq d/2 \\ C_o \sin(ux_s/d) & \text{for } -d/2 \leq x_s \leq d/2 \\ -C_o \sin(u/2)\exp\{v(x_s+d/2)/d\} & \text{for } x_s \leq -d/2. \end{cases} \quad (2.83)$$

The field expressions for all the guided modes enable us to evaluate the power which these modes carry, and relate this power to the excitation coefficients. From the z-component S_z of the complex Poynting vector $\vec{S} = \vec{E} \times \vec{H}^*$, with $\vec{H}^*$ as the complex conjugate of the magnetic field vector $\vec{H}$, the power in the z-direction, carried per unit width of the film in the y-direction follows as

$$P = \int_{-\infty}^{\infty} S_z \, dx.$$

For H-modes with $S_z = -E_y H_x^*$ and H_x from equation (2.60), the power per unit width is

$$P = \frac{\beta}{\omega\mu_0} \int_{-\infty}^{\infty} |E_y|^2 \, dx. \quad (2.84)$$

Substituting for E_y in the different cross-sectional regions from equations (2.65), (2.66) and (2.67), respectively, we can evaluate the integral, and taking equation (2.72) as well as the characteristic equation (2.11) into consideration, we find

$$P = |A|^2 \; (\beta d/(2\omega\mu_0))(1 + w^2/u^2)(1 + 1/v + 1/w). \qquad (2.85)$$

For E-modes with $S_x = E_x H_y^*$ and E_x from equation (2.62), we obtain for the power per unit width

$$P = \frac{\beta}{\omega\varepsilon_0} \int_{-\infty}^{\infty} \frac{1}{n^2} |H_y|^2 \; dx. \qquad (2.86)$$

If here we substitute for H_y from equations (2.65), (2.66), (2.67), respectively, and also take n as n_0, n_1 or n_2 depending on the cross-sectional region, we find with equations (2.12) and (2.77)

$$P = |C|^2 \frac{\beta d}{2\omega\varepsilon_0} \left(1 + \frac{n_1^4 w^2}{n_0^4 u^2}\right) \left[\frac{1}{n_1^2} + \frac{(u^2+v^2)n_2^2/v}{n_2^4 u^2 + n_1^4 v^2} + \frac{(u^2+w^2)n_0^2/w}{n_0^4 u^2 + n_1^4 w^2}\right]. \qquad (2.87)$$

For real and positive excitation coefficients A or C, equations (2.85) or (2.87) may serve to express all field components in terms of power guided per unit width. With $P = 1$ the fields may then also be normalized with respect to this power per unit width.

For the distribution of transverse fields in weakly guiding films according to equations (2.78) and (2.79) the factors B and D rather than A and C were regarded as excitation coefficients of H-modes and E-modes, respectively. If we substitute B for A from equation (2.72) in equation (2.85) and let $w \to \infty$, the expression for power guided per unit width in H-modes of weakly guiding films reduces to

$$P = |B|^2 \; [\beta d/(2\omega\mu_0)](1 + 1/v). \qquad (2.88)$$

Similarly substituting D from equation (2.77) for C in equation (2.87) and again letting $w \to \infty$, we obtain for the power per unit width of E-modes in weakly guiding films

$$P = |D|^2 \frac{\beta d}{2\omega\varepsilon_0} \left(\frac{1}{n_1^2} + v^2 \frac{n_2^2/v}{n_2^4 u^2 + n_1^4 v^2} \right). \qquad (2.89)$$

Guided modes in symmetric slabs have the even or odd field distributions of equations (2.80) to (2.83), where now A_e and A_o as well as C_e and C_o are regarded as excitation coefficients. Evaluating equation (2.84), we find for the power guided per unit width by even and also by odd H-modes one and the same expression, namely

$$P = |A|^2 \, [\beta d / (2\omega\mu_0)](1 + 2/v), \qquad (2.90)$$

where $A = A_e$ for even H-modes and $A = A_o$ for odd H-modes. Also for E-modes, with even as well as with odd field distributions equation (2.86) yields one and the same expression in the form

$$P = |C|^2 \frac{\beta d}{2\omega\varepsilon_0} \left[\frac{1}{n_1^2} + \frac{2 n_2^2 v^2}{v(n_2^4 u^2 + n_1^4 v^2)} \right]. \qquad (2.91)$$

The power formulae for H- and E-modes, respectively, in symmetrical slabs appear quite similar to those of corresponding modes in weakly guiding asymmetrical films. Considering that any mode of the weakly guiding film represents in its field distribution just one half of the field of the corresponding mode in the symmetrical slab of twice the width, these power formulae actually are identical.

The fields of guided modes in dielectric films or slabs are only partly confined to the actual guiding layer. In the transverse direction perpendicular to these layers they form standing waves, with $k_{1x} = u/d$ as the phase constant and $\lambda_t = 2\pi/k_{1x}$ as the wavelength of the standing wave pattern. Beyond the boundaries all field components decay exponentially with $\alpha_2 = v/d$ and $\alpha_0 = w/d$, the transverse attenuation constants into the substrate and the cover region (the half-space above the film), respectively. In the case of $n_0 < n_2$ and near the cut-off of any particular mode, α_2 is quite small, and the fields extend very far into the substrate. Far above cut-off, on the other hand,

α_2 increases to larger and larger values just as α_0, and the fields confine themselves more and more to the film.

Any measure of the effective width of a particular film mode should take into consideration these evanescent fields beyond the boundaries. A simple means of measuring the effective field extension into the regions of evanescence is obtained with the subtangents of the exponential decay:

$$d_0 = 1/\alpha_0 = d/w \qquad (2.92)$$

and

$$d_2 = 1/\alpha_2 = d/v \;. \qquad (2.93)$$

These subtangents should be added to d for the effective width of the mode:

$$d_e = d + d_0 + d_2 = d(1 + \frac{1}{v} + \frac{1}{w})\;. \qquad (2.94)$$

Expressing w according to equation (2.38) in terms of u and v and one of the asymmetry parameters a_H or a_E, respectively, the ratio d_e/d for any particular mode appears as a function of only the film parameter V and of a_H or a_E, respectively. The numerical solution of the eigenvalue equation for H-waves and, in the case of $n_1 \simeq n_2$, also for E-waves, is plotted in Fig.2.7 with $B = v^2/V^2$ as a function of $V = (u^2 + v^2)^{\frac{1}{2}}$. The effective width may be expressed with these parameters according to

$$d_e = d\{1 + [B^{-\frac{1}{2}} + (B + a)^{-\frac{1}{2}}]/V\}. \qquad (2.95)$$

Note, however, that in Fig.2.7 the asymmetry parameter $a = a_H$ applies only for H-modes, and a_E must be used for E-modes. In equation (2.38) on the other hand, the parameter a_H holds for H- as well as for E-modes. Fig.2.14 shows the effective width of the fundamental H_0-mode normalized in the form

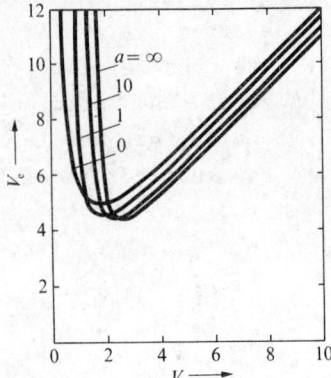

Fig.2.14. Normalized effective width of the H_0-mode in film guides.

$$V_e = kd_e \, (n_1^2 - n_2^2)^{\frac{1}{2}} \qquad (2.96)$$

just as d is normalized with $V = kd(n_1^2 - n_2^2)^{\frac{1}{2}}$.

$$V_e = V + B^{-\frac{1}{2}} + (B + a)^{-\frac{1}{2}} \qquad (2.97)$$

is plotted as a function of V for different values of the asymmetry parameter a. V_e starts with infinitely large values at cut-off, goes through a relatively broad minimum and eventually increases linearly with V far above cut-off. For the weakly guiding film with $a \to \infty$, equation (2.97) reduces to

$$V_e = V + B^{-\frac{1}{2}}, \qquad (2.98)$$

with

$$V_e = V + 1$$

far above cut-off. In this case the minimum value of V_e for the dominant H_0-wave occurs at $V = 2 \cdot 55$ and amounts to $V_{e\,min} = 4 \cdot 4$. This implies a minimum achievable effective width of $d_e = 0 \cdot 7 \, \lambda \, (n_1^2 - n_2^2)^{-\frac{1}{2}}$, which obtains when the weakly guiding film has a thickness of $d = 0 \cdot 41 \lambda \, (n_1^2 - n_2^2)^{-\frac{1}{2}}$.

The symmetric slab has $a = 0$, and equation (2.97) reduces to

$$V_e = V + 2B^{-\frac{1}{2}} \qquad (2.99)$$

with

$$V_e = V + 2$$

far above cut-off.

For all H-modes, including the lowest-order dominant mode for which Fig.2.14 has been plotted, the Goos-Haenchen shift offers another interpretation of the effective mode width. The beam shift in total internal reflection of the plane-waves which, in zig-zagging down the film, form the guided modes, may be accounted for by apparent reflectors displaced from the boundary by the distance d_H from equation (1.30) for H-waves and by the distance d_E from equation (1.31) for E-waves. In the case of H-waves these distances are identical to the subtangents d_0 and d_2 (equations (2.92) and (2.93)) of the exponential decay for the evanescent fields in cover and substrate. The effective width d_e for any particular H-wave therefore corresponds directly to the total spacing of the apparent reflectors, which account for the Goos-Haenchen beam shift in the zig-zag wave picture of Fig.2.15. This correspondence does not appear quite as direct in the case of E-modes. Here the apparent reflectors are spaced at distances which according to equation (1.31) differ somewhat from the respective subtangents d_0 and

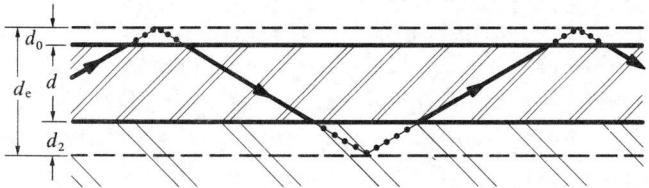

Fig.2.15. Effective width d_e of the film guide with apparent reflectors accounting for Goos-Haenchen shift of zig-zag beam for H-modes.

d_2 of the exponential decay.

The difference at the substrate boundary ranges, however, only between

$$d_E = (n_1^2/n_2^2) \; d_2$$

near cut-off and

$$d_E = (n_2^2/n_1^2) \; d_2$$

far above cut-off, so that the apparent reflectors always remain spaced in the order of d_0 and d_2, respectively, for any particular E-mode. The interpretation of the effective width as the total spacing of the apparent reflectors for the Goos-Haenchen shift therefore also applies within a certain approximation to E-modes.

2.4. GUIDED MODE ABSORPTION

The previous three sections have assumed the film and slab guides to consist of perfectly uniform materials without any absorption. As a consequence these materials have real wave numbers nk at any frequency of consideration and the refractive indices n differ only from substrate to film and to space above, but remain constant within any of the three guide regions. In addition the boundaries of the film or slab have been assumed plane and parallel to each other. Practical film and slab guides differ from such idealized models in all three respects. The guide materials have absorption, and their refractive index varies locally. Also, due to manufacturing imperfections, the guide boundaries deviate from the perfect geometry.

Material absorption is most easily accounted for by replacing the real wave numbers $n_i k$ in each cross-sectional region i by the complex quantities $n_i k - j\alpha_i$ with α_i, the negative of their imaginary components, representing the attenuation constant of a uniform plane wave in region i. With all $\alpha_i = 0$ the guided modes propagate unattenuated with a purely imaginary propagation constant $\gamma = j\beta$. The absorption will cause the modes to be attenuated and their propagation constant to assume complex values $\gamma = \alpha + j\beta$.

Any material which is suitable for low-loss film guides should have low absorption and therefore $\alpha_i \ll n_i k$. Due to such low absorption the transverse field distribution of guided modes changes only little from its loss-free distribution. We can therefore expect the transverse phase- and attenuation parameters u, v, and w to deviate only slightly from their real and positive values u_0, v_0, and w_0 in loss-free guides, and let

$$u = u_0 + j\Delta u, \quad v = v_0 + j\Delta v, \quad w = w_0 + j\Delta w, \quad (2.100)$$

with

$$|\Delta u| \ll u_0, \quad |\Delta v| \ll v_0, \quad |\Delta w| \ll w_0. \quad (2.101)$$

The guided modes will under these low-loss conditions suffer only little attenuation, rendering their attenuation constant

$$\alpha \ll \beta. \quad (2.102)$$

Upon substituting $n_i k - j\alpha_i$ for the different k_i into equations (2.13), and also replacing β by $\beta - j\alpha$, these equations become complex; but under conditions (2.101) and (2.102) their real parts relate u_0, v_0, and w_0 to the $n_i k$ and to β as in the loss-free case, while their imaginary parts appear as

$$u_0 \Delta u = (\beta\alpha - n_1 k \alpha_1) d^2$$
$$v_0 \Delta v = (n_2 k \alpha_2 - \beta\alpha) d^2 \quad (2.103)$$
$$w_0 \Delta w = (n_0 k \alpha_0 - \beta\alpha) d^2.$$

Hence the absorption effects a purely imaginary perturbation of u_0, v_0, and w_0.

The eigenvalue equation (2.11) for H-modes depends only on u, v, w. Equation (2.12) for E-modes, however, depends in addition on n_0, n_1, and n_2. In the case of absorption u, v, w turn complex according to equation (2.100), and in equation (2.12) the n_i must be replaced by $n_i - j\alpha_i/k$. Writing the real eigenvalue equation for the loss-free film guide in the form

$$D(u_0, v_0, w_0, n_i) = 0 \qquad (2.104)$$

where n_i represents the dependence on n_0, n_1, and n_2, this eigenvalue equation turns complex in the form

$$D(u, v, w, n_i - j\alpha_i/k) = 0,$$

when the guide regions have absorption. Under low-loss conditions, however, the real part of this complex eigenvalue equation maintains the form (2.104) and the perturbation due to absorption manifests itself only in the following imaginary part of the complex eigenvalue equation

$$D_u \Delta u + D_v \Delta v + D_w \Delta w - \sum_i D_i \alpha_i/k = 0. \qquad (2.105)$$

Here the indices of D denote the following partial derivatives of D in equation (2.104), evaluated for the loss-free solution u_0, v_0, w_0 and n_i, for the respective guided mode:

$$D_u = \frac{\partial D}{\partial u}; \quad D_v = \frac{\partial D}{\partial v}; \quad D_w = \frac{\partial D}{\partial w}; \quad D_i = \frac{\partial D}{\partial n_i}. \qquad (2.106)$$

Equation (2.105) together with equations (2.103) form a system of four equations for the four perturbations Δu, Δv, Δw, and α. Eliminating Δu, Δv, Δw leads to the following first-order perturbational expression for the attenuation constant:

$$\alpha\beta\, kd^2 = \frac{(vwD_u n_1\alpha_1 - uwD_v n_2\alpha_2 - uvD_w n_0\alpha_0)k^2 d^2 - uvw(D_1\alpha_1 + D_2\alpha_2 + D_0\alpha_0)}{vw\, D_u - uw\, D_v - uv\, D_w}. \qquad (2.107)$$

u, v, and w in this formula stand for their values u_0, v_0 and w_0 of the respective mode in the loss-less structure. For convenience and without any risk of confusion, we omit the index 0 in this and the following formulae.

For most practical film guides equation (2.107) is unnecessarily general. Usually the space above the film is air with $n_0 = 1$ and $\alpha_0 = 0$ or at least $\alpha_0 \ll \alpha_1, \alpha_2$. Equation (2.107) then reduces to

$$\alpha\beta k d^2 = w \frac{(v D_u n_1 \alpha_1 - u D_v n_2 \alpha_2)k^2 d^2 - uv(D_1\alpha_1 + D_2\alpha_2)}{vw D_u - uw D_v - uv D_w}. \qquad (2.108)$$

A further simplification applies for modes with D in the eigenvalue equation (2.104) independent of any of the n_i. The H-modes have such an eigenvalue equation in the form of equation (2.11). In this case

$$\alpha\beta = kw \frac{v D_u n_1 \alpha_1 - u D_v n_2 \alpha_2}{vw D_u - uw D_v - uv D_w}. \qquad (2.109)$$

An even simpler expression holds for the weakly guiding film with the refractive index n_1 only slightly larger than the refractive index n_2 of the substrate, but a pronounced difference between n_1 and n_0. Here we have $w \gg v, u$, reflecting the strongly evanescent character of the fields above the film. Furthermore the approximate form (2.16) of the eigenvalue equations for weakly guiding films does not depend on any of the n_i. Therefore even with absorption in the space above the film, we obtain for the attenuation constant

$$\alpha = \frac{k}{\beta} \frac{v D_u n_1 \alpha_1 - u D_v n_2 \alpha_2}{v D_u - u D_v}. \qquad (2.110)$$

Because of the eigenvalue equation $D(u,v) = 0$, the partial derivatives D_u and D_v in equation (2.110) are related to the derivative dv/du as prescribed for solutions of $D(u,v) = 0$ by

$$\frac{dv}{du} = -\frac{D_u}{D_v}. \qquad (2.111)$$

This allows us to write the attenuation constant in terms of

v/u and dv/du:

$$\alpha = \frac{k}{\beta} \frac{(dv/du)\, n_1\, \alpha_1 + (u/v)\, n_2\, \alpha_2}{dv/du + u/v}. \qquad (2.112)$$

So far we have abstained from evaluating any of the partial derivatives of D or dv/du in the attenuation formulae for the particular eigenvalue equations of film guides. We also intend to apply these formulae to other forms of optical waveguides. With no specific eigenvalue equation introduced into them, they also hold for guided modes of other axially uniform structures consisting of up to three different media with low enough absorption. The only other prerequisite for the attenuation formulae to apply is that the transverse phase and attenuation parameters must be related to the wave numbers of the media and to the axial phase constant by expressions such as (2.13). A number of optical waveguides with circular symmetry fall into this category, and for them some of the above general attenuation formulae prove quite useful.

Certain general features of the attenuation characteristic of modes in optical waveguides may directly be read from equation (2.112). In film guides or similar guides with two relevant layers, we have near cut-off of any guided mode

$$v \ll u, \qquad \beta \simeq n_2 k. \qquad (2.113)$$

These conditions also hold for modes with zero cut-off. Here v and u are both small near cut-off, but, as in Figs. 2.9 and 2.7, they start from zero with $v \ll u$ and $du/dv \gg 1$. Furthermore our perturbation analysis based on the inequalities (2.101) also holds near cut-off for any of the guided modes, although v or even v and u are quite small. For low enough absorption these inequalities are also maintained near cut-off, as may be inferred from equation (2.103). As a consequence of the conditions (2.113) introduced into equation (2.110) or (2.112), any guided mode propagates near cut-off with the absorption $\alpha \simeq \alpha_2$ of the medium with lower refractive index (which is the substrate in case of the weakly guiding film). An attenuation $\alpha = \alpha_2$ is, of course, to be expected near cut-off, because at the limit of total internal reflection, the fields extend infinitely wide into the medium with lower refractive index.

Far above cut-off any guided mode has

$$v \gg u, \quad du/dv \ll 1, \quad \beta \simeq n_1 k .$$

Under these circumstances, from equation (2.112) the mode propagates with the attenuation α nearly equal to the absorption α_i of the medium with higher refractive index. An attenuation $\alpha \simeq \alpha_1$ is in turn to be expected far above cut-off because then the fields concentrate almost entirely within the medium with n_1 and α_1, where they propagate as if this medium extends to infinity.

The weakly guiding film has the simple and universal characteristic equation (2.16) for all its guided modes. If this relation between u and v is used to evaluate equation (2.112), the following expression results for the attenuation of its guided modes:

$$\frac{\alpha \beta}{n_1 \alpha_1 k} = 1 + \frac{n_2 \alpha_2/(n_1 \alpha_1)-1}{1 + v} \frac{u^2}{v^2} \qquad (2.114)$$

Fig.2.16 shows the attenuation according to this formula for

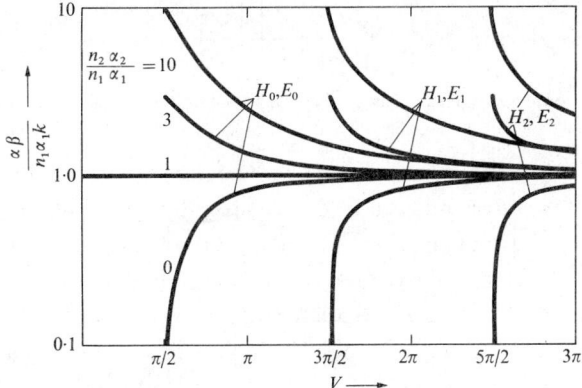

Fig.2.16. Film mode attenuation of weakly guiding films due to bulk losses α_1 and α_2 in film and substrate, respectively.

a number of low order modes in weakly guiding films. The curves reflect the general attenuation characteristic of film modes, when bulk absorption or scattering in film and substrate cause the loss. At cut-off the attenuation equals the bulk loss α_2 of the substrate. Far above cut-off the attenuation approaches the bulk loss α_1 of the film.

The various attenuation formulae of this section do not only take account of absorption in the bulk of the different regions of a dielectric waveguide; they may also serve to analyse how emission instead of absorption in any of the layers reduces the attenuation, and may even lead to amplification of a film mode. This problem occurs when waveguide lasers for amplification or generation of coherent light are constructed in the form of film guides, and film or substrate are brought into a state for stimulated emission of radiation. Fig.2.17 shows as an example of considerable practical interest the relevant layers of a double hetero laser epitaxially grown from GaAs and GaAlAs. The central GaAs-layer forms the active layer into which elec-

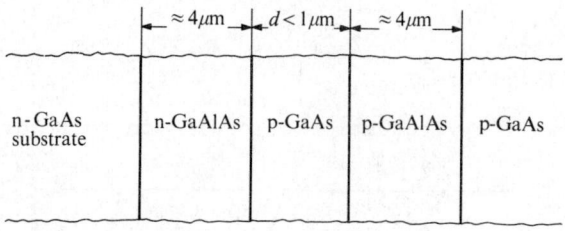

Fig.2.17. p-GaAs waveguide layer in double hetero-injection laser.

trons are injected from one side and holes from the other side, and are confined to this region by the energy barriers at the hetero-junctions to the GaAlAs of wider energy gap. These densely concentrated carriers recombine by radiative transitions and, when stimulated by the electric field of a light wave, add their photons in phase to this wave for amplification. The gain which this stimulated emission produces is to a first approximation given by (Unger 1976)

$$g_1 = \frac{\eta \lambda^2 J}{16\pi q n_1^2 \Delta f\, d} \quad , \qquad (2.115)$$

where J is the density of injection current, η the quantum efficiency (the degree to which the recombination is radiative), and Δf the bandwidth of the gain characteristic. q denotes the electron charge, and d the width of the active layer. Associated with a given injection current density are a certain number of injected carriers. They distribute uniformly over the width d of the GaAs-layer so that carrier concentration and gain g_1 are proportional to d^{-1}.

The active GaAs-layer has not only the gain g_1 but also a higher refractive index n_1 than the index n_2 of the GaAlAs-layers on both sides. It therefore forms a symmetrical slab, which guides waves according to Section 2.2. A gain constant g_1 such as in equation (2.115) acts as a negative attenuation constant $\alpha_1 = -g_1$, which reduces the slab mode attenuation α or even leads to negative values of $\alpha = -g$, which mean amplification for the particular slab mode. To analyse the effect of g_1 on the gain of the slab guide, we neglect any absorption in the embedding medium and let $\alpha_2 = 0$. This corresponds to the double hetero structure of Fig.2.17, where the GaAlAs-regions with their wider energy gap are transparent at wavelengths for which the GaAs-active region shows gain by stimulated emission. We then restrict attention to the dominant H_0-mode, which always is best confined to the slab, and therefore uses g_1 most efficiently for amplification. The H_0-mode is the lowest-order even H-mode of the slab, and its characteristics follow from the corresponding solution of equation (2.45). Substituting from this equation for u and dv/du into equation (2.112), we obtain for the gain $g = -\alpha$ of any even-order H-mode in the slab

$$\frac{g\beta}{g_1 n_1 k} = \frac{B + v/2}{1 + v/2} \; .$$

By abbreviating g_1 from equation (2.115) as $g_1 n_1 = O/d$ and introducing the opto-electronic coefficient

$$O = \frac{\eta \lambda^2 J}{16\pi q n_1 \Delta f},$$

the even H-mode gain appears as

$$\frac{g\beta}{\partial k} = \frac{1}{d} \frac{B + v/2}{1 + v/2}. \qquad (2.116)$$

We will now vary the thickness d of the active layer to find the highest possible gain of the dominant H_0-mode of the active slab guide. For very small values of d the right-hand side of equation (2.116) goes to zero because according to equation (2.52), v approaches zero as d^2 and $B = v^2/V^2$ therefore also as d^2. For larger and larger values of d, on the other hand, the right-hand side of equation (2.116) goes again to zero because then B approaches unity while v goes to infinity as d. Over this full range of d the effective refractive index β/k of the H_0-mode changes only from n_1 to n_2.

For a weakly guiding slab, as in case of the GaAs-GaAlAs-structure with n_2 only slightly smaller than n_1, the variation of β/k with d may be neglected and the maximum of the right-hand side regarded as the optimum condition with respect to H_0-mode amplification. This maximum occurs when

$$d/\lambda = 0 \cdot 227 \, (n_2^2 - n_1^2)^{-\frac{1}{2}}$$

and amounts to

$$\frac{g\beta}{\partial k} = 0 \cdot 356 \, \beta \, (n_2^2/n_1^2 - 1)^{\frac{1}{2}}.$$

Double hetero laser amplifiers or oscillators should have this thickness d of their active layer in order to obtain the maximum possible gain for a given injection current J.

2.5. GUIDED MODE SCATTERING

The guided modes of film or slab guides suffer loss not only from absorption, but also from scattering at local variations of the refractive index, and at deviations of their boundaries from a perfectly plane geometry. The local variations of the refractive index originating from the random molecular structure of the amorphous film and substrate materials are of microscopic dimensions, small compared to the wavelength of light. Their scattering loss for plane uniform waves in the bulk of the respective material follows the Rayleigh law and contributes to the imaginary component of the complex wave number with a term proportional to λ^{-4}, as in equation (1.35).

For guided modes of films and slabs the scattering loss due to such microscopic index variations is somewhat modified by the presence of the film or slab boundaries. Scattered wave components from inside the film, which strike the boundary in the range of total internal reflection, are trapped within the film and may excite guided modes of the film in practically all directions, including the direction in which the primary guided mode travels. In the latter case they superimpose themselves on this wave. With microscopic index variations of random nature, and short correlation distance, only very little power will scatter back into the primary mode or any of the other guided modes in the same direction. Most of the scattered power will be radiated and some of it will be scattered into guided modes which travel in other directions. The radiated part of the scattered power will be larger compared to the power which is scattered into other modes the smaller the index differences between film and substrate, and the weaker the guiding characteristics of the other modes are. In most practical film and slab guides, it therefore suffices to account for scattering at microscopic index variations by including the imaginary component of the wave vector for bulk scattering loss according to equation (1.35) in the α_i-values of the different cross-sectional regions of the guide. The guided mode attenuation due to this Rayleigh scattering then follows from the same expressions as the attenuation due to absorption. All the loss characteristics of guided modes depend under these conditions on the guide parameters and on the bulk attenuation constants due to scattering as they depend on these parameters in the case of absorption.

Deviations of the film or slab boundaries from their

116   DIELECTRIC FILMS

nominally plane geometry represent another source of scattering and loss. Such deviations result at both boundaries from the fabrication process. Usually, however, they are more likely to occur at the boundary of larger index difference and also cause more scattering at this boundary. Sputtered films on a glass substrate are a typical case, where more imperfections must be expected at the boundary of the film with the space over it which has a larger index difference.

Fig.2.18 shows such an imperfect boundary with random displacements $f(y,z)$ of the actual boundary from its nominal plane at $x = 0$. With n as the actual index near the boundary, and $\bar{n}$ as the index at the same point for a perfect

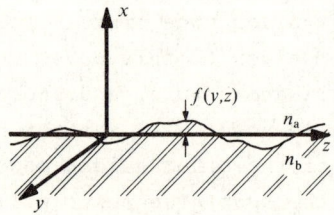

Fig.2.18. Nominally plane boundary between regions of refractive index n_a and n_b with displacements $f(y,z)$.

boundary, the displacement may be accounted for in the field equations by introducing the effective current density

$$\vec{J} = j\omega\varepsilon_0 \, (n^2 - \bar{n}^2) \, \vec{E}.$$

Here $\vec{E}$ represents the electric field which is associated with the particular guided mode in the presence of the perturbation. More specifically, we have

$$\vec{J}_{a \atop b} = j\omega\varepsilon_0 \, (n_{b \atop a}^2 - n_{a \atop b}^2) \, \vec{E}_{ab \atop ba} , \qquad (2.117)$$

if the region with $n_{b \atop a}$ extends beyond the nominal boundary into the region with $n_{a\atop b}$. $\vec{E}_{ab\atop ba}$ represents the actual field in the particular dent. To a good approximation we have

$$\vec{E}_{ab\atop ba} = \vec{E}_{b\atop a} \, ,$$

with $\vec{E}_{b\atop a}$ as the field of the guided mode of the perfect guide in the region with index $n_{b\atop a}$ at the boundary, i.e. at $x = \mp 0$. Decomposing these boundary fields into their Cartesian components, we may write

$$\vec{E}_a = \begin{bmatrix} E_{xa} \\ E_{ya} \\ E_{za} \end{bmatrix} = E_0 \, \exp(-j\beta z) \begin{bmatrix} r_x \\ r_y \\ r_z \end{bmatrix} ;$$

(2.118)

$$\vec{E}_b = \begin{bmatrix} E_{xb} \\ E_{yb} \\ E_{zb} \end{bmatrix} = E_0 \, \exp(-j\beta z) \begin{bmatrix} (n_a^2/n_b^2) r_x \\ r_y \\ r_z \end{bmatrix} .$$

For H-modes we have $r_x = r_z = 0$ and, with $E_0 = A$ being the y-directed electric field at $x = 0$ in this representation, obtain from equations (2.65) and (2.72) $r_y = r_{y01} = 1$ at the boundary between n_0 and n_1, and $r_y = r_{y12} = [(n_1^2 - n_0^2)/(n_1^2 - n_2^2)]$ at the boundary between n_1 and n_2.

For E-modes we have $r_y = 0$ and, with

$$E_0 = \frac{\beta}{\omega \varepsilon_0 \, n_1^2} \, C$$

representing the x-component of the electric field inside the film at $x = 0$, obtain from equations (2.62) and (2.66) $r_{z01} = n_1^2/n_0^2$, $r_{z01} = j(n_1^2 \, w)/(n_0^2 \, \beta \, d)$ at the boundary between n_0 and n_1, while

$$r_{x12} = \frac{n_1^2}{n_2^2} \left(\cos u + \frac{n_1^2 w}{n_0^2 u} \sin u \right); \; r_{z12} = -j \, \frac{u}{\beta d} \left(\sin u - \frac{n_1^2 w}{n_0^2 u} \cos u \right)$$

at the boundary between n_1 and n_2.

With the effective currents of density $\vec{J}$ as given by equaton (2.117), any boundary element $dydz$ which is displaced by $f(y,z)$ has the electric current element of moment

$$\vec{Il} = \vec{J}.f(y,z) \; dydz \tag{2.119}$$

associated with it. Depending on the direction of displacement, and therefore the sign of $f(y,z)$, we have such an electric dipole moment either on side a or on side b of the boundary, but always directly adjacent to it.

For a more unified representation of these effective dipole moments, we replace any dipole moment in region b by an equivalent source in region a. Consider the distribution of dipole moments in Fig.2.19 with $(\vec{Il})_a$ and $(\vec{Il})_b$ directly

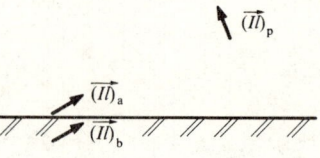

Fig.2.19. Equivalent dipole moments on both sides next to the boundary.

above and below the boundary, respectively, and $(\vec{Il})_p$ of any orientation located anywhere in space. Denoting by E_{pa} the field component of source p in the direction of source a and at its location, we have with corresponding notations for the other field components, and by the law of reciprocity,

$$E_{pa} (Il)_a = E_{ap} (Il)_p \; ,$$

as well as

$$E_{pb} (Il)_b = E_{bp} (Il)_p \; .$$

For $(\vec{Il})_a$ and $(\vec{Il})_b$ to be equivalent to each other, they must generate the same fields $E_{ap} = E_{bp}$ anywhere in space. This requirement results in the following condition for equivalence:

$$E_{pa} (Il)_a = E_{pb} (Il)_b .$$

In the case of dipoles a and b oriented parallel to the boundary, we have $E_{pa} = E_{pb}$ and consequently $(Il)_b = (Il)_a$. In the case of dipoles perpendicular to the boundary, we have $n_a^2 E_{pa} = n_b^2 E_{pb}$. With all these relations we may now take account of any displacement $f(y,z)$, positive or negative, of the boundary element $dy\,dz$ by the electric dipole moment

$$\vec{J}(y,z) \; dV = j\omega\varepsilon_0 \; (n_b^2 - n_a^2) \; E_0 \begin{bmatrix} r_x \\ r_y \\ r_z \end{bmatrix} \exp(-j\beta z) \; f(y,z) \; dy\,dz \quad (2.120)$$

on the a-side next to the boundary.

To evaluate the radiation loss from such dipoles, we need to calculate their radiation field in the presence of both film boundaries. The effect of any boundary on the radiation field, and on the radiated power, will be more pronounced the closer this dipole is located to the boundary. Any boundary at a sufficient distance will still modify the radiated field by reflecting and refracting the incident radiation, but will not react so much on the source, and therefore will not change its radiated power very much. For a loss analysis we are primarily interested in this power and will therefore take account only of the boundary to which the dipole is immediately adjacent.

The assertion that the film boundary to which the dipole is not directly adjacent does not react on the dipole to change its total radiated power significantly is corroborated by an accurate computation of the dipole radiation in the presence of both film boundaries (Hinken 1977). If the film guides any modes which the dipole excites, then not all the power is radiated into the space above and substrate below the film, but part of it is guided by these film modes which propagate radially away from the dipole with circularly cylindrical phase fronts. The total power, however, as the sum of radiated power and the power into film modes, remains nearly the same as the radiated dipole power when only the adjacent film boundary is present.

The radiation field of electric current elements at the plane boundary between two different media was first given by Sommerfeld (1926) in an integral representation, which, when evaluated for its far field by a transformation of wave-functions, has the following magnetic field vector components in the coordinates of Fig.2.20, with the dipole at their origin (Hinken 1976):

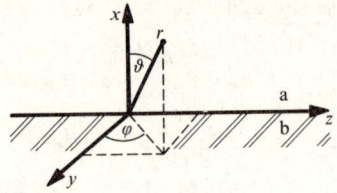

Fig.2.20. Coordinates of dipole at the boundary between a and b for its radiation field.

$$\vec{H} = \begin{bmatrix} H_r \\ H_\vartheta \\ H_\varphi \end{bmatrix} = \begin{bmatrix} 0 & 0 & 0 \\ 0 & g_{\vartheta y} & g_{\vartheta z} \\ g_{\varphi x} & g_{\varphi y} & g_{\varphi z} \end{bmatrix} \begin{bmatrix} (Il)_x \\ (Il)_y \\ (Il)_z \end{bmatrix} . \tag{2.121}$$

For the components of the Green's-function dyadic in this expression, we must distinguish, in which of the two regions the field is to be determined. With the wave numbers

$$k_a = n_a \omega (\mu_0 \epsilon_0)^{\frac{1}{2}} \qquad k = n_b \omega (\mu_0 \epsilon_0)^{\frac{1}{2}} .$$

in each region and

$$\delta = \frac{n_b^2 - n_a^2}{n_a^2} \tag{2.122}$$

as the relative difference in permittivities, the dyadic has the following components in the respective regions:

DIELECTRIC FILMS

$$g_{\varphi x}^{a} = j \frac{k_a}{2\pi r} \exp(-jk_a r) \frac{\sin \vartheta}{1 + \frac{1}{1+\delta}(1 + \delta \cos^{-2}\vartheta)^{\frac{1}{2}}}$$

$$g_{\varphi x}^{b} = j \frac{k_b}{2\pi r} \exp(-jk_b r) \frac{\sin \vartheta}{1 + (1+\delta)(1 - \frac{\delta}{1+\delta} \cos^{-2}\vartheta)^{\frac{1}{2}}}$$

$$g_{\vartheta y}^{a} = j \frac{k_a}{2\pi r} \exp(-jk_a r) \frac{\sin \varphi}{1 + (1 + \delta \cos^{-2}\vartheta)^{\frac{1}{2}}} \qquad (2.123)$$

$$g_{\vartheta y}^{b} = j \frac{k_b}{2\pi r} \exp(-jk_b r) \frac{\sin \varphi}{1 + (1 - \frac{\delta}{1+\delta} \cos^{-2}\vartheta)^{\frac{1}{2}}}$$

$$g_{\varphi y}^{a} = j \frac{k_a}{2\pi r} \exp(-jk_a r) \frac{\cos \varphi \,(\delta + \cos^2\vartheta)^{\frac{1}{2}}}{1 + \delta + (1 + \delta \cos^{-2}\vartheta)^{\frac{1}{2}}}$$

$$g_{\varphi y}^{b} = j \frac{k_b}{2\pi r} \exp(-jk_b r) \frac{\cos\varphi\{(1+\delta)\cos^2\vartheta - \delta\}^{\frac{1}{2}}}{1 + (1 + \delta - \delta \cos^{-2}\vartheta)^{\frac{1}{2}}}$$

$g_{\vartheta z}^{a,b}$ and $g_{\varphi z}^{a,b}$ follow from $g_{\vartheta y}^{a,b}$ and $g_{\varphi y}^{a,b}$, respectively, when φ is replaced by $\varphi + \frac{\pi}{2}$, therefore

$$g_{\vartheta z}^{a} = \cot\varphi \; g_{\vartheta y}^{a} \; ; \qquad g_{\vartheta z}^{b} = \cot\varphi \; g_{\vartheta y}^{b} \; ;$$

$$g_{\varphi z}^{a} = -\tan\varphi \; g_{\varphi y}^{a} \; ; \qquad g_{\varphi z}^{b} = -\tan\varphi \; g_{\varphi y}^{b} \; . \qquad (2.124)$$

Note that the square roots of $1 - \delta/[(1+\delta)\cos^2\vartheta]$ in the expressions for the g^b go to zero at

$$\vartheta_c = \arcsin(n_a/n_b),$$

the limiting angle of total reflection at the boundary from n_b to n_a. Note also that in the boundary plane for $\vartheta = \pi/2$ the $g_{\varphi y}$ as well as the $g_{\varphi z}$ vanish.

The power which the single dipole at the boundary radiates into the space above and below the boundary is obtained, when we integrate the far-field Poynting vector

$$S = \frac{1}{n} \left(\frac{\mu_0}{\varepsilon_0}\right)^{\frac{1}{2}} |H|^2 \qquad (2.125)$$

over all directions Ω in space

$$P_r = \iint r^2 \, S(\Omega) \, d\Omega \, . \tag{2.126}$$

The solid angle Ω contains the polar angle ϑ as well as the azimuth φ, and $d\Omega = \sin\vartheta \, d\vartheta \, d\varphi$ is the infinitesimal element of this solid angle. Depending on the region, the refractive index n in equation (2.125) is either n_a or n_b.

When we separate the contributions from dipoles which are horizontal, that is parallel to the boundary from those which are vertical or oriented perpendicular to it, the radiated power may be written as

$$P = P_a \left\{ \begin{array}{c} h(\delta) \\ v(\delta) \end{array} \right\}, \tag{2.127}$$

where

$$P_a = \frac{2\pi}{3} I^2 \frac{l^2}{\lambda^2} n_a \, , \tag{2.128}$$

in accordance with equation (1.33), is the power radiated by a dipole of moment Il into an unlimited space of refractive index n_a. The factors

$$h(\delta) = h_a(\delta) + h_b(\delta)$$
$$v(\delta) = v_a(\delta) + v_b(\delta) \tag{2.129}$$

account for the change in power radiation due to the boundary, when the dipole at the boundary is oriented either parallel or perpendicular to it. They depend only on the relative permittivity difference δ from equation (2.122) and consist of two terms each of which, according to their index a or b, describe the power fractions which are radiated into the half space a or b, respectively.

Fig. 2.21 shows the factor $h(\delta)$ for a horizontal dipole, and Fig. 2.22 shows $v(\delta)$ for a vertical dipole both together with their constituent parts for region a and b. The radiated power from an electric current dipole that lies parallel to an interface changes only slightly when the index difference increases. It increases somewhat when

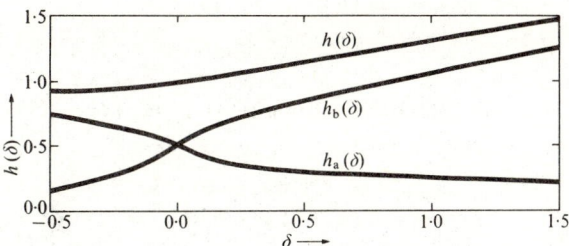

Fig.2.21. Power radiated from an electric dipole which lies in a half-space with n_a parallel and on the interface to a half-space with n_b, relative to the dipole power radiated into homogeneous space with n_a. h_a and h_b are the relative powers radiated into half-spaces a and b, respectively.

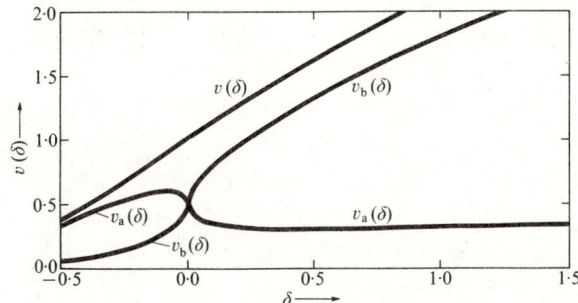

Fig.2.22. Power radiated from an electric dipole that sits in half-space a vertical to the interface to half-space b, relative to the dipole radiation in homogeneous space with n_a.

the region b below increases its index with respect to region a. It decreases, but at an even smaller rate, when the region below assumes an index n_b smaller than n_a. In this range $h(\delta)$ goes through a minimum. The constituent parts $h_a(\delta)$ and $h_b(\delta)$ show a more pronounced change. For $n_a < n_b$ most of the power radiates into the lower half-space with n_b, while for $n_a > n_b$ it radiates more and more into the upper half-space with n_a.

Fig.2.22 shows similar characteristics for the power that a dipole perpendicular to the interface radiates relative to the power radiated into free space. They are, however, much more pronounced in that they depend strongly on δ, the relative difference in permittivities.

With the far-field magnetic vector of a single dipole available in the form of equation (2.121), we can now return to the film guide with its boundary distortions and their effective current distributions. If these distortions extend from $0 \le y \le L_y$ in y-direction and from $0 \le z \le L_z$ in z-direction, they generate a far-field magnetic vector according to

$$\vec{H} = \sum_{t=x,y,z} \vec{g}_t \int_{y=0}^{L_y} \int_{z=0}^{L_z} J_t(y,z) \exp(j\psi(y,z)) dV. \quad (2.130)$$

Any vector $\vec{g}_t$ in this expression has the spherical components $g_{\vartheta t}$ and $g_{\varphi t}$ and, from equation (2.121), gives the magnetic field, which the Cartesian component $(\vec{Il})_t = J_t\, dV$ (with $t = x$, y, or z) of the effective current element $\vec{J}\, dV$ generates. The difference $\psi(y,z)$ in phase retardation between a source at $y = z = 0$ and one at (y,z) on the boundary to any far-field point (r, ϑ, φ) in space is given by

$$\psi(y,z) = nk \sin\vartheta\, (y \cos\varphi + z \sin\varphi). \quad (2.131)$$

n_a or n_b must be substituted for the refractive index n in this expression, depending on the region in which the far-field is to be evaluated.

If we introduce the component representation of effective current dipoles from equation (2.120), the magnetic radiation field appears as

$$\vec{H} = j\omega\varepsilon_0 (n_b^2 - n_a^2) E_0 \sum_{t=x,y,z} r_t \vec{g}_t \int_{y=0}^{L_y} \int_{z=0}^{L_z} f(y,z) \exp(j[\psi(y,z) - \beta z]) \, dy\, dz. \quad (2.132)$$

For the total power which this field radiates, the Poynting vector according to equation (2.125), with $\vec{H}$ substituted from equation (2.132), must be integrated over all directions in space just as in equation (2.126). In order to evaluate the integrals from equation (2.132), a particular boundary displacement $f(y,z)$ must be specified.

In the case of random distributions for $f(y,z)$, only certain statistical characteristics are known. Here we will evaluate the average of radiated power and the average mode loss associated with it. To this end we take the ensemble average of equation (2.126),

$$\langle P \rangle = \int_\Omega r^2 \langle S(\Omega) \rangle \, d\Omega. \qquad (2.133)$$

Substituting

$$\langle S(\Omega) \rangle = \eta(\omega\varepsilon_0 n_a^2 \delta E_0)^2 \, [|\sum_{t=x,y,z} r_t g_{\vartheta t}|^2 + |\sum_{t=x,y,z} r_t g_{\varphi t}|^2] \langle |F_{1,2}|^2 \rangle, \qquad (2.134)$$

we have

$$\langle |F_{1,2}|^2 \rangle = \int_{y_1=0}^{L_y} \int_{z_1=0}^{L_z} \int_{y_2=0}^{L_y} \int_{z_2=0}^{L_z} \langle f(y_1,z_1) f(y_2,z_2) \rangle$$

$$\exp\{-j[\beta(z_1-z_2) - \psi(y_1-y_2, z_1-z_2)]\} \, dz_2 \, dy_2 \, dz_1 \, dy_1. \qquad (2.135)$$

For the latter expression we consider first the one-dimensional example:

$$\langle |F_1|^2 \rangle = \int_{x_1=0}^{L} \int_{x_2=0}^{L} \langle f(x_1) f(x_2) \rangle \exp\{-j \, \gamma(x_1-x_2)\} \, dx_2 \, dx_1. \qquad (2.136)$$

We assume the random distribution $f(x)$ to be stationary and ergodic with a mean square

$$\overline{f^2} = \frac{1}{L} \int_0^L f^2(x) \, dx. \qquad (2.137)$$

For such a distribution, the autocorrelation $\langle f(x_1) f(x_2) \rangle$ depends only on the difference $u = x_1 - x_2$ in location, but is independent of the particular location x_1. We may therefore introduce

$$\overline{f^2} \varphi(u) = \langle f(x_1) f(x_2) \rangle \qquad (2.138)$$

into equation (2.136), and change to u and x_2 as the new variables of integration:

$$\langle |F_1|^2 \rangle = \overline{f^2} \, [\int_{-L}^{0} \int_{-u}^{L} \varphi(u) \exp(-j\gamma u) \, dx_2 \, du$$

$$+ \int_{0}^{L} \int_{0}^{L-u} \varphi(u) \exp(-j\gamma u) \, dx_2 \, du].$$

(2.139)

The new limits of integration are illustrated in Fig.2.23 in the (x_1-x_2)-plane, in which u = const describes straight

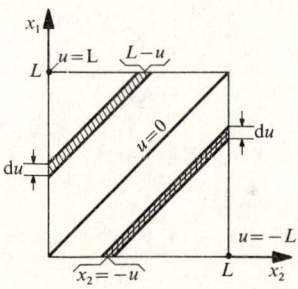

Fig.2.23. Variables and limits of integration for equation (2.139).

lines inclined at 45°. The first integral extends over the lower right triangle of the square in Fig.2.23 and the second integral over the upper left. We integrate over x_2 and combine both integrals over u into

$$\langle |F_1|^2 \rangle = 2\overline{f^2} \int_{0}^{L} (L-u) \cos \gamma u \, \varphi(u) \, du. \qquad (2.140)$$

It can here be assumed that random boundary displacements are correlated only over a small fraction of the total extent L of the film in any direction. Under this assumption we have $u \ll L$ for any u at which $\varphi(u)$ contributes to the integral. The integral then reduces to

$$\langle |F_1|^2 \rangle = 2L \, \overline{f^2} \int_{0}^{L} \varphi(u) \cos \gamma u \, du. \qquad (2.141)$$

Because of the limited correlation distance, the upper limit of integration may be extended to infinity:

$$\langle |F_1|^2 \rangle = L \, \overline{f^2} \int_{-\infty}^{+\infty} \varphi(u) \cos \gamma u \, du. \qquad (2.142)$$

The Fourier transform of the autocorrelation function, which appears in this expression, is for the stationary distribution $f(x)$ equal to its power spectrum

$$\phi(\gamma) = \overline{f^2} \int_{-\infty}^{\infty} \varphi(u) \exp(-j\gamma u) \, du. \qquad (2.143)$$

Here the quantity γ for the space-domain function $f(x)$ corresponds to the angular frequency ω in time-domain functions. It is the angular frequency in space of a particular spectral component of $f(x)$. In terms of this power spectrum, we have

$$\langle |F_1|^2 \rangle = L \, \phi(\gamma). \qquad (2.144)$$

These results may now be applied to the two-dimensional distribution and lead to the following expression

$$\langle |F|^2 \rangle = 4\overline{f^2} \, L_y \, L_x \int_0^{L_y} \int_0^{L_z} \varphi(u,v) \cos[\beta v - \psi(u,v)] \, dv \, du, \qquad (2.145)$$

in which u is the variable in the y-direction, transverse to the propagation direction of the film mode, and v is the variable in the z-direction, in which the wave propagates. The mean square of the boundary displacement, $\overline{f^2}$, now takes account of the two-dimensional distribution and follows from

$$\overline{f^2} = \frac{1}{L_y L_z} \int_0^{L_y} \int_0^{L_z} f^2(y,z) \, dz \, dy. \qquad (2.146)$$

Because of limited correlation, the integrals in equation (2.145) may also be extended to infinity, and the two-dimensional autocorrelation function expressed by the power spectral distributions in both directions, according to

$$\phi(\xi,\zeta) = \overline{f^2} \iint_{-\infty}^{\infty} \varphi(u,v) \exp\{-j(\xi u + \zeta v)\} \, du \, dv. \qquad (2.147)$$

Instead of equation (2.145), we then obtain

$$\langle |F|^2 \rangle = L_y L_z \, \phi(\xi,\zeta) \qquad (2.148)$$

with the power spectrum at the spatial frequency

$$\xi = -nk \sin\vartheta \cos\varphi \qquad (2.149)$$

of the y-distribution in $f(y,z)$, and at the spatial frequency

$$\zeta = \beta - nk \sin\vartheta \sin\varphi \qquad (2.150)$$

of the z-distribution in $f(y,z)$. n is equal to n_a or n_b, depending in which region (a or b) the radiation is considered.

With the factor $\langle |F|^2 \rangle$ taken from equation (2.145) for a specified autocorrelation function, or from equation (2.148) in terms of the power spectrum of boundary displacements, the average radiated power follows from equations (2.133) and (2.134). This power subtracts from the power P of the guided mode, so that

$$\frac{dP}{dz} = -\frac{\langle P \rangle}{L_y L_z}$$

and P decays exponentially according to

$$P = P_0 \exp(-2\bar{\alpha}z).$$

The mean attenuation constant $\bar{\alpha}$ follows from these two relations as the power ratio

$$\bar{\alpha} = \frac{\langle P \rangle}{2 L_y L_z P} \,. \qquad (2.151)$$

If the average Poynting vector of the radiation is expressed in terms of the power spectrum $\phi(\xi,\zeta)$ of boundary displacements, we obtain the following expression for the mean attenuation constant:

$$\bar{\alpha} = \frac{1}{2} \frac{E_0^2}{P} \omega^2 \varepsilon_0^2 (n_b^2 - n_a^2)^2 \left(\frac{\mu_0}{\varepsilon_0}\right)^{\frac{1}{2}}$$

$$\iint_\Omega \frac{r^2}{n} [|\sum_{t=x,y,z} r_t g \vartheta_t|^2 + |\sum_{t=x,y,z} r_t g\varphi_t|^2] \phi(\xi,\zeta) d\Omega. \qquad (2.152)$$

Note that the spatial frequencies ξ and ζ at which the power spectrum must be evaluated are given by equations (2.149) and (2.150). Through them the power spectrum depends on the variables of integration ϑ and φ in the solid angle Ω.

To further evaluate the scattering loss, we need to specify the power spectrum of boundary displacement or its autocorrelation function. For want of better information on the nature of surface roughness and its statistics, we resort to a particularly simple example of autocorrelation in the form of the exponential function

$$\varphi(u,v) = \exp\left(-\frac{|u|}{B_y} - \frac{|v|}{B_z}\right). \qquad (2.153)$$

The quantities B_y and B_z in the exponential give the distances at which the autocorrelation drops from unity at $u = v = 0$ to $1/e$ in the y- or z-directions, respectively. As such characteristic quantities they are called autocorrelation distances of the random distribution, B_y in the y-direction, and B_z in the z-direction.

The power spectrum in the case of the exponential autocorrelation has the two-dimensional distribution

$$\phi(\xi,\zeta) = 4\overline{f^2} B_y B_z \frac{1 + B_y B_z \xi \zeta}{(1+\xi^2 B_y^2)(1+\zeta^2 B_z^2)} . \qquad (2.154)$$

If the correlation in surface roughness extends only over distances

$$B_y \ll \frac{1}{nk} \qquad (2.155)$$

transverse to the direction of wave propagation, and only over distances

$$B_z \ll \frac{1}{nk + \beta} \qquad (2.156)$$

parallel to the direction of wave propagation, the power spectrum remains flat up to the highest spatial frequencies of $f(y,z)$ which contribute to radiation. Under these conditions, we can set

$$\phi(\xi,\zeta) = 4\overline{f^2} \, B_y \, B_z. \qquad (2.157)$$

For such short correlation distances, the effective sources of all displaced surface elements radiate independently of each other, and the total scattered power is simply the sum of contributions from all individual surface scatterers.

Relatively simple expressions for the scattering loss result under these conditions for H_m-waves in the film. If the surface roughness occurs at the boundary between a film with n_1 and upper space with n_0, the mean attenuation constant may be expressed as

$$\overline{\alpha}_{01} = \frac{2}{3\pi} \overline{f^2} \, B_y \, B_z \, \frac{n_0 k^3 V^2}{\beta d_e d^2} (1-B)(n_1^2 - n_0^2) \, h\left(\frac{n_1^2 - n_0^2}{n_0^2}\right). \qquad (2.158)$$

If the boundary between film and substrate with n_2 shows such roughness, it causes scattering loss to H_m-waves according to

$$\overline{\alpha}_{12} = \overline{\alpha}_{01} \frac{n_2}{n_0} \frac{n_1^2 - n_2^2}{n_1^2 - n_0^2} \frac{h\left(\frac{n_1^2 - n_2^2}{n_2^2}\right)}{h\left(\frac{n_1^2 - n_0^2}{n_0^2}\right)} \qquad (2.159)$$

d_e in equation (2.158) is the effective width of the particular H_m-mode which, according to equation (2.94), not only takes account of the film thickness d but also of the penetration of fields into the substrate and covering medium. V is the film parameter while β and B are phase constant and phase parameter (equation (2.18)) respectively of the particular H_m-mode under consideration. The function $h(\delta)$ as as plotted in Fig.2.21 represents the change in radiated power of the horizontally oriented dipole due to the presence of the respective boundaries.

Because of the factor $k^3 v^2/\beta$ in equation (2.158), the scattered loss depends essentially as $1/\lambda^4$ on wavelength. This corresponds to the Rayleigh law of scattering and is typical for scatterers which are small compared to the wavelength. For surface roughness with the same value for the product $\overline{f^2} B_y B_z$ at both film boundaries, the ratio of scattering loss is essentially that of the differences in permittivities $(n_1^2 - n_2^2)/(n_1^2 - n_0^2)$. The boundary with larger index difference therefore scatters more power according to this ratio. For weakly guiding films with a small index difference between film and substrate, scattering from the upper film surface dominates, and this surface must be particularly smooth for low scattering loss.

For a particular H_m-mode, the scattering loss depends as $1/d_e$ on the effective mode width. Near cut-off, the fields extend into the substrate, and surface roughness, which then is relatively small compared to the mode width, causes only little scattering. As we move away from cut-off, d_e goes through a minimum according to Fig.2.14. At this minimum surface roughness becomes most effective in scattering guided mode power. Beyond this minimum, d_e increases monotonically and approaches d. In this region of thicker and thicker films, a particular film mode consists of rays, which propagate at steadily decreasing angles θ (see Fig.2.2) and undergo fewer and fewer reflections at the boundaries. The scattering at rough boundaries will then also decrease and with it the scattering loss.

Fig.2.24 shows the scattering loss of the fundamental H_0-mode in the form $W = \overline{\alpha} \, d^5/(\overline{f^2} B_y B_z)$ for three different values of the frequency parameter V plotted as a function of the asymmetry parameter a_H from equation (2.39). For small a_H-values in nearly symmetrical film guides, W_{01} differs only little from W_{12}. When a_H increases, we approach the case of weakly guiding films. W_{12} then stays nearly constant while W_{01} increases almost linearly with a_H.

The range of correlation distances B_y and B_z for which these results apply is limited by the assumptions (2.155) and (2.156). Of these two conditions, the inequality (2.156)

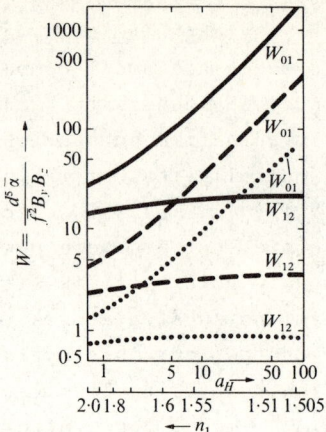

Fig.2.24. Scattering loss $\overline{\alpha_{01}}$ from surface roughness of the upper film boundary and $\overline{\alpha_{12}}$ from roughness of the film substrate boundary. $a_H = (n_2^2 \ n_0^2)/(n_1^2 \ n_2^2)$; $n_0 = 1$; $n_2 = 1.5$.

$$V = kd \ (n_1^2 - n_2^2)^{\frac{1}{2}} = \begin{matrix} 1.87 & \cdots \\ 2.57 & \text{---} \\ 4.57 & \text{———} \end{matrix}$$

is more restrictive than (2.155). For boundary perturbations of increasing correlation distance, this condition will therefore be the first one that is no longer satisfied.

In some cases of practical interest, the correlation extends much further in one direction than it does in the other. As one such case, and for application in later sections, we will now consider the limit $B_y \to \infty$. This corresponds to boundary displacements which are constant transverse to the direction of wave propagation and depend only on the coordinate z in the propagation direction. The scattering from an elementary line of width dz and displacement f may, in the case of excitation by an H_m-wave, be considered to come from an effective line source of electric current:

$$I = j\omega\varepsilon_0 \ (n_a^2 - n_b^2) \ E_0 \ r_y \ f(z) \ \exp(-j\beta z) \ dz. \qquad (2.160)$$

For the radiation field of this line source, we again take account only of the boundary next to this line source and neglect the effect of the other film boundary on this field. Its far field then has a magnetic vector with only a θ-component in the cylindrical coordinates of Fig.2.25. For this far-field magnetic component, the following expression holds (Hinken 1976):

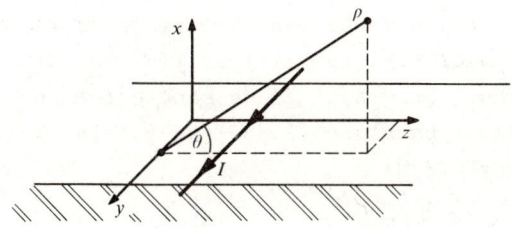

Fig.2.25. Effective line source for an elementary boundary displacement transverse to the direction of wave propagation.

$$H_\theta = \frac{jI}{1 + \sqrt{(1 + \delta_\theta \sin^{-2}\theta)}} \sqrt{\left(\frac{nk}{2\pi\rho}\right)} \exp\{jnk(z \cos \theta - \rho) + j\phi_0\}, \quad (2.161)$$

where $n = n_a$ and $\delta_\theta = (n_b^2 - n_a^2)/(n_a^2)$ for $0 \le \theta \le \pi$, but
$n = n_b$ and $\delta_\theta = (n_a^2 - n_b^2)/(n_b^2)$ for $\pi \le \theta \le 2\pi$. (2.162)

The phase constant ϕ_0 in equation (2.161) has no consequences in the present context. Any distribution $f(z)$ of boundary displacement and the associated distribution $I(z)$ of effective line sources along z generate a magnetic component in the far field according to

$$H_\theta = -\frac{\omega\varepsilon_0(n_a^2 - n_b^2)E_0\, r_y}{1 + (1+\delta_\theta \sin^{-2}\theta)^{\frac{1}{2}}} \left(\frac{nk}{2\pi\rho}\right)^{\frac{1}{2}} \exp\{j(\phi_0 - nk\rho)\}$$
$$\int_0^{L_z} f(z) \exp\{-jz(\beta - nk \cos \theta)\}\, dz. \quad (2.163)$$

It has the intensity

$$S = \frac{1}{n}\left(\frac{\mu_0}{\varepsilon_0}\right)^{\frac{1}{2}} |H_\theta|^2 \quad (2.164)$$

and radiates the power

$$P = L_y \int_0^{2\pi} S(\theta)\, \rho\, d\theta \tag{2.165}$$

in a film guide of width L_y in the y-direction.

For a random boundary displacement, we find the scattered power to be expected from it by assuming, as we did before, that $f(z)$ is stationary and ergodic. The ensemble average of scattered power can then again be expressed in terms of the power spectrum $\phi(\gamma)$ of $f(z)$, or its autocorrelation function $\varphi(u)$. If we choose the power spectrum $\phi(\gamma)$ according to equation (2.143) we obtain the average scattered power $\langle P \rangle$ as

$$\langle P \rangle = \frac{kL_y L_z}{2\pi} \left(\frac{\mu_0}{\varepsilon_0}\right)^{\frac{1}{2}} [\omega\varepsilon_0 (n_a^2 - n_b^2) E_0 r_y]^2 \int_0^{2\pi} \frac{\phi(\beta - nk \cos\theta)}{|1 + (1+\delta_\theta \sin^{-2}\theta)^{\frac{1}{2}}|^2} d\theta, \tag{2.166}$$

where n and δ_θ in the integrand depend on θ as in equation (2.162). The mean attenuation constant due to the average power P may now again be evaluated as in equation (2.151). For the more important practical case of scattering from a rough top surface of the film, the average loss is

$$\bar{\alpha}_{01} = \frac{k^2 v^2 (1-B)(n_1^2 - n_0^2)}{2\pi \beta d_e d^2} \int_0^{2\pi} \frac{\phi(\beta - nk \cos\theta)}{|1 + (1+\delta_\theta \sin^{-2}\theta)^{\frac{1}{2}}|^2} d\theta. \tag{2.167}$$

From these expressions for two-dimensional radiation and its guided mode loss, we can see more readily, than from the more complicated forms of three-dimensional radiation, how the spectral components of the surface perturbation contribute to radiation into different directions θ. For $0 \leq \theta \leq \pi$, the spatial frequencies at which $\phi(\gamma)$ must be evaluated in equation (2.167) range from $\beta - n_0 k < \gamma < \beta + n_0 k$. Whatever the power spectrum contributes at these particular γ-values will be radiated into the corresponding directions θ of the upper space. Of the remaining θ-range only power scattered into directions $\theta_c - \pi \leq \theta \leq -\theta_c$ will radiate into the substrate; $\theta_c = \arccos(n_1/n_2)$ is the limiting angle of total reflection at the substrate. Power which scatters into $-\theta_c \leq \theta \leq 0$ will convert to forward-travelling

H_m-modes, while power scattered into $-\pi < \theta < \theta_c - \pi$ will excite backward-travelling modes.

If the film guides only the fundamental H_0-mode, the forward-scattered power within $-\theta_c \leq \theta \leq 0$ will nearly all remain in the H_0-mode, so that this range then should be excluded from the integration in equations (2.166) and (2.167). If, in addition to the H_0-mode, the film also guides higher-order modes, we need to exclude from the integration only that range of small negative θ-values, which corresponds to the H_0-mode. The power which scatters into higher-order modes will then also be accounted for in the H_0-loss.

As special cases of practical interest, we consider first a power spectrum of the surface roughness that is wide enough to remain flat for all spatial frequencies over which the integration in equation (2.167) extends. Such a wide power spectrum corresponds to a very small correlation distance, which then satisfies a condition such as (2.156). Under this condition, the integration of equation (2.167) with $\phi(\gamma) \simeq \phi(0)$ results in

$$\bar{\alpha}_{01} = \frac{k^2 V^2 (1-B)(n_1^2 - n_0^2)}{4\beta d_e d^2} \phi(0) \, h_L(\delta). \tag{2.168}$$

The factor

$$h_L = \frac{2}{\pi} \int_0^{2\pi} \left| \frac{1}{1 + (1+\delta_\theta \sin^{-2}\theta)^{\frac{1}{2}}} \right|^2 d\theta \tag{2.169}$$

in this expression represents the power, which an electric current line source on the interface radiates into both half-spaces relative to the power which it would radiate into free space. Integration of equation (2.169) leads to $h_L = 1$. Any electric current line next to a dielectric interface therefore radiates just as much power as it would radiate into free space. The ratio in which this radiation distributes into both half-spaces depends on the difference in permittivities δ_θ.

Next we assume again the particularly simple exponential correlation

$$\varphi(u) = \exp\{-|u|/B\},$$

which has associated with it the power spectrum

$$\phi(\gamma) = \frac{2\,\overline{f^2}\,B}{1 + \gamma^2 B^2}$$

corresponding to a Lorentz curve in its dependence on spatial frequency γ. If

$$B \ll 1/(\beta + nk),$$

this particular power spectrum remains flat for all spatial frequencies over which the integration in equation (2.167) extends. The scattering loss of equation (2.167) is then, according to

$$\overline{\alpha}_{01} = \frac{k^2 v^2 (1-B)(n_1^2 - n_0^2)}{2\beta d_e d^2} \overline{f^2} B, \qquad (2.170)$$

proportional to the variance $\overline{f^2}$ of surface roughness and its correlation distance B.

The approximate theory of scattering at a film boundary with two-dimensional roughness, as well as its one-dimensional specialization for transverse grooves neglects to certain extent the effect of the respective second film boundary on the scattered field and power. Even if, in case of the transverse grooves, we exclude the range of θ-values from the integration that belongs to the primary film mode, we still expect differences from a more exact calculation. Scattering at transverse grooves of film guides has been analysed by Marcuse (1974), who set up the complete set of guided and radiation modes of the film and found the coupling between these modes due to interface displacements in the form of transverse grooves. In his numerical evaluations he considers symmetrical slabs and assumes random grooves of exponential correlation. Figs.2.26 and 2.27 compare some of his results (Marcuse 1969) with results from our present more approximate analysis. The representation adopted from Marcuse gives the attenuation constant $\overline{\alpha}$ normalized with respect to $4\overline{f^2}/d^3$ as a function of the correlation distance B relative to half the slab width. Fig.2.26 is for a weakly

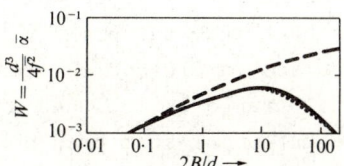

Fig.2.26. H_0-mode scattering loss of a symmetric slab with random transverse grooves of exponential autocorrelation in one boundary with correlation distance B with $n_2 = 1$, $n_1 = 1\cdot01$, $V = 2\cdot27$.

——— Integration in equation (2.167) extended over $0 < \theta < 2\pi - \theta_c$

---- Integration in equation (2.167) extended over $0 < \theta < 2\pi$

.... results from complete coupled mode theory (Marcuse 1969)

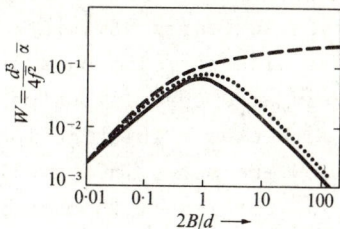

Fig.2.27. H_0-mode scattering loss of a symmetric slab as in Fig.2.26, but with $n_2 = 1$, $n_1 = 1\cdot5$, $V = 2\cdot91$.

guiding slab and Fig.2.27 for a more pronounced index difference as, for example, a slab in free space would have.

In both cases, monomode conditions prevail so that the slabs guide only the H_0- and E_0-modes, and the scattered power from the H_0-mode goes into the radiation field.

From these examples the results of our approximate analysis show good accuracy, in particular for short correlation. To maintain sufficient accuracy for longer correlation as well, the integration over the scattered power should exclude the range $-\theta_c < \theta < 0$, in which power of the monomode slab scatters back into the primary H_0-wave. The solid and the broken curves in Figs. 2.26 and 2.27 compare scattering loss when this range is included and when it is excluded from the integration of scattered power.

The loss due to surface scattering appears proportional to the correlation distance B only for very short correlation in Figs. 2.26 and 2.27. As the correlation widens, the loss goes through a maximum and decreases again for very wide correlation. Such a dependence on correlation distance is typical for distributions of scatterers with a continuous power spectrum. For very short correlation, their power spectrum is very wide, and power scatters into all directions of space. As the correlation widens, the power spectrum narrows down to lower and lower spatial frequencies. Spectral components of low frequency scatter power into forward directions with small angles θ. For very small θ, however, the denominator in the integrand (2.167) increases and thus reduces the scattered power. In addition, for $-\theta_c < \theta < 0$, any power is scattered back into guided modes. We therefore observe for wide correlation a steady decrease in scattering loss. Very wide correlation corresponds to gradual changes in film thickness which no longer cause much scattering but allow the mode, in travelling through the film, to gradually adjust itself to changing dimensions and maintain its independent normal mode character.

2.6. GRATING COUPLERS

The surface scattering of power from guided modes of films and slabs into the radiation field as well as into other guided modes appears as a detrimental effect that causes mode attenuation and interference. This surface scattering may, however, also be put to practical use for coupling in and out of guided modes, and to produce coupling between certain guided modes for specific purposes. Instead of the random surface imperfections, we need to produce boundary displacements of a more regular nature and, more specifically, in the form of a periodic grating for such couplers. The lines of this grating should run transverse to the direction of wave propagation, and should have no change of boundary displacement in this transverse direction, but space periodic changes only in the direction of wave propagation. Fig.2.28 shows such a grating of rectangular grooves that could be etched into the film by planar

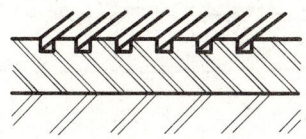

Fig.2.28. Transverse periodic grating of rectangular grooves in a film guide.

photolithographic techniques. When excited by the field of a guided mode of the film, each line of the grating, with its effective line source of electric current, will radiate into the space above and also through the film into the substrate below. In addition it will scatter some power into other guided modes of the film as well as into the primary mode in the forward as well as the backward direction. The forward-scattered part into the primary mode superimposes on to this mode and causes some phase change, while the backward-scattered component appears as reflection of the incident mode.

In a periodic grating of many such lines the contributions from all line sources interfere with each other. In some directions of upper space and into the substrate the interference can be constructive, and it will then lead to beams of plane-waves, which radiate into these particular directions of the covering space or into the substrate. Such beams are able to gather a substantial part of the power which the primary film mode guides. They thus offer a convenient means to couple power out of film or slab modes into beams and, by reciprocity, launch beams into film modes.

To identify the origin of specific scattered beams and facilitate the analysis, we examine first the scattering at a sinusoidal variation of the upper film boundary as shown in Fig.2.29. In the case of H_m-wave excitation, the effective sources of electric currents from equation (2.160) have the surface density distribution

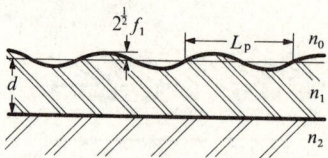

Fig.2.29. Transverse grating of sinusoidal surface variation in a film guide.

$$J_s = j\omega\varepsilon_0 \ (n_0^2 - n_1^2) \ E_0 \ r_y \ 2^{\frac{1}{2}} \ f_1 \ \sin(\gamma_1 z) \ \exp(-j\beta z). \quad (2.171)$$

where $2^{\frac{1}{2}} f_1$ constitutes the maximum boundary displacement of the sinusoidal variation, and $\gamma_1 = 2\pi/L_p$ its angular spatial frequency, corresponding to a period of length L_p. The power spectrum of the sinusoidal distribution $f(z) = 2^{\frac{1}{2}} f_1 \sin(\gamma_1 z)$ consists only of one line at $\gamma = \gamma_1$ and may be written with the help of Dirac's delta function

$$\phi(\gamma) = f_1^2 \ \delta(\gamma - \gamma_1). \quad (2.172)$$

Such a single line contributes to the scattered power of equation (2.166) only in that direction $0 < \theta < \pi$ of the space above the film, for which

$$\beta - n_0 \ k \ \cos\theta = \gamma_1. \quad (2.173)$$

Into the film, it scatters in that direction $-\pi < \theta < 0$, for

which

$$\beta - n_1 k \cos \theta = \gamma_1.$$

Upon refraction at the film-substrate interface this direction changes and according to Snell's law follows now from

$$\beta - n_2 k \cos \theta = \gamma_1. \tag{2.174}$$

Scattered fields from the sinusoidal surface variation interfere constructively only in these particular directions of space above and substrate below. Radiation into all other directions extinguishes itself by destructive interference.

Fig.2.30 shows θ as a function of γ_1 from equations (2.173) and (2.174). For $\gamma_1 < \beta - n_2 k$ and $\gamma_1 > \beta + n_2 k$ power scatters neither into the substrate, nor into the covering space. For

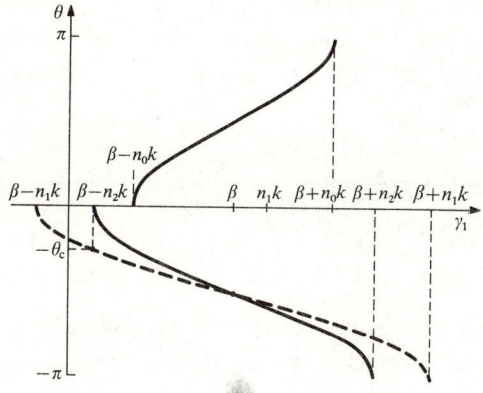

Fig.2.30. Beam angle θ for beams into the covering space and substrate from a film mode with phase constant β at a sinusoidal grating with angular spatial frequency γ.

$$\beta - n_2 k < \gamma_1 < \beta - n_0 k \tag{2.175}$$

only a forward beam into the substrate, and for

$$\beta + n_0 k < \gamma_1 < \beta + n_2 k \tag{2.176}$$

only a reverse beam into the substrate is excited. For the remaining range of spatial frequencies,

$$\beta - n_0 k < \gamma_1 < \beta + n_0 k, \tag{2.177}$$

any sinusoidal variation excites a beam into the substrate as well as into the covering space.

A grating coupler to launch from a film mode into a beam or vice versa should preferably excite only one beam. A simple sinusoidal variation can do this only for spatial frequencies in the γ_1-ranges of (2.175) or (2.176); in these spatial frequency ranges it will then only excite a substrate beam.

If with $\gamma_1 = \beta - n_2 k \cos\theta_2$ in the range of either (2.175) or (2.176) only a substrate beam is excited, it extracts power from the primary wave and, in this process, renders it an exponentially decaying or leaky mode. The power from the primary wave leaks out into the susbtrate beam; the attenuation constant connected with this power leakage follows from (2.167) if we substitute Dirac's function from equation (2.172) for the power spectrum $\phi(\gamma)$:

$$\alpha = \frac{kV^2(1-B)(n_1^2-n_0^2)f_1^2}{2\pi\beta d_e \, d^2 n_1 \sin\theta_1 \left|1 + (1 - \frac{n_1^2-n_0^2}{n_1^2}\sin^{-2}\theta_1)^{\frac{1}{2}}\right|^2} \tag{2.178}$$

θ_1 in this expression represents the beam angle before refraction at the film-substrate interface and is given by $n_1 k \cos\theta_1 = \beta - \gamma_1$.

In leaking power to the substrate beam, the primary wave has now the complex propagation constant $\alpha + j\beta$; its amplitude decays exponentially along z according to $e^{-\alpha z}$. The beam amplitude will then likewise change, but transverse to its direction of propagation, as Fig.2.31 indicates. Its excitation is only as strong as the amplitude of the primary wave at the particular location z, where the respective part

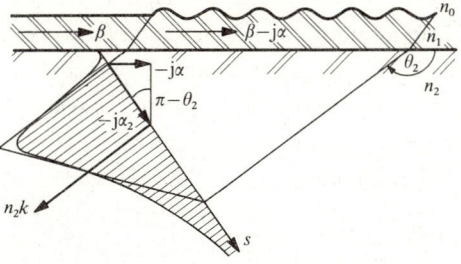

Fig.2.31. Reverse substrate beam excited by a film mode in a sinusoidal grating coupler.

of the substrate beam originates. With the primary wave decaying exponentially along z, the substrate beam will also decay exponentially along s, transverse to its propagation direction. For the transverse rate of decay in the beam to match the longitudinal rate of decay of the mode, we must have $\alpha_2 s = \alpha z$ for distances s and z that correspond to each other in Fig.2.31. With $s = z \sin(\theta_2 - \pi)$ for the substrate beam in reverse direction, we obtain

$$\alpha_2 = \alpha/\sin(\theta_2 - \pi).$$

The beam which couples out of the film mode at the sinusoidal surface grooves has the form of a transversely evanescent wave. It propagates with a complex wave vector $\vec{k}_2 - j\vec{\alpha}_2$ of magnitude equal to the wave number $n_2 k$ of the substrate. Its real part $\vec{k}_2$ has the propagation direction of the beam and for $|\vec{\alpha}_2| \ll |\vec{k}_2|$ is nearly equal to $n_2 k$. Its imaginary component has the transverse direction parallel to the beam coordinate s in Fig.2.31. For a long enough grating of the film, most of the film mode power will couple into this transversely decaying beam. The length of the coupler for nearly complete power transfer depends critically on the depths of grooves and the amount of scattering which they effect.

In most of the practical situations a small sinusoidal variation will not suffice to obtain enough coupling per length into a beam. More coupling occurs when rectangular grooves of sufficient depth are etched into the film surface or when a periodic grating of possibly quite a different refractive index is deposited on to the film, as in Fig. 2.32. The scattering from such more general and effective gratings can to a certain approximation still be described by effective current sources. The periodic function $f(z)$,

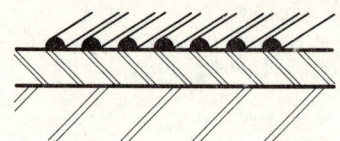

Fig.2.32. Grating coupler with high-index lines deposited on a film for more efficient coupling.

on which this source distribution depends, can be expanded into a Fourier series with coefficients

$$2^{\frac{1}{2}} f_\nu = \frac{2}{L_p} \int_0^{L_p} f(z) \cos(\nu \, \gamma_1 \, z) \, dz. \qquad (2.179)$$

Each Fourier component with amplitude f_ν contributes a line at frequency $\gamma_\nu = \nu \, \gamma_1$ to the power spectrum $\phi(\gamma)$ in equation (2.167). In addition to the one or two beams radiating into the substrate and covering space with directions according to equations (2.173) and (2.174), we will then obtain additional beams for whatever θ-values satisfy

$$\beta - n_0 k \cos \theta = \nu \, \gamma_2$$

or $\qquad\qquad\qquad\qquad\qquad\qquad\qquad\qquad\qquad\qquad (2.180)$

$$\beta - n_1 k \cos \theta = \nu \, \gamma_1$$

for beams into covering space and substrate, respectively. Such beams of higher diffraction orders extract additional power from the film mode and reduce the useful output power of the main beam. To avoid these parasitic beams and the power loss associated with them, the fundamental frequency γ_1 of the grating should be chosen such that all its space harmonics at $\nu\,\gamma_1$ remain outside the range in which their spectral components can excite any beam into the covering space or substrate. The only range of γ_1 that allows to satisfy this condition is given by equation (2.176). A grating coupler with space harmonics f_ν in its distributions should therefore be tuned to the range (2.176) where it excites only the reverse substrate beam, but no other beam from higher-order harmonics either into the substrate or into the covering space. Fig.2.31 therefore also represents the proper conditions for a more general grating coupler with space harmonics in its grating.

An obvious disadvantage of coupling by a grating into a substrate beam instead of coupling into a space beam is that the substrate beam can leave the coupler only through the bottom of the substrate or at a side face. Quite often, a prism may even be placed underneath the substrate to extract the beam as in Fig.2.33. Such output beams, and even

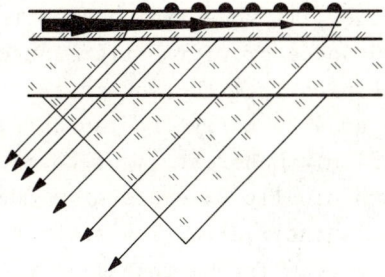

Fig.2.33. Grating coupler with prism to extract a reverse substrate beam.

more so the prism, destroy the pure planar geometry, which is a desirable feature of grating couplers for applications in planar optical circuits.

To avoid a substrate beam and couple only to one space beam, a periodic grating must be composed of elements that scatter individually only into a limited range of directions of the covering space. Its period length L_p of fundamental spatial frequency $\gamma_1 = 2\pi/L_p$ must then be so adjusted that the corresponding θ-value from equation (2.173) falls into the main lobe of the individual radiation pattern. Kogelnik and Sosnowski (1970) employed a thick dichromatic gelatine layer on top of the film guide, which was exposed holographically so as to produce a fringe pattern inclined at an angle of about 45° with respect to the film surface. In such a thick grating layer all but the fundamental diffraction order into the covering space are virtually suppressed by the Bragg effect. Only one beam corresponding to this fundamental diffraction radiates under these conditions near the Bragg angle, and into the covering space.

A perfect grating coupler with only one output beam excites this beam with its transversely evanescent amplitude. To couple all the film mode power into this beam, the grating must be very long, and the beam is correspondingly wide. However, since the exponential tail beyond a beam width of $s = 1/\alpha_2$ carries only little power, the grating coupler may be terminated at this point, or somewhat beyond, and the remaining power cut off.

By way of reciprocity, such a grating coupler can also serve to launch the film mode from a beam incident either from the covering space or, for better launching efficiency, in the reverse fashion from the substrate. Such a grating coupler presents an alternative to launching film modes at the edge of the film. Coherent light beams out of gas or solid-state lasers usually extend over a considerable beam width transverse to their direction of propagation. To launch such a wide beam into a thin film would pose considerable problems if the beam were to be focused on to the film edge. It would not only be difficult to transform the incident beam to the extremely narrow field distribution of a particular film mode but also any roughness of the film edge, which can hardly be avoided, would scatter much of the incident power into radiation fields.

Instead of forcing a relatively wide beam into the narrow distribution of a film mode, the grating coupler transforms the laser beam gradually into the film mode over its full coupling length. To achieve the same efficiency in launching as in coupling out of the film mode, the input beam must have plane phase fronts and the same direction as the outgoing beam; it must also have the same transverse amplitude distribution with exponential decay as the outgoing beam. Power from an incoming beam with any other amplitude distribution $A_i(s)$ will only convert to the film mode to the degree to which this amplitude distribution overlaps with the exponential distribution

$$A_0(s) = (2\alpha_2)^{\frac{1}{2}} \exp\{-\alpha_2 s\} \qquad (2.181)$$

of the corresponding outgoing beam. The overlap integral (Tien 1971)

$$p = [\int_s A_i(s) A_0(s) ds]^2 / [\int_s |A_i(s)|^2 ds \int_s |A_0(s)|^2 ds] \qquad (2.182)$$

specifies the correlation between the amplitude distribution $A_i(s)$ and that required for a perfect coupler. It also represents the launching efficiency for an input beam of amplitude distribution $A_i(s)$. The integrals in equation (2.182) extend over the transverse beam coordinate, as shown in Fig.2.31. They should cover the full ranges over which the respective amplitude distributions contribute anything. Only in the case $A_i(s) = A_0(s)$ will the launching efficiency equal the output efficiency, and approach 100 per cent if s and the coupler length extend sufficiently far beyond $1/\alpha_2$. Fig.2.34 shows this exponential amplitude distribution as the uppermost example.

An input beam with a uniform intensity and amplitude distribution of width s_i launches its power into the film mode with an efficiency that follows from equation (2.182) according to

$$p = \frac{2}{\alpha_2 s_i} [1 - \exp(-\alpha_2 s_i)]^2. \qquad (2.183)$$

As a function of $\alpha_2 s_i$ this launching efficiency has a maximum of $p = 0.81$ at $\alpha_2 s_i = 5/4$. Fig.2.34 shows as the second example this beam of uniform intensity distribution and op-

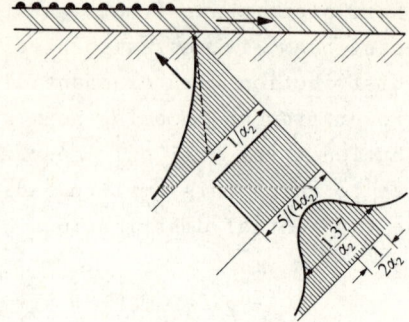

Fig.2.34. Input beams for a reverse grating coupler with exponential, uniform, and Gaussian transverse amplitude distribution. The beam positions and beam widths give maximum launching efficiencies.

timum width with respect to launching efficiency.

The third example in Fig.2.34 represents a beam of Gaussian intensity and amplitude distribution according to

$$A_i(s) = \exp\{-(s-s_o)^2/w^2\}.$$

Such Gaussian beams emerge from lasers, which oscillate in the mode of lowest transverse order of their optical cavity. For best launching efficiency, the beam waist should be moved close to the grating coupler in order that its phase fronts be as plane as possible in the region of the coupler. The launching efficiency which is then obtained from equation (2.182) assumes its maximum value for a spot size of the Gaussian beam of $w = 0.684/\alpha_2$ and a displacement of the beam axis from the edge of the exponentially decaying reference beam of $s_o = 1/(2\alpha_2)$. If all these requirements are met, the Gaussian beam launches into the film mode with the maximum efficiency of $p = 80$ per cent. Optimum spot size and displacement of the Gaussian beam are indicated in Fig.2.34.

2.7. PRISM COUPLERS

The prism which appears schematically in Fig.2.35 serves the same purpose as the grating coupler namely to launch light beams into film modes and couple them out into beams. It also works on a principle similar to that of the grating coupler.

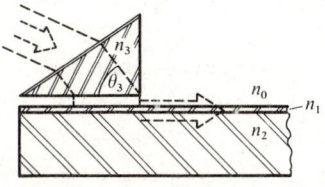

Fig.2.35. Prism film coupler.

The prism coupler solves the problem of launching light beams into film modes by partially frustrating the total internal reflection and by distributed coupling. The prism in Fig.2.35 is positioned above the film with its base parallel and quite close to the film. The uniform plane wave of a light beam, which is incident through the prism on to its base, experiences total internal reflection at the base interface to the air gap. Its fields are evanescent in this gap and decay exponentially transverse to the gap. If this air gap is narrow enough, the exponential tails of evanescent fields will extend down to and into the film. Depending on the angle of incidence, these fields can then excite a wave into the substrate or also one of the film modes. The total reflection at the prism base is then partially frustrated, and power converts from the incident beam into a substrate or film wave. Designating the refractive index of the prism by n_3 and the incident beam angle in the prism by θ_3, this beam excites substrate waves as long as

$$\cos \theta_3 < n_2/n_3 \ .$$

The excitation of film waves requires incident beam angles in the range of

$$n_2/n_3 < \cos \theta_3 < n_1/n_3 \ , \tag{2.184}$$

while for

$$\cos \theta_3 > n_1/n_3$$

no frustration of the total internal reflection at the prism base occurs. For incident beam angles of suitable values the beam excites a film mode and transfers power to it over the full width of the beam along the base of the prism. While this power is continuously extracted from the beam, it builds up the film mode, until eventually, most of the beam power converts to the film mode.

By way of reciprocity the prism coupler also transfers power from a film mode into a beam inside the prism. As in the case of the grating coupler, the conversion of a film mode into an outgoing beam is somewhat simpler to explain and to analyse. We will therefore first discuss the prism coupler in this mode of operation. From the results for the output coupler we may then, by the principle of reciprocity, draw all conclusions of practical interest for the input coupler.

Fig.2.36 shows the front end of a prism coupler for coupling a guided mode of the film out into a beam through

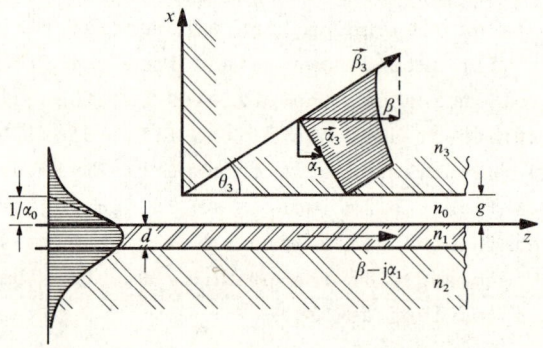

Fig.2.36. Front end of a prism coupler with wave vectors and transverse field distribution of film wave and output beam.

the prism. Up to this front end the film mode propagates without attenuation along the film and its evanescent fields in the space above the film decay exponentially in transverse direction with an attenuation constant α_0. At a distance g above the film surface, which corresponds to the gap of the prism coupler, its field may be assumed small enough that the film mode continues to travel underneath the prism with nearly the same phase constant β as before entering the coupler. The evanescent field of the film mode at the base of the prism, although quite small, nevertheless excites a plane wave inside the prism, which travels with a phase constant equal to the wave number $n_3 k$. Originating from the film mode with phase constant β the plane wave in the prism must have the same phase change in z-direction, so it therefore travels into the prism at an angle

$$\theta_3 = \arccos [\beta/(n_3 k)] \qquad (2.185)$$

with respect to the z-direction. Its phase fronts are perpendicular to this direction. For a wide gap and correspondingly weak coupling, this plane wave will be quite low in intensity and extract power from the film wave only at a low rate. However, as this beam continues to gather power from the film wave the latter will be attenuated exponentially with some attenuation constant α_1. Therefore, underneath the prism the film mode changes its character from a truly guided wave before it entered the prism region, to a leaky wave that continuously leaks power into the plane wave travelling in the direction θ_3 into the prism. With the attenuation of the leaky film mode, the plane wave in the prism will likewise be attenuated, but transverse to its direction of propagation. It therefore propagates with the complex wave vector $\vec{k}_3 = \vec{\beta}_3 + j\vec{\alpha}_3$, where $\vec{\beta}_3$ makes the angle θ_3 with the z-direction and $\vec{\alpha}_3$ is at right angle to $\vec{\beta}_3$, in accordance with equation (1.11), for a transversely evanescent wave in the loss-less prism region. Parallel to the film this prism wave has the component $\alpha_1 = \alpha_3 \sin\theta_3$, which represents the attenuation constant of the leaky film mode to which the prism wave is attached.

To determine the leaky mode attenuation α_1, and with it α_3 for the form and width of the beam in the prism, we need to solve the boundary-value problem of the four-layer region underneath the prism. We require this solution for modes of propagation which correspond to the respective guided film modes without the prism. By way of example, we will treat here the case of an H_m-wave incident in the film. H_m-waves underneath the prism are not only more convenient to analyse but they also deserve more practical interest than E_m-waves. Among the H_m-waves the lowest-order H_0-wave is the fundamental mode of the film and is normally utilized for film guide applications.

To set up a solution for H_m-waves in the prism coupler, we use the same expressions for the field components E_y and H_z inside the film and in the substrate as in equations (2.66), (2.67), and (2.68). For the gap we let

$$E_y = A \cosh(\alpha_0 x) + F \sinh(\alpha_0 x) \quad \text{for} \quad 0 \leq x \leq g \quad (2.186)$$

which matches this field component at $x = 0$ to the E_y from equation (2.66) inside the film. Inside the prism we let

$$E_y = (A \cosh(\alpha_0 g) + F \sinh(\alpha_0 g)) \exp\text{-}jk_{3x}(x-g) \quad \text{for} \quad x \geq g \quad (2.187)$$

which matches E_y from equation (2.186) at $x = g$. The corresponding magnetic field components follow from equation (2.60). In particular, we obtain for the tangential magnetic field above and below the three interfaces, at $x = g$

$$H_z^+ = \frac{k_{3x}}{\omega\mu} (A \cosh(\alpha_0 g) + F \sinh(\alpha_0 g))$$

$$H_z^- = \frac{j\alpha_0}{\omega\mu} (A \sinh(\alpha_0 g) + F \cosh(\alpha_0 g)),$$

at $x = 0$

$$H_z^+ = \frac{j\alpha_0}{\omega\mu} F$$

$$H_z^- = \frac{jk_{1x}}{\omega\mu} B,$$

and at $x = -d$

$$H_z^+ = \frac{jk_{1x}}{\omega\mu} (A \sin u + B \cos u)$$

$$H_z^- = \frac{j\alpha_2}{\omega\mu} (A \cos u - B \sin u).$$

Requiring these components to be continuous at the three boundaries leads to a homogeneous system of three linear equations for the factors A, B, and F. For non-trivial solutions of this system its coefficient determinant must be zero; this constitutes the characteristic or eigenvalue equation for H_m-modes of the four-layer structure. Using the notations u, v, w as previously for the transverse phase and attenuation parameters, and in addition

$$t = k_{3x} d \quad \text{as well as} \quad \delta = g/d, \qquad (2.188)$$

the characteristic equation may be written as

$$\begin{aligned}&w(u \cos u + v \sin u) - u(u \sin u - v \cos u) \\ &= -\exp(-2\delta w) \frac{t+jw}{t-jw} [w(u \cos u + v \sin u) + u(u \sin u - v \cos u)].\end{aligned} \qquad (2.189)$$

For a wide gap between film and prism we have $\delta w \gg 1$, and the exponential on the right-hand side of equation (2.189) renders this side of the equation quite small. In the limit of $\delta w \to \infty$, equation (2.189) reduces to the characteristic equation (2.71) of guided H_m-modes in the film.

If, due to the exponential, the right-hand side of equation (2.189) remains small but finite, the prism perturbs the film modes only slightly and a perturbation analysis can start from unperturbed values u, v, and w of film modes with small perturbations added to them in the form

$$u + u_1, \quad v + v_1, \quad w + w_1$$

to account for the effect of the prism. Such perturbations u_1, v_1, w_1 of u, v, and w cause a perturbation β_1 of the unperturbed phase parameter β in the form $\beta + \beta_1$. Equations

(2.13) relate the perturbation β_1 of β to the perturbations u_1, v_1, w_1 of u, v, and w:

$$\beta_1 \beta d^2 = -u_1 u = v_1 v = w_1 w. \qquad (2.190)$$

The characteristic equation (2.189) now yields the following expression for β_1 due to its right-hand side perturbation:

$$\beta_1 \beta d^2 = \exp(-2\delta w) \frac{w-jt}{w+jt} \frac{2u^2 vw^2}{(u^2+w^2)(v+w+vw)}. \qquad (2.191)$$

This perturbation is complex. Its real part indicates a slight shift in phase constant of the film mode underneath the prism. The negative of the imaginary part of equation (2.191) is

$$\alpha_1 \beta d^2 = \exp(-2\delta w) \frac{4t\, u^2 w^2}{(t^2+w^2)(u^2+w^2)(1+\frac{1}{w}+\frac{1}{v})} ; \qquad (2.192)$$

it gives the attenuation constant α_1 of the film mode underneath the prism due to its power leaking into the prism beam. If we introduce the effective width d_e of the film mode from equation (2.94) and replace the remaining factors of the right-hand side denominator by $u^2 + w^2 = k^2 d^2 (n_1^2 - n_0^2)$ and $t^2 + w^2 = k^2 d^2 (n_3^2 - n_0^2)$, the leaky mode attenuation is given by

$$\alpha_1 = \exp(-2\delta w) \frac{4t\, u^2 w^2}{(n_1^2-n_0^2)(n_3^2-n_0^2) k^4 \beta d^5 d_e}. \qquad (2.193)$$

Due to the exponential factor it depends most critically on the relative gap width $\delta = g/d$ and the transverse attenuation parameter $w = \alpha_0 d$ of the evanescent fields in the gap. For moderate to large values of δw the exponential tails of gap fields are quite weak at the prism base and transfer power of the film mode to the prism beam only at a very low rate.

The factor associated with $w^2 \exp(-2\delta w)$ in the formula for α_1 may alternatively be expressed in terms of the ray angles, θ_1 for the film mode, and θ_3 for the prism beam:

$$\alpha_1 = 4\omega^2 \exp(-2\delta w) \frac{n_3 \, n_1 \, \sin\theta_1 \, \tan\theta_1 \, \sin\theta_3}{(n_1^2-n_0^2)(n_3^2-n_0^2) \, k^2 \, d^2 \, d_e} \,. \quad (2.194)$$

For weakly guiding films with $n_1 \simeq n_2$ but a large asymmetry parameter a_H, the more explicit form

$$\alpha_1 = \frac{4n_3 \, n_1}{(n_3^2-n_0^2)d_e} \, \theta_1^2 \, \sin\theta_3 \, \exp\{-2kg(n_1^2-n_0^2)^{\frac{1}{2}}\} \quad (2.195)$$

approximates the leaky-mode attenuation quite well.

The leaky-mode attenuation α_1 of the film wave causes the transverse attenuation $\alpha_3 = \alpha_1/\sin\theta_3$ of the prism beam. The output beam therefore has its highest intensity in that part of its cross-section that originates near the front end of the prism. The beam components from subsequent sections of the coupling gap continue to decrease in accordance with the attenuation of the leaky film mode. The output beam with its one-sided exponential intensity distribution gathers more and more power with increasing coupling length. For a given length L of the prism coupler, the output coupling efficiency amounts to

$$p_o = 1 - \exp(-2\alpha_1 L)$$

For $L = 1/\alpha_1$ it reaches $p_o = 86\cdot 4$ per cent.

To obtain an output beam with a more uniform or symmetric transverse intensity distribution we need to change the coupling strength and the ensuing leaky-mode attenuation along the coupler. One way to accomplish this is by tapering the gap between the film and prism base. We start with a wide gap at the front end of the prism, where the incident film wave still has its full power, and then narrow the gap gradually so that the increasing rate α_1 of power transfer balances the decrease in film mode power as it transfers to the prism beam. With a suitably tapered gap the output beam could be shaped to any transverse intensity distribution. It could also approximate the Gaussian intensity distribution that matches beams out of gas or solid-state lasers.

Fig.2.37 shows schematically the practical implementation of such a prism coupler with a tapered gap. The substrate has the form of a slide, which is covered by the thin film. A clamp presses the slide against the prism near the

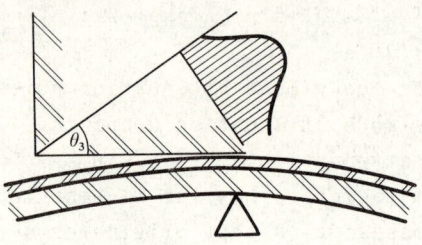

Fig.2.37. Prism coupler with tapered gap for shaping the transverse field distribution of the output beam.

end of the coupler. Toward the front of the coupler the slide bends down elastically and thus increases the width of the gap between film and prism. By varying the pressure of clamping, the prism-slide configuration can be adjusted to shape the output beam to a near-Gaussian intensity distribution.

We return now to the operation of a prism coupler as an input coupler for launching a beam into a film mode. Fig.2.35 shows this mode of operation. If the beam is incident under the same angle θ_3 as the outgoing beam is excited in the other direction, and if it has plane phase fronts and the same transverse intensity distribution, then, by way of reciprocity, the incident beam will launch its power into the film mode with the same efficiency as the output coupler. If, for example, the prism coupler has a constant gap width and uniform leaky-mode loss, then the input beam should match the one-sided exponential distribution of the corresponding output beam in order to achieve maximum launching efficiency. Any deviation from this distribution will reduce the efficiency. This launching efficiency may again be determined from the overlap integral in equation (2.182), which we discussed for the grating coupler.

DIELECTRIC FILMS

Optimum values with respect to launching efficiency for beam position and width in the case of uniform and Gaussian intensity distributions may again be read from Fig.2.34 with α_2 replaced by the transverse attenuation constant α_3 of the output beam in the prism coupler. Tapering the gap as in Fig.2.37 to shape the transverse intensity distribution of the output beam opens the possibility to match it to any input beam and thereby, in principle, increase the launching efficiency, to 100 per cent.

2.8. CURVED SLABS

Film guides are mostly used in planar optics and integrated optics. In these applications they are fabricated on plane substrates and have no curvature in any direction. The same is normally true for symmetrical slabs, as they occur in buried form in optical devices and circuits or as films suspended in free space.

We will, nevertheless, discuss wave propagation in curved slabs, because a number of optical guiding structures, as we derive them in subsequent chapters from film or slab guides, actually need to be curved in specific applications for optical circuits. We can restrict our attention to curved slabs, that is symmetrical films with curvature, because only the curvature effects of such symmetrical structures occur in waveguides that we derive from films or slabs. What we learn from wave propagation in curved slabs will help us to understand the effects of bends in a number of other waveguides, and even allow us to estimate quantitatively curvature loss and other detrimental effects due to curvature.

We consider the slab with uniform curvature in Fig. 2.38. The guided modes which we found for the same slab without curvature will, to a certain extent and with some modification, also exist in the curved slab. If the slab is curved gently enough, we may consider the modes to consist of uniform plane waves again, which travel on a zig-zag path through the bend and experience reflection at both boundaries. When they satisfy a phase condition corresponding to equation (2.5), which we derived with the help of

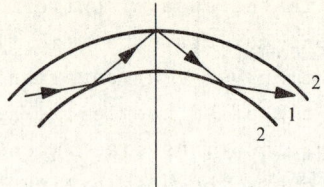

Fig.2.38. Curved slab with zig-zag ray of a guided mode.

Fig.2.2, they repeat in phase after two successive reflections, and superimpose to form the self-consistent field distribution of a guided mode.

The most important consequence of the curvature in the boundary shows up as a change in the magnitude of the reflection coefficient for a plane uniform wave. In the case of total reflection at a plane boundary, this reflection coefficient has unit magnitude, while in the case of concave curvature of the upper boundary of the curved slab this reflection coefficient becomes smaller than unity. The fractional power reflection at the concave boundary signifies radiation of some power into space and as a consequence curvature loss to the guided mode.

The following approximate analysis of wave reflection at a curved dielectric interface will explain the mechanism of radiation and lead to expressions for the radiation field, for the reflection coefficient, and eventually also to formulas for guided-mode attenuation in curved slabs. To this end we consider the cylindrical interface with curvature radius ρ between regions with n_1 and n_2 in Fig.2.39. We let a plane uniform wave be incident from region 1 at an angle θ with respect to the z-direction. The z-axis forms the tangent to the boundary at the point $z = x = 0$, where the ray which represents our plane uniform wave is incident on the boundary. By assuming

$$n_1 k \rho \gg 1, \qquad (2.197)$$

we may take the liberty to consider all fields to be locally of plane-wave character. Not only the incident wave, but

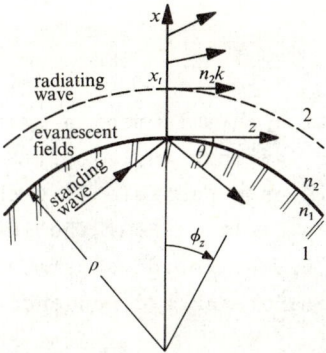

Fig.2.39. Plane-wave reflection at a curved interface.

also its reflection can then be approximated by a plane uniform wave. The significant change due to curvature occurs not in the reflected field distribution but in the fields of medium 2 beyond the boundary.

We begin with an incident plane wave, which corresponds to an H-mode and has its electric field vector parallel to the curved boundary. The field components of its incident and reflected parts superimpose to a distribution similar to equations (2.66) and (2.68), but for our present purposes must be separated into incident and reflected terms

$$E_y = (a \exp(-jk_{1x}x) + b \exp(jk_{1x}x)) \exp(-j\beta z)$$
$$H_z = \frac{k_{1x}}{\omega\mu_0} (a \exp(-jk_{1x}x) - b \exp(jk_{1x}x)) \exp(-j\beta z) \ .$$
(2.198)

In region 2, we expect a significant change of the field distribution and therefore let

$$E_y = E(x) \exp(-j\beta z), \qquad (2.199)$$

where $E(x)$ must follow as a solution of equation (2.59), which has the form

$$\left(\frac{\partial^2}{\partial x^2} + k_{2x}^2 \right) E(x) = 0 \qquad (2.200)$$

with

$$k_{2x}^2 = n_2^2 k^2 - \beta^2. \qquad (2.201)$$

The field distribution in this region must depend as $\exp(-j\, l_z\, \phi_z)$ on the azimuth angle ϕ_z. For equation (2.199), such a dependence is only possible if β changes with x according to $l_z\, \phi_z = \beta(x) z$. Near the interface and the origin of our rectangular coordinates x and z, we have $z \simeq (\rho+x)\phi_z$ and therefore an x-dependence of β according to

$$\beta(x) \simeq l_z/(\rho+x). \qquad (2.202)$$

At the interface, we match $\beta(x)$ to the constant β of equation (2.198), which leads to

$$l_z = \rho\beta.$$

Since these considerations remain restricted to $x \ll \rho$, we have

$$\beta^2(x) = \beta^2 (1-2x/\rho) \qquad (2.203)$$

and

$$k_{2x}^2 = n_2^2 k^2 - \beta^2 (1-2x/\rho). \qquad (2.204)$$

Such an x-dependence of k_{2x}^2 suggests the following change of variables (Snyder and Love 1975) for the differential equation (2.200)

$$\xi = \left(\frac{\rho}{2\beta^2}\right)^{2/3} [\beta^2 - n_2^2 k^2 - 2\beta^2\, x/\rho], \qquad (2.205)$$

which transforms equation (2.200) into Airy's equation

$$\left(\frac{d^2}{d\xi^2} - \xi\right) E(\xi) = 0. \qquad (2.206)$$

It has the two Airy functions $Ai(\xi)$ and $Bi(\xi)$ as linearly independent solutions. But only the combination

$$Ai(\xi) + jBi(\xi) = 2\exp(j\pi/3)\, Ai(\xi\exp(-j2\pi/3)) \qquad (2.207)$$

represents an outgoing wave for $x \to \infty$, which lets $\xi \to -\infty$. With its argument going to this limit, the combination of Airy functions in equation (2.207) has the following asymptotic form:

$$Ai(\xi\exp(-j2\pi/3)) = [2\pi^{1/2}(-\xi)^{1/4}]^{-1}\exp{-j[(2/3)(-\xi)^{3/2} + \pi/12]}. \qquad (2.208)$$

It is therefore the appropriate solution to describe a radiating wave from the curved interface into medium 2.

In the range of total reflection at the plane interface, that is for $\theta < \theta_c$, with $\cos\theta_c = n_2/n_1$, we have

$$\beta^2 - n_2^2 k^2 = n_1^2 k^2 (\cos^2\theta - \cos^2\theta_c) > 0.$$

Hence at the interface for $x = 0$, the Airy function argument is $\xi > 0$. Under these conditions $Ai[\xi\exp(-j2\pi/3)]$ decreases monotonically with increasing distance from the interface. For $\xi \gg 1$, the asymptotic form

$$Ai(\xi\exp(-j2\pi/3)) \simeq [2\pi^{1/2}\xi^{1/4}]^{-1}\exp[(2/3)\xi^{3/2} + j\pi/6] \qquad (2.209)$$

even indicates an exponential decay with decreasing ξ away from the interface. At

$$x = x_t \equiv (\rho/2)(1-\cos^2\theta_c/\cos^2\theta), \qquad (2.210)$$

ξ goes through zero and becomes negative. With ξ turning negative in Airy's differential equation (2.206), its solutions change from a decaying form to an oscillating form. Beyond the turning point x_t we therefore have an outward travelling wave, radiating power away from the interface.

For a physical explanation of this phenomenon, we consider the space periodic excitation with period $\lambda_m = 2\pi/\beta$ which the incident wave from region 1 provides at the interface to region 2. Its period is so short that region 2 is excited in a mode below cut-off so that its fields are evanescent and decay exponentially into region 2. Due to the curvature, however, the excitation widens in period length as we move away from the interface. At $x = x_t$, we reach the cut-off point, where

$$(\rho + x_t)/\rho = \lambda/(n_2 \lambda_m). \tag{2.211}$$

Here the excitation period has widened to the free wavelength λ/n_2 of region 2, and can now begin to excite a travelling wave in the direction $\theta_2 = 0$ parallel to the interface. Moving still further away from the interface, the excitation widens even more and the outgoing waves turn in directions $\theta_2 > 0$ so that always

$$(\rho + x)/\rho = \lambda/(n_2 \lambda_m \cos \theta_2).$$

The energy in region 2, which is associated with the incident wave of a guided mode in region 1, moves slower than the speed of light $c_2 = c/n_2$ in region 2, as long as x remains smaller than x_t from equation (2.210). At $x = x_t$, however, it reaches c_2 and for $x > x_t$ would have to move faster than c_2 in order to remain attached to the mode around the bend. It therefore detaches itself from this mode and radiates out into region 2 with ever-increasing angles θ_2, in order to maintain the speed c_2 of free propagation in this medium.

The situation reminds us of the prism coupler for film guides where the prism of index $n_3 > n_1$ frustrates the total reflection and lets energy tunnel through the evanescent field region of the gap between film and prism. At the curved interface the region between $x = 0$ and $x = x_t$ corresponds to the gap of the prism coupler. Through this evanescent field region some energy tunnels into the region beyond the turning point x_t and from there radiates out into region 2. The widening of space in region 2, which is due to curvature of the interface, therefore frustrates the total reflection and lets energy leak out into a radiation field.

On closer examination we find that the condition (2.211)

DIELECTRIC FILMS 163

leads to the expression

$$x_t = \rho(1 - \cos\theta_c/\cos\theta), \tag{2.212}$$

which differs from equation (2.210) for $\xi = 0$ in Airy's equation. This discrepancy dissolves, however, when we consider the assumptions which led to equation (2.210). For equation (2.203) in particular, we assumed $x \ll \rho$ for all significant distances from the interface and also, of course, for x_t. Requiring $x_t \ll \rho$ in equation (2.210) leads to

$$\cos\theta \simeq \cos\theta_c. \tag{2.213}$$

Airy's equation and its solution therefore hold only if for $\theta < \theta_c$ we have either a sufficiently small limiting angle θ_c, or as long as θ differs not too much from θ_c. Within this approximation the difference between equations (2.210) and (2.212) disappears.

With the meaning of the Airy function solution and its limitations fully understood, we can now write for the field distribution in region 2

$$E_y = c\, Ai(\xi\, \exp(-j2\pi/3))\, \exp(-j\beta z)$$

$$H_z = j\, \frac{c}{\omega\mu_0}\, \frac{\partial}{\partial x}\, Ai(\xi\, \exp(-j2\pi/3))\, \exp(-j\beta z).$$

Matching these field components at the interface to those of region 1 from equation (2.198) leads to

$$\frac{b}{a} = \frac{\psi - 1}{\psi + 1}, \tag{2.214}$$

where

$$\psi = j\ Ai(\xi_0\ \exp(-j2\pi/3))/[\zeta\ Ai'(\xi_0\ \exp(-j2\pi/3))]$$

$$\zeta = (2\beta^2/\rho)^{1/3}/k_{1x} = (2\cos^2\theta/n_1 k\ \rho)^{1/3}/\sin\theta \qquad (2.215)$$

$$\xi_0 = \xi\ (x=0) = \left(\frac{n_1 k \rho}{2\cos^2\theta}\right)^{2/3} (\cos^2\theta - \cos^2\theta_c)$$

and the prime at the Airy function denotes differentiation with respect to ξ.

The power transmitted and radiated into region 2 is the difference between incident and reflected power. We call its ratio to the incident power the power transmission coefficient and calculate it from

$$T = 1 - |b/a|^2 = 4\mathrm{Re}(\psi)/[1 + |\psi|^2 + 2\mathrm{Re}(\psi)]. \qquad (2.216)$$

For this power transmission coefficient, the real part of ψ follows from equation (2.215) as

$$\mathrm{Re}(\psi) = |Ai'(\xi_0\ \exp(-j2\pi/3))|^{-2}/4\pi\zeta. \qquad (2.217)$$

A simpler but still adequate approximation for the power transmission coefficient has the form (Snyder and Love 1975)

$$T = |T_F^H|\ |Ai(\xi_0\ \exp(-j2\pi/3))|^{-2}/[4\pi\ |\xi_0|^{1/2}], \qquad (2.218)$$

in which

$$T_F^H = \frac{4\ [1 - (\sin\theta_c/\sin\theta)^2]^{1/2}}{\{1 + [1 - \sin\theta_c/\sin\theta]^{1/2}\}^2} \qquad (2.219)$$

represents the Fresnel coefficient of power transmission for H-waves through the plane equivalent of the curved dielectric interface in Fig.2.39. For this plane interface we would actually obtain $T_F^H = 0$ in the range $\theta \le \theta_c$ of total reflection. Nevertheless, the finite values which equation (2.219) yields for $\theta < \theta_c$, together with the remaining factors in equation (2.218), account quite accurately for the power transmission through the curved interface. Further-

DIELECTRIC FILMS 165

more equation (2.218) can also be applied to the range $\theta > \theta_c$, where even the plane interface reflects only partially. For $\xi_0 \gg 1$, the approximation (2.218) approaches the same asymptotic expression

$$T = 4 \frac{\sin\theta(\cos^2\theta - \cos^2\theta_c)^{1/2}}{\sin^2\theta_c} \exp(-4\xi_0^{3/2}/3),$$

as does T from the more accurate equation (2.216). Also, for the other limiting case of $\xi_0 = 0$, we obtain from equation (2.218)

$$T = T_0 = 3 \cdot 182 \ (\cos^2\theta_c/n_1 k \ \rho)^{1/3}/\sin\theta_c, \qquad (2.220)$$

while the more accurate expression (2.216) in the case $\xi_0 = 0$ leads to

$$T = T_0/[1 + T_0/2 + [T_0/(2\sqrt{3})]^2]. \qquad (2.221)$$

At this limit, therefore, the transmission coefficient in its simple form gives only an adequate approximation when it is sufficiently small itself, that is, when in equation (2.220) we have

$$n_1 k \ \rho \gg \cot^2\theta_c/\sin\theta_c. \qquad (2.222)$$

Actually the greatest departure of equation (2.218) from the more accurate form (2.216) occurs in this limit of $\xi_0 = 0$. Hence, if only the condition (2.222) is satisfied and thus T_0 small enough, the expression (2.218) gives an adequate approximation for the power transmission coefficient of H-waves over the full range of ξ_0.

For E-waves with their magnetic field parallel to the boundary and normal to the plane of incidence, the same procedure as for H-waves results in a power transmission coefficient, that may likewise be approximated by

$$T = |T_F^E| \ |Ai(\xi_0 \exp(-j2\pi/3))|^{-2}/(4\pi \ |\xi_0|^{1/2}). \qquad (2.223)$$

T_F^E in this expression is

$$T_F^E = \frac{4\,[1 - (\sin\theta_c/\sin\theta)^2]^{1/2}/\cos^2\theta_c}{\{1 + [1 - (\sin\theta_c/\sin\theta)^2]^{1/2}/\cos^2\theta_c\}^2} \qquad (2.224)$$

and represents the Fresnel coefficient of power transmission for E-waves through the plane equivalent of the curved dielectric interface in Fig.2.39. Here again, as for T_F^H, we would actually have $T_F^E = 0$ in the range of total reflection. However, the finite values which equation (2.224) formally yields for $\theta < \theta_c$ render the expression (2.223) a good approximation for the power transmission of E-waves through the curved interface, not only for $\theta \leq \theta_c$ but also in the range $\theta > \theta_c$, where even the plane interface reflects only partially.

For the approximation (2.223) to be adequate at the point of largest departure from the accurate value, that is at $\xi_0 = 0$, we must now require that

$$n_1 k\,\rho \gg \cos^{-4}\theta_c \sin^{-3}\theta_c. \qquad (2.225)$$

This condition, as well as the corresponding condition (2.222) for H-waves, restricts the product $k\rho$ more than the initial assumption $k\rho \gg 1$. However, θ_c usually remains well below 90° and approaches this limit only for extremely large index differences $n_1 - n_2$. Under normal circumstances, therefore the conditions (2.222) and (2.225) are not really much more restrictive than the initial assumption $k\rho \gg 1$.

For small index differences $n_1 - n_2$, as in weakly guiding slabs, the parameter ξ_0 from equation (2.215) is approximately given by

$$\xi_0 = (n_1 k\rho/2)^{2/3}(\theta_c^2 - \theta^2), \qquad (2.226)$$

and we then have $T_F^E \simeq T_F^H$. The product $(\theta_c^2/\theta)T$ with T either from equation (2.216) or equation (2.218) then depends only on $\theta_c^2 - \theta^2$ and on $n_1 k\rho$. Fig.2.40 shows for $n_2 \simeq n_1$ curves of $(\theta_c^2/\theta)T$ from equation (2.218) or equation (2.223) as a function of $\theta_c^2 - \theta^2$ for a number of values of the parameter $n_1 k\rho$. These curves decrease monotonically when the angle of incidence decreases from the limit $\theta = \theta_c$ of total inter-

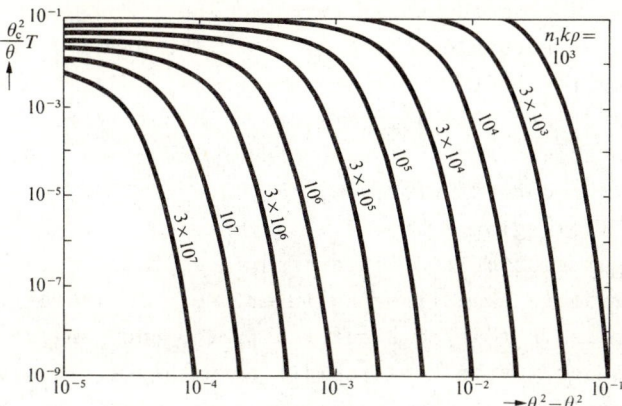

Fig.2.40. Power transmission coefficient T for a uniform plane wave incident on a curved dielectric interface at angle θ. θ_c is the limiting angle of total reflection and ρ the radius of curvature of the interface.

nal reflection to $\theta = 0$. Near $\theta = \theta_c$, the fields in region 2 decay only slowly and extend very far into region 2; the total internal reflection is therefore most effectively frustrated near this limit due to the curvature and widening of space in region 2. With decreasing θ, the fields decay ever more rapidly and less frustration of total internal reflection occurs, so that less power tunnels into region 2. Fig.2.40 also displays monotonically decreasing power transmission for increasing values of the parameter $n_1 k\rho$. Increasing k or ρ effects the same reduction in power transmission. In the limit of $\xi_0 \gg 1$, but for small index differences, the power transmission coefficient for H- and E-waves is given by

$$T \simeq 4(\theta/\theta_c^2)(\theta_c^2-\theta^2)^{1/2} \exp[-\frac{2}{3} n_1 k\rho \,(\theta_c^2-\theta^2)^{3/2}]. \qquad (2.227)$$

The exponential factor in this expression governs the dependence of T on $\theta_c^2 - \theta^2$ and lets T fall nearly exponentially with $(\theta_c^2-\theta^2)^{3/2}$. The power transmission also drops exponentially with $k\rho$. Both factors in the exponent reflect the tunnelling of power through the exponentially decaying evanescent fields of region 2.

To determine the loss which guided modes of a curved slab suffer due to the power transmission through the curved interface, we return now to Fig.2.38. The plane wave, which the zig-zag ray in this figure represents, and which forms a guided mode of the slab, reflects alternately from the outer concave interface and the inner convex interface. In reflecting from the convex interface, the wave excites evanescent fields in the region 2 beyond. In contrast to the region 2 on the other side of the concave surface, this region narrows due to curvature away from the interface. The total internal reflection is in this case not frustrated but, on the contrary, the fields decay even more rapidly than they would upon reflection at the plane equivalent of this interface. No tunnelling of wave energy occurs at such a curvature of the interface, and the reflection is indeed total.

In reflecting from the concave outer surface, however, the wave suffers loss due to power transmission into region 2. A mode which guides the total power P loses the power fraction TP per reflection at the concave interface. It undergoes this reflection and suffers this power loss once in a distance

$$\Delta z = 2d_e \cot \theta \qquad (2.228)$$

along the guide. d_e in equation (2.228) is the effective width and takes account of the Goos-Haenchen shift as in Fig. 2.15. Between two successive reflections at the outer boundary, the ray power of any particular mode travels not only the distances $\Delta z_1 = 2d \cot \theta$ within the slab, but, upon reflection at both boundaries, shifts by an additional distance $\Delta z_2 = 2d_2 \cot \theta$. The situation is somewhat complicated by the fact that d_2 and hence also d_e depend on the angle θ and in the case of E-waves differ somewhat from the relatively simple expressions in equations (2.94) and (2.95), which hold only for H-waves. Note also from Fig.2.38 that due to the curvature, the angle θ at the concave boundary is larger than the corresponding angle θ at the other convex boundary.

To avoid all these complications, we restrict the curvature radius to large enough values, so that

$$\rho \geqslant d/(1-\cos\theta) \qquad (2.229)$$

for the particular mode and its ray angle θ that we wish to consider. Under this condition, both angles θ at the upper and lower boundary are close to each other, and we keep a safe distance from the situation in Fig.2.41 where the ray just grazes the convex boundary, and the guided mode changes from a slab mode to a whispering-gallery mode of the outer concave interface.

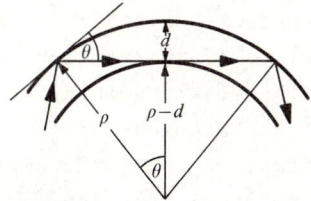

Fig.2.41. Transition from slab mode to whispering-gallery mode of the concave interface at grazing incidence of the ray on the convex interface.

In the general expression for the attenuation constant of a guided mode,

$$\alpha = -(dP/dz)/(2P),$$

we have $-(dP/dz) = TP/\Delta z$ with Δz from equation (2.228), and hence obtain a formula for α in terms of T:

$$\alpha = T/(4d_e \cot\theta). \qquad (2.230)$$

Any of the expressions for T, the more accurate form of equation (2.216), or approximations such as equation (2.218) or equation (2.223), may be substituted in equation (2.230) to obtain more or less accurate expressions for α. In the case of small index differences, the angles θ and θ_c likewise stay small, so that for weakly guiding films

$$\alpha = T\theta/(4d_e). \tag{2.231}$$

Under these conditions, the quantity

$$(\theta_c^2/\theta)T = (\theta_c/\theta)^2 \, 4\alpha d_e = 4 \, \frac{n_1^2 - n_2^2}{n_1^2 - (\beta/k)^2} \, \alpha d_e \tag{2.232}$$

may for specific values of the parameter $n_1 k\rho$ be read directly from Fig.2.40 as a function of

$$\theta_c^2 - \theta^2 = (\beta/(n_1 k))^2 - (n_2/n_1)^2 \, ,$$

and the attenuation constant α extracted from it. β in this case is the phase constant of the particular mode in the plane slab. If the mode in the weakly guiding slab is not too close to cut-off and $\theta_c^2 - \theta^2$ is large enough for ξ_0 according to equation (2.215) to satisfy $\xi_0 \gg 1$, the explicit expression (2.227) approximates the power loss, and α may be determined from

$$\alpha \, d_e = (\theta/\theta_c)^2 \, (\theta_c^2 - \theta^2)^{1/2} \, \exp[-\tfrac{2}{3} n_1 k\rho (\theta_c^2 - \theta^2)^{3/2}]. \tag{2.233}$$

In terms of index and wave numbers as well as β this formula appears as

$$\alpha \, d_e = [1 - (\beta/(n_1 k))^2] \frac{B^{1/2}}{(1 - (n_2/n_1)^2)^{\frac{1}{2}}}$$
$$\cdot \exp\{-\tfrac{2}{3} n_1 k\rho [(\beta/(n_1 k))^2 - (n_2/n_1)^2]^{3/2}\}, \tag{2.234}$$

where B is the phase parameter of the particular mode from equation (2.18). We infer from Fig.2.40 that a particular mode in a slab of specific curvature suffers its highest bending loss near cut-off. As we move away from cut-off, this loss decreases steadily and disappears completely when $\theta \to 0$ or $\beta \to n_1 k$ far above cut-off. In this limit, the evanescent fields in region 2 decay so rapidly that curvature cannot frustrate the total reflection any more and any tunnelling of energy through the evanescent field region ceases.

Care must be exercised, however, that this approximate theory is only applied to slab guide modes that obey all the assumptions which were made in deriving the expressions for curvature loss. From equation (2.213) the angles θ and θ_c should either both be small or not too different from each other. The slab waveguide should therefore have only small index differences, or else the approximations hold only for slab guide modes not too high above their respective cut-off.

Furthermore, the condition (2.229) holds in a bent slab only for gentle enough curvature. From Fig.2.41 we see that it breaks down completely when θ falls below the value that follows from

$$\rho \cos \theta = \rho - d.$$

Despite all these restrictions, however, most cases of practical interest are actually covered by this approximate theory. In the first place, we are only interested in curvature loss where it really can be excessive, and that is near or not too high above cut-off for any particular mode. For similar reasons there is not much practical interest in such small θ-values or large curvature as would be excluded by the condition (2.229).

2.9. SLABS AND FILMS WITH GRADED INDEX

The film and slab guides, which we considered in previous sections, were all assumed to have well-defined surfaces or interfaces with abrupt transitions in refractive index from one medium to the other and constant values for the refractive index in each particular region. Even in cases where we consider the volume scattering due to small index variaitons, we assumed these variations to extend only over distances which were small compared to the wavelength of light. The average refractive index was also taken as constant for these cases.

By the same token then, our previous analysis applies not only to films and slab guides which have abrupt transitions of refractive index at their boundaries, but also when these transitions are gradual. The total change of index from one homogeneous region to the other must only occur in a distance which is small compared to the wavelength. This condition prevails at film surfaces with free space and, depending on the materials and their formation into films, quite often also at interfaces between different film and substrate or covering materials.

However, some fabrication methods, especially those which involve diffusion processes, lead to smoothly graded transitions in index, which extend over a substantial fraction of the light wavelength or even over several wavelengths. The index may even be graded across the full width of the guide. Such grading of index transitions and index profiles changes the cut-off and dispersion characteristics of the guides and may also influence the scattering loss. Whether these changes are objectionable or not depends on the particular application. We even foresee the possibility that suitably graded transitions may yield desirable characteristics for guided modes.

For the analysis of such films and slabs with graded index profiles, we will find it useful to distinguish between symmetric profiles, such as in Fig.2.42(a), and strongly asymmetric index distributions with one abrupt transition such as Fig.2.42(b). The first case occurs intentionally, but also unintentionally, when a buried layer is produced by some diffusion process. The second case occurs when a

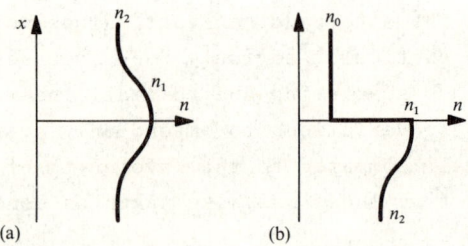

Fig.2.42. Refractive index profiles of graded index films: (a) symmetric profile of buried layer; (b) asymmetric profile of film on substrate.

guiding film is formed near a substrate surface by in- or out-diffusion. The very surface of the substrate or film has the abrupt index transition. To a certain approximation the set of modes of the symmetric profile contains the modes which are guided by one half of the corresponding profile in the strongly asymmetric case.

We consider first the symmetric profile of Fig.2.42(a) and try the ray optics approach. The index profile is a special case of an inhomogeneous medium. We assume the refractive index to change only gradually over the distance of a wavelength. A uniform plane wave, for example, will then only be slightly modified, so that we may write

$$\vec{E} = \vec{e}(\vec{r}) \exp(-jnkS(\vec{r})) \qquad \vec{H} = \vec{h}(\vec{r}) \exp(-jnkS(\vec{r})) \qquad (2.235)$$

for its electric and magnetic field vectors, with slowly varying functions $\vec{e}(\vec{r})$ and $\vec{h}(\vec{r})$ of the vector $\vec{r}$ to a point of observation in space. In the case of wave propagation in the z-direction we have $S(\vec{r}) = z$; but in general $S(\vec{r}) = $ constant gives the local phase front and ∇S its direction of propagation, while $\vec{e}(\vec{r})$ and $\vec{h}(\vec{r})$ are essentially field distributions transverse to this direction.

If we substitute equation (2.235) into Maxwell's equations (1.3), we obtain

$$(\nabla S) \times \vec{h} + n^2 \left(\frac{\varepsilon_0}{\mu_0}\right)^{\frac{1}{2}} \vec{e} = \frac{1}{jk} \nabla \times \vec{h}$$

$$(\nabla S) \times \vec{e} - \left(\frac{\mu_0}{\varepsilon_0}\right)^{\frac{1}{2}} \vec{h} = \frac{1}{jk} \nabla \times \vec{e} \qquad (2.236)$$

while $\nabla(n^2 \vec{E}) = 0$ leads to

$$\vec{e} \cdot \nabla S = \frac{1}{jnk} [\nabla \vec{e} + \vec{e} \cdot \nabla(\log_e n)] . \qquad (2.237)$$

Here we go to the limit $k \to \infty$, which represents the zero-order optics or geometric optics approximation, and lets the right-hand sides of all three equations vanish. Equation (2.237) then reduces to

$$\vec{e} \cdot \nabla S = 0 \qquad (2.238)$$

and equations (2.236) to

$$(\nabla S) \times \vec{h} + n^2 \left(\frac{\varepsilon_0}{\mu_0}\right)^{\frac{1}{2}} \vec{e} = 0$$

$$(\nabla S) \times \vec{e} - \left(\frac{\mu_0}{\varepsilon_0}\right)^{\frac{1}{2}} \vec{h} = 0.$$

(2.239)

Non-trivial solutions of this homogeneous system of linear equations for the six components of $\vec{e}$ and $\vec{h}$ require its coefficient determinant to be zero, which, in view of equation (2.238), leads to

$$(\nabla S)^2 = n^2.$$

(2.240)

This is the so-called *eikonal equation*. It represents the basic equation of geometric optics and relates wavefronts $S(\vec{r})$ = constant to the index distribution, thus describing the motion of waves.

Better suited for our purposes is another form of equation (2.240), which describes light rays more directly. These light rays form orthogonal trajectories to the wave fronts; they therefore have the direction of ∇S. Fig.2.43 shows one such light ray and two phase fronts. If we let $\vec{r}$ be a position vector along a particular ray and s the length of this ray from a fixed reference point, then

$$\vec{u} = \frac{d\vec{r}}{ds}$$

(2.241)

is the unit vector tangential to this ray. Since $\vec{u}$ is parallel to ∇S and $|\nabla S| = n$, from the eikonal equation (2.240), we arrive at

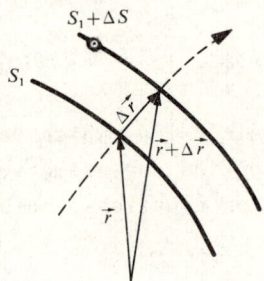

Fig.2.43. Phase fronts and light ray in an inhomogeneous medium.

DIELECTRIC FILMS

$$n \frac{d\vec{r}}{ds} = \nabla S , \qquad (2.242)$$

which specifies rays by means of the wave-front function S.

We would like, however, to specify rays directly in terms of $n(\vec{r})$. To this end we differentiate equation (2.242) with respect to s and, on its right-hand side, express this operation as

$$\frac{d}{ds} = \sum_i \frac{dx_i}{ds} \cdot \frac{\partial}{\partial x_i} = \frac{d\vec{r}}{ds} \nabla .$$

We then substitute for $d\vec{r}/ds$ from equation (2.242). The result

$$\frac{d}{ds}(n \frac{d\vec{r}}{ds}) = \frac{1}{n}(\nabla S) \nabla(\nabla S) = \frac{1}{2n} \nabla[(\nabla S)^2]$$

allows us to substitute for $(\nabla S)^2$ from the eikonal equation and obtain the differential equation for light rays in its general form:

$$\frac{d}{ds}(n \frac{d\vec{r}}{ds}) = \nabla n. \qquad (2.243)$$

According to this equation light rays essentially curve in the direction of ∇n.

For a homogeneous medium with $\nabla n = 0$, equation (2.243) integrates to $\vec{r} = s\vec{a}+\vec{b}$, with constant vectors $\vec{a}$ and $\vec{b}$ representing straight rays.

Light rays which are guided by graded index films run in planes y = constant of the coordinate system in Fig.2.44.

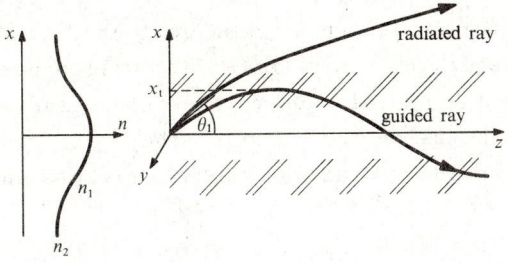

Fig.2.44. Symmetric index profile and rays of guided and radiation modes.

They are also nearly parallel to the z-axis with the transverse component dx/ds of their unit direction vector $d\vec{r}/ds$ always small compared to the longitudinal component dz/ds. Under these conditions we have $ds \simeq dz$. Also for such films, n does not change in the z-direction. The general ray equation (2.243) then reduces to the paraxial approximation

$$\frac{d^2 x}{dz^2} = \frac{d}{dx} \log_e n . \qquad (2.244)$$

This equation allows the deviation of a ray from the centre plane along the film to be traced.

To identify different modes of propagation in the layer with symmetric index distribution, we look at rays that radiate from a line source in the z-axis of Fig.2.44 at an angle θ_1 with respect to the centre plane. In the ray equation (2.243) the z-component of ∇n vanishes, because n remains constant in this direction. We therefore have for the ray in Fig.2.44 a z-component of equation (2.243) which integrates along s to $n(d\vec{r}/ds) = n \cos \theta = $ const. Snell's law (equation (1.19)) generalizes for this ray to

$$n \cos \theta = n_1 \cos \theta_1 . \qquad (2.245)$$

Any ray, starting with θ_1 at the centre plane, curves towards this plane so that the longitudinal component $nk \cos \theta$ of its wavevector $n\vec{k}$ remains constant. If only $\theta_1 < \arccos (n_2/n_1)$, with n_2 as the lower limit of the refractive index beyond the tail of the profile, this ray will bend back towards the centre plane, and the profile will guide it. Its turning point x_t occurs within the profile. Such rays can form guided modes of the profile, which we also call profile modes. Rays with initial angles $\theta_1 > \arccos(n_2/n_1)$ will experience some bending within the profile due to the index gradient, but not enough to remain inside the profile; they will radiate out into the surrounding medium and therefore represent radiation modes.

Not all rays with angles $\theta_1 < \arccos (n_2/n_1)$ actually form guided modes of propagation; to do this, their locally plane waves need to satisfy a phase condition for self-consistent fields within the profile. This phase condition corresponds to equation (2.5) for the film with abrupt boundaries, which postulates that after two successive reflections at the upper and lower boundary the wave repeats in phase, in order to fit into the film. We formulate this phase condition for the profile modes by utilizing the transverse component k_x of the wave vector $n\vec{k}$. The phase integral between a lower and an upper turning point must be an integer multiple $m\pi$ of π:

$$\int_{-x_t}^{x_t} k_x \, dx = (m + \tfrac{1}{2})\pi \ . \tag{2.246}$$

Only under this condition will we obtain the required standing wave pattern in the transverse field distribution with m modes. The phase shift $\pi/2$ on the right-hand side takes account of the peculiar situation at the turning point. Looking in the x-direction transverse to the direction of propagation, the wave impedances are $Z = \omega\mu_0/k_x$ for H-waves and $Z = k_x/(\omega\varepsilon_0 n^2)$ for E-waves. At the turning point we have $k_x = 0$, and the wave impedances change from real values for $|x| < |x_t|$ to purely imaginary values for $|x| > |x_t|$. They have equal magnitude at small but equal distances to both sides from the turning point. With $Z^{(-)} = j\, Z^{(+)}$, therefore, the waves experience a reflection at the turning point with a reflection coefficient $r = (j-1)/(j+1)$ of unit magnitude but $\pi/2$ phase shift.

To further evaluate equation (2.246), we express k_x according to the local separation condition by the local wave number $n(x)k$ and the axial phase constant $\beta = n_1 k \cos\theta_1$. This leads to the universal form

$$\int_{-x_t}^{x_t} (n^2(x)\, k^2 - \beta^2)^{\tfrac{1}{2}}\, dx = (m + \tfrac{1}{2})\pi \tag{2.247}$$

of the characteristic equation for any symmetric index profile with sufficiently small index gradient. It holds for any ray that has turning points $\pm x_t$, where the integrand of equation (2.247) vanishes. Any β-value in the range $n_2 < \beta/k < n_1$ will yield such turning points. If, in addition, this β-value satisfies equation (2.247) for an integer m, it represents the phase constant β_m of the respective H_m- and E_m-modes. Equation (2.246) therefore provides us with a general method of determining the phase constant of guided modes in symmetric index profiles.

For an evaluation of the phase integral in equation (2.247), we need to specify a particular index profile $n(x)$. As a first example, we take the truncated parabolic profile

$$n = n_1 [1 - \Delta(x/d)^2] \qquad \text{for } |x| < d$$
$$n = n_2 \qquad \text{for } |x| > d \qquad (2.248)$$

of Fig.2.45.

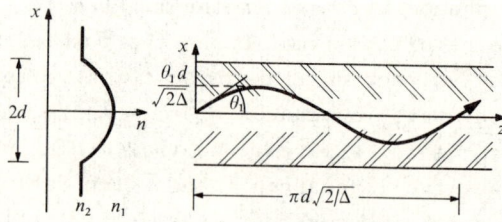

Fig.2.45. Truncated parabolic index profile and ray path of a guided mode.

$$\Delta = \frac{n_1 - n_2}{n_1} \qquad (2.249)$$

represents the maximum relative index difference of this profile. The paraxial ray equation (2.244) for this profile,

$$\frac{d^2 x}{dz^2} + 2(\Delta/d^2)x = 0, \qquad (2.250)$$

has the particular solution

$$\frac{x}{d} = \{\theta_1/(2\Delta)^{\frac{1}{2}}\}\sin\{(2\Delta)^{\frac{1}{2}}z/d\}. \qquad (2.251)$$

Rays starting at $z = 0$ with $x = 0$ and θ_1 undulate as sine waves with period length $\pi d(2/\Delta)^{\frac{1}{2}}$ and turning points $x_t = \pm d\theta_1/(2\Delta)^{\frac{1}{2}}$. The phase integral (2.247) leads to a finite number of modes with phase constants

$$\beta = \left[n_1^2 k^2 - (2m+1) \, n_1 k (2\Delta)^{\frac{1}{2}}/d\right]^{\frac{1}{2}}. \qquad (2.252)$$

The highest-order mode has $|x_t| = d$ and just grazes the edge of the profile. Its phase constant is $\beta = n_1 k \cos(2\Delta)^{\frac{1}{2}}$, which when substituted into equation (2.252) yields the highest order m of a guided mode as the largest integer smaller than

$$n_1 k \, d(\Delta/2)^{\frac{1}{2}} - 1/2.$$

This quantity, as well as equation (2.252), suggests a finite cut-off even for the lowest order $m = 0$.

A more accurate field analysis shows, however, that the H_0- and E_0-modes of the truncated parabolic profile in Fig.2.45 have zero cut-off. Such a field analysis also reveals the transverse field distribution of modes in layers with graded index profiles. The field analysis of H-waves with $E_x = E_z = H_y = 0$ starts from the Cartesian components (2.53), (2.55), and (2.57) of Maxwell's equations and, by eliminating H_x and H_z, arrives at

$$\frac{\partial E_y}{\partial x^2} + (n^2(x)k^2 - \beta^2)E_y = 0. \qquad (2.253)$$

This wave equation differs from equation (2.59) only in that n now depends on the transverse coordinate x. For E-waves with $H_x = H_z = E_y = 0$, the wave equation which follows from (2.54), (2.56), and (2.58) has the form

$$n^2 \frac{\partial}{\partial x} \left[\frac{1}{n^2} \frac{\partial H_y}{\partial x}\right] + (n^2 k^2 - \beta^2) H_y = 0. \qquad (2.254)$$

It reduces to

$$\frac{\partial^2 H_y}{\partial x^2} + (n^2 k^2 - \beta^2) H_y = 0 \qquad (2.255)$$

only when the index is graded smoothly enough for

$$\left|\frac{\partial n}{\partial x}\right| \ll \left|(n/2)(n^2k^2-\beta^2)H_y/(\partial H_y/\partial x)\right| \qquad (2.256)$$

to be satisfied. Under these conditions E- and H-waves obey the same transverse wave equation for the y-components of their fields and therefore have corresponding propagation characteristics and field distributions.

As a first example (Kirchhoff 1972), we consider again the truncated parabolic profile of equation (2.248). Changing to the new variable

$$\xi = \{2n_1 k(2\Delta)^{\frac{1}{2}}/d\}^{\frac{1}{2}} x \qquad (2.257)$$

and substituting

$$\alpha = d(\beta^2 - n_1^2 \cdot k^2)/(2n_1 k(2\Delta)^{\frac{1}{2}}) \qquad (2.258)$$

transform equation (2.253) into

$$\frac{d^2 E_y}{d\xi^2} - (\xi^2/4 + \alpha)E_y = 0, \qquad (2.259)$$

which is one of the standard forms of differential equations of the parabolic cylinder (Abramowitz and Stegun 1964). Because of the symmetry of the structure, the solutions split into even and odd modes corresponding in the ray picture to even and odd integers m, respectively.

The even mode solutions of equation (2.259) are

$$E_y^{(e)} = E_0 \exp(-\xi^2/4) \, M(\frac{\alpha}{2} + \frac{1}{4}, \frac{1}{2}, \frac{\xi^2}{2}) , \qquad (2.260)$$

where $M(a,b,\xi^2/2)$ represents the confluent hypergeometric function

$$M(a,b,\frac{\xi^2}{2}) = \sum_{m=0}^{\infty} \frac{(a)_m (\xi^2/2)^m}{(b)_m m!} \qquad (2.261)$$

with $(a)_m$ and $(b)_m$, respectively, defined by

$$(a)_m = a(a+1)(a+2)\ldots(a+m-1)$$

$$(a)_0 = 1. \tag{2.262}$$

Equation (2.260) gives the transverse field distribution within the profile. Outside it, for $|x| > d$, the evanescent field solution such as in equation (2.80) must be taken and the tangential fields matched at the boundaries $x = \pm d$. This leads to the characteristic equation for even modes, which determines their phase constant β.

Any such phase constant from this procedure, which satisfies the turning point condition

$$n(x_t)k = \beta \tag{2.263}$$

well within the profile, has evanescent fields beyond this point. They decay to negligible values at the edge of the profile. To ensure that equation (2.260) shows this evanescent decay beyond x_t, the series (2.261) must be a finite polynomial, which in turn requires the parameter a to be a negative integer $-l$:

$$a = \frac{\alpha}{2} + \frac{1}{4} = -l. \tag{2.264}$$

This condition represents the characteristic equation for even H-modes. Their phase constants follow from it according to

$$\beta^2 = n_1^2 k^2 \left[1 - (2\Delta)^{\frac{1}{2}}(4l+1)/(n_1 kd)\right]. \tag{2.265}$$

Except for the condition that x_t from equation (2.263) be not too close to d, the characteristic equation (2.264) is independent of the profile size. It only holds for modes which are well enough guided within the profile. The finite polynomial to which the hypergeometric function reduces in this case is, by

$$M(-l, \tfrac{1}{2}, \tfrac{\xi^2}{2}) = (-1)^l \frac{l!}{(2l)!} H_{2l}(\tfrac{\xi}{\sqrt{2}}), \tag{2.266}$$

related to the Hermite polynomial

$$H_m(x) = (-1)^m \exp(x^2) \frac{d^m}{dx^m} \exp(-x^2). \tag{2.267}$$

Its lowest orders are

$$H_0(x) = 1$$
$$H_1(x) = 2x \tag{2.268}$$
$$H_2(x) = 4x^2 - 2.$$

We turn back now to the differential equation (2.259) of the parabolic cylinder and examine its odd mode solutions. They are given by

$$E_y^{(0)} = E_0 \, \xi \, \exp(-\xi^2/4) \, M(\frac{\alpha}{2} + \frac{3}{4}, \frac{3}{2}, \frac{\xi^2}{2}). \tag{2.269}$$

Odd modes with only small evanescent fields at the edge of the profile ($x_t < d$) have, by similar considerations to the above, a phase constant given by

$$\beta^2 = n_1^2 k^2 \, [1 - (2\Delta)^{\frac{1}{2}}(4l+3)/(n_1 kd)], \tag{2.270}$$

while their hypergeometric function in equation (2.269) reduces under these conditions to an odd-order Hermite polynomial

$$M(-l, \frac{3}{2}, \frac{\xi^2}{2}) = (-1)^l \, \frac{l!}{(2l+1)!} \, \frac{\sqrt{2}}{2\xi} \, H_{2l+1}(\frac{\xi}{\sqrt{2}}). \tag{2.271}$$

In summary then, we find even and odd order H_m-waves which have the same phase constants (2.265) as follow from the ray optics approach and field distributions

$$E_y = E_0 \, H_m(\sqrt{2} \, x/w) \, \exp{-(x/w)^2} \tag{2.272}$$

where

$$w = \left[\frac{d}{n_1 k}\left(\frac{2}{\Delta}\right)^{\frac{1}{2}}\right]^{\frac{1}{2}}. \tag{2.273}$$

These relations are valid as long as the modal fields are well confined to the profile. They also hold for E_m-waves as long as equation (2.256) is satisfied, such as in smoothly graded profiles. In the case of E-waves, the expressions (2.260), (2.269), and (2.272) describe the transverse distribution of H_y.

As a second example, we list results for the squared hyperbolic-secant profile (Kogelnik 1975)

$$n(x) = n_2(1 + \Delta \cosh^{-2}(x/d)) \tag{2.274}$$

with

$$\Delta = (n_1 - n_2)/n_2. \tag{2.275}$$

The index gradient of this profile changes only gradually, so that it approximates what we obtain as profiles in buried layers by diffusion processes.

The guided modes of this profile have phase constants according to

$$\beta^2 = n_2^2 k^2 + (M-m)^2/d^2 \tag{2.276}$$

and field distributions

$$E_y = u_m(x/d) \cosh^2(x/d). \tag{2.277}$$

The parameter

$$M = \frac{1}{2}\{(1 + 4k^2 d^2 (n_1^2 - n_2^2))^{\frac{1}{2}} - 1\} \tag{2.278}$$

in equation (2.276) gives the number of modes which the profile guides at $\lambda = 2\pi/k$. The function u_m follows again from hypergeometric functions with

$$u_m = 1 - \frac{m}{2}(2M-m)\sinh^2(x/d)/(1\times1!) + \frac{1}{4}m(m-2)(2M-m)(2M-m-2)$$
$$\sinh^4(x/d)/(1\times3\times2!) + \ldots \quad (2.279)$$

for even mode numbers m, and

$$u_m = \sinh(x/d)\,[1 - \frac{1}{2}(m-1)(2M-m-1)\sinh^2(x/d)/(3\times1!) +$$
$$+ \frac{1}{4}(m-1)(m-3)(2M-m-1)(2M-m-3)\sinh^4(x/d)/(3\times5\times2!)\ldots] \quad (2.280)$$

for odd mode numbers.

For some of the lower-order modes these functions are

$$u_0 = 1$$
$$u_1 = \sinh(x/d) \quad (2.281)$$
$$u_2 = 1 - 2(M-1)\sinh^2(x/d).$$

A third example of some practical significance is the exponential profile (Conwell 1973)

$$n(x) = n_2(1+\Delta\exp(-|x|/d)). \quad (2.282)$$

One half of this profile, such as in Fig.2.46, represents an index distribution which, due to a diffusion process, decays exponentially from a surface into the substrate.

The solution of equation (2.253) for the exponential

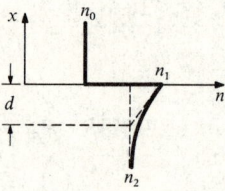

Fig.2.46. Asymmetric film with exponential index profile.

distribution can conveniently be written in terms of the parameter

$$V = kd\,(n_1^2 - n_2^2)^{\frac{1}{2}}, \qquad (2.283)$$

which, in correspondence with equation (2.17), accounts for the width and maximum index difference of the layer as well as the wavelength. The modes have transverse field distributions according to

$$E_y = J_p(2V\exp[-|x|/(2d)]) \qquad (2.284)$$

with J_p as the Bessel function of the non-integral order p. This order p assumes discrete values p_m for each of the modes which the exponential profile guides, and determines the phase constant of the respective mode by

$$\beta^2 = n_2^2 k^2 + p_m^2/(2d)^2. \qquad (2.285)$$

The ratio

$$\frac{p_m^2}{V^2} = \frac{\beta^2/k^2 - n_2^2}{n_1^2 - n_2^2} \qquad (2.286)$$

corresponds to the phase parameter B of equation (2.18) and is plotted in Fig.2.47 as a function of V for a number of odd-order modes. V corresponds similarly to the previously defined film parameter of equation (2.17).

If the index profile has the strong asymmetry which Fig.2.42(b) and Fig.2.46 indicate, the large index step from n_1 down to n_0 will have a relatively large limiting angle $\theta_{0c} = \arccos(n_0/n_1)$ for total reflection. Any modes which are still guided by the comparatively weak index change in the profile will be associated with rays that have angles $|\theta_1| < \arccos(n_2/n_1)$, that are quite small compared to θ_{0c}. Their total internal reflection leads to a very low field at the boundary between n_1 and n_0. In the limit of infinite asymmetry, there is even a field node at the boundary. In the corresponding symmetric profile all odd-order H- and E-modes have this field node in the centre plane. One

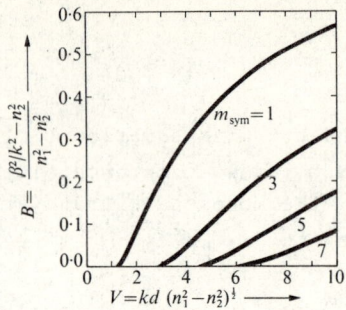

Fig.2.47. Phase parameter of odd-order modes in the symmetric profile with exponential index distribution and of modes of order $m_{asym} = (m_{sym} -1)/2$ in the corresponding asymmetric profile (Carruthers et al. 1974).

half of their field distributions therefore also fits into the strongly asymmetric guide and propagates with the same phase constant as in the symmetric profile. All odd-order modes of any symmetric profile are therefore also modes of propagation in the corresponding asymmetric structure.

As a particular example of practical significance, Fig. 2.47 shows the phase parameter $B = p_m^2/V^2$ of the odd-order modes in the symmetric profile of exponential index distribution. These are also the modes of the strongly asymmetric case with an exponential profile. Since they are the only modes in the asymmetric case, their mode order should be renumbered. Any odd mode number m_{sym} of the symmetric profile gives the mode of number

$$m_{asym} = (m_{sym}-1)/2 \qquad (2.287)$$

in the asymmetric guide. The lowest-order mode in the symmetric profile has zero cut-off, but it is of even order with $m_{sym} = 0$. The lowest-order mode of the asymmetric profile has the order $m_{asym} = 0$, but it corresponds to $m_{sym} = 1$ and obeys the finite cut-off of this mode.

REFERENCES

Abramowitz, M. and Stegun, I.A. (1964). *Handbook of mathematical functions*. Dover Publications, Inc., New York.

Carruthers, J.R., Kaminow, I.P. and Stulz, L.W. (1974). Diffusion kinetics and optical waveguiding properties of outdiffused layers in lithium niobate and lithium tantalate. *Appl. Opt.* 13, 2333-42.

Conwell, E.M. (1973). Modes in optical waveguides formed by diffusion. *Appl. Phys. Lett.* 23, 328-9.

Hinken, J.H. (1976). *Film mode attenuation due to random surface irregularities*. Conference Proceedings of the 6th European Microwave Conference. Microwave Exhibition and Publishers Ltd., Sevenoaks (England), 513-7.

——— (1977). Dielectric film mode excitation by Hertzian dipoles. *Arch. Elektron. & Übertragungstech.* 31 (to be published).

Kirchhoff, H. (1972). The solution of Maxwell's equations for inhomogeneous dielectric slabs. *Arch. Elektron. & Übertragungstech.* 26, 537-41.

Kogelnik, H. (1975). *Theory of dielectric waveguides in 'integrated optics'* (ed. T. Tamir), p.55. Springer Verlag, Berlin.

——— and Ramaswamy, V. (1974). Scaling rules for thin-film optical waveguides. *Appl. Opt.* 13, 1857-62.

——— and Sosnowski, H.P. (1970). Holographic thin film couplers. *Bell Syst. tech. J.* 49, 1602-08.

Marcuse, D. (1969). Mode conversion caused by surface imperfections of a dielectric slab waveguide. *Bell Syst. Tech. J.* 48, 3187-216.

——— (1974). *Theory of dielectric optical waveguides*. Academic Press, New York.

Snyder, A.W. and Love, J.D. (1975). Reflection at a curved dielectric interface-electromagnetic tunneling. *IEEE Trans. MTT* 23, 134-41.

Sommerfeld, A. (1926). Über die Ausbreitung der Wellen in der drahtlosen Telegraphie. *Ann. Phys.* 81, 1135-53.

Tien, P.K. (1971). Light waves in thin films and integrated optics. *Appl. Opt.* 10, 2395-413.

Ulrich, R. (1974). Thin dielectric films for integrated optics. *J. Vac. Sci. & Technol.* 11, 156-62.

Unger, H.-G. (1976). *Optische Nachrichtentechnik*. Elitera Verlag, Berlin.

3
PLANAR GUIDES WITH TRANSVERSE CONFINEMENT

The film and slab guides of the previous sections were assumed to extend without limitation in the direction transverse to wave propagation. We also assumed the fields of all modes of propagation to be uniform in this transverse direction. As the only exception, we treated scattering from surface imperfections which change in the longitudinal as well as in the transverse direction.

Practical arrangements of films or slabs in planar optics always involve some limitation or change in transverse direction parallel to the film. Also, the beams which excite guided modes in films or slabs through prism or grating couplers have only a finite width. The films and slabs, as such, provide no transverse confinement for this excitation of finite width. A number of planar devices, such as the thin-film light deflector for optical data processing, need no such confinement. But even in these applications, the film guides are of finite width. They need to be excited by light beams which extend even less in the transverse direction. The transverse confinement which this excitation provides, may be maintained in an iterative fashion by periodic focusing. It is also possible to confine light beams by continuous focusing. A still more effective method of confining the light transversely in a continuous fashion is by limiting the film or slab sideways to a strip or at least reducing its height accordingly.

3.1. FILM LENSES AND LENS GUIDES

We discuss the different arrangements for confining film waves sideways by starting with the method of periodic focusing, which changes the character of the film waves only slightly. By exciting the film wave through a prism or grating coupler with the Gaussian intensity distribution of a laser beam, the fields will no longer be constant in the y-direction transverse to their propagation direction. Rather, they will have the y-dependence of the Gaussian excitation. This y-dependence is, however, quite weak compared to the rapid change of fields in the x- and z-directions. If, for this situation, the separation condition (1.9) is written with Cartesian components of the wave vector,

$$k_x^2 + k_y^2 + k_z^2 = n^2 k^2,$$

we have $|k_y| \ll |k_x|$, $|k_z|$ in any of the film guide regions. The field distribution in the x-direction as well as the laws for propagation in the z-direction will remain unchanged for this case of weak y-dependence. Any film waves, which are thus confined in the y-direction by excitation, will travel for some distance along the film without much change in their original y-distribution. Eventually, however, the dispersing effect of diffraction will widen the y-distribution, in much the same way as a beam widens in free space due to diffraction.

To limit such spreading in the y-direction, we can employ the same method of refocusing as we used for wave beams in free space. Refocusing of wave beams is conveniently accomplished with converging lenses. We therefore need the equivalent of a converging lens in the film guide.

The dispersion diagrams of film modes show that the phase constant of any particular mode increases with film thickness. Considering

$$N = \beta/k , \qquad (3.1)$$

the effective index of refraction of a film mode, this increase with film thickness also applies to N. The lens-like region of Fig.3.1 with its raised film thickness will there-

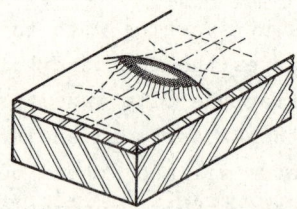

Fig.3.1. Converging film lens formed by a lens-like region of increased film thickness and effective refractive index.

for provide the same refocusing of film-wave beams as a converging lens provides for a spreading wave beam in free space. To be more specific, the cylindrically curved phase front of a spreading beam of a film wave will be transformed by a suitable film lens into phase fronts of opposite curvature, and thus the diverging beam of the film wave will be changed into a converging beam. For such a film lens to perform properly and not cause too much scattering at its edges, the transition in thickness between film and lens regions must be gradual. If the film wave beam is to be transmitted through a long section of film guide, refocusing can be repeated in a periodic fashion giving rise to a lens guide for film-wave beams.

The modes of propagation in such a lens guide can be analysed by extending the analogy between regular lenses in free space and film lenses to lens guides in free space (Kogelnik and Li 1966) and film lens guides as well. With this analogy we apply Huygens principle to the section of a film-lens guide in Fig.3.2. We let this section begin in the

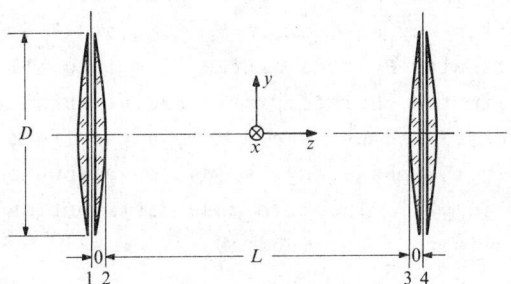

Fig.3.2. Section of periodic sequence of film lenses between two centre planes 1 and 4 of neighbouring lenses.

centre plane 1 of one lens and end in the centre plane 4 of the next lens. The lens-guide section therefore contains one half of each neighbouring lens.

A mode of propagation in such a beam waveguide repeats its y-distribution of field from centre plane to centre plane and its phase fronts coincide with these centre planes. Half the film lens has twice the focal length f of the complete film lens. The plane phase front of a beam mode in the centre plane 1 is therefore transformed by half the film lens to a cylindrically curved phase front with curvature radius $R = 2f$. The distribution $E_y = E_1(y)$ of an H_m-beam mode with a plane phase front changes to

$$E_2 = E_1 \exp\{j(\beta/(4f))y^2\} \qquad (3.2)$$

in the plane 2 next to the first lens. β is the phase constant of the particular H_m-film mode which forms the beam. The film lens is here assumed to be an ideal phase-front transformer, and its extent in the z-direction is neglected. It only transforms the phase fronts; any additional phase shift is irrelevant to our present considerations and therefore omitted. The phase angle in equation (3.2) increases with the square of distance y from the beam-guide axis. Hence, to a first approximation, it forms a cylindrical phase front, which converges in the direction of propagation. To find the film mode distribution to which E_2 transforms in front of the next lens, we employ Huygens principle. Each element of plane 2 with its Huygens source $E_y = E_2$ and $H_x \simeq -E_2/\eta$ radiates the particular H_m-mode of the film into all forward directions with the characteristic radiation pattern of such Huygens sources. Because $L \gg D \gg \lambda$, a far-field evaluation suffices; only the phase angles must be accounted for to second order in y/L. The film mode distribution in front of the next lens is thus obtained as

$$E_3 = \left(\frac{j\beta}{2\pi L}\right)^{\frac{1}{2}} \exp(-j\beta L) \int_{-D/2}^{D/2} E_2 \exp\{-j\frac{\beta}{2L}[y_2^2+y_3^2-2y_2y_3]\}dy_2. \quad (3.3)$$

The first half of the next lens transforms the phase of this distribution according to

$$E_4 = E_3 \exp j(\beta/4f)(y')^2. \qquad (3.4)$$

In order to form a beam mode of this lens guide, the field $E_4(y')$ must have the same y'-distribution as the field $E_1(y)$ has as a y-distribution in the first centre plane. It should differ from E_1 only by a complex factor $g = \exp{-(a+jb)}$ to account for an overall phase shift b and for the loss a. By letting $E_4 = gE_1$, this factor g is the transmission factor of the lens guide section, with loss a and phase shift b. Lumping equations (3.2), (3.3), and (3.4) into one equation leads to the following condition for film-beam modes:

$$gE(y) = \left(\frac{j\beta}{2\pi L}\right)^{\frac{1}{2}} \int_{-D/2}^{D/2} E(y) \exp\{\frac{j\beta}{2L}[(y^2+y'^2)(\frac{L}{2f} - 1)+2yy']\}dy.$$
(3.5)

Any field distributions which solve this integral equation form beam modes of the film lens guide in the particular H_m- or E_m-mode of the film.

If the film lenses focus these beam modes sufficiently, their fields decay to very small values at the distance $y = \pm D/2$ from the centre, which corresponds to the lens width. Under these conditions, we may extend the integration in equation (3.5) to $|y| \to \infty$. For any beam mode which concentrates its field near enough to the lens guide axis, such an extension of the limits of integration will not change the right-hand side of equation (3.5). With these new limits of integration, we now consider equation (3.5) as the operator equation

$$\overline{G} E(y) = g E(y) \qquad (3.6)$$

with the operator

$$\overline{G} = \left(\frac{j\beta}{2\pi L}\right)^{\frac{1}{2}} \int_{-\infty}^{\infty} \exp\{\frac{j\beta}{2L} (y^2+y'^2)(\frac{L}{2f} - 1) + 2yy'\} dy$$

and eigenvalues g yet to be determined as solutions of equation (3.6) for particular beam modes. The operator $\overline{G}$ may be shown to commute with the Hamiltonian $\overline{H} = (d/dy^2) - \alpha^4 y^2$ of the linear harmonic oscillator, which, in a normalized form, appears also as the operator of the parabolic cylinder equation (2.259). Under equivalent boundary conditions, that is for fields that are confined near the axis to small enough y-values, equation (3.6) therefore has the same set of eigenfunctions as equation (2.259) for the parabolic cylinder. In the case of the parabolic-cylinder equation for a parabolically graded index profile, the eigenfunctions for those fields which are confined to the index profile are given by equation (2.272) as a product of the Gaussian distribution and the Hermite polynomials.

In the present situation, these eigenfunctions have the form

$$E_l(y) = E_0 \left(\frac{w_0}{w}\right)^{\frac{1}{2}} H_l (2^{\frac{1}{2}} \tfrac{y}{w}) \exp{-(y/w)^2} \tag{3.7}$$

with

$$w = [w_0^2 + L^2/(w_0 \beta)^2]^{1/2} \tag{3.8}$$

and

$$w_0^2 \beta = [L(4f-L)]^{1/2} . \tag{3.9}$$

They describe the field distribution of beam modes over the lens cross-sections for which equation (3.5) was set up. The eigenvalues g_l for these eigenfunctions all have unit amplitude. For infinitely wide film lenses or well-confined beam modes, such eigenvalues of unit magnitude are to be expected, because these beam modes suffer no diffraction loss at lens edges and maintain their intensity from lens to lens. The eigenvalues show only the phase shift

$$b_l = \beta L - 2(l+1) \arctan [L/(w_0^2 \beta)] \tag{3.10}$$

for the field distribution from lens to lens.

The transverse field distribution of beam modes is given by equation (3.7) only in the centre plane of the film lenses and on cylinder surfaces of radius $R = 2f$ directly adjacent to these lenses. This expression may, however, be generalized to also describe the field distribution at all points between lenses. For this generalization, we replace the distance L between lenses by the longitudinal coordinate z, and let $L = 2z$, as well as

$$w = [w_0^2 + (2z/(w_0 \beta))^2]^{1/2} . \tag{3.11}$$

The parameter w_0 remains unchanged, so that the field distribution depends on z only through $w(z)$. Fig.3.3 indicates how w and the transverse field distribution of the lowest-order beam mode depend on the axial distance z. With w_0 ac-

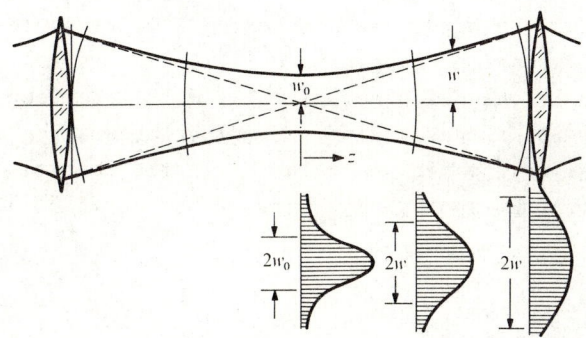

Fig.3.3. Spot size w and transverse field distribution of the lowest-order beam mode in a lens guide section.

cording to equation (3.9) for any periodic sequence of film lenses, the radius of curvature $R = 2f$ of phase fronts at these lenses is given by

$$R = L/2 + w_0^4 \beta^2/(2L). \qquad (3.12)$$

At any distance z from the centre between two lenses, the radius of curvature of the phase fronts may also be determined from equation (3.12) when we generalize this expression, and again replace L by $2z$:

$$R(z) = z + w_0^4 \beta^2/(4z). \qquad (3.13)$$

Right of the centre plane between the lenses, for $z > 0$, the radius vector points in the positive z-direction and $R(z)$ from equation (3.13) is likewise positive. For $z < 0$, the radius vector points in the opposite direction, and $R(z)$ is also negative. The phase shift of beam modes in their direction of propagation along the z-axis is obtained from the corresponding generalization of equation (3.10), again by substituting $2z$ for L.

Across any of the phase fronts, the transverse field distribution $E(y)$ of any beam mode corresponds to the field distribution in equation (2.272) of the same order mode, which is guided by the parabolically graded index profile. Fig.3.3 shows such distributions for the lowest order mode. The parameter w according to equation (3.11) measures the transverse extension of beam-mode fields and is called the spot size. It changes hyperbolically with z between the lenses from its largest value according to equation (3.8) at both lenses, to the minimum value w_0 in the centre.

The minimum spot size w_0 from equation (3.9) has real values only for

$$L < 4f. \tag{3.14}$$

For the periodic sequence of film lenses to have any truly guided modes, the distance between lenses must be within $4f$. For larger lens distances, film modes cannot form beam modes of the lens sequence and will diverge in the transverse y-direction as they propagate through the sequence of lenses. For a given phase constant β of a film mode and spacing L between lenses, the maximum spot size w from equations (3.8) and (3.9) still depends on the focal length f of the lenses. Choosing this focal length according to

$$2f = L \tag{3.15}$$

leads to a confocal arrangement, which minimizes this spot size to

$$w = 2^{\frac{1}{2}} w_0 = (2L/\beta)^{\frac{1}{2}}. \tag{3.16}$$

A confocal lens sequence therefore guides beam modes most efficiently and confines their fields closest to the guide axis.

If, for example, the film mode has a phase constant $\beta = 3\pi \cdot 10^4$ cm^{-1}, corresponding to an effective index $N = 1\cdot 5$

at λ = 1 µm, and if the confocal spacing of lenses amounts
to L = 1 cm, the spot size is only w = 46 µm. Film lenses
need then be only 0·2 mm wide to guide beam modes without too
much diffraction loss. To facilitate fabrication but also
to ease the launching of film modes, lens guides on films will
have much wider lenses under practical conditions. They will
then also be designed to guide film waves in beam modes of
larger spot size.

3.2. STRIP GUIDES

Better confinement of film waves than by iterative refocusing in film lenses may be provided, transverse to their
direction of propagation and parallel to the film, in a continuous fashion. The most obvious way to obtain such continuous confinement is to limit the film to a finite width as
in Fig.3.4. The film with refractive index n_1 now extends in
the transverse direction only over the limited width b and

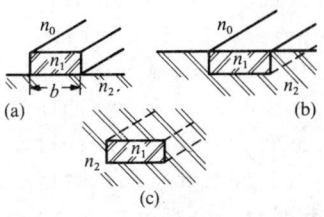

Fig.3.4. Basic strip-guide structures: (a) raised strip; (b) embedded strip; (c) buried strip.

forms a strip. This strip may be deposited on a substrate
surface in form of the raised strip in Fig.3.4(a); it may
also be embedded in the substrate with its cover boundary
flush with the substrate surface as in Fig.3.4(b). As a
third version, it may even be buried completely inside the
substrate medium with the same refractive index n_2 surrounding it on all four sides, as in Fig.3.4(c). All three basic
forms of strip guide, the raised strip, the embedded strip,
and the buried strip, occur as guiding structures in optical
devices and as connecting lines in planar optical circuits.
By confining the fields of guided modes very much to their

cross-sections, they reduce the size of these devices and circuits and offer nearly the ultimate in space saving. Wherever optical devices need drive power, as, for example, lasers and modulators do, the close confinement of light and the reduction in size lowers power consumption and increases the efficiency. The simple rectangular geometry of strip-guide cross-sections, moreover, is easy to fabricate and lends itself readily to those production technologies that are well established for planar integrated solid-state circuits. If we maintain the same rectangular cross-section in all components and waveguides, they may conveniently be assembled in planar circuits and connected to each other without any transformation of wave forms. Such simple and uniform optical assemblies avoid any scattering loss at transitions and reduce the interference from reflections to a minimum.

The basic strip guide structures of Fig.3.4 are assumed to have abrupt transitions of the refractive index and perfectly rectangular cross-sections. They represent only idealized models of actual strip guides which due to fabrication techniques, deviate in their cross-section from the perfect rectangle. In some cases, actual strip guides also have graded transitions due, for example, to diffusion processes during their fabrication. Despite all these deviations, the idealized models of rectangular strip guides serve well to analyse the field distribution and propagation characteristics of their guided modes. All three forms of strip guide, the raised strip, the embedded strip, and the buried strip, work on the same principle of transverse confinement, and their guided modes show similar characteristics. The general strip guide structure of Fig.3.5 includes all three cases of Fig.3.4 and serves us as a sufficiently general model to

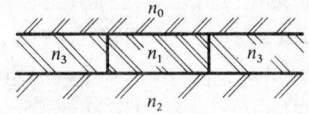

Fig.3.5. General model for raised, embedded, and buried strips.

analyse guided modes in all three structures. It represents the raised strip for $n_3 = n_0$, the embedded strip for $n_3 = n_2$, and the buried strip for $n_3 = n_2 = n_0$.

The confinement may be regarded as total reflection of a film wave inside the strip at the side walls of the strip. For any of the guided modes of the strip, we start with the corresponding H_m- or E_m-wave of the film with index n_1 between the substrate of index n_2 and the cover of index n_0. We let this film wave with its phase constant $\beta = N_s k$ propagate at an angle θ with respect to the strip axis as indicated by Fig.3.6. For this film wave, the strip region has the effective index of refraction N_s. When incident on the side

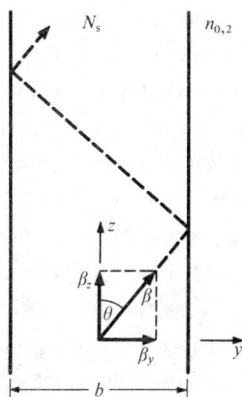

Fig.3.6. Film mode on zig-zag path in strip guide forms strip mode.

walls of the strip, this film wave excites a field distribution in the region beyond, which is similar to the field of the corresponding film wave in a film with index n_3 between the same substrate and cover as in the strip region. In our situation, however, we have either $n_3 = n_0$, or $n_3 = n_2$, or $n_3 = n_2 = n_0$ and hence no film waves exist in the regions with n_3. The film wave which is incident from the strip will therefore excite a radiation field and will then not be guided by the strip. If, however,

$$\beta_z = \beta \cos \theta > n_{0,2} k, \qquad (3.17)$$

the incident film wave from the strip excites only evanescent fields in the region beyond the strip. It will then be totally reflected by the side walls and travel on a zigzag path inside the strip. If now, in addition, it satisfies a phase condition corresponding to equation (2.6) for film waves, it repeats in phase after two successive reflections. Its field distribution then fits into the film and forms guided modes of propagation of the particular strip. Such guided modes of propagation we call strip modes.

If the particular strip mode grows out of an H_m-mode of the film in the strip region, we designate this strip mode as HE_{ml}-mode; if it grows out of an E_m-mode, we designate it as EH_{ml}-mode. This designation with both letters H and E accounts for the hybrid character of strip modes. While the H_m- and E_m-modes of films have only either a magnetic or an electric field component in the direction of propagation, any of the strip modes has both, magnetic as well as electric field components in propagation direction. This hybrid character can also be inferred from the zigzag propagation of film modes in the strip. An H_m-mode of the film with its electric field E transverse to its direction of propagation and parallel to the film has the z-component $E_z = E \sin\theta$ when it travels on the zigzag path of Fig.3.6. An E_m-mode of the film has the corresponding H_z-component when travelling on the zigzag path of Fig.3.6. When an H_m-film mode undergoes total reflection at the side walls of the strip, it has the polarization of an E-wave with respect to these reflecting side walls. An E_m-film mode on the other hand, has the polarization of an H-wave with respect to the reflecting side wall.

The first index m in the HE_{ml}- and EH_{ml}-designation of strip modes continues to count the order of the film mode out of which it grows. It represents the number of nodes in the transverse field distribution perpendicular to the film. The second index l is also an integer number and designates the transverse order of the respective strip mode parallel to the bottom and cover of the strip. The index l counts the number of periods 2π in the phase condition corresponding to equation (2.6). It is the number of periods by which the direct shift of a phase front differs from the detour via two successive reflections at the side walls with their additional phase shifts. In zigzagging along the strip, the film-mode fields superimpose to standing waves in the transverse direction, which is also parallel to the bottom and cover of the strip. The index l counts the number of nodes of their transverse standing wave pattern.

To actually formulate the phase condition for strip modes that would correspond to equation (2.6) for film modes, we need to know the phase shift of the particular film mode upon total reflection at the side wall. To determine this phase shift would involve us in a complicated field analysis, in particular, when the strip is so narrow, that local fringing fields at the side walls reach into the centre of the strip, and let both reflections at the opposite side walls interact directly with each other. We will not pursue this problem here, but rely for accurate results on a more direct field analysis of the complete strip. We will discuss this field analysis later and right now use the ray picture of film modes on a zigzag path in the strip to obtain simple approximations for strip-mode characteristics. To this end, we neglect any direct interaction between total reflections at the opposite side walls of the strip. In addition, we will use equations (1.25) and (1.26) to evaluate the phase shift for total reflection. These phase-shift expressions were derived for plane uniform waves and total reflection at a plane interface. To apply them for the total reflection of a film wave at the side wall of the strip, we replace k_{1x} by $\beta_y = \beta \sin \theta$ and k_{2x} by $-j[\beta_z^2 + (u/d)^2 - n_3^2 k^2]^{1/2}$. The first of these substitutions introduces the transverse y-component of the relevant wave vector in the strip region, while the second substitution introduces the corresponding quantity in the evanescent field region beyond the side walls. This latter substitution follows from the separation condition (1.9) with the wave vector components parallel to the sidewall matched by

$$k_{3x} = u/d \qquad k_{3z} = \beta_z$$

to the wave vector in the strip region. To facilitate formulation of the characteristic equations for strip modes, we abbreviate these substitutions by

$$u_s/b = \beta_y = (n_1^2 k^2 - (u/d)^2 - \beta_z^2)^{\frac{1}{2}} \tag{3.18}$$

$$v_s/b = ((u/d)^2 + \beta_z^2 - n_3^2 k^2)^{\frac{1}{2}} ; \tag{3.19}$$

the quantities u_s and v_s then represent the transverse phase parameter inside the strip and the transverse attenuation parameter beyond the strip, respectively, both in the y-direction. They correspond to the transverse parameters u and v in the y-direction of the constituent film modes.

The phase angles of total reflection follow from equations (1.25) and (1.26). With the above substitutions they appear as

$$\phi_m = 2 \arctan[n_1^2 v_s/(n_3^2 u_s)] \tag{3.20}$$

for an H_m-film mode and

$$\phi_e = 2 \arctan(v_s/u_s) \tag{3.21}$$

for an E_m-film mode. Note again that an H_m-film mode has a magnetic field H_x as the dominant component parallel to the sidewall, for an E_m-film mode it is the electric field component E_x.

The characteristic equations of the various strip modes can now be formulated by starting from the general equation (2.6) and proceeding directly to its special forms for symmetric films. The strip of Fig.3.5 shows this symmetry with respect to its centre plane $y = 0$. We therefore obtain from equation (2.47) the characteristic equation for HE_{ml}-modes with even order l:

$$u_s \tan(u_s/2) = (n_1/n_3)^2 v_s, \tag{3.22}$$

while equation (2.48) must be applied for HE_{ml}-modes with odd orders l:

$$-u_s \cot(u_s/2) = (n_1/n_3)^2 v_s. \qquad (3.23)$$

In the case of EH_{ml}-modes with even order l, equation (2.45) applies and leads to

$$u_s \tan(u_s/2) = v_s, \qquad (3.24)$$

while for EH_{ml}-modes with odd order l, we obtain from equation (2.46)

$$-u_s \cot(u_s/2) = v_s. \qquad (3.25)$$

To evaluate these characteristic equations for any of the strip modes, we specify the film parameter $V = kd(n_1^2 - n_2^2)^{\frac{1}{2}}$ and determine first the transverse phase parameter u of the particular film mode that forms the strip mode. We next take the corresponding strip parameter $V_{st} = kb(n_1^2 - n_3^2)^{\frac{1}{2}}$ of the strip and by solving the respective characteristic equation out of equations (3.22) to (3.25), determine the transverse phase parameter u_s in the strip. The phase constant of the strip mode then follows from equation (3.18) as

$$\beta_z = (n_1^2 k^2 - (u/d)^2 - (u_s/b)^2)^{\frac{1}{2}}. \qquad (3.26)$$

For an HE_{ml}-strip mode in this approximation, u represents the transverse phase parameter of the H_m-mode in a film of thickness d and index n_1 between cover and substrate with index n_0 and n_2, respectively, while u_s represents the transverse phase parameter of the E_l-mode in a symmetrical slab of thickness b and index n_1 between regions with index n_3. For an EH_{ml}-strip mode, u belongs to the E_m-film mode and u_s to the H_l-mode of the symmetrical slab.

Practical strip guides have quite often only a slightly higher refractive index n_1 than the index n_2 of the substrate. However, both n_1 and n_2 differ substantially from $n_0 = 1$ if the cover is free space. Under these weak guiding conditions, the above approximate evaluation of the characteristic equations simplifies still further, and the results may be represented in terms of the universal phase and film parameters

B and V that were also applied to represent film mode characteristics.

For the raised strip we have $n_3 = n_0$ and in the case of weak guidance $n_1 \simeq n_2$ but a pronounced difference between n_1 and $n_0 = 1$. The large index step from n_1 to n_0 renders the fields nearly zero at the side walls of the strip, so that

$$u_s = (l+1)\pi. \qquad (3.27)$$

The other transverse phase parameter u follows from the universal characteristic equation (2.16), which when written in terms of u and v appears as

$$u = \pm V \sin u. \qquad (3.28)$$

With solutions for u from this equation, and u_s according to equation (3.27), the phase constant of the raised strip with weak guidance may now be determined from equation (3.26). Fig. 3.7 shows from this approximation the phase parameter B as previously defined by equation (2.18) plotted against the film parameter V for a few of the low-order modes of the weak-

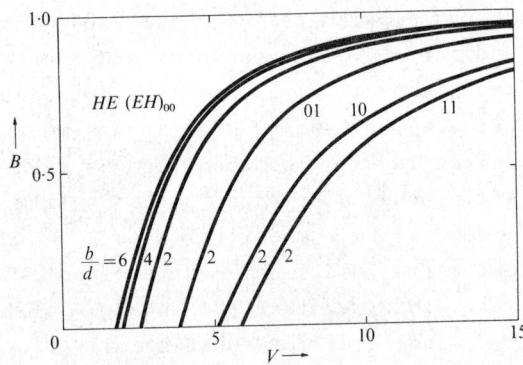

Fig.3.7. Phase parameter $B = (\beta/k-n_2)/(n_1-n_2)$ of low-order modes in a raised strip with $(n_1-n_2) \ll (n_1-n_0)$ versus the film parameter $V = kd(n_1^2-n_2^2)^{\frac{1}{2}}$.

ly guiding raised strip. The approximate expressions (3.27) and (3.28) do not differentiate between H- and E-film modes of the same transverse orders. Any HE_{ml}-mode from this approximation has therefore the same phase constant as the EH_{ml}-mode of the same transverse orders m and l.

For the embedded strip, we have $n_3 = n_2$. If, in addition, we assume again weak guidance with $n_1 \simeq n_2$, but a pronounced difference between n_1 and n_0, the transverse phase parameter u again follows from equation (3.28), while u_s as the transverse phase parameter of the weakly guiding symmetrical slab follows from equation (3.24) for even orders l and from equation (3.25) for odd orders l. The transverse attenuation parameter v_s in these equations is related to u_s and the strip parameter V_{st} by $u_s^2 + v_s^2 = V_{st}^2$, so that instead of equations (3.24) and (3.25), the following forms of the characteristic equation

$$u_s = \pm V_{st} \cos(u_s/2) \qquad (3.29)$$

$$u_s = \pm V_{st} \sin(u_s/2) \qquad (3.30)$$

give u_s directly though not explicitly in terms V_{st}. Fig.3.8

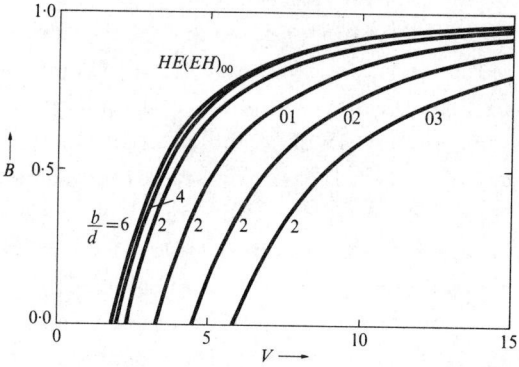

Fig.3.8. Phase parameter $B = (\beta/k - n_2)/(n_1 - n_2)$ of low-order modes in an embedded strip with $(n_1-n_2) \ll (n_1-n_0)$ versus the film parameter $V = kd(n_1^2 - n_2^2)^{\frac{1}{2}}$.

shows the phase parameter B as calculated according to this approximation for embedded strips of weak guidance. Any particular mode of the embedded strip has always a lower cut-off and a larger phase constant than the same mode in the raised strip.

When the film and strip parameters V and V_{st} are sufficiently far above cut-off for the particular film or slab mode, equation (2.35) together with equations (2.36) and (2.37) provide us with explicit approximations for the transverse phase parameters u and u_s. The approximations hold for any differences $n_1 - n_0$ and $n_1 - n_2$. If we substitute such approximations for u and u_s into equation (3.26), we obtain an explicit expression for the phase constant of strip modes in terms of the frequency parameters V and V_{st} and the ratio $(n_1^2 - n_0^2)/(n_1^2 - n_2^2)$ of permittivity differences. Since the strip mode cut-off is always somewhat higher than the respective cut-off values of V_{st} and V for the constituent film and slab modes, this explicit expression approximates the phase constant quite well, even close to cut-off of any particular strip mode.

All the approximate solutions for strip modes, which we have derived so far by starting with film modes and considering their zigzag propagation down the strip, give us not only the dispersion characteristics, but provide us also with the transverse phase parameters u and u_s inside the strip as well as the attenuation parameters v, w, and v_s for the evanescent fields outside of the strip. These parameters enable us to plot the transverse field distribution to the same approximation as we obtained u and u_S for the phase constant. Inside the strip, the fields form standing waves in the x- and y-directions. In the x-direction, they have the same distribution as the constituent film wave, in the y-direction, they distribute as $\cos(u_s\, y/b)$ for even orders l and as $\sin(u_s\, y/b)$ for odd orders l. Outside the film, the fields decay exponentially with attenuation constants v/d into the substrate, w/d into the cover, and v_s/b beyond the side walls. Fig.3.9 illustrates the field distribution schematically for a few low-order modes. The arrows indicate the direction of the transverse electric field and their length gives an

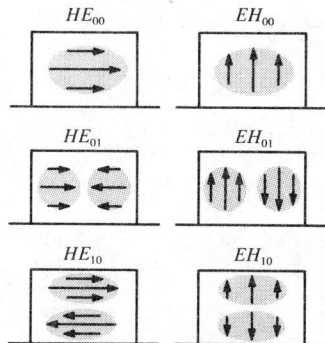

Fig.3.9. Transverse electric field configurations and intensity distributions of low-order strip-guide modes.

idea of the electric field strength. The shading marks the cross-sectional areas with high power-flow density.

The field distributions which we obtained by considering film modes inside the strip and their total internal reflection at the side walls do not solve the boundary value problem of the layered structures of Figs. 3.4 and 3.5 exactly. The phase constant from equation (3.26), which is associated with these field distributions, is likewise only an approximation to the actual phase constant of strip modes. The approximation was made when calculating the reflection at the side walls as if the strip was of infinite height with the standing wave pattern of the film mode according to $\cos(ux/d)$ extending to infinity. In determining the reflection phase, we only matched field components at the side walls for the height d of the strip but neglected completely the fields at $|y| = b/2$ for $x > 0$ and $x < -d$. We can therefore expect this solution to give accurate results only when the particular film mode is sufficiently far above cut-off and its cover and substrate fields low and strongly evanescent. Any mismatch in these small and localized evanescent fields will then cause only little error.

If the strip mode is close to cut-off, however, the constituent film mode is possibly also not too much above cut-off. The reflection phase from equations (3.20) and (3.21) may then not be accurate enough and may cause the approximations for the transverse phase and attenuation parameters and for the longitudinal phase constant to differ significantly from the actual values.

A field solution for the rectangular dielectric waveguide which can be made as accurate as desired was first derived by Schlosser (1964). He also found an approximation (Schlosser and Unger 1966) for the phase constant and field distribution which is actually equivalent to our approximate analysis of strip modes from film modes. In his complete field solution for guided modes of the rectangular dielectric waveguide Schlosser (1964) considered a rectangular metallic waveguide with the rectangular dielectric strip inserted in its centre. When we apply this method to our general strip guide model of Fig.3.5, we would consider the rectangular waveguide in Fig.3.10, which is inhomogeneously filled with half the

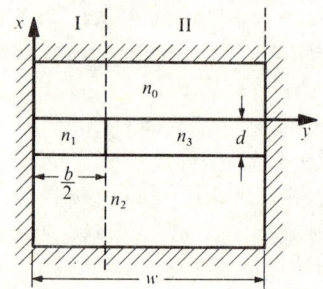

Fig.3.10. Closed waveguide model inhomogeneously filled with half a strip guide structure.

strip structure of Fig.3.5. For HE_{ml}-strip modes with even l, and EH_{ml}-strip modes with odd l, we would let its walls, in particular the wall through the strip centre, be ideal magnetic conductors, while for HE_{ml}-strip modes with odd l as well as EH_{ml}-strip modes with even l, we would let them be ideal electric conductors. This choice of conductors will allow the same field configuration in one half of the symmetric guiding structure as the modes of even and odd symmetry have in the complete waveguide. By enclosing the complete structure with ideally conducting walls, we gain the advantage of discrete sets of eigenfunctions, which are complete enough to represent any source-free field distribution, including all guided modes.

The electromagnetic field of a particular waveguide mode in the different cross-sectional areas is now expanded in terms of eigenfunctions of the respective area. By matching tangential field components at the boundaries, we obtain a homogeneous system of linear equations for the expansion coefficients. To reduce numerical work for the solution of the equation system calls for a suitable subdivision of the waveguide cross-section. Fig.3.10 shows such a subdivision into regions (I) and (II) with their common interface in the longitudinal section $y = b/2$. Looking from this interface in the transverse direction, each subarea represents a parallel-plate waveguide which has three dielectric layers and terminates in conducting planes at $y = 0$ and $y = w$. For these parallel-plate transverse waveguides, we can set up complete standing wave solutions that satisfy all boundary conditions at the conducting walls. Matching their tangential fields at the interface $y = b/2$ provides us with the homogeneous system of linear equations for the expansion coefficients. For a nontrivial solution, its coefficient determinant must vanish, which condition represents the characteristic equation. Its solutions correspond to modes of propagation of the complete waveguide. We are interested here only in those solutions which have evanescent fields outside the strip of index n_1, when the conducting walls at the bottom, the cover, and the side move further and further out. These solutions represent strip modes.

The exact characteristic equation in the form of the coefficient determinant of an infinite system of equations has infinitely many terms. For its numerical evaluation, only a limited number of terms can be taken into account. For the phase constant of a low-order strip mode to be accurately determined from the finite characteristic equation, it was found adequate, under most practical conditions, to represent the fields in subregions I and II by the standing and evanescent waves of the six lowest-order modes in each region. With four tangential field components to match at the interface, this leads to a 24 × 24 determinant.

The closed waveguide model in this analysis must have its conducting walls at the bottom, cover and side wall far enough removed from the strip in order that the evanescent fields near them decay to insignificant values. The numerical evaluations have shown that, depending on the index differences, bottom and cover should at least be twice the strip thickness d removed from the strip, while the sidewall should be at least twice the strip width b removed from the strip. To obtain accurate solutions in a short time with this method requires a high speed computer.

Another computer-generated solution for strip modes is based on an expansion of the electromagnetic fields in terms of circular harmonics. Goell (1969) has developed this method for the buried strip in Fig.3.11. Because of the symmetry of

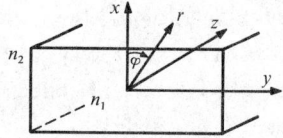

Fig.3.11. Buried strip with cylinder coordinates for circular harmonic expansion.

the structure with respect to the longitudinal centre sections at $x = 0$ and $y = 0$, we need to consider only one quadrant of the cross-section, e.g. the quadrant $0 \leq \varphi \leq \pi/2$. Depending on the symmetry of the particular strip mode, we can then let the planes $\varphi = 0$ and $\varphi = \pi/2$ be ideal electric or magnetic conductors.

The longitudinal components of the electric and magnetic fields inside the strip are set up in the form:

$$E_{z1} = \sum_{p=0}^{N} a_p J_p(k_r r) [\sin(p\varphi+\varphi_p)]\exp(-j\beta z)$$
$$H_{z1} = \sum_{p=0}^{N} b_p J_p(k_r r) [\sin(p\varphi+\psi_p)]\exp(-j\beta z)$$
(3.31)

and outside the strip as:

$$E_{z2} = \sum_{p=0}^{N} c_p K_p(\alpha_r r) [\sin(p\varphi+\varphi_p)]\exp(-j\beta z)$$

$$H_{z2} = \sum_{p=0}^{N} d_p K_p(\alpha_r r) [\sin(p\varphi+\psi_p)]\exp(-j\beta z)$$

(3.32)

with $k_r = (n_1^2 k^2 - \beta^2)^{1/2}$ as well as $\alpha_r = (\beta^2 - n_2^2 k^2)^{1/2}$ and J_p and K_p as Bessel functions and modified Hankel functions, respectively, of the integer order p. The transverse fields for these longitudinal components follow readily from Maxwell's equations. The circular harmonic expansion of equation (3.31) when extended to infinitely many terms forms a complete set, which allows any source-free field distribution inside the strip to be represented with $\exp(-j\beta z)$ as its z-dependence. It therefore also lends itself to any strip mode with phase constant β. The expansions (3.32), when extended to infinitely many terms, are likewise complete enough to represent the evanescent fields of any strip mode outside the strip. To determine the expansion coefficients a_p, b_p, c_p, and d_p we choose N points along the boundary of the strip and match all four tangential field components inside and outside the strip at these points. This provides us with just as many equations as there are unknown quantities in the expansions (3.31) and (3.32). They form a homogeneous and linear system for the expansion coefficients. For non-trivial solutions its coefficient determinant must vanish, which condition is the characteristic equation for strip modes. Values for β which solve this equation are phase constants of strip modes.

Numerical computations show that for moderate aspect (or dimensional) ratios b/d of the strip, only a limited number of circular harmonics are needed to achieve sufficient accuracy for the phase constant. For larger aspect ratios, however, the convergence is not very good.

Fig.3.12 shows accurate results for the phase parameter B of a buried strip as a function of the film parameter V. Such results are obtained from the numerical analysis of either Schlosser (1964) or Goell (1969). The broken lines in

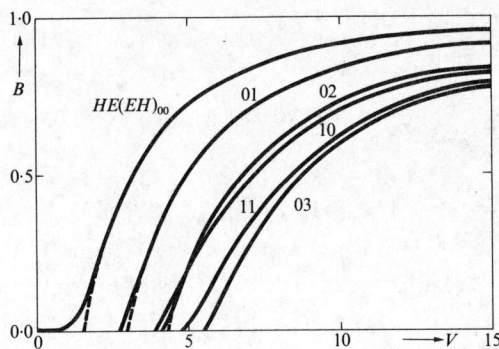

Fig.3.12. Phase parameter $B = (\beta/k - n_2)/(n_1 - n_2)$ versus the film parameter $V = kd(n_1^2 - n_2^2)^{\frac{1}{2}}$ for low-order modes in a buried strip with $(n_1 - n_2) \ll n_1$ and dimensional ratio $b/d = 2$. Broken lines represent the approximation from equation (3.26) with film modes in the strip.

Fig.3.12 represent results from equation (3.26), which are based on the film mode approximation for the strip mode. These approximations agree so well with the accurate numerical results that the two types of curves cannot be distinguished from each other over most of their range. A noticeable deviation occurs only close to their respective cut-offs. This discrepancy between the accurate numerical solution and the film-mode approximations appears most pronounced for the lowest-order modes in their two polarizations HE_{00} and EH_{00}. While the accurate numerical solutions display the dispersion curve down to their zero cut-off, the film-mode approximation ends with $B = 0$ at a finite cut-off value for V.

The raised strip and the embedded strip have fundamental modes with finite cut-off. For them, the film-mode approximation will not show this relatively large discrepancy at cut-off. We can therefore pronounce the film-mode approximation to be sufficiently accurate for most practical purposes in all three types of strip guides of Fig.3.4, except near cut-off, in particular for the lowest-order modes of the buried strip with their zero cut-off.

Fig.3.12 gives the phase parameter of the buried strip

only for the aspect ratio $b/d = 2$. For other aspect ratios, strip modes have different phase constants. The total range over which the phase constant varies is illustrated by Fig. 3.13 for the two lowest-order modes of the buried strip. It

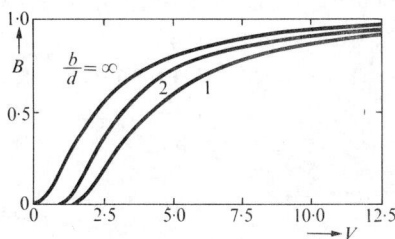

Fig.3.13. Phase parameter $B = (\beta/k-n_2)/(n_1-n_2)$ of the lowest-order strip modes HE_{00} and EH_{00} in a buried strip with different dimensional ratio (Goell 1969); $(n_1-n_2) \ll n_1$.

compares the phase constant for the limiting cases of a square strip and an infinitely wide strip to the phase constant of a strip with aspect ratio $b/d = 2$. The curves for $b/d = 1$ and 2 are from an exact numerical evaluation of Goell's computer analysis while the curve for $b/d \to \infty$ reproduces the phase constant for the H_0- and E_0-modes of a weakly guiding symmetrical film. The latter case actually represents a symmetrical film.

All the dispersion curves which have been presented so far are for weakly guiding strips with $(n_1-n_2) \ll (n_1-n_0)$ or $(n_1-n_2) \ll n_1$. In this limiting case any HE_{ml}-mode of a particular strip has the same phase constant as the EH_{ml}-mode of the same transverse orders m and l. As in case of film modes, any substantial difference in refractive index of strip and substrate breaks this degeneracy between HE_{ml}- and EH_{ml}-modes. Fig.3.14 shows the dispersion curves of the lowest-order modes in their two polarizations HE_{00} and EH_{00} for a buried strip of aspect ratio $b/d = 2$ and various relative index differences Δ between the strip and the embedding material. The phase parameter in this general case follows from the exact definition $B = (\beta^2/k^2-n_2^2)/(n_1^2-n_2^2)$, which holds for any value of

$(n_1 - n_2)$. The curves represent accurate results, again from Goell's computer analysis. A significant difference between the phase parameter of the HE_{00}-mode and the EH_{00}-mode already appears at $n_1 - n_2 = 0.5$. The EH_{00}-phase constant decreases more from its limiting value for the weakly guiding strip than does the HE_{00}-phase constant. The HE_{00}-mode with the highest of all phase constants is best confined to the strip at all wavelengths and for any aspect ratio $b/d > 1$. It therefore forms the fundamental or dominant mode of propagation of the strip.

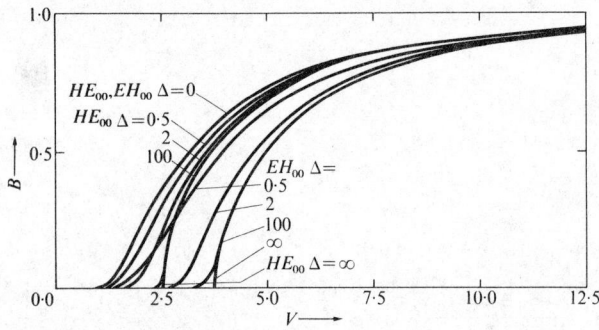

Fig.3.14. Phase parameter $B = (\beta^2/k^2 - n_2^2)/(n_1^2 - n_2^2)$ of the lowest-order strip mode in a buried strip of dimensional ratio $b/d = 2$ for different relative index differences $\Delta = (n_1 - n_2)/n_2$.

3.3. STRIP-LOADED FILM GUIDES

When films are limited to a finite width in the form of strips, they provide transverse confinement of film waves by total reflection at the side walls of the strip. The sidewalls of the raised strip with their large index step reflect film waves more effectively, and fields concentrate more inside the strip than in case of the embedded strips with their weakly guiding side walls. This characteristic favours the raised strip. The side walls of such a raised strip must, however, be very smooth in order not to cause too much scattering loss. The embedded strip may be allowed to have more side wall imperfections without excessive scattering loss. From equations (2.158) and (2.159) the scattering loss at

rough interfaces with different index steps increases with
the difference in relative permittivities; it is therefore
much larger for the raised strip than for the embedded strip.

Another advantage of the embedded strip as compared
to the raised strip of the same size may be its wider extension of transverse fields. The evanescent fields beyond
both sidewalls of the strip for any particular mode are much
stronger in the case of the embedded strip and extend further sideways than in case of the raised strip. The larger
mode size in the embedded strip facilitates launching of these
modes and joining strip guides.

To gain still more in mode size, and further reduce
the scattering at rough edges of the structure, the transverse confinement should be provided by an even smaller effective difference in refractive index at the side walls of
the structure. The strip-loaded film of Fig.3.15 offers this

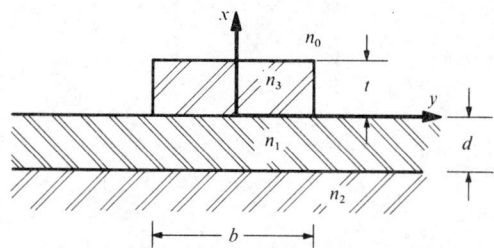

Fig.3.15. Strip-loaded film guide.

possibility. It allows the effective change in refractive
index to be controlled quite accurately down to very small
differences. According to Fig.3.15, the strip-loaded film
consists of a thin film of index n_1, deposited on a substrate
of index n_2, loaded with a raised strip of index n_3. The
index of the covering region n_0 is normally that of free
space with $n_0 = 1$. The strip can have the same index as the
substrate; it could also have an index between substrate
and film index. However, to maintain the favourable characteristics of the strip-loaded film guide, the strip index
should neither be much lower than the substrate index nor

much higher than the film index.

We learn to understand wave guidance and field confinement in the strip-loaded film by considering the strip region as a four-layer film guide. A particular film mode of this four-layer structure has a phase constant β_s and an effective index $N_s = \beta_s/k$ which are always larger than the phase constant β_f and effective index N_f, respectively, of the corresponding film mode in the unloaded side regions of the film. If the four-layer film mode propagates at an angle θ with respect to the strip axis, as shown in Fig.3.16, it

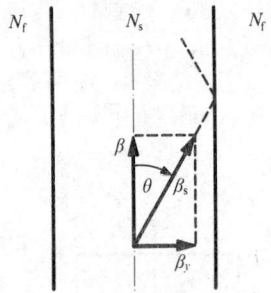

Fig.3.16. Ray at angle θ representing a film mode in the four-layer strip-loaded region.

will experience total reflection at the edge of the four-layer region when

$$N_s \cos \theta > N_f. \qquad (3.33)$$

If, in addition, this film mode satisfies a phase condition corresponding to equation (2.5), it repeats in phase after two successive reflections at the opposite strip edges. It then fits into the strip-loaded film region and forms self-consistent field solutions for guided modes of propagation in the strip-loaded film.

To evaluate the phase condition for these guided modes, we need to know the phase shift ϕ which the film mode of the strip-loaded region experiences upon total reflection at the edges of the four-layer region. An analytical solution to

this problem is not available. Any numerical method to find
this phase shift would involve us in extensive computations
and hardly make the influence of strip and film parameters
clear. These computations would be particularly complex if
the strip is not very wide and interactions between reflections at the opposite edges need to be taken into account.
For the general structure of Fig.3.15, we will therefore resort to the same approximate procedure as we applied to strip
guides. A more accurate numerical solution and some representative results will be discussed in the next section for a
special case of a strip-loaded structure.

For our present approximation of film-mode reflection
at the edge of the four-layer structure, we consider the impedances of the film modes in the transverse direction parallel to the film. Let an H_m-film mode of the strip-loaded
region propagate with phase constant

$$\beta_s = (n_1^2 k^2 - (u_s/d)^2)^{\frac{1}{2}} \tag{3.34}$$

at an angle θ with respect to the strip axis. In expression
(3.34) we have related β_s to the wave number $n_1 k$ and the
transverse phase parameter u_s of the film mode inside the
film. This film mode then has the axial phase constant

$$\beta = \beta_s \cos \theta \tag{3.35}$$

and the transverse phase constant

$$\beta_y = \beta_s \sin \theta . \tag{3.36}$$

From Maxwell's equation its transverse magnetic field component H_x relates to the electric field components as

$$-j\omega\mu_0 H_x = -j\beta_y E_z + j\beta E_y . \tag{3.37}$$

From $\nabla \vec{E} = 0$ it follows that $E_y = -(\beta/\beta_y)E_z$, so that the
transverse wave admittance $Y = H_x/E_z$ in the strip-loaded
region is.

$$Y_s = \beta_s^2/(\omega\mu_0\beta_y). \tag{3.38}$$

The same considerations hold for an H_m-film mode in the unloaded side-regions, except that in the case of total internal reflection at the edge of the four-layer region, its fields decay exponentially in the y-direction with an attenuation constant

$$\alpha_y = (\beta^2 - \beta_f^2)^{\frac{1}{2}}. \tag{3.39}$$

The phase constant β_f in this expression is that of the respective H_m-film mode in the unloaded region. It is related by

$$\beta_f^2 = n_1^2 k^2 - (u_f/d)^2 \tag{3.40}$$

to the wave number $n_1 k$ of the film and the transverse phase parameter u_f of this film mode. Given β_f and α_y of the transversely evanescent film mode in the unloaded film region, its transverse wave admittance is

$$Y_f = j\,\beta_f^2/(\omega\mu_0\alpha_y). \tag{3.41}$$

Before we proceed to evaluate the total reflection at the wave admittance step from Y_s to Y_f, we will first determine the corresponding wave impedances for E_m-film modes. Noting the duality between H_m- and E_m-waves in dual structures, the wave impedances of E_m-waves constitute dual quantities to the wave admittances of corresponding H_m-waves. They therefore follow from equations (3.38) and (3.41) with the permeability μ replaced by the permittivity $\varepsilon = n^2 \varepsilon_0$. The transverse E_m-wave impedance in the strip-loaded region is

$$Z_s = \beta_s^2/(\omega n^2 \varepsilon_0 \beta_y), \tag{3.42}$$

while for the evanescent fields of an E_m-wave in the unloaded region, it is

$$Z_f = j\,\beta_f^2/(\omega n^2 \varepsilon_0 \alpha_y). \tag{3.43}$$

For the reflection of an E_m-wave at the edge of the four-layer region, we have the E_x-component perpendicular to the plane of incidence and therefore take the reflection coefficient r_e corresponding to equation (1.22)

$$r_e = \frac{Z_f - Z_s}{Z_f + Z_s} \ . \qquad (3.44)$$

For the reflection of an H_m-wave on the other hand, we have the H_x-component perpendicular to the plane of incidence, and the reflection coefficient is

$$r_m = \frac{Y_f - Y_s}{Y_f + Y_s} \ . \qquad (3.45)$$

In determining the reflection from transverse wave impedances, we achieve a fairly good match of field components inside the film, and almost as good inside the substrate as well. The fields above the film in the loading strip on one side of the edge, and in the cover space on the other side, are, however, not accounted for in this impedance match. The reflection coefficients will therefore be accurate only as long as the fields are well confined to the film.

When we substitute the transverse wave impedances and admittances into the respective reflection coefficient, we obtain one and the same expression for the phase shift of total internal reflection for E_m-film waves and H_m-film waves. This phase shift is

$$\phi = 2 \arctan[\alpha_y \beta_s^2 / (\beta_y \beta_f^2)] . \qquad (3.46)$$

It shows similarity with the phase angles in equations (1.25) and (1.26) for total reflection of a plane uniform wave at a plane interface. If we replace the transverse attenuation constant jk_{2x} in equation (1.26) by our present transverse attenuation constant α_y and the transverse phase constant k_{1x} by our present transverse phase constant β_y, and if furthermore the ratio n_1/n_2 of refractive indices in equation (1.26) is replaced by the ratio $N_s/N_f = \beta_s/\beta_f$ of our present effective indices of refraction, we obtain equation (3.46) for the phase angle of total reflection at the edge of the four-layer region. With these substitutions, therefore, the reflection corresponds to that of a plane uniform E-wave at the plane interface.

This correspondence leads us directly to the characteristic equations of guided modes in the strip-loaded film. The structure is symmetric with respect to the longitudinal centre-section at $y = 0$. With equation (3.46) corresponding to the reflection of E-waves, we therefore take the characteristic equations (2.47) and (2.48) of even and odd order E_m-waves of the symmetrical slab, and, with the substitutions

$$u \rightarrow \beta_y b$$

$$v \rightarrow \alpha_y b \qquad (3.47)$$

$$n_1/n_2 \rightarrow \beta_s/\beta_f$$

obtain the characteristic equation

$$\beta_y \tan(\beta_y b/2) = \alpha_y (\beta_s/\beta_f)^2 \qquad (3.48)$$

for HE_{ml}- and EH_{ml}-modes of even order l and

$$-\beta_y \cot(\beta_y b/2) = \alpha_y (\beta_s/\beta_f)^2 \qquad (3.49)$$

for HE_{ml}- and EH_{ml}-modes of odd order l. The designations HE_{ml} and EH_{ml} correspond in all details to those of strip modes: HE_{ml}-modes come from H_m-film modes and EH_{ml} from E_m-film modes. The transverse order l again represents the integer multiple of 2π by which the phase shift along the direct path of a film mode differs from the detour via the two successive reflections at the opposite edges of the strip-loaded region. l also equals the number of nodes of the standing wave pattern which the transverse field distribution forms under the loading strip in the y-direction.

The characteristic equations (3.48) and (3.49) for guided modes of the strip-loaded film are equivalent to those for E_m-modes of the symmetrical slab when we substitute according to equation (3.47). The same substitutions in their solutions will therefore yield solutions for the HE_{ml}- and EH_{ml}-modes of the strip-loaded film.

For a universal representation of dispersion characteristics of film modes, we have used the film parameter $V = (u^2 + v^2)^{\frac{1}{2}}$ and the phase parameter $B = (\beta^2/k^2 - n_2^2)/(n_1^2 - n_2^2)$. We have then obtained B as a function of V for each particular film mode with n_1/n_2 as an additional parameter. In the present case of the strip-loaded film, the substitutions (3.47) lead to the new waveguide parameter

$$V_s = b(\beta_y^2 + \alpha_y^2)^{\frac{1}{2}} . \qquad (3.50)$$

This parameter can also be written in terms of the effective indices of refraction for the film mode in the strip-loaded region

$$N_s = \beta_s/k = (\beta^2 + \beta_y^2)^{\frac{1}{2}}/k \qquad (3.51)$$

and in the unloaded region

$$N_f = \beta_f/k = (\beta^2 - \alpha_y^2)^{\frac{1}{2}}/k, \qquad (3.52)$$

so that

$$V_s = kb(N_s^2 - N_f^2)^{\frac{1}{2}} . \qquad (3.53)$$

Substituting N_s for n_1 and N_f for n_2 in B leads to the new phase parameter

$$B_s = (\beta^2/k^2 - N_f^2)/(N_s^2 - N_f^2) . \qquad (3.54)$$

For any HE_{ml}- or EH_{ml}-mode in the strip-loaded film, B_s now depends on V_s and on N_s/N_f, just as B depends on V and on n_1/n_2 for the E_l-mode of the same order l as the strip-loaded film mode has in the y-direction. Weakly guiding films with $n_1 \simeq n_2$ show the additional simplification that with $n_1/n_2 \simeq 1$, B becomes a unique function of V. By the same token, for the strip-loaded film, if the loading is weak enough to let $\beta_s \simeq \beta_f$ and consequently $N_s \simeq N_f$, the phase parameter B_s is the same unique function of V_s as $B(V)$ for the weakly guiding symmetrical film.

To actually evaluate the dispersion characteristic B_s as a function of V_s and N_s/N_f for a particular mode in a specific strip-loaded film, we first need to know the effective indices N_s and N_f of the constituent film mode in the loaded and unloaded regions. The effective index N_f of the film mode in the unloaded region may be read from Fig.2.7 or computed more accurately as a solution of the characteristic equations (2.11) or (2.12), respectively, of the asymmetrical film.

The effective index N_s, however, belongs to the film mode in the four-layer structure. The characteristic equation for H-modes in such a structure has already been derived for the prism coupler. Its approximate solution in the case of the prism coupler was, however, oriented towards the leaky wave loss, when the prism in the covering region frustrates the total reflection. Under present conditions, we have evanescent fields in the covering region and possibly also in the strip region of the four-layer structure. The interface at $x = 0$ between film and strip region will then show total reflection for any plane uniform wave which forms film waves in its zigzag propagation down the film. For an H-wave incident on this interface at $x = 0$, the reflection coefficient r_e according to equation (1.22) is determined by the impedance which the strip layer of thickness t and transverse wave impedance $Z_3 = \omega\mu_0/k_{3x}$ presents at the interface, when it is backed up by the covering region of transverse wave impedance $Z_0 = \omega\mu_0/k_{0x}$. The strip layer acts as a transmission line of wave impedance Z_3 that transforms its load impedance Z_0 into

$$Z_{13} = Z_3 \frac{Z_0 + Z_3 \tanh jk_{3x}t}{Z_3 + Z_0 \tanh jk_{3x}t} . \qquad (3.55)$$

If, for total reflection, both the covering region and the strip region are excited with evanescent fields, we have $k_{0x} = -j\alpha_0$ and $k_{3x} = -j\alpha_3$ as well as $Z_0 = j\omega\mu_0/\alpha_0$ and $Z_3 = j\omega\mu_0/\alpha_3$. The reflection coefficient

PLANAR GUIDES WITH TRANSVERSE CONFINEMENT 223

$$r_e = \frac{Z_{13} - Z_1}{Z_{13} + Z_1} \qquad (3.56)$$

at the film-strip interface has unit amplitude under these conditions, and with $Z_1 = \omega\mu_0/k_{1x}$ its phase angle amounts to

$$\phi_e = 2\arctan\left[\frac{\alpha_3(\alpha_0 + \alpha_3 \tanh \alpha_3 t)}{k_{1x}(\alpha_3 + \alpha_0 \tanh \alpha_3 t)}\right] \qquad (3.57)$$

with the transverse phase constant

$$k_{1x} = (n_1^2 k^2 - \beta_s^2)^{\frac{1}{2}} \qquad (3.58)$$

inside the film, and transverse attenuation constants

$$\begin{aligned}\alpha_3 &= (\beta_s^2 - n_3^2 k^2)^{\frac{1}{2}} \\ \alpha_0 &= (\beta_s^2 - n_0^2 k^2)^{\frac{1}{2}}\end{aligned} \qquad (3.59)$$

in the strip and covering regions respectively. At the film-substrate interface the total reflection has the phase

$$\phi_e = 2\arctan \frac{\alpha_2}{k_{1x}} \qquad (3.60)$$

with

$$\alpha_2 = (\beta_s^2 - n_2^2 k^2)^{\frac{1}{2}} . \qquad (3.61)$$

Substituting these phase shifts for ϕ_{10} and ϕ_{12} into equation (2.6) leads to the characteristic equation for H-waves

$$\tan k_{1x} d = \frac{k_{1x}\left[\alpha_2 + \alpha_3 \dfrac{\alpha_0 + \alpha_3 \tanh \alpha_3 t}{\alpha_3 + \alpha_0 \tanh \alpha_3 t}\right]}{k_{1x}^2 - \alpha_2 \alpha_3 \dfrac{\alpha_0 + \alpha_3 \tanh \alpha_3 t}{\alpha_3 + \alpha_0 \tanh \alpha_3 t}} . \qquad (3.62)$$

Its solutions for a given four-layer structure at a specified wave number k yield the effective index $N_s = \beta_s/k$ of H_m-modes in the four-layer region.

For a loading strip of sufficient thickness t, so that

the argument $\alpha_3 t$ is large enough for $\tanh \alpha_3 t \simeq 1$, equation (3.62) reduces to the characteristic equation

$$\tan k_{1x} d = \frac{k_{1x}[\alpha_2 + \alpha_3]}{k_{1x}^2 - \alpha_2 \alpha_3} \qquad (3.63)$$

of the asymmetric film, for which the strip material with index n_3 forms the cover. Under these conditions, the effective index may again be read directly from Fig.2.7 or obtained from a more accurate numerical solution of equation (2.11).

The special case of $n_3 = n_2$ deserves some attention because with the limited choice of suitable materials, we might want to use the same material for substrate and loading strip. If for $n_3 = n_2$, the film is weakly guiding with $n_1 \simeq n_2$, but the strip is thick enough for equation (3.63) to apply in the loaded region, we have a symmetric slab for the N_s of the film modes in the loaded region, and a strongly asymmetric film for the N_f of film modes in the unloaded side regions. Under these conditions, the phase parameter $B \simeq (\beta/k - n_2)/(n_1 - n_2)$ follows for each mode of the strip-loaded film as a universal function of the film parameter $V = kd(n_1^2 - n_2^2)^{\frac{1}{2}}$ with the dimensional ratio b/d as the only additional parameter. Fig.3.17 shows this phase parameter for the two polarizations of the fundamental mode of this

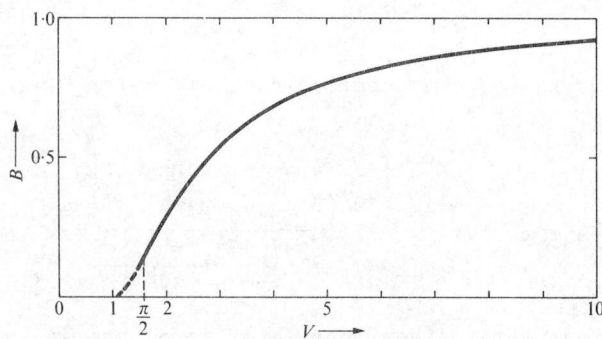

Fig.3.17. Phase parameter $B = (\beta/k - n_2)/(n_1 - n_2)$ versus film parameter $V = kd(n_1^2 - n_2^2)^{\frac{1}{2}}$ of the lowest-order mode in a strip-loaded film with $n_1 - n_2 = n_1 - n_3 \ll n_1 - n_0$ and $\tanh \alpha_3 t \simeq 1$, $b/d = 2$.

weakly guiding structure with substrate and strip of the same index. Under the conditions of Fig.3.17 the strip-loaded film guides only the fundamental mode. The next higher-order mode in its two polarizations HE_{01} and EH_{01} has its cut-off at $b/d = 7 \cdot 2$.

3.4. RIB GUIDES

Another special guiding structure is obtained from the strip-loaded film in Fig.3.15 when film and loading strip have the same index $n_3 = n_1 > n_2 > n_0$. Such a structure results when we first deposit a film on to a substrate, and then remove part of this film to make a lower thickness, except for the strip region. The strip-like region of raised film thickness remains as a ridge or rib on top of the film of reduced thickness; it lends the name ridge or rib guide to this type of planar optical waveguide. The rib guide of-s the same advantages as the strip-loaded guide to nearly the 'me extent. It relaxes the stringent requirements on the smu*hness of sidewalls, which low scattering losses impose on st.'ip guides, particularly on the raised strip.

Fig.3.18 shows a rib guide with a rib of height h and width b and the film of thickness d extending symmetrically to both sides of the rib. The modes which are guided by the rib may again be considered to consist of film modes of the rib region which experience total reflection at the opposite edges of the rib and propagate on a zigzag path down the rib. The rib itself forms a simple film guide which is asymmetrical if the covering layer index n_0 differs from the substrate index n_2. At the edge of the rib, the film

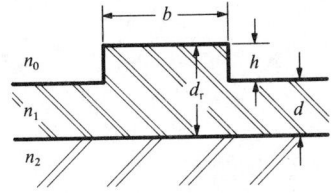

Fig.3.18. Rib guide of index n_1 on a substrate of index n_2.

steps down to a thinner layer. For any of the film modes, this step means also that its effective index steps down from its value $N_r = \beta_r/k$ inside the rib to $N_f = \beta_f/k$ in the thinner film to both sides of the rib. The phase shift for total reflection of a film mode in the rib at its edges obtains approximately from the same considerations, that led to equation (3.46) for the phase shift of total reflection at the edge of the strip-loaded film. We again match transverse impedances of a particular film mode inside the film at the rib's edge. By matching these impedances we also ensure a fairly good, but not complete, match for tangential field components at this interface. With this impedance match, we account for field continuity inside the film, but neglect the evanescent fields in cover and substrate. We therefore get a better approximation to the actual situation the more the fields of the constituent film modes are confined to the film. Under these conditions, and with β_s replaced by β_r, the same expression (3.46) applies also for the phase shift of total reflection at the rib's edges. As a further consequence, the condition for film modes to repeat in phase after two successive reflections leads to the same characteristic equations (3.48) and (3.49) for rib modes of even and odd symmetry with respect to the rib's longitudinal plane of symmetry. As the only changes in equations (3.48) and (3.49), we replace β_s by β_r and interpret β_y as the transverse phase constant, $\beta_y = \beta_r \sin\theta$, of the film mode in the rib region on its zigzag path down the rib. We thus obtain the characteristic equation

$$\beta_y \tan(\beta_y b/2) = \alpha_y (\beta_r/\beta_f)^2 \qquad (3.64)$$

for HE_{ml}- and EH_{ml}-modes of even order l, and the characteristic equation

$$-\beta_y \cot(\beta_y b/2) = \alpha_y (\beta_r/\beta_f)^2 \qquad (3.65)$$

for HE_{ml}- and EH_{ml}-modes of odd order l.

To solve these equations, we proceed very much as in case of the strip-loaded film. We first determine $N_r = \beta_r/k$

and $N_f = \beta_f/k$ of the constituent H_m- or E_m-mode in the films of thickness d_r and d, respectively. We then obtain the phase parameter

$$B_{rf} = (\beta^2/k^2 - N_f^2)/(N_r^2 - N_f^2) \qquad (3.66)$$

as a function of the rib parameter

$$V_{rf} = kb\,(N_r^2 - N_f^2)^{\frac{1}{2}} \qquad (3.67)$$

from the solution of the equivalent symmetrical slab problem.

For a rib guide which consists of weakly guiding or strongly asymmetric films with $(n_1-n_2) \ll (n_1-n_0)$, the phase parameter

$$B = (\beta^2/k^2 - n_2^2)/(n_1^2 - n_2^2) \qquad (3.68)$$

may be presented as a function of the film parameter

$$V_r = kd_r\,(n_1^2 - n_2^2)^{\frac{1}{2}} \qquad (3.69)$$

in the rib region of the film, with the dimensional ratio b/d_r and the film thickness ratio d/d_r as the only additional parameters. To show that such a representation is possible, we note that for any constituent H_m- or E_m-mode in weakly guiding films, we have always $N_r \simeq N_f$ for their effective indices, so that $\beta_r/\beta_f \simeq 1$ in the characteristic equations (3.64) and (3.65). The phase parameter B_{rf} depends under these conditions only on the rib parameter V_{rf}. Furthermore for such weakly guiding films, the phase parameter

$$B_r = (N_r^2 - n_2^2)/(n_1^2 - n_2^2) \qquad (3.70)$$

and

$$B_f = (N_f^2 - n_2^2)/(n_1^2 - n_2^2) \qquad (3.71)$$

for any particular film modes depend only on the respective film parameters

$$V_r = kd_r(n_1^2 - n_2^2)^{\frac{1}{2}} \quad \text{and} \quad V_f = kd(n_1^2 - n_2^2)^{\frac{1}{2}}.$$

With $B_r(V_r)$, we therefore have

$$B_f(V_f) = B_r(\frac{d}{d_r}V_r).$$

Expressing B_{rf} and V_{rf} by these phase parameters, we obtain

$$B_{rf} = (B - B_f)/(B_r - B_f)$$

and (3.72)

$$V_{rf} = (b/d_r)(B_r - B_f)^{\frac{1}{2}}V_r.$$

We solve equation (3.72) for B and obtain

$$B = B_{rf}(B_r - B_f) + B_f.$$

Hence with V_{rf} as well as B_r and B_f as unique functions of V_r, b/d_r, and d/d_r, the phase parameter B depends likewise only on these three parameters.

By way of example, Fig.3.19 shows this phase parameter

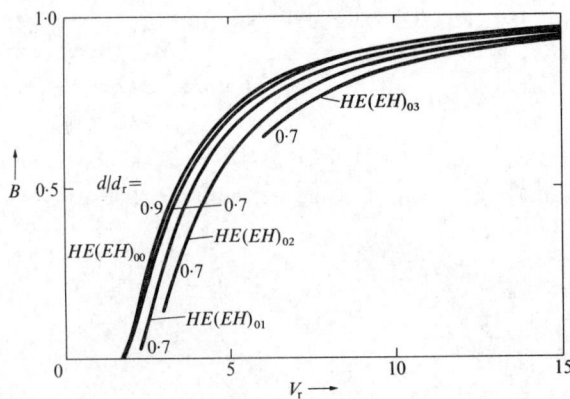

Fig.3.19. Phase parameter $B = (\beta^2/k^2 - n_2^2)/(n_1^2 - n_2^2)$ of rib-guide modes versus rib parameter $V_r = kd_r(n_1^2 - n_2^2)^{\frac{1}{2}}$ for the dimensional ratio $b/d_r = 4$ and $(n_1 - n_2) \ll (n_1 - n_0)$.

B for a number of low-order modes in a rib guide of dimensional ratio $b/d_r = 4$ and two different ratios of film thickness. We note, that in this weakly guiding film approximation, HE_{ml}-modes are degenerate with EH_{ml}-modes of equal transverse orders m and l. This degeneracy already exists for the constituent H_m- and E_m-modes of the weakly guiding film and transfers to the corresponding rib guide modes. Any rib guide mode in Fig.3.19 with a transverse order l different from zero starts with a finite value B at its respective cut-off point.

A rib guide which, with $n_0 = n_2$ in Fig.3.18, consists of symmetrical films in the rib region and beyond its edges, deserves special interest as a model for optical waveguides that are made of only one material of uniform refractive index n_1, surrounded by free space of index $n_0 = 1$ above and below the rib guide. To support such a single-material waveguide, the symmetrical slab with its centre rib is enclosed in a capillary as shown in Fig.3.20. The slab is attached on both

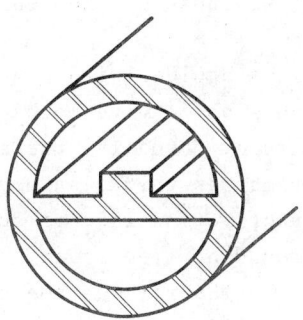

Fig.3.20. Single-material rib guide supported in capillary.

sides to the inner wall of the capillary. Fields of modes which are guided by the rib confine themselves to this rib and decay exponentially in the transverse direction into the slab. Near the walls of the capillary they must be vanishingly small. If not, these walls frustrate the total reflection of film modes at the rib's edges. Energy then leaks out into the capillary and renders the rib mode a leaky wave.

For an approximate representation of rib modes in this single-material waveguide, we start again with film modes, but now of the symmetrical slabs of thickness d_r in the rib region, and d for the supporting slab. We consider their total reflection at the rib's edges, obeying the phase condition for a slab mode to repeat in phase after two successive reflections. We then arrive at the same characteristic equations (3.64) and (3.65) for rib modes in the single-material waveguide. These equations may be solved by determining first the effective indices N_r and N_f of the constituent slab mode in the rib region and in the supporting slab, respectively. The phase parameter B_{rf} according to equation (3.66) is then the same function of the rib parameter V_{rf} from equation (3.67) as B is a function of V for the E_l-mode in a symmetrical slab of refractive index $n_1 = N_r$ surrounded by a medium of refractive index $n_2 = N_f$. The index l represents the transverse order of HE_{ml}- and EH_{ml}-modes parallel to the slab. If the rib differs only slightly in height from the supporting slab, we have $N_r \simeq N_f$ and the solutions of equations (3.64) and (3.65) correspond to those of the weakly guiding symmetrical slab.

Single-material waveguides with slab-supported ribs require a thick enough slab to give mechanical strength and rigidity to the structure. Usually, therefore, the slab is at least several micrometres thick, and even thicker in the rib region. Its refractive index n_1 is substantially larger than that of the surrounding free space. Typically, we have $n_1 > 1\cdot 4$, while $n_2 = 1$. The film parameter $V = kd(n_1^2 - n_2^2)^{\frac{1}{2}}$ for such slabs is $V > 2\pi d/\lambda$, and, for $d > \lambda$, substantially larger than unity. Under these conditions, at least the lower-order slab modes are quite high above cut-off, and their transverse phase parameters may then be approximated by

$$u_r \simeq u_f \simeq (m+1)\pi \qquad (3.73)$$

in the rib region as well as for the supporting slab.

In order to utilize these approximations, we express the phase parameter B_{rf} and rib parameter V_{rf} in terms of u_r and u_f:

$$B_{rf} = [\beta^2 - n_1^2 k^2 + (u_f/d)^2]/[(u_f/d)^2 - (u_r/d_r)^2], \qquad (3.74)$$

$$V_{rf} = b((u_f/d)^2 - (u_r/d_r)^2)^{\frac{1}{2}}. \qquad (3.75)$$

With the approximation (3.73) for thick enough slabs, these parameters reduce to

$$B_{rf} = [1 + d^2(\beta^2 - n_1^2 k^2)/(m+1)^2 \pi^2]/[1 - (d/d_r)^2] \qquad (3.76)$$

$$V_{rf} = (m+1)\pi (b/d_r)((d_r/d)^2 - 1)^{\frac{1}{2}}. \qquad (3.77)$$

To this approximation the parameter V_{rf} depends only on the dimensional ratio b/d_r, the thickness ratio d_r/d, and the slab-mode order m. It is independent of wavelength. Since the cut-off for HE_{ml}- and EH_{ml}-modes occurs at $V_{rf} = l\pi$, the approximation (3.77) leads to the cut-off condition

$$\pi(m+1)(b/d_r)((d_r/d)^2 - 1)^{\frac{1}{2}} = l\pi, \qquad (3.78)$$

which is also independent of wavelength. For $V_{rf} < \pi$, or equivalently

$$d^{-2} - d_r^{-2} < (m+1)^{-2} b^{-2}, \qquad (3.79)$$

the rib guides only HE_{m0}- and EH_{m0}-modes.

A condition for the rib to guide only the lowest-order mode in its two polarizations HE_{00} and EH_{00} is obtained from equation (3.79) when we let $m = 0$:

$$d^{-2} - d_r^{-2} < b^{-2}. \qquad (3.80)$$

This inequality constitutes a necessary but not a sufficient condition for the rib to guide only the HE_{00}- and EH_{00}-modes. To obtain sufficient conditions for the exclusive guidance of only these lowest-order modes, we need to consider the interaction between slab modes of different orders m at the rib's edges. For HE_{m0}- and EH_{m0}-modes of order $m = 1$, we note that their constituent H_m- and E_m-slab modes, upon reflection at the rib's edges, do not only excite the same order H_m- and

E_m-modes in the supporting slab. Rather, they also excite all other orders m which the supporting slab guides. In particular, any of these H_m- or E_m-modes excites also the H_0- or E_0-mode in the supporting slab. For total reflection their fields must decay exponentially in the transverse direction parallel to the slab. Otherwise power would leak out of the H_m- or E_m-mode in the rib and render this mode a leaky wave. To prevent this power leakage into H_0- or E_0-modes of the supporting slab, we must have

$$\beta^2 > n_1^2 k^2 - \pi^2/d^2 \ . \tag{3.81}$$

The phase constant β in this inequality for our present 'thick-slab' approximation follows from

$$\beta^2 = n_1^2 k^2 - (m+1)^2 \pi^2/d_r^2 - \beta_y^2 \ . \tag{3.82}$$

When the particular HE_{ml}- or EH_{ml}-mode is at cut-off, its transverse phase constant is $\beta_y = l\pi/b$, and the axial phase constant has the cut-off value

$$\beta = \{n_1^2 k^2 - (m+1)^2 \pi^2/d_r^2 - l^2 \pi^2/b^2\}^{\frac{1}{2}} \ . \tag{3.83}$$

With this cut-off value in the inequality (3.81), the cut-off condition for higher-order modes is

$$d > d_r [(m+1)^2 + (l \ d_r/b)^2]^{-1/2} \ . \tag{3.84}$$

For $m = 1$ and $l = 0$ this condition reduces to

$$d > d_r/2 \ . \tag{3.85}$$

Under this condition the HE_{10}- and EH_{10}-modes leak power into the H_0- and E_0- modes of the supporting slab and are not guided by the rib. For higher orders m or l, this power leakage starts at even smaller thickness ratios d/d_r. Conditions (3.80) and (3.85) therefore represent necessary and sufficient conditions for the single-material rib to guide only the lowest-order mode in its two polarizations HE_{00}- and EH_{00}.

The possibility of a single-material waveguide which supports only the lowest order mode of propagation has aroused considerable practical interest (Kaiser and Ash 1974). This interest received additional impetus from the unique conditions for the exclusive guidance of only these lowest-order modes. These conditions are independent of wavelength. A rib even on a fairly thick supporting slab will guide just the lowest-order mode only if, according to equations (3.80) and (3.85), it is shallow and narrow enough.

For many applications in optical communications single-mode propagation in a waveguide is a desirable feature or even an important prerequisite. The single-material rib guide assumes the single mode character under conditions for its cross-sectional form and size which are independent of wavelength, and furthermore not too difficult to fulfil in an actual guiding structure. All these features count in favour of this particular waveguide. Unfortunately, however, the single mode character is not quite perfect in the case of the single-material rib guide, because the lowest-order mode exists in its two polarizations HE_{00} and EH_{00}. These two polarizations propagate with different phase constants and suffer different time delays. A signal which excites and propagates in both these polarizations will be distorted due to the different phase constants and delay times.

The above approximation for rib guide modes yields the same dispersion characteristics for both polarizations. To determine the actual differences between both modes, a more complete and accurate analysis is necessary. A numerical method which can achieve any desired degree of accuracy proceeds, as in the case of the strip guide, by enclosing the structure with conducting walls (Rütze 1977). As a representative example, we consider the single-material structure of Fig.3.21 with two planes of symmetry. Because of this two-fold symmetry, we need to consider only one quadrant of the total cross-section as in Fig.3.22, with the centre planes as ideal electric or magnetic conductors depending on which mode we want to analyse.

We divide the quadrant into regions I and II with their interface at the transverse junction between rib and slab.

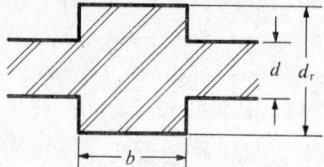

Fig.3.21. Single-material waveguide in the form of a double rib with a two-fold cross-sectional symmetry.

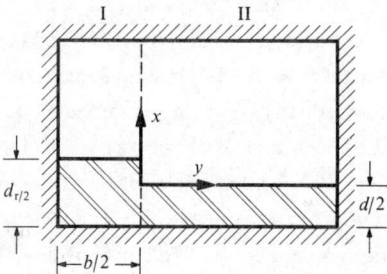

Fig.3.22. Quadrant of double rib guide cross-section bounded by conducting planes.

In y-direction both regions represent partially filled parallel-plate transmission lines which terminate in conducting planes. Their discrete spectra of modes, when set up as standing waves to satisfy boundary conditions at the conducting planes, form complete sets to represent any source-free fields of the composite structure. By matching tangential fields at the interface, we obtain equations to determine the coefficients of mode expansions as well as the characteristic equation for the phase constant of modes in the composite structure.

From a numerical evaluation of this characteristic equation, Fig.3.23 shows the effective index of refraction $N = \beta/k$ of the lowest-order mode in its two polarizations HE_{00} and EH_{00} plotted versus $kd_r = 2\pi d_r/\lambda$ for a typical set of values for the aspect ratio b/d_r, the rib to slab ratio d_r/d as well as the refractive index n_1. In this more accurate evaluation, the two polarizations differ significantly in their phase constants. The HE_{00}-mode has the larger phase constant, just as its constituent H_0-slab mode has also a larger phase constant than the E_0-slab mode, from which the EH_{00}-mode of the rib guide originates. The difference in phase constants becomes smaller and smaller as kd_r increases, and the fields of both polarizations confine themselves more and more to the rib region.

The difference in dispersion characteristics of both polarizations of the dominant rib guide mode causes the time delay to differ also, when a signal travels in both polarizations. Any signal that is band-limited to a narrow enough frequency band will be delayed per unit axial distance along the guide according to the group delay per unit axial distance

$$\tau = d\beta/d\omega. \qquad (3.86)$$

Signals which are carried by optical waves have a bandwidth which is always quite small compared to the frequency of the optical carrier. The group delay serves therefore as an accurate measure for their delay time. When the signals travel in both polarizations, they suffer different group delays and spread apart. An impulse signal, for example, splits into two, which separate in time according to the difference in group delay between the two polarizations. Fig.3.24 shows this group delay difference for the same example of double-rib guide for which Fig.3.23 gives the corresponding dispersion characteristics. The group delay difference in Fig.3.24 was obtained by determining the derivatives dN/dk of the curves in Fig.3.23, evaluating the group delay for each polarization according to

$$\frac{d\beta}{d\omega} = [N + k\, dN/dk]/c \qquad (3.87)$$

and taking their difference. The broken line in Fig.3.24 represents the third-order hyperbola

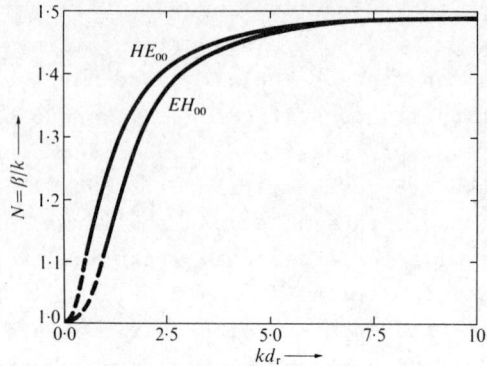

Fig.3.23. Fundamental mode phase constants in a single-material double rib guide of rib to slab ratio $d_r/d = 1\cdot 25$ and dimensional ratio $b/d_r = 1$ with $n_1 = 1\cdot 5$ (Rütze 1977).

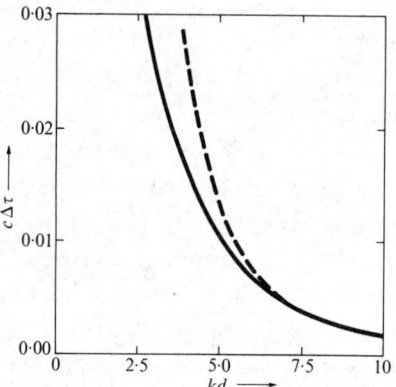

Fig.3.24. Group delay difference per unit axial distance of the two fundamental mode polarizations in the single-material double rib guide of Fig.3.23 (Rütze 1977). Solid line from an accurate numerical solution; broken line according to equation (3.88).

$$c\Delta\tau = K(k\, d_r)^{-3} \qquad (3.88)$$

with its coefficient K chosen so that this hyperbola matches the group delay difference at large values of kd_r. According to this approximation the group delay difference diminishes quite rapidly when the rib grows higher. The single-material rib guide can therefore have very little delay difference between its two fundamental mode polarizations if only its rib is high enough. Such a high rib must, of course, also be narrow enough and supported by a thick enough slab in order to maintain the single-mode character of the waveguide.

The conditions (3.80) and (3.85) represent only approximations for the single-mode range of a single-material rib guide. They are based on slab-mode approximations which hold only far enough from cut-off for the constituent slab modes. The present numerical solution for the single-material double-rib guide allows a more accurate evaluation of the single-mode limits for this particular single-material guide. Fig.3.25 compares the single-mode limits of the symmetrical

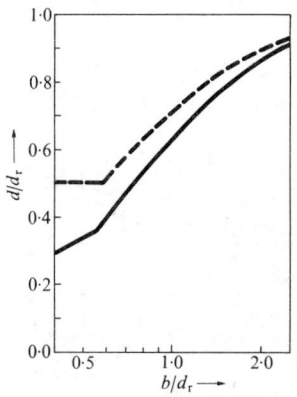

Fig.3.25. Single-mode limits of single-material rib guides for $n_1 = 1\cdot 5$ and $kd_r = 20$ (Rütze 1977). Solid line from an accurate numerical solution for the symmetrical, double-rib guide; broken line from the approximate conditions for the asymmetrical, single-rib guide.

double-rib guide from the accurate numerical solution with the single-mode limits from the approximate conditions (3.80) and (3.85) for the asymmetrical rib guide with equal sizes of rib and supporting slab. Below a certain dimensional ratio b/d_r the approximate conditions require the supporting slab to only have a thickness $d > d_r/2$. The numerical solution yields a more relaxed single-mode condition, particularly for small dimensional ratios b/d_r. The smaller this dimensional ratio, the more the thickness ratio d/d_r may drop below 1/2 and still maintain the single-mode character of the single-material rib guide.

3.5. BULGE GUIDES

The rib waveguide on a substrate in Fig.3.18, as well as the single-material rib guide with a supporting slab in Fig.3.20 have an abrupt step in their film or slab thickness at the junction between the rib region and the film or supporting slab. Such abrupt transitions are difficult to manufacture In actual waveguides, these transitions will be more or less gradual, and the structures in Figs. 3.18 and 3.20 would then only be mathematical models for the actual waveguides, which have gradual transverse changes in film or slab thickness in the rib region.

A more realistic model for these structures is shown in Fig.3.26. It takes account of these gradual transitions

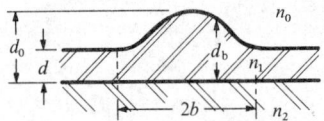

Fig.3.26. Bulge guide on a substrate.

and, instead of a rectangular rib, has a longitudinal centre region of the film or slab that bulges gradually to a larger film or slab thickness. The maximum bulge thickness in Fig. 3.26 is d_0, $d_b(y)$ designates the varying film thickness in the bulge region which tapers out to the uniform thickness d of the film at both sides of the bulge. The total width of the bulge is $2b$. Fields and energy of waves that are guided by such a structure are confined at or near the bulge. Hence, the structure may be called a bulge guide. The gradual transverse transition in film or slab thickness in rib or bulge guides does not only occur due to manufacturing deficiencies.

Such gradual transitions are also introduced intentionally because they may offer certain advantages. As one desirable characteristic, the scattering at rough edges of a rectangular rib guide is reduced substantially when, instead of having an edge with its inherent roughness, the rib or bulge thickness changes gradually in the transverse direction. Any longitudinal imperfections remaining in an otherwise smooth bulge will not cause nearly as much scattering as the same imperfections in a rectangular rib. Such smooth bulges may also be much easier to fabricate and, when tailored to specific bulge forms, they may offer certain desirable dispersion characteristics for their guided modes.

To analyse wave propagation in the bulge guide, we start, as in case of the rib and strip guides, with film or slab modes in the bulging region of the film or slab. We assume the local bulge thickness to change so gradually in the transverse direction that this slight change in thickness causes only little modification of the local film modes.

We represent this local film mode by the ray in Fig. 3.27 and let this ray run through the longitudinal centre

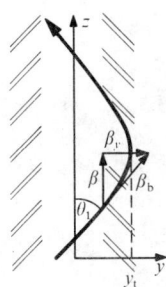

Fig.3.27. Undulating path of film-mode ray in a bulge guide.

section of the bulge at an angle θ_1 with respect to its axis. This ray then penetrates into bulge regions with decreasing bulge thickness; here it propagates with a decreasing phase constant β_b or, equivalently, decreasing effective index of refraction $N_b = \beta_b/k$, according to the decreasing bulge thickness. The change in effective index has the same effect on the ray of the film mode as the index change in a graded index medium has on a locally plane wave. According to the ray equation (2.243), it causes the ray to curve in the direction of increasing effective index N_b. As a result, the ray bends towards the axis and turns back to it, as long as a turning point $y = y_t$ occurs within the bulge. In this case, it runs along an undulating path and can form a guided mode of the bulge.

As an additional requirement for the local film mode to actually form a guided mode of the bulge, it must satisfy a phase condition for self-consistent fields in the bulge. For the analogous case of the graded-index slab, the phase condition was formulated in equation (2.246) with the phase integral in the transverse direction. The same procedure may be adopted here.

$$\beta_y = (\beta_b^2 - \beta^2)^{\frac{1}{2}} \qquad (3.89)$$

is the transverse phase constant, where β designates the phase constant of the bulge-guide mode, which is yet to be determined. The phase constant β_b of the constituent film or slab mode depends on the transverse coordinate y and follows from

$$\beta_b^2 = n_1^2 k^2 - u_b^2/d_b^2(y) \qquad (3.90)$$

with u_b as the transverse phase parameter of the constituent film or slab mode according to equation (2.13).

To formulate the characteristic equation for bulge-guide modes from a phase condition such as equation (2.246), we replace k_x in equation (2.246) by β_y, integrate over y as the relevant transverse coordinate instead of over x, and replace m by the integer l, the transverse order in the y-direction of the particular HE_{ml}- or EH_{ml}-mode of the bulge. We thus obtain

$$\int_{-y_t}^{y_t} (n_1^2 k^2 - \beta^2 - u_b^2/d_b^2)^{\frac{1}{2}} dy = (l + \frac{1}{2})\pi . \qquad (3.91)$$

The limits of integration $\pm y_t$ in the phase integral of this characteristic equation represent the turning points of the undulating ray path. At these points, we have $\beta_y = 0$, so that they follow from

$$u_b^2/d_b^2(y_t) = n_1^2 k^2 - \beta^2. \tag{3.92}$$

The phase integral in the characteristic equation (3.91) accounts only approximately for the phase shift in total internal reflection at the turning points $\pm y_t$. It therefore gives accurate results for the dispersion characteristic of a particular guided mode only sufficiently far above its cut-off point. It is hence well suited to analysing bulge guides, which have many guided modes, most of which are high above cut-off. For a bulge guide with only few, and possibly only the fundamental guided mode, another method of analysis must be used. Also another method of analysis is required to find the transverse field distribution of guided modes.

For a complete field analysis, we would actually need to start from Maxwell's equations and solve them for guided modes under the boundary conditions of the bulging film or slab. The concept of effective refractive index of film modes, and the analogy between graded index films and bulge guides to which this concept leads, allows us to formulate directly, though only approximately, a transverse wave equation for modal fields from a particular film wave.

In the case of graded-index films, equations (2.253) and (2.255) represent the wave equations for the transverse field distribution of H- and E-waves respectively. When, in these equations, we replace the refractive index $n(x)$ by the effective index

$$N_b = [n_1^2 - u_b^2/(k^2 d_b^2)]^{\frac{1}{2}} \tag{3.93}$$

of a particular film mode in the bulge region, they hold for the transverse distribution in the y-direction of Fig.3.27 for either the H_x-component of HE-modes or the E_x-component of EH-modes. Denoting these x-directed field components by F_x, we obtain the following differential equation for their transverse distribution:

$$\frac{\partial^2 F_x}{\partial y^2} + [N_b^2(y)k^2 - \beta^2]F_x = 0. \qquad (3.94)$$

β is the longitudinal phase constant and, for any guided mode solution of equation (3.94), will follow as its eigenvalue. The effective index N_b depends on the transverse coordinate y, not only because of the varying bulge thickness $d_b(y)$, but also because u_b, the transverse phase parameter of the constituent film mode, changes with d_b, and hence with y.

For film and slab guides the guide parameter $V = kd(n_1^2-n_2^2)^{\frac{1}{2}}$ appears as a useful quantity to uniquely represent propagation characteristics, particularly for weakly guiding structures. It also counts the number of guided modes in the film or slab. The corresponding parameter for rib guides is given by equation (3.67) and, in its approximate form (3.77), it is independent of wavelength.

By taking the effective index

$$N_0 = [n_1^2 - u_0^2/(k^2 d_0^2)]^{\frac{1}{2}} \qquad (3.95)$$

of the constituent film mode in the centre of the bulge and its index

$$N = [n_1^2 - u^2/(k^2 d^2)]^{\frac{1}{2}} \qquad (3.96)$$

in the supporting slab or uniform film outside the bulge, a bulge parameter may be defined along the same line as equation (3.67):

$$V_b = kb(N_0^2-N^2)^{\frac{1}{2}}. \qquad (3.97)$$

In terms of the transverse phase parameters u_0 and u of the constituent film mode in the bulge centre, and outside the bulge, respectively, this bulge parameter appears as

$$V_b = (b/d_0)(u^2(d_0/d)^2 - u_0^2)^{\frac{1}{2}}. \qquad (3.98)$$

All formulae which we have obtained for the bulge guide so far apply to the bulge on a substrate as well as to the single-material bulge with $n_2 = n_0$, which the adjoining slabs on both sides support in free space. As the only prerequisite for these formulae, the bulge must change its thickness gradually enough in the transverse direction so that the local fields still derive from only the one respective local film or slab mode in the bulge region and beyond.

In the case of the single-material bulge guide with $n_2 = n_0$, and n_1 substantially larger than n_0, we may again utilize the same approximations as for the single-material rib guide. With the slab parameter $V = kd_b(n_1^2 - n_2^2)^{\frac{1}{2}}$ much larger than unity, at least the lower-order slab modes are far enough above cut-off for their transverse phase parameter to be approximated by

$$u_b \approx (m+1)\pi . \tag{3.99}$$

The effective index and bulge parameter are than likewise approximated by

$$N_b = [n_1^2 - (m+1)^2 \pi^2/(k^2 d_b^2)]^{\frac{1}{2}} \tag{3.100}$$

$$V_b = (m+1)\pi (b/d_0)[(d_0/d)^2 - 1]^{\frac{1}{2}} , \tag{3.101}$$

and N_b depends on y only through the varying bulge thickness $d_b(y)$.

To further evaluate either the phase integral in equation (3.91), or solve the wave equation (3.94), we need to specify the shape of the bulge. In order to utilize results from the analysis of graded-index slabs, we let

$$d_b = d_0/(1 + \delta(y/b)^2)^{\frac{1}{2}} \quad \text{for} \quad |y| \leq b$$
$$d_b = d = d_0/(1+\delta)^{\frac{1}{2}} \quad \text{for} \quad |y| \geq b. \tag{3.102}$$

In this form of 'parabolic' bulge, the parameter δ measures the relative thickness of the bulge and, with the single-material or thick-film approximation (3.99), the bulge parameter from equation (3.101) is given by

$$V_b = (m+1)\pi(b/d_0)\,\delta^{\frac{1}{2}} . \tag{3.103}$$

Also in this approximation, the effective index according to

$$N_b^2 = n_1^2 - (m+1)^2(\pi k/d_0)^2\,[1 + \delta(y/b)^2] \tag{3.104}$$

corresponds to the truncated parabolic profile of equation (2.248), which further justifies the designation parabolic bulge. Continuing in this approximation, we transform from the transverse coordinate y to

$$\eta = V_b^{\frac{1}{2}}(y/b) \tag{3.105}$$

and obtain instead of equation (3.94), the transverse wave equations for modes with $m = 0$

$$\frac{d^2 F_x}{d\eta^2} + \left\{ \begin{array}{c} \chi - \eta^2 \\ \chi - V_b \end{array} \right\} F_x = 0 \quad \text{for} \quad \left\{ \begin{array}{c} \eta^2 \leq V_b \\ \eta^2 \geq V_b \end{array} \right. \tag{3.106}$$

where χ relates to the longitudinal phase constant β according to

$$\chi = \frac{b^2}{V_b}[n_1^2 k^2 - \beta^2 - (\pi/d_0)^2] . \tag{3.107}$$

The first of equation (3.106) corresponds to equation (2.259), the differential equation of the parabolic cylinder. Its even- and odd-mode solutions may be written in terms of hypergeometric functions just as in equations (2.260) and (2.269).

The second of equations (3.106) for the slab outside the bulge is solved by $\exp\{-|\eta|(V_b-\chi)^{\frac{1}{2}}\}$. This solution describes evanescent fields which decay exponentially away from the bulge. At the junction $\eta^2 = V_b$ between bulge and slab, the field F_x and its derivative dF_x/dx from the solutions in both regions must be matched. These conditions yield the characteristic equation. The values of χ which solve this equation represent eigenvalues of the problem and correspond to guided modes of the bulge. Their phase constants β may then be determined from equation (3.107).

From a numerical evaluation, based on the lowest-order slab modes with $m = 0$, Fig.3.28 shows the eigenvalues for the guided modes of zero and first order in the y-direction. With $m = 0$, the zero order represents the fundamental mode of the bulge in its two polarizations HE_{00}- and EH_{00}, while

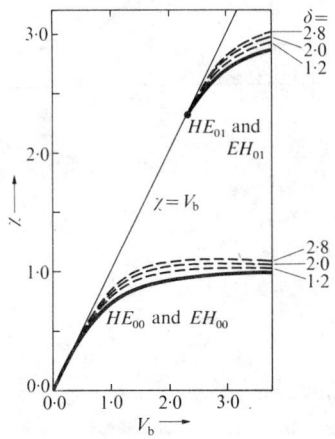

Fig.3.28. Eigenvalue χ of low-order modes in a parabolic bulge (Petermann 1976). Solid line, zero-order approximation; broken line first-order approximation.

the first order in y represents the HE_{01}- and EH_{01}-modes.

In the present approximation for very slight bulges, the results in Fig.3.28 are accurate only for small values of the relative bulge thickness δ. For more pronounced bulges the local fields in the bulge region lose their slab-mode character. Instead of the uniform polarization of the single slab-mode approximation in Fig.3.29a, they will then have the more complex field configuration of Fig.3.29b, which needs for its accurate representation not only one, but many slab modes. The change in slab thickness in the bulge region couples local slab modes with each other so that, instead of just one transverse wave equation (3.94), the problem is more accurately described by a set of coupled wave equations. This system of coupled wave equations has been derived by Petermann (1976) and evaluated with the thick-slab

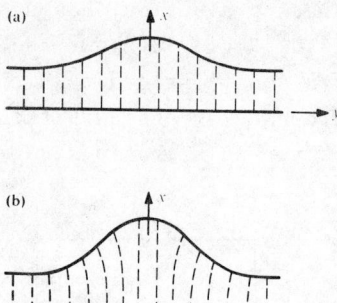

Fig.3.29. Transverse fields of bulge modes: a) approximation with uniformly polarized fields in a slight bulge; b) curved field lines in a pronounced bulge.

approximation (3.99), including coupling between the zero- and first-order slab modes. The broken lines in Fig.3.28 show the eigenvalue χ from this more accurate solution for a number of δ-values. The zero-order approximation from the uncoupled wave equation (3.94) differs only little from the more accurate solution which includes the interaction between two local slab modes. Petermann (1976) has estimated also that eigenvalues of the exact solution differ twice as much from the zero-order solution as the first-order solution in the form of the broken lines in Fig.3.28 differs from the solid lines.

All the curves for eigenvalues χ as a function of the bulge parameter V_b in Fig.3.28 end at the straight line $\chi = V_b$. When χ approaches V_b, the fields in the slab outside the bulge, with their y-dependence according to $\exp\{-|\eta|(V_b-\chi)^{\frac{1}{2}}\}$, extend further and further into the slabs and at $\chi = V_b$ cease to decay altogether. $\chi = V_b$ is the cutoff point, where the guided modes turn into leaky waves that leak power into modes of the supporting slab. Hence for

$$V_b < 2 \cdot 3 , \qquad (3.108)$$

the HE_{01}- and EH_{01}-modes as well as all other HE_{0l}- and EH_{0l}-modes of orders $l > 1$ are not guided by the slab. The condition (3.108) represents a necessary but not a sufficient requirement for the bulge to guide only the fundamental mode in its two polarizations. To find sufficient conditions for single-mode guidance, we need to consider also bulge modes which derive from higher-order film or slab modes. For modes of order $m = 1$, and with the thick-slab approximation, the transverse wave equation appears as

$$\frac{d^2 F_x}{dy^2} + [n_1^2 k^2 - \beta^2 - (2\pi/d_b)^2] F_x = 0. \qquad (3.109)$$

For the parabolic bulge according to equation (3.102), we obtain

$$\frac{d^2 F_x}{d\eta^2} + \left\{ \begin{array}{l} \chi - 3V_b/\delta - 4\eta^2 \\ \chi - 3V_b/\delta - 4V_b \end{array} \right\} F_x = 0 \qquad \text{for} \quad \left\{ \begin{array}{l} \eta^2 \leq V_b \\ \eta^2 \geq V_b \end{array} \right. , \qquad (3.110)$$

where η has the same meaning as in equation (3.105) and χ as in equation (3.107). By defining a modified bulge parameter

$$V' = 2V_b \qquad (3.111)$$

and a modified parameter for the eigenvalues

$$\chi' = \frac{1}{2} (\chi - 3V_b/\delta), \qquad (3.112)$$

equations (3.110) assume the same form as equations (3.106) with V_b replaced by V' and χ by χ'. The eigenvalues χ for HE_{00}- and EH_{00}-waves as well as for HE_{01}- and EH_{01}-waves, the solid lines in Fig.3.28, are therefore also modified eigenvalues χ' plotted versus the modified bulge parameter V' for HE_{10}- and EH_{10}-waves as well as for HE_{11}- and EH_{11}-waves, respectively.

However, this zero-order approximation for HE_{1l}- and EH_{1l}-waves neglects any coupling between H_1- and E_1-slab waves and the lower-order H_0- and E_0-waves. Such coupling occurs at any variation of the slab thickness in the one-sided bulge, and will transfer power from the first-order slab modes to their zero-order neighbours. If in the presence of such coupling, we have $\chi > V_b$ or, equivalently

$$\chi' > (V_b/2)(1-3/\delta), \qquad (3.113)$$

the H_0- and E_0-slab fields will no longer decay exponentially away from the bulge in the supporting slab, but will propagate, and the power coupled into these fields will leak out of the bulge into the slab. The HE_{1l}- and EH_{1l}-modes are then no longer guided modes of the bulge. The condition (3.113) therefore limits the single-mode character with respect to HE_{1l}- and EH_{1l}-modes. For these modes not to be guided by the bulge, its relative thickness must be

$$\delta < 3/(1-2\chi'/V_b) \ . \qquad (3.114)$$

In order for these modes not to be guided at large values of V_b, we must require $\delta < 3$ or, equivalently

$$d_0 < 2d \ . \qquad (3.115)$$

This is the same condition as equation (3.85) for the single-material rib guide, which was also obtained by excluding the corresponding modes to be guided by the rectangular rib.

We now have the two conditions (3.108) and (3.115) for the bulge to guide only the fundamental mode. But still, even these two conditions are not quite sufficient. To arrive at the third and last condition for single-mode guidance, we consider HE_{ml}- and EH_{ml}-modes of very high order m in the thick slab approximation. For their transverse field distribution a single-material bulge, with large index steps at its boundaries as in Fig.3.30, looks very much like an optical cavity with a plane and a curved reflector. Adding its mirror image we obtain the symmetrical optical cavity with two curved mirrors in Fig.3.30 where the image reflector is indicated by the broken line. Such an optical cavity forms the equivalent of the periodic lens guide of Figs.3.2 and 3.3. Whereas in the lens guide, beam modes propagate by periodic refocusing at each lens, the optical cavity refocuses these beam modes upon reflection at the curved mirrors, and

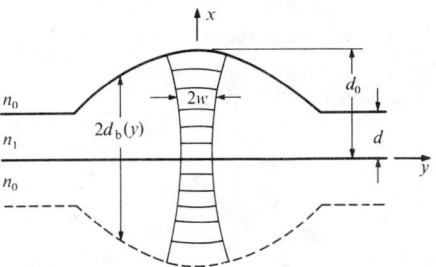

Fig.3.30. Transverse beam mode pattern of bulge mode with high transverse order m.

a beam mode travels back and forth between the mirrors. A lens of focal length f corresponds in focusing power to a reflector of radius of curvature

$$R = 2f \ . \tag{3.116}$$

If only these beam modes fit into the cavity with any integer multiple of half their wavelength, they form modes of oscillation in it. High-order bulge modes correspond in their transverse standing wave pattern to these beam modes. Fig.3.30 illustrates one such bulge mode by showing the transverse beam or spot size w as it widens from the beam waist in the centre plane to its largest spot size at the reflector.

To have stable beam modes, the lens guide must satisfy condition (3.14). By the same token, the optical cavity can only have stable modes of oscillation in the beam of Fig. 3.30 when $d_0 < R$. To exclude such stable modes and thereby prevent any bulge modes of high transverse order m being guided, we must require

$$d_0 > R \ (y=0) \ . \tag{3.117}$$

The centre of curvature at the bulge centre must hence lie within the slab above its lower plane boundary.

There is yet another beam orientation possible for high-order bulge modes. This slanted beam is indicated in

Fig.3.31, and can occur, whenever the normal to the bulge boundary cuts the lower boundary in the bulge centre. To prevent any high-order slab modes with this beam orientation

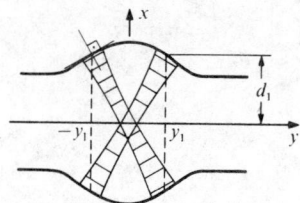

Fig.3.31. High-order bulge modes with transverse beam mode pattern of slanted orientation.

to be guided, we must require

$$(d_1^2 + y_1^2)^{\frac{1}{2}} > R \ (y=y_1) \ . \tag{3.118}$$

According to this condition the centre of curvature must again lie within the slab.

For the parabolic bulge of equation (3.102), the condition (3.117) leads to

$$V_b < \pi\delta \ , \tag{3.119}$$

while the condition (3.118) leads to

$$V_b < (2/3)\pi\delta \ . \tag{3.120}$$

Condition (3.120) is more restrictive than equation (3.119) and must therefore be satisfied in addition to equations (3.108) and (3.115) to ensure single-mode guidance of the parabolic bulge. All three conditions together form a sufficient set for the bulge to be single-mode.

If, by obeying the three single-mode conditions, the bulge guides only the fundamental mode, a signal may still travel in either one or both of its polarizations HE_{00} and EH_{00} and suffer different delays. However, all numerical results which have been obtained so far for the eigenvalues show no difference between these polarizations. In the above approximate solutions, their phase constants and dispersion characteristics appear to be identical. Any difference between them was neglected when the thick-slab approximation (3.99) was introduced, with identical transverse phase parameters u_b for the constituent slab modes in the two different polarizations. HE_{ml}- and EH_{ml}-bulge modes of equal transverse orders m and l appear degenerate both in the zero-order approximation, as represented by the solid lines in Fig.3.28, as well as in the first-order approximation shown by the broken lines of Fig.3.28. To remove this degeneracy we need to base the analysis on more accurate values for the transverse phase parameter u_b. A better approximation for u_b is afforded by equation (2.35). Although this formula, to be accurate, requires the slab modes to be still sufficiently far above cut-off, it at least accounts to first order for the difference in H_m- and E_m-slab modes of equal transverse order m. A perturbation calculation which starts from the solution of the wave equation (3.94) with u_b according to equation (3.99) and introduces the differences in u_b between equations (2.35) and (3.99) as a perturbation leads to the following expression for the relative difference in group delay between the HE_{00}- and EH_{00}-modes of the bulge (Petermann 1976)

$$\Delta\tau/\tau = 4\pi^2 \delta_3 (n_1 k d_0)^{-3} [1 - (n_0/n_1)^2]^{\frac{1}{2}} . \qquad (3.121)$$

In this expression the factor δ_3 represents a weighted average of the relative bulge thickness $d_0/d_b(y)$ to the third power, in which the zero-order solution for the transverse field distribution $F_x(y)$ is applied as the weighting function. δ_3 may also be written as

$$\delta_3 = (d_0/\bar{d}_b)^3 \qquad (3.122)$$

with an appropriate value $\bar{d}_b$ for the weighted average of the bulge thickness. Fig.3.32 plots δ_3 as a function of V_b with δ as a parameter for the parabolic bulge according to equations (3.102).

The difference in group delay between the two fundamental mode polarizations in its approximate form (3.121) depends on the bulge shape only through the factor δ_3. Single-

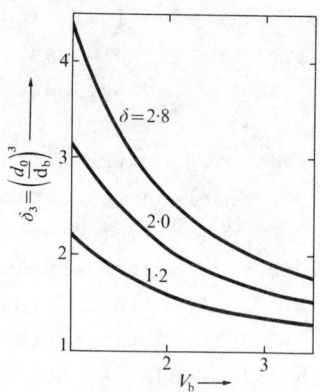

Fig.3.32. Weighted average δ_3 of the relative bulge thickness to the third power, for the parabolic bulge (Petermann 1976).

mode bulges in thick supporting slabs must be relatively slight. δ_3 is then not too much larger than unity and $\Delta\tau/\tau$ proportional to $(\lambda/d_0)^3$. This dependence on wavelength and bulge thickness was also observed in Fig.3.24 for the single-material rib guide. Hence for a single-mode bulge or rib guide we should make $d_0 \gg \lambda$ in order to have little delay difference between the two polarizations. At the same time, however, all three conditions (3.108), (3.115), and (3.120), for single-mode guidance must be obeyed.

3.6. COUPLED WAVEGUIDES AND DIRECTIONAL COUPLERS

Optical waveguides which run parallel to each other interact with one another by way of their evanescent fields. The interaction between two parallel waveguides becomes significant when the evanescent field of a mode in one of the waveguides reaches far enough into the vicinity of the other waveguide to overlap with the evanescent field of a mode in the other waveguide. The interaction leads to power transfer from the mode in the first waveguide to the overlapping mode in the other waveguide. In planar optical circuits containing waveguides which run close and parallel to each other, such power transfer may degrade the desired performance of the circuit. It then needs to be reduced to a low enough value by spacing parallel waveguides far enough apart or by otherwise reducing the interaction.

In some applications, however, waveguides in optical circuits are brought close together and parallel to each other to intentionally transfer a specific amount of power between particular modes of both guides. Such waveguide couplers are used to monitor the power in one waveguide by way of a secondary waveguide or to launch a specific mode into a waveguide from the mode of another waveguide. These represent only two examples of the numerous and varied applications which waveguide couplers find in optical circuits.

Fig.3.33. Coupled waveguides

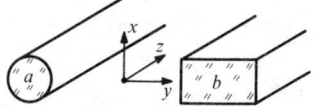

To avoid parasitic coupling between parallel waveguides or to design waveguide couplers we need to analyse the interaction between any two modes of such guides and find its dependence on waveguide and mode parameters. For this purpose we consider the two parallel waveguides a and b in Fig.3.33 with one mode a in waveguide a whose evanescent fields overlap the evanescent fields of a mode b in waveguide b. Without the other waveguide present each mode would propagate according to the differential equation

$$dA/dz = -j\beta A \qquad (3.123)$$

for its amplitude A. $\beta = \beta_a$ or β_b represents the phase constant of each mode without the other waveguide present. We let the amplitude A be normalized with respect to power, so that

$$P = AA^* \qquad (3.124)$$

represents the power in this mode. The axial dependence of A that follows from equation (3.123) is simply

$A(z) = A(0) \exp(-j\beta z)$. Interaction between the two modes of both waveguides will couple the differential equations for their mode amplitudes A_a and A_b, so that we have the following system of differential equations:

$$dA_a/dz = -j\beta_a A_a - jc\, A_b$$
$$dA_b/dz = -j\beta_b A_b - jc\, A_a \; . \tag{3.125}$$

The rate of change with z for the amplitude of the mode in one waveguide now depends in addition on the amplitude of the mode in the other waveguide. The coupling coefficient c gives the extent to which one mode determines the rate of change of the other mode with z.

Before we can determine this coupling coefficient, we need first to derive the solution of the system (3.125) of linear differential equations with constant coefficients. By trying $A_a = A_n \exp(-j\beta_n z)$ as well as $A_b = B_n \exp(-j\beta_n z)$ we postulate a wave in the system of coupled waveguides that propagates with a phase constant β_n and consists of the superposition of mode a with amplitude A_n and mode b with amplitude B_n. Substituting this trial solution into equations (3.125) transforms these equations into a homogeneous system of linear equations for the amplitudes A_n and B_n. For non-trivial solutions its coefficient determinant must vanish, which condition is the characteristic equation for the system (3.125), and has the form

$$(\beta_a - \beta_n)(\beta_b - \beta_n) = c^2 \; . \tag{3.126}$$

This quadratic equation for the phase constant β has the two solutions

$$\beta_{1,2} = \tfrac{1}{2}(\beta_a + \beta_b) \pm [\tfrac{1}{4}(\beta_a - \beta_b)^2 + c^2]^{\tfrac{1}{2}} \; . \tag{3.127}$$

In the coupling region, therefore, the two modes a and b of the uncoupled waveguides form two new modes which propagate independently from each other, one with β_1, the other with β_2. The general solution of equation (3.125) is a combination of these two modes, resulting in

$$A_a = A_1 \exp(-j\beta_1 z) + A_2 \exp(-j\beta_2 z) \qquad (3.128)$$

for the amplitude of mode a and in

$$A_b = \frac{\beta_1 - \beta_a}{c} A_1 \exp(-j\beta_1 z) + \frac{\beta_2 - \beta_a}{c} A_2 \exp(-j\beta_2 z) \qquad (3.129)$$

for the amplitude of mode b. If, at $z = 0$, the front end of the coupling region, mode a is launched with unit power but there is no power in mode b, we have

$$A_a(0) = A_1 + A_2 = 1$$
$$A_b(0) = \frac{\beta_1 - \beta_a}{c} A_1 + \frac{\beta_2 - \beta_a}{c} A_2 = 0 \; . \qquad (3.130)$$

With amplitudes A_1 and A_2 of the normal modes in the coupling region from these initial conditions, we obtain for the amplitudes in each of the waveguides modes along the coupling region

$$A_a = \tfrac{1}{2}[1+\Delta\beta/(\Delta\beta^2+c^2)^{\frac{1}{2}}]\exp(-j\beta_1 z) + \tfrac{1}{2}[1-\Delta\beta/(\Delta\beta^2+c^2)^{\frac{1}{2}}]\exp(-j\beta_2 z)$$
$$\qquad (3.131)$$
$$A_b = c/[2(\Delta\beta^2+c^2)^{\frac{1}{2}}]\exp(-j\beta_1 z) - c/[2(\Delta\beta^2+c^2)^{\frac{1}{2}}]\exp(-j\beta_2 z) ,$$

where $\Delta\beta = (\beta_a - \beta_b)/2$ represents half the difference in phase constants between modes a and b. The power P_a which the incident mode a on waveguide a carries, changes along the coupling region according to

$$P_a = \cos^2[z(c^2+\Delta\beta)^{\frac{1}{2}}] + [\Delta\beta^2/(c^2+\Delta\beta^2)]\sin^2[z(c^2+\Delta\beta^2)^{\frac{1}{2}}], (3.132)$$

while it transfers the power

$$P_b = [c^2/(c^2+\Delta\beta^2)]\sin^2[z(c^2+\Delta\beta^2)^{\frac{1}{2}}] \qquad (3.133)$$

over into mode b on waveguide b. Fig.3.34 illustrates this power transfer by plotting P_a and P_b versus the effective coupling length cz for three values of the ratio $\Delta\beta/c$ of half the phase difference to the coupling coefficient. The condition

$$c \ll \Delta\beta \tag{3.134}$$

characterizes weak coupling. Under this condition, the power of the incident mode is

$$P_a \simeq 1 - (c/\Delta\beta)^2 \sin^2(\Delta\beta z), \tag{3.135}$$

while the power coupled into mode b fluctuates according to

$$P_b \simeq (c/\Delta\beta)^2 \sin^2(\Delta\beta z). \tag{3.136}$$

At $\Delta\beta z = (p + 1/2)\pi$ with $p = 0, 1, 2...$ the incident mode has the minimum power

$$P_{a\min} \simeq 1 - (c/\Delta\beta)^2, \tag{3.137}$$

with the difference from the incident power transferred to mode b:

$$P_{b\max} \simeq (c/\Delta\beta)^2. \tag{3.138}$$

This case of weak coupling is typical for interaction between waveguides that occurs unintentionally and leads to undesired power loss and cross talk between waveguides. Equations (3.137) and (3.138) represent maximum power loss and cross talk; they can serve to specify limits on $c/\Delta\beta$ for whatever power loss or cross talk is tolerated.

Phase synchronism between both coupled modes,

$$\Delta\beta = 0, \tag{3.139}$$

leads to

$$\begin{aligned} P_a &= \cos^2 cz \\ P_b &= \sin^2 cz. \end{aligned} \tag{3.140}$$

In this case, no matter how weak the coupling, all the power incident in mode a transfers to mode b at distances

$z = (p+1/2)\pi/c$ from the front end of the coupler with $p = 0, 1, 2...$. At distances $z = p\pi/c$ the power transfers completely back to the incident mode. Waveguide couplers for complete power transfer should therefore have exact phase synchronism between both modes and be $l = \pi/(2c)$ long.

Any difference in phase constants of the coupled modes frustrates the complete power transfer. For the tight coupling condition

$$c \gg \Delta\beta \qquad (3.141)$$

the power fluctuates according to

$$P_a \simeq \cos^2(cz) + (\Delta\beta/c)^2 \sin^2(cz)$$
$$P_b \simeq [1 - (\Delta\beta/c)^2] \sin^2(cz) \qquad (3.142)$$

between both waveguides. Maximum power transfer from one waveguide to the other still occurs approximately at $z = (p+1/2)\pi/c$, but only the fraction

$$P_{b\max} \simeq 1 - (\Delta\beta/c)^2 \qquad (3.143)$$

of the total power converts maximally to waveguide b. If, in a waveguide coupler of length $l = \pi/(2c)$ for nominally complete power transfer, we want to convert at least 90% of the incident power from one waveguide to the other, there must be less than $2\Delta\beta l < 60°$ of phase difference between both modes over the full length of the coupler.

In a number of applications waveguide couplers are required to transfer only a specified fraction of the total power from one waveguide to the other. Quite often this power fraction or the associated coupling loss should, in addition, be insensitive to any changes in coupling parameters. This would also facilitate the design for certain specifications. In particular, for such couplers the coupling loss should not change too much with wavelength. Such couplers will operate satisfactorily over a wide spectral range.

To meet these requirements we examine equation (3.133)

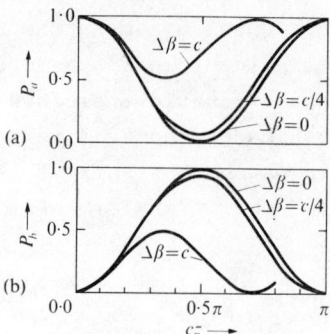

Fig.3.34. Power conversion along two uniformly coupled waveguides. c = coupling coefficient, $2\Delta\beta$ = difference in phase constants between coupled modes.

and Fig.3.34. P_b becomes insensitive to any changes of z, c, and $\Delta\beta$ if we choose $z(c^2+\Delta\beta^2)^{\frac{1}{2}} = \pi/2$ and thereby operate in the maximum of P_b. For a specified coupling loss $a_b = 1/P_b$ we should then adjust the phase difference and coupling coefficient according to

$$(\Delta\beta/c)^{\frac{1}{2}} = a_b - 1 \tag{3.144}$$

and make the coupler length

$$l = \pi/[2\,(\Delta\beta^2+c^2)^{\frac{1}{2}}]. \tag{3.145}$$

Waveguide couplers for the optical frequency range have as a desirable characteristic very pronounced directivity. We explain this directivity with reference to the waveguide coupler in Fig.3.35. A mode a, which is incident at

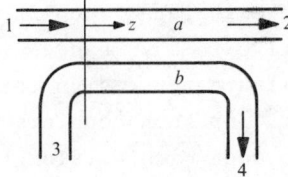

Fig.3.35. Waveguide directional coupler.

port 1, couples mainly to a mode b, which propagates in the other waveguide in the same direction as mode a. For the coupling to a mode b travelling in opposite direction, we have $\beta_b^{(-)} = -\beta_b$ and a rather larger difference $2\Delta\beta = \beta_a + \beta_b$ in phase constants. The power transfer due to this coupling obeys the weak coupling approximation (3.134), and, according to equation (3.138), only very little power transfers to the backward travelling mode. We define the ratio of power that converts to the forward travelling mode, and appears at port 4, to that of the backward mode which appears at port 3, as the directivity

$$D = P_4/P_3 = P_b/P_b^{(-)} \ . \tag{3.146}$$

In the case of equal coupling coefficients for forward and backward coupling, and for weak coupling, but maximum power transfer according to equation (3.138), the directivity is always larger than

$$D = (\beta_a+\beta_b)^2/(\beta_a-\beta_b)^2 \ . \tag{3.147}$$

Coupled modes of optical waveguides in waveguide couplers have phase constants β which differ little from each other but are much larger than the coupling coefficient c. Such couplers possess a high directivity and deserve the designation directional coupler. In the directional coupler of Fig.3.35 all power that enters port 1 splits between ports 2 and 4, and no power appears at port 3. Such a directional coupler serves well for monitoring the power that travels in each direction of the primary waveguide. With the ratio of output power at 3 and 4 it measures the reflection in the termination of port 2.

Besides the difference in phase constants $2\Delta\beta$, the all-determining factor in waveguide coupling is the coupling coefficient c. To obtain an adequate approximation for this coupling coefficient (Arnaud 1974) we consider the section Δz of coupled waveguides in Fig.3.36 and in particular the circular cylinder volume which both waveguides intersect. We let $\vec{E}_a$ and $\vec{H}_a$ be the electric and magnetic fields of the

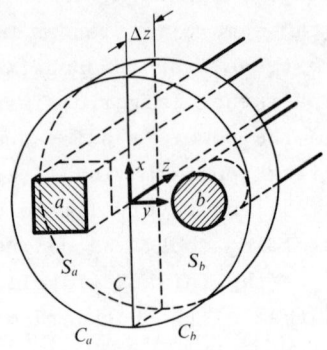

Fig.3.36. Section of coupled waveguides to which the Lorentz reciprocity theorem is applied in order to determine the coupling coefficient.

mode a on waveguide a with the waveguide b absent. These fields depend on z according to $\exp(-j\beta_a z)$. Furthermore we let $\vec{E}_1^{(-)}$ and $\vec{H}_1^{(-)}$ be the fields of one of the two normal modes of the coupled waveguides travelling in the opposite, i.e. the negative z-direction. These fields depend on z according to $\exp(j\beta_1 z)$. For weak coupling $\vec{E}_a$ and $\vec{H}_a$ solve Maxwell's equations in the vicinity of waveguide a, even in the presence of waveguide b, at least approximately. $\vec{E}_a$ and $\vec{H}_a$ on one hand and $\vec{E}_1^{(-)}$ and $\vec{H}_1^{(-)}$ on the other then constitute different solutions of Maxwell's equations near waveguide a for which the Lorentz reciprocity theorem (Harrington 1961) holds:

$$\nabla(\vec{E}_a \times \vec{H}_1^{(-)} - \vec{E}_1^{(-)} \times \vec{H}_a) = 0. \qquad (3.148)$$

We integrate this equation over the one half cylinder section in Fig.3.36, which waveguide a intersects, and let $\Delta z \to 0$; this leads us to

$$j(\beta_a - \beta_1) \int_{S_a} (\vec{E}_a \times \vec{H}_1^{(-)} - \vec{E}_1^{(-)} \times \vec{H}_a) \overrightarrow{dS}_a$$
$$= \int_{C_a} (\vec{E}_a \times \vec{H}_1^{(-)} - \vec{E}_1^{(-)} \times \vec{H}_a) \overrightarrow{dC}_a , \qquad (3.149)$$

where $\overrightarrow{dS}_a = \vec{u}_z \, dS_a$ denotes a vector of magnitude dS_a pointing

PLANAR GUIDES WITH TRANSVERSE CONFINEMENT   261

normal to the cross-section S_a in the positive z-direction, while $\vec{dc}_a$ is a vector perpendicular to the contour C_a, pointing outward. The same considerations applied to a mode b on waveguide b, and integration of the Lorentz reciprocity theorem over the other half of the cylinder section in Fig.3.36 leads to

$$j(\beta_b - \beta_1) \int_{S_b} (\vec{E}_b \times \vec{H}_1^{(-)} - \vec{E}_1^{(-)} \times \vec{H}_b) \vec{ds}_b$$
$$= \int_{C_b} (\vec{E}_b \times \vec{H}_1^{(-)} - \vec{E}_1^{(-)} \times \vec{H}_b) \vec{dc}_b . \quad (3.150)$$

The product of (3.149) and (3.150) provides us with $(\beta_a - \beta_1)(\beta_b - \beta_1)$ on its left-hand side. This quantity, according to the characteristic equation (3.126), equals the square of the coupling coefficient c. To determine c we need to evaluate the surface and line integrals of equations (3.149) and (3.150). For weak coupling the fields $\vec{E}_1^{(-)}$ and $\vec{H}_1^{(-)}$ of a normal mode of the coupling region may, at $z = 0$, be written as a linear combination of the fields of both modes of the separate waveguides

$$\vec{E}_1^{(-)} = \vec{E}_a^{(-)} + b\,\vec{E}_b^{(-)}$$
$$\vec{H}_1^{(-)} = \vec{H}_a^{(-)} + b\,\vec{H}_b^{(-)} . \quad (3.151)$$

This representation involves the same error as we allowed when we postulated the Lorentz reciprocity theorem over each half of the cross-section of the coupled waveguide with the respective modal fields. The factor b in the linear combination (3.151) need not be determined for our present purposes. When we substitute from equations (3.151) into the surface integrals of equations (3.149) and (3.150) the cross terms can be neglected, because $\vec{E}_b$ and $\vec{H}_b$ are small where $\vec{E}_a$ and $\vec{H}_a$ have significant values, and vice versa. Hence these surface integrals reduce to

$$P'_a = \int_{S_a} (\vec{E}_a \times \vec{H}_a^{(-)} - \vec{E}_a^{(-)} \times \vec{H}_a) \vec{ds}_a \quad (3.152)$$

in equation (3.149) and to

$$P_b' = b \int_{S_b} (\vec{E_b} \times \vec{H}_b^{(-)} - \vec{E}_b^{(-)} \times \vec{H_b}) \, d\vec{s}_b \qquad (3.153)$$

in equation (3.150). We simplify these integrals even further by noting that, if the fields of a forward travelling mode at $z = 0$ are written in terms of its transverse vectors and longitudinal components according to

$$\vec{E} = \vec{E_t} + \vec{u_z} E_z$$
$$\vec{H} = \vec{H_t} + \vec{u_z} H_z , \qquad (3.154)$$

the same mode, when travelling in the opposite direction, has

$$\vec{E}^{(-)} = \vec{E_t} - \vec{u_z} E_z$$
$$\vec{H}^{(-)} = -\vec{H_t} + \vec{u_z} H_z . \qquad (3.155)$$

With these relations between forward and backward travelling modal fields, we obtain for the surface integrals

$$P_a' = 2 \int_S (\vec{E}_a \times \vec{H}_a)_z \, dS$$
$$P_b' = 2b \int_S (\vec{E_b} \times \vec{H_b})_z \, dS . \qquad (3.156)$$

We have extended each of these integrals over the full cross-section S of the coupled waveguide, because the fields of mode a are small enough over S_b not to contribute anything there, and vice versa. We furthermore extend the limits of S out to infinity. The fields in the line integrals on the right-hand sides of equations (3.149) and (3.150) will then contribute to these integrals only on the dividing line C which is common to both contours. We evaluate these line integrals by substuting for $\vec{E}_1^{(-)}$ and $\vec{H}_1^{(-)}$ from equation (3.151). When we apply the Lorentz reciprocity theorem to each waveguide, we note that for products of $\vec{E}_a$, $\vec{H}_a$, and $\vec{E}_a^{(-)}$, $\vec{H}_a^{(-)}$ as well as of $\vec{E}_b$, $\vec{H}_b$, and $\vec{E}_b^{(-)}$, $\vec{H}_b^{(-)}$, the line integrals vanish because with $\beta_1 = \beta_a$ and, respectively, $\beta_1 = \beta_b$ in these cases, the left-hand sides of equations (3.149) and (3.150) are zero. Hence, only the cross terms contribute to these line integrals, and we obtain for the right-hand side of equation (3.149)

$$c_a = b \int_C (E_{bx}H_{az} + E_{bz}H_{ax} - E_{ax}H_{bz} - E_{az}H_{bx})\,dC \qquad (3.157)$$

as well as

$$c_b = \int_C (E_{bx}H_{az} + E_{bz}H_{ax} - E_{ax}H_{bz} - E_{az}H_{bx})\,dC \qquad (3.158)$$

for the right-hand side of equation (3.150).

If we now solve the characteristic equation (3.126) for the coupling coefficient

$$c = [(\beta_a - \beta_1)(\beta_b - \beta_1)]^{\frac{1}{2}}$$

and substitute for the product $(\beta_a - \beta_1)(\beta_b - \beta_1)$ of differences in phase constants from the product of equations (3.149) and (3.150), we obtain

$$c = c_b / [4j(P_a P_b)^{\frac{1}{2}}], \qquad (3.159)$$

where P_a and P_b follow from

$$P_{a,b} = \int_S (\vec{E}_{a,b} \times \vec{H}_{a,b})_z \, dS \qquad (3.160)$$

and for all practical purposes in optical waveguides represent the power flow in each mode of the two separate waveguides. Once the field distributions $\vec{E}_a$, $\vec{H}_a$ as well as $\vec{E}_b$, $\vec{H}_b$ of each of the two modes in the two separate waveguides a and b are known, c_b may easily be evaluated from the integral along the intersection C between the two waveguides. If, for the evaluation of the line integral (3.158), we choose modal fields that are normalized with respect to power flow, then we have $P_a = P_b = 1$ and

$$c = c_b / (4j). \qquad (3.161)$$

The coupling coefficient depends critically on the relative strength of the evanescent fields on the line C, which divides the cross-section between both waveguides. To obtain sufficient coupling for the purposes of partial or even complete power transfer, we need a certain magnitude of these evanescent fields. Optical waveguides in waveguide couplers must therefore be placed at a relatively close distance.

As a representative example of an optical waveguide coupler we consider the strip-guide coupler in Fig.3.37a. EH_{ml}-modes in each of the strips originate from E_m-film modes in the strip region, which experience total reflection at the sidewalls of the strip, and satisfy a phase condition on their

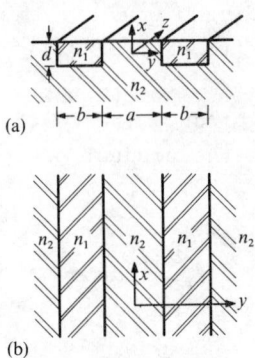

Fig.3.37. Coupled strips and equivalent slab coupling: (a) Coupling of embedded strip guides; (b) coupled symmetrical slabs.

zigzag path, to repeat in phase after two successive reflections. This phase condition leads to their characteristic equations (2.45) and (2.46). With respect to these sidewall reflections their fields as well as their transverse propagation constants correspond to those of H_l-waves in a symmetric slab of thickness b and index n_1, embedded in a medium of index n_2. We will therefore first evaluate the coupling between H_l-waves in two symmetric slabs, which, according to Fig.3.37b, are spaced by a distance a. Their field components follow from equations (2.80) and (2.81) as well as from equation (2.60), with the appropriate transformation of coordinates. Equation (2.90) gives the power which such an H_l-mode guides in the z-direction per unit width in the x-

direction of Fig.3.37b. If we take these field components at a distance $a/2$ from the slab walls and substitute them into equation (3.158) and if P from equation (2.90) is substituted for P_a and P_b, we obtain from equation (3.159) one and the same expression for the coupling coefficient between even and odd order H_l-modes of the same order l

$$c = \frac{2vu^2 \exp(-va/b)}{\beta b^2 (u^2+v^2)(1+2v)} \ . \tag{3.162}$$

u and v represent the transverse phase and attenuation parameters respectively, for the particular H_l-mode in the slab of width b, and β its longitudinal phase constant.

We return now to EH_{ml}-modes of the coupled strips in Fig.3.37a, which correspond in their y-dependence to the above H_l-modes of the slab. We replace the transverse attenuation constant v/b of the slab mode by the corresponding attenuation constant

$$\alpha_y = [k^2(n_1^2-n_2^2) - \beta_y^2]^{1/2} \tag{3.163}$$

of the strip mode, and the transverse phase constant u/b of the slab mode by the β_y of the strip mode, as it also appears in equation (3.18). With these substitutions and with β as the longitudinal phase constant of the strip mode, the coupling coefficient between two EH_{ml}-modes of equal orders m and l is found from

$$c = \frac{2\alpha_y \beta_y^2 \exp(-\alpha_y a)}{\beta b (\alpha_y^2 + \beta_y^2)(1+2\alpha_y b)} \ . \tag{3.164}$$

This formula as well as the expression (3.162) for slab-mode coupling show clearly the exponential decay of c with α_y and a, as it is due to the evanescent fields which likewise decay exponentially with α_y and a. The coupling coefficient according to equation (3.164) shows no explicit dependence on the thickness d of the strips. It changes with d, however, because the longitudinal phase constant β of the particular strip mode depends on d. Fig.3.38 shows the coupling coefficient according to equation (3.164) for interaction between

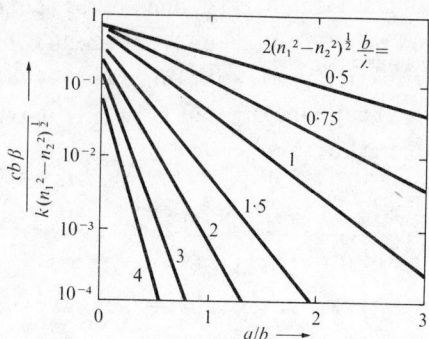

Fig.3.38. Coupling between EH_{m0}-modes of equal order m in embedded strips with $(n_1^2 - n_2^2) \ll 1$.

EH_{m0}-modes of equal order m. By plotting $cb\beta/[k(n_1^2-n_2^2)^{\frac{1}{2}}]$ versus a/b the curves hold for guided EH_{m0}-modes of any order m. Any dependence of c on m in this equivalent-slab approximation appears only in the phase constant β. Coupling between these strip modes decreases drastically not only when the spacing between both strips increases, but also when the strip parameter $V_b = kb(n_1^2-n_2^2)^{\frac{1}{2}}$ grows. With growing V_b, any particular mode moves further away from cut-off, so that with increasing α_y its evanescent fields decay more rapidly in the y-direction.

As a representative example we let $n_1 = 1 \cdot 5$, $n_1/n_2 = 1 \cdot 01$, and $b = 2d$. At $V_b = kb(n_1^2-n_2^2)^{\frac{1}{2}} = 3\pi/4$ each of the strips guides only the fundamental mode in its two polarizations HE_{00} and EH_{00}. $V_b = 3\pi/4$ and $b = 2d$ require the strips to be $d = 1 \cdot 77 \lambda$ thick and $b = 3 \cdot 54 \lambda$ wide. We let these strips be spaced by half the strip width $a = b/2$. Under these circumstances the EH_{00}-coupling amounts to $c = 0.002/\lambda$. To completely transfer the power from one strip to the other, such a strip-guide coupler must be $l_c = 770 \lambda$ long.

Waveguide couplers for a specified low coupling loss require near phase synchronism of coupled modes. For zero coupling loss and complete power transfer they must have exact phase synchronism. Also the waveguides must be spaced at a close distance, and this spacing maintained constant over the full length of the coupler, to ensure sufficiently strong and uniform coupling. All these requirements impose small dimensions and strict tolerances on waveguide couplers. Raised strip couplers are very critical in these respects. Their evanescent fields decay rapidly beyond the sidewalls of the strips, and even for only moderate coupling, they must be extremely close together. Embedded strips have evanescent fields which extend much further beyond their sidewalls; the same coupling allows much more spacing between both strips. The above example has shown, however, that even for embedded strips, the spacing should not be any larger than the other strip dimensions.

Strip-loaded guides as well as rib and bulge guides can be designed to have evanescent fields of the their fundamental modes which extend even further beyond their edges. They allow even more spacing between guides in couplers. Relative tolerances on such wider spacings are easier to maintain for the same uniformity of coupling.

3.7. PLANAR WAVEGUIDE TECHNOLOGY

Planar waveguides such as films and strip guides, or any of the waveguides that derive from the basic strip structure, serve in planar optical circuits to form components or connect circuit components with each other. The choice of materials for these waveguides and their fabrication depend on the kind of component and circuit for which they are used. We distinguish broadly between passive components and passive circuits on one hand, and active components, or circuits in which active components are to be integrated on the other. Waveguide materials for passive components and circuits need to have only suitable values of their refractive index for guidance by total reflection and must be low enough in absorption and scattering to allow the fabrication of low-loss components and circuits. Active components consist at least partially of laser-active materials such as semiconductors with direct and hence radiative recombination, or of yttrium-aluminium garnets (YAG) or ultraphosphates with neodymium as a laser-active additive. Electro-optic materials, such as lithium niobate for optical modulators and magnetooptic materials, or photoelectric materials, such as semiconductors for photodetectors, are also included among the optically active materials. Usually these active components depend for their operation on the monocrystalline structure of the optically active material.

While planar waveguides for passive components and circuits are mostly made from amorphous or polycrystalline materials, the monocrystalline regions of active components require the whole waveguide structure to be fabricated as a single crystal. For monolithic integration of such active components into a planar circuit, all its components and connecting waveguides must be built up on the same single-crystal substrate, or built into it. Active components for different functions require different active materials to work efficiently; monolithic integration is therefore hardly feasible for them. Instead we obtain more efficient optical devices when the best materials for the respective functions are combined. This leads to the integration of active and passive components into hybrid planar circuits. Since usually only such hybrid integration results in optical circuits with satisfactory performance, we will discuss separately first the fabrication of passive waveguides and planar passive circuits (Zernike 1975), and then the fabrication of single-crystal components with active waveguides and their possible monolithic integration (Garmire 1975).

In general, for passive planar waveguides we need transparent films of different refractive indices on substrates, or transparent substrates into which layers of higher indices are embedded or buried. For transverse confinement, these films or layers must be patterned into strips, or into the more complex configurations of various planar components. We treat first the formation of films or guiding layers and subsequently the pattern fabrication.

The substrate for planar passive circuits must be plane and smooth and have a surface layer, which is free of any defects as far into the substrate as the evanescent fields of guided modes will reach. If mechanical and chemical polishing or a combination of both fail to produce such a defect-free surface layer, then a layer of lower index may first be deposited on the substrate before the actual guide is grown. Before any film is deposited, the substrate surface must be cleaned thoroughly. Care should be taken that no cleaning agents are used which degrade the surface quality by etching. Surface cleaning methods which were developed for semiconductors also prove quite effective for optical substrates of fused silica or other glasses.

There are different methods of depositing thin films for planar guides on substrates. The classical method of vacuum evaporation is not so well suited here, because it usually produces films with relatively high scattering loss for guided modes. Rather well proven methods for the deposition of low-loss films are sputtering, plasma polymerization, and the application of a liquid by dipping or by spinning. Which of these methods is to be used depends on the combination of materials, and the materials in turn depend on the characteristics which we require the waveguide to have.

For film deposition by sputtering, material is removed by ion bombardment from a target and collected on the substrate as a thin film. In the vacuum chamber (here called a sputtering tank), the substrate is mounted on a water-cooled substrate holder opposite and parallel to the target cathode. The tank is filled at low pressure ($2 - 20 \times 10^{-3}$ Torr) with a suitable gas (usually oxygen) and a negative r.f. voltage is applied to the cathode. This r.f. voltage generates the plasma discharge for ion bombardment of the target. For a compound target, the film on the substrate has not, in general, the same composition as the target material. The sputtering yield differs for the atoms of the compound as well as the rate at which different atoms find suitable bonding sites for deposition. The same target material can hence form films of different compositions on different substrates. Metal compounds such as oxides or nitrides can also be deposited by sputtering from a metal target with an appropriate gas for the chemical reaction. In the sputtering tank not only is the target sputtered but, depending on the charge distribution in the r.f. plasma, some sputtering also occurs on other exposed surfaces and could contaminate the film. The substrate is therefore first covered by a shutter until all these surfaces are coated with target material, and all contaminations are removed from them.

Film deposition by sputtering in general yields quite durable and low-loss films. Under stable conditions, the growth rate is constant though somewhat slow (3 nm/min); but films of well defined and uniform thickness are produced. The method is well suited for depositing low-loss glass films on glass substrates with the film index ranging between $n \simeq 1 \cdot 48 - 1 \cdot 61$, depending on the composition of the film glass. Films of higher refractive index ($n > 2$) are made by sputtering of tantalum or niobium to form Ta_2O_5- and Nb_2O_5-films, respectively. However, these oxide films have higher losses.

The method of plasma polymerization deposits films on substrates by passing an electrical discharge through a gas which contains an organic compound of low molecular weight. The compound polymerizes in the discharge and deposits as a smooth and transparent film with a thickness ranging from 20 nm to several micrometres. A large number of monomers can be used to form these films, and by mixing two monomers, the refractive index of the film can be made to have any value between 1·48 and 1·70. The film thickness grows at a constant rate of typically 100 nm/min. During the time of deposition, the mixture of monomers that is admitted into the discharge may be varied and thus the index across the film be formed to any specified profile. Organo-silicon films which are deposited by this method on glass susbtrates have less than 0·1 dB/cm loss.

A number of varnishes and epoxy resins are suitable as film guide materials and can be applied in liquid form. The substrate is covered with the liquid material and then turned upright to allow the excess material to drip off. The excess material can also be removed by spinning the substrate around an axis normal to its surface. Alternatively, the substrate is dipped into the solution and withdrawn slowly for a uniform film to remain on its surface. Depending on the material and possibly on its solvent, the film is air-dried or heated for baking or curing. Films can be produced quite quickly with this method but not controlled very accurately in their thickness. Polyurethane and epoxy films as well as lead silicate films have been deposited on glass substrates in this way with losses as low as 0·3 dB/cm for epoxy films (Ulrich and Weber 1972). Photoresist films can also be solution-deposited; any guide or circuit patterns can be directly written on them by exposing them to ultraviolet light and thereby changing their refractive index (Weber et al. 1972).

Instead of depositing films on the substrate, layers of higher refractive index can also be produced inside the substrate by chemical or physical reactions. These reactions lead to gradual index changes and hence produce graded-index layers. Ion migration or ion exchange raises the index of a glass substrate near its surface by changing its composition there. The glass substrate is heated and an electrode in a molten salt applies an electric field in which ions migrate out of the glass to be exchanged for ions from the molten salt. Thus a borosilicate substrate is immersed in molten potassium nitrate, sodium nitrate, and thallium nitrate, which contains platinum electrodes and is heated to more than 500°C (Izawe and Nakagome 1972). Na^+-ions in the glass are exchanged with Tl^+-ions, the latter forming a high-index layer near the surface. By continuing the process without thallium nitrate in the molten salts, Na^+- and K^+-ions diffuse from the melt into the substrate surface, and the Tl^+-ions migrate further into the substrate. They now form a buried layer of higher index a short distance under the substrate surface.

Surface or buried layers of higher index are also produced by proton bombardment and ion implantation. By bombarding the substrate with ions, these ions penetrate into the substrate and are implanted there with a Gaussian distribution of concentration and depth of penetration that depend on the substrate material, the type of ions, and their energy. In fused silica substrates, bombardment with lithium ions is particularly effective in raising the index by implanting Li^{7+}-ions (Standley, Gibson, and Rodgers 1972). The index increases by $n = 2·1 \times 10^{-21} D$, where D is the density of Li^{7+}-ions per cm^3. For index differences as high as $\Delta n = 0·035$, the guided mode loss is 2 dB/cm. Annealing the substrate heals some of the glass matrix defects from the bombardment and reduces the guided mode loss substantially, but it also reduces the index again, so that only a smaller index difference remains in the guiding layer. In the above example of Li^{7+}-implantation in a fused silica substrate, annealing reduces the index difference from 0·035 initially to 0·01. However, the loss decreases by more than an order of magnitude to below 0·2 db cm^{-1}.

Planar strips and other planar waveguides with continuous transverse confinement as well as passive components and circuit patterns are fabricated with methods summarized in Fig. 3.39 (Smith 1974). The substrate surface is first coated with a radiation-sensitive polymer and is then exposed to the appro-

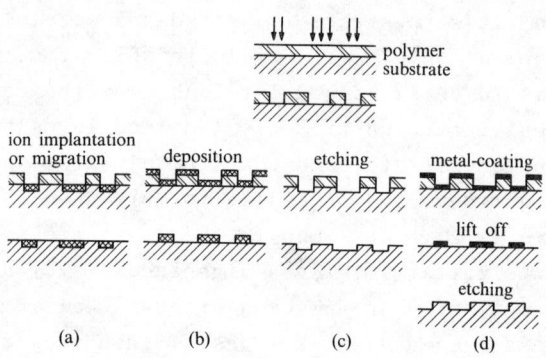

Fig.3.39. Pattern fabrication with polymer or metal masks.

priate radiation in the specified pattern. Subsequent to exposure, development of the polymer removes either its exposed or its unexposed regions and leaves a pattern of the polymer on the substrate. This pattern serves as a barrier or mask for ion implantation or migration of ions into the exposed regions of the substrate, or for depositing the waveguide material into these regions between the polymer relief pattern. It is also used as a barrier in an etching process, which removes a surface layer of the exposed substrate regions and thus transfers the polymer relief pattern on to the substrate. If in the etching process, the polymer cannot itself resist the etching, a metal coating is applied to the entire surface (Fig.3.39d), and the polymer pattern as well as the metal covering it are lifted off. This leaves only a metal pattern to serve as a barrier in an etching procedure.

In the case of suitable photo-resist polymer films that are deposited from a solution, the polymer itself increases its refractive index as a result of exposure to radiation. The guide or circuit pattern is here written directly into the polymer film, as in Fig.3.40a, and no further process steps are needed. If a negative photo-resist is used that, without exposure, is suitable as a waveguide material, then the negative pattern may be written into it, and development leads directly to the circuit pattern in Fig.3.40b.

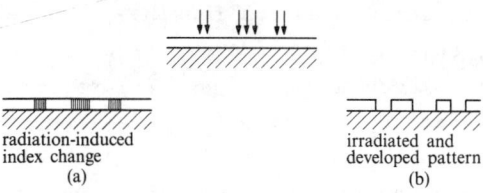

radiation-induced
index change
(a)

irradiated and
developed pattern
(b)

Fig.3.40. Pattern fabrication with radiation-sensitive waveguide material.

The first critical step in all these methods of pattern fabrication is the exposure of the pattern in the radiation sensitive polymer film. The patterns for planar optical circuits must be produced over relatively long distances to extremely tight tolerances and must have very smooth edges. Some of the subsequent process steps, as for example the lift-off, require the developed polymer pattern to have vertical sidewalls. Conventional contact printing does not fill these requirements. The photo-mask and the polymer-coated substrate are both not plane enough and the diffraction of the exposing light, which occurs in the spacing between both, limits the minimum line-width to $\gtrsim 2$ μm; it also rounds off the sidewalls of the polymer contour. Better results obtain with contact printing, when a conformable photo-mask is used. Such a conformable mask has the pattern printed in a thin film of chromium or Fe_2O_3 on a glass slide which is only 0·2 mm thick. This photo-mask is sucked by vacuum, face-down on to the polymer-coated substrate. It thus conforms intimately to the substrate surface and even deforms around any particles. Minimum line-widths of less than 0·4 µm thus become feasible and can be controlled to within 0·1 µm. The exposure and development results in vertical or even slightly undercut photo-resist profiles, which are ideal for metal lift-off and certain etching methods (Smith 1974).

Scanning electron beam lithography is another method to write patterns with high resolution and precision on a radiation sensitive polymer film. Both positive and negative resists are available for this method. It employs a scanning electron microscop whose electron beam can be focussed to less than 10 nm in diameter. Under favourable conditions, linewidths of less than 0·1 µm are feasible. The electron beam can, however, only be scanned over a few millimeters before lens aberrations become too large. For patterns which are larger than a rectangle of about 1 mm × 1 mm, the substrate must be moved from rectangle to rectangle. To maintain enough accuracy between individual rectangles, any errors in the substrate motion are detected interferometrically and corrected by adjusting the beam position. In a positive photo-resist, the scattering of the beam electrons in the resist layer, and their back scattering from the substrate write a line, which is wider at the bottom of the resist than at the top. Although this undercutting of the resist sidewalls is to the advantage of metal lift-off and certain etching methods, it nevertheless limits the minimum line-width to more than twice the extent of undercutting.

A third method of writing a pattern in the radiation-sensitive polymer film uses a focused light beam from an argon laser. In order to trace the desired pattern, the substrate is translated perpendicular to the laser beam axis. Precision and smoothness of the edges therefore depend not only on the uniformity of the resist sensitivity and its development but also on the quality of the translating carriage. The laser beam exposes a region which is wider at the top of the resist layer than at its bottom. Only a negative photo-resist will therefore develop with undercut edges after laser beam exposure. A groove in a positive resist is wider at its top than at the bottom and is therefore not suited for metal lift-off. Laser beam exposure with the UV beam of 364 nm wavelength from an argon laser is used for direct writing of circuit patterns into photosensitive waveguide materials. It works particularly well when high-resolution, so-called photo-locking materials are employed for the film. In these materials, a moderately volatile dopant of higher refractive index is dissolved in the lower-index polymer and this mixture deposited as a film. The photochemical reaction upon exposure to the laser beam reduces the dopant mobility and virtually locks it in the film. Unreacted dopant molecules are then removed by heating (Tomlinson et al. 1975). Index changes of up to 1·3 per cent and optical losses of typically 0·3 dB cm^{-1} are achieved in the photo-locked material. It readily lends itself to making short directional couplers with complete power transfer and grating couplers of high coupling strength.

When the polymer film is used as a mask or barrier in subsequent steps of pattern fabrication, the substrate is ready after exposure and development for any of the production steps of the third stage in Fig.3.39. Ion implantation or ion migration proceed for the masked pattern in the same way as they do for continuous films and as they were explained in that context. After implantation or migration and after the polymer pattern has been removed, we obtain embedded strip guides as the basic circuit elements. Deposition of waveguide material on the exposed substrate regions of the pattern and on the polymer pattern can likewise be done by any of the film-deposition methods such as sputtering, plasma polymerization, and solution deposition. After removal of the polymer pattern and of the material which has been deposited on it, we obtain raised strips as the basic element. Alternatively, with a continuous film underneath, these strips form the loading element of strip-loaded guides.

The etching process in Fig.3.39c and Fig.3.39d employs sputtering to remove substrate material from the exposed surface regions between the polymer pattern. In a sputtering tank similar to the arrangement for depositing films on substrates, the properly masked substrate now takes the place of the target directly on the cathode. To prevent the organic material of the polymer mask oxidizing and burning up during sputter etching, an argon plasma is employed. For the cathode, a metal such as titanium is preferred from which no oxygen is freed and which even acts as a getter. Redeposited titanium from such a cathode is removed by briefly sputter etching it on a glass cathode.

Sputter etching is performed with better control and more precision in an ion-beam machine. In it an ion gun produces a collimated beam of ions and directs it against the substrate which is now placed in high vacuum and kept at lower temperatures than during ordinary sputter etching.

The ion bombardment during sputter etching not only removes substrate material but also the mask. Polymer masks are removed at about the same rate as glass. Relief can be etched in the substrate with such a mask material only as deep as the mask is thick. The sidewalls of the relief will be faceted at 45° and will only be smooth if the mask has smooth sidewalls from top to bottom. Metals usually sputter more slowly than glasses. Manganese, for example, sputters at only 1/5 the rate of glass (Goell 1973); also its edges remain smooth during sputter etching and its residues can be removed after sputtering without degrading the remaining surface. Manganese is therefore well suited to producing patterns by the lift-off process in Fig.3.39d and by subsequent sputter etching. Both etching methods, with and without metal coating and lift-off, result in substrate relief patterns that with lift-off reproduce the polymer mask pattern and without lift-off represent its negative. Depending on the underlying substrate layers, with which the process starts, the basic circuit element is either a raised strip, a strip loaded guide, or a rib guide.

The methods for fabrication of planar waveguides and circuit patterns which we have so far discussed in this section use amorphous or polycrystalline materials. In general, it is difficult to perform any active functions with these materials. Thin-film dye lasers use liquid guides, and neodymium-doped glass films also show light amplification. But they must be pumped optically and have little efficiency. Modulators can likewise be made with liquid films or use acousto-optic interaction in amorphous layers; but in size, efficiency, and speed, they cannot compete with modulators in single-crystal form.

The more efficient active components for light generation, amplification, modulation, and switching as well as for detection use single-crystal structures. Semiconductors with direct and therefore predominantly radiative recombination offer the possibility of performing the three basic functions of active components - amplification, modulation, and detection - with one basic material. In principle, such a semiconductor allows monolithic integration of active components for all these functions, together with passive components on a single-crystal chip. Such a monolithic circuit will not perform as efficiently as if the best materials for the particular functions are combined in a hybrid circuit; but the advantage of monolithic integration may outweigh any reduction in efficiency that this compromise brings.

Of the direct semiconductors, gallium arsenide and its combination with aluminium arsenide in the ternary GaAlAs system perform best when all three basic functions of optically active components are considered together. In particular, the injection laser made from GaAs and GaAlAs works so well that for any monolithic integration these semiconductors are usually the first choice. We therefore take GaAs and GaAlAs as examples to discuss methods of planar waveguide fabrication in single-crystal semiconductors.

The refractive index of a semiconductor decreases when the concentration of free carriers increases. Decreasing the free-carrier concentration in layers, for example by doping, will hence raise the refractive index there and create planar waveguides. Free carriers of sufficient mobility lend to the semiconductor the character of a plasma. The electric field $\vec{E}$ accelerates a free electron in the crystal lattice according to

$$m^*(\mathrm{d}\vec{v}/\mathrm{d}t) = -q\,\vec{E}\,, \qquad (3.165)$$

where m^* designates the effective electron mass, $-q$ its charge, and $\vec{v}$ its velocity. Under steady-state conditions, these electrons oscillate according to

$$\vec{v} = \mathrm{j}(q/(\omega m^*))\vec{E}\,, \qquad (3.166)$$

so that N electrons per unit volume cause the a.c.-density

$$\vec{J} = -j(Nq^2/(\omega m^*))\vec{E} . \tag{3.167}$$

This current adds to the displacement current and results in a total current density

$$j\omega\varepsilon_s \vec{E} + \vec{J} = j\omega\varepsilon_s [1 - \omega_p^2/\omega^2]\vec{E} , \tag{3.168}$$

where

$$\omega_p = (Nq^2/(\varepsilon_s m^*))^{1/2} \tag{3.169}$$

is called the plasma frequency of the material and $\varepsilon_s = n_s^2 \varepsilon_0$ is its permittivity. Due to the plasma effect of freely moving carriers, the effective index of refraction of the semiconductor depends as

$$n = n_s [1 - (\omega_p/\omega)^2]^{\frac{1}{2}} \tag{3.170}$$

on ω_p and the angular frequency ω. The free carriers therefore lower the index by

$$\Delta n = -\frac{\mu_0 q^2 N \lambda^2}{8\pi^2 m^* n_s} . \tag{3.171}$$

For n-GaAs with $N = 10^{19}$ cm^{-3}, the index changes by $\Delta n = -0 \cdot 02$ compared to the intrinsic material. Such an index difference is entirely sufficient to form guiding layers in the semiconductor. This relatively large index difference due to free electrons in GaAs is mainly due to their small effective mass which amounts to only 8 per cent of the free electron mass. The holes in p-GaAs have a much larger effective mass and so do free carriers in other semiconductors.

If on an n-GaAs substrate with N_2 free electrons per unit volume, a film guide is formed with an electron concentration $N_1 < N_2$, then the cut-off condition (2.21) for film modes of order m requires the electron concentrations in substrate and film to differ by

$$N_2 - N_1 = \frac{(2m+1)^2 \pi^2 m^*}{4\mu_0 q^2 d^2} . \qquad (3.172)$$

In this cut-off condition, the wavelength dependence of the plasma effect in equation (3.171) cancels the λ-factor in V_m of equation (2.21); it hence becomes independent of wavelength. For $N_1 \ll N_2 = 5 \times 10^{18}$ cm^{-3} in n-GaAs, the film must be 1 μm thick to guide the fundamental film mode in its two orthogonal polarizations.

The film modes which are guided by the plasma effect of free carriers suffer loss from free carrier absorption. Under forced oscillations in the electric field, the free carriers interact with lattice defects and lattice vibrations. They transfer energy from the wave to the lattice. This absorption increases with carrier concentration and is hence much stronger in the embedding medium than in the guiding layer. For well-guided modes which confine most of their fields and energy in the guiding layer, the free carrier absorption loss remains quite low. Typically in n-GaAs with $N = 10^{18}$ cm^{-3}, free electrons absorb with 40 dB/cm at $\lambda = 1$ μm (Garmire 1975). If the substrate material has this much absorption, but the layer only has 4 dB cm^{-1}, then, according to equation (2.114) and Fig.2.16, the fundamental mode will still suffer a loss of 8 dB cm^{-1} at the cut-off of the next higher-order mode.

Guiding layers of reduced electron concentration on top or near the surface of n-type semiconductor substrates can be produced by a number of the standard methods of semiconductor device fabrication. Epitaxial growth of an intrinsic layer on an n-type substrate allows steep transitions in carrier concentrations and hence in refractive index and also very accurate control of the layer thickness. Epitaxial layers on n-GaAs for single-mode guidance in the near infrared are several microns thick and guide the fundamental mode with losses less than 4 dB cm^{-1} (Garmire 1975).

Surface layers or buried layers of low electron concentration near the surface of n-GaAs are also produced by ion implantation. The energetic ions cause lattice displacements which trap free carriers and thereby lower the free-carrier concentration. The layer thickness and its depth depend on the proton energy. In GaAs, protons penetrate nearly 1 μm per 10^5 eV. The lattice defects from proton bombardment not only trap free carriers, but also cause optical absorption. Annealing subsequent to implantation heals some of the damage but maintains sufficient index difference. With 300 keV protons, guiding layers are fabricated starting from n-GaAs substrates of 2×10^{18} cm^{-3} free-electron concentration. Annealing reduces the loss in these layers to below 10 dB cm^{-1} (Garmire *et al.* 1972).

Guiding layers on an n-GaAs substrate near its surface are also produced by diffusion of acceptor impurities into the n-type material. Such diffusion creates a p-type layer that reaches the diffusion depth from the surface into the substrate. It also causes a depletion layer between the p-layer and the n-substrate. However, this depletion layer extends only over a few tenth of a micron and is hence too thin to guide waves. The free holes in GaAs have an effective mass which is about ten times the effective electron mass. The same concentration of holes therefore lowers the index only a tenth of the index difference due to free electrons. Both the depletion region and the p-type surface layer together form the film guide. The diffused pn-junctions of homojunction lasers represent an example of such diffused waveguides. They improve the laser performance by guiding and confining the generated light.

The maximum index difference that can be obtained by a change in free-carrier concentration is limited directly by the maximum concentration that free carriers can have in a semiconductor. At very high electron concentrations in n-GaAs, the material becomes a conductor, and acts not so much as a plasma for the light to be guided but more and more as an absorber.

Larger index differences can be achieved when the material composition is changed. To combine two different materials in a single crystal requires both to be similar in lattice structure and match very closely in their lattice constants. GaAs has such a close lattice match to AlAs. The ternary system $Ga_{1-x}Al_xAs$ can therefore be combined in any AlAs concentration without altering the lattice constant to any significant extent. The refractive index, however, decreases substantially when the concentration of AlAs is raised. Fig. 3.41 shows this refractive index as a function of the mole fraction x of aluminium in $Ga_{1-x}Al_xAs$ for the two wavelengths 0·9 and 0·8 μm that limit the range in which active GaAlAs-components normally operate. According to Fig.3.41, a film

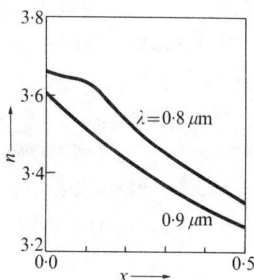

Fig.3.41. Refractive index of $Ga_{1-x}Al_xAs$ versus aluminium mole fraction (Casey, Sell, and Panish 1974).

guide of GaAs or GaAlAs regions must be built on or embedded in GaAlAs-regions of higher AlAs concentration.

The usual procedure is to epitaxially grow on the GaAs substrate first a layer of GaAlAs of a certain AlAs concentration, and then a layer of GaAlAs of lower AlAs concentration, or also pure GaAs. The intermediate layer serves to optically isolate the top-layer waveguide from the substrate. For this purpose, it must be thick enough for the evanescent fields of modes in the top layer to decay through the intermediate layer to insignificant values at the substrate boundary. Since in this three-layer structure, the top-layer index can never be larger than the substrate index, it only guides leaky modes. They correspond to the leaky modes underneath the prism coupler in Fig.2.36. With proper substitution of parameters, their leaky mode attenuation is given by equations (2.192) and (2.193). As long as the intermediate layer spaces the top layer far above the substrate, the leaky mode attenuation remains quite small, and for all practical purposes, these leaky modes then qualify as guided modes.

A single-layer film guide on top of a GaAs-substrate is obtained when the epitaxial growth starts with a high AlAs concentration and then decreases this concentration as the growth proceeds. Within the epitaxial layer of changing AlAs concentration, the refractive index decreases from top to bottom, and it forms a graded-index film similar to Fig.2.46. In liquid-phase epitaxial growth of $Ga_{1-x}Al_xAs$, aluminium grows out from the melt more rapidly because of its high segregation coefficient. The melt is gradually depleted of aluminium during growth, and towards the top of the layer, the refractive index increases as the AlAs concentration diminishes. The index step at the substrate boundary will render all modes of this graded-index film leaky. The index difference within the index profile must be large enough and the profile sufficiently wide to keep the leaky-mode attenuation below specified limits.

Otherwise the modes in these epitaxially grown GaAlAs guides are not so much attenuated by free-carrier absorption as by scattering at layer and boundary imperfections. The free carriers are not needed here for their guiding effect and can therefore be kept at their low intrinsic concentrations, unless, of course, an active function as, for example, the laser action in heterojunction lasers, requires a certain free-carrier concentration. Scattering loss is, however, difficult to avoid because epitaxial layers and their heterojunctions are just not perfect enough.

For active components such as lasers, luminescent diodes, and modulators, multilayer structures based on the ternary GaAlAs system are quite versatile. Since for field confinement, they rely on the index change due to the AlAs concentration, they can be doped to whatever free-carrier concentration the active function requires without degrading their guiding characteristics.

Single-crystal semiconductor layers with different refractive indices cannot only be grown epitaxially on semiconductor substrates; epitaxial growth of such layers is also possible on other single-crystal substrates. It requires, of course, a close lattice match between both crystals. As one particularly important example, single-crystal zinc oxide films are epitaxially deposited on sapphire substrates (Hammer *et al.* 1972). ZnO is transparent in the optical range from 0·4 to 2 μm wavelength; it exhibits the Pockel's electro-optic effect, is piezoelectric, has sizable nonlinear polarizability coefficients, and can be doped to be an efficient photoconductor. A low-loss ZnO waveguide can therefore perform a variety of active operations on guided modes in integrated planar circuits. For the epitaxial growth of single-crystal ZnO-films, the heated Al_2O_3-substrate is suspended close to a heated ZnO powder-source in a flow of hydrogen gas. This close-spaced chemical vapour transport produces films with rough surfaces. After polishing, however, the fundamental mode loss is as low as 4 dB cm^{-1} at 633 nm wavelength. The large refractive index difference between ZnO with $n = 1 \cdot 99$ and Al_2O_3 with $n = 1 \cdot 69$ at $\lambda = 633$ nm limits the film thickness to between 150 and 450 nm, if only the fundamental modes are to be guided at this wavelength.

Single-crystal films of insulating materials also deserve special interest for active components in planar circuits, because some of the most efficient electro-optic, magneto-optic, or acousto-optic, and optically non-linear materials are electric insulators. Lithium niobate ($LiNbO_3$) and lithium tantalate ($LiTaO_3$) are notable examples of insulating but very effective electro- and acousto-optic as well as optically non-linear crystals. They have a variety of applications in optical circuits. One method of forming a waveguide near the surface of a $LiNbO_3$-crystal is by out-diffusion of Li (Kaminow and Carruthers 1973). $LiNbO_3$ can be crystallized in the non-stoichiometric form $(Li_2O)_x(Nb_2O_3)_{1-x}$. The extraordinary refractive index, i.e. the index for a polarization parallel to the axis of the uniaxial crystal changes with x according to

$$dn_e/dx = -1 \cdot 63 . \qquad (3.173)$$

If x decreases towards the surface, n_e increases in this direction and a film guide is formed. Li is smaller than Nb and not as strongly bound to the lattice. Heating the crystal in vacuum to above 1000°C for several hours lets Li diffuse out of the crystal and leads to a surface layer nearly 10 μm thick with less than 10^{-2} index difference. Such thick surface layers guide a number of film modes. Indiffusion of Nb into

LiTaO$_3$ also produces a guiding surface layer (Hammer and Philips 1974). Nb is sputtered on the polished LiTaO$_3$-crystal which is then heated to above 1000°C for several hours. During this process, Nb replaces Ta in LiTaO$_3$ and raises the ordinary index by Δn_o = 0·113 and the extraordinary by Δn_e = 0·021. The surface layer of raised indices is about 1 µm thick and guides only the fundamental film mode.

Actually not only Nb but a variety of metals may be diffused into both LiTaO$_3$ and LiNbO$_3$ to raise their refractive indices and form guiding layers. Notable examples of transition elements suitable for indiffusion are Ti, V, and Ni (Schmidt and Kaminow 1974). Thin films of these metals with 20 to 80 nm thickness are evaporated first on to the LiNbO$_3$ or LiTaO$_3$ single-crystal substrates. These substrates are then heated to temperatures in the range of 850 - 1000°C for indiffusion of the metal in an Ar-flow to prevent oxidation of the metal. The adjustment of diffusion parameters in this process allows waveguide parameters to be controlled. Single-mode low-loss film guides have been obtained in particular by indiffusion of Ti into LiNbO$_3$. They confine the guided light to within 1 µm of the surface.

The refractive index of LiNbO$_3$ crystals may also be changed by ion implantation (Wei, Lee, and Bloom 1974). Implantation of argon and neon ions with energies near 60 keV and a dose of 10^{16} cm^{-2} or more lowers the refractive index by as much as 10 per cent. This negative index change occurs in a thin layer whose distance from the crystal surface depends on the implantation energy. By implanting a buried layer of lower refractive index, the top layer with unchanged index and with only few radiation defects forms a low-loss waveguide with good electro-optic characteristics.

Single-crystal films of LiNbO$_3$ are also grown epitaxially on LiTaO$_3$-substrates by melting (Miyazawa 1973). LiNbO$_3$ melts nearly 300°C below the melting point of LiTaO$_3$. A flat layer of LiNbO$_3$-lacquer suspension is painted on to the LiTaO$_3$-substrate, heated to near the melting point of LiNbO$_3$, and then cooled very slowly. The films from this epitaxial growth by melting have a graded index profile (Tien *et al.* 1974), they are a few tenths of a millimetre thick and guide a very large number of modes, yet with very low loss. Single-crystal $LiNbO_3$-films on $LiTaO_3$-substrates with an abrupt index change at the substrate boundary are produced by epitaxial growth in flux. This method dips the substrate into a flux that contains $LiNbO_3$ together with other lithium compounds (Tien, Riva-Sanseverino, and Ballman 1974).

The dipping method of liquid phase epitaxy is also employed to grow single-crystal garnet films on garnet substrates (Tien et al. 1972). $Eu_3Ga_5O_{12}$-films, for example, with a refractive index of $n = 1 \cdot 967$ on a $Gd_3Sc_2Al_3O_{12}$-substrate with $n = 1 \cdot 901$ are smooth and uniform and have losses as low as 1 dB cm^{-1}. Hardly any other single-crystal film guide has so little loss.

In active components for planar circuits as well as in monolithically integrated circuits with active components, single-crystal films must be fabricated in strips or other patterns of the respective component or circuit. The basic strip of rectangular cross-section occurs in the three different surroundings of Fig.3.4, as a raised, embedded, or a buried strip. They require different methods of pattern fabrication. Strip-loaded and rib guides can then also be fabricated with one of these methods. Some of these pattern fabrication methods are identical or at least similar to those for amorphous waveguide materials. Others derive from the methods for fabrication of single-crystal films.

A relief pattern of raised strips and ridges may be fabricated as shown in Fig.3.39c. First the pattern is produced in photo-resist that serves as a mask or barrier; or the photo-resist pattern is metal coated and, after lift-off, the remaining metal-coat pattern serves as mask for material removal. Sputter etching or ion-beam machining is then employed to remove the unmasked parts of the single-crystal layer underneath and possibly some of the substrate crystal as well. Ion-beam etching affords better control over the sections of planar layers that are to be removed. By orienting its beam at appropriate angles, raised strips can even be tailored to specific cross-sectional shapes. Chemical etching is also used, in particular, when parts of semiconductor layers are to be removed. The rate of material removable depends on the type of chemical reaction and on crystal orientation. Different ridge and rib contours can thus be etched chemically. In general, however, chemical etching is less controllable than sputter etching and ion-beam machining.

Instead of removing material as in Fig.3.39c and d, relief patterns can also be formed by the deposition of material as in Fig.3.39b. To grow such patterns in single crystal semiconductors, vapour phase epitaxy is used, in particular for GaAs and GaAlAs. Oxide masks serve to cover the channel areas of the pattern (Blum et al. 1974).

If waveguide patterns are to be embedded into the substrate as in Fig.3.39a, the method of fabrication depends on the crystal material which is to form the waveguide. Ion implantation is used, for example, for GaAs to raise the refractive index by compensation of the free-carrier concentration (Somekh et al. 1973). The pattern is formed by a gold mask which must be several microns thick to keep the protons or ions from penetrating the masked areas. This gold mask in turn is fabricated directly on the GaAs by ion-beam machining through a photo-resist mask. Also employed for making embedded guide patterns is diffusion of compensating ions of a higher index material through a mask.

A less direct but still quite precise method of fabricating embedded guide patterns starts with the single-crystal epitaxial film guide on the crystal substrate. It first removes part of the film by etching and leaves only those film strips that are to form the embedded guide. Between these strips, the substrate material is then filled in by epitaxial regrowth. The mask that served in etching the relief pattern may now be used to prevent any growth of substrate material on top of the guides. Without this mask, the epitaxial growth of substrate material all around the guide leads to a buried waveguide pattern. Liquid-phase epitaxy may be used to grow the embedding layer around the waveguide pattern. However, better results obtain with molecular beam epitaxy. The machine for molecular beam epitaxy consists of an ultra high-vacuum enclosure inside which an oven emits a beam of molecules that deposit epitaxially in thin uniform films on the substrate. Shadow masks between the molecular beam oven and the substrate also allow the deposition rate and concentration to vary laterally, and thus produce embedded guides more directly.

If strip loading on single-crystal films is to be used to form waveguides in active components, the loading strip itself need not to be a single-crystal. Its main function is to confine the light to the film region underneath by raising the effective index of the particular film mode. For this purpose its material must only have the proper index of refraction but otherwise can be polycrystalline or amorphous. Loading-strip patterns in active components can therefore be deposited by sputtering or any of the other simpler methods of film deposition and pattern fabrication for passive circuits.

REFERENCES

Arnaud, J.A. (1974). Transverse coupling in fibre optics. *Bell Syst. tech. J.* 53, 217-24.

Blum, F.A., Lawley, K.L., Scott, W.C., and Holton, W.C. (1974). Optically pumped grown GaAs mesa surface laser. *Appl. Phys. Lett.* 24, 430-32.

Casey, H.C., Sell, D.D., and Panish, M.B. (1974). The refractive index of $Al_xGa_{1-x}As$ between 1·2 and 1·8 eV. *Appl. Phys. Lett.* 24, 63-5.

Garmire, E. (1975). Semiconductor components for monolithic applications in integrated optics. In *Topics in Applied Physics*, Vol. 7. Springer Verlag, Berlin, 243-304.

——— Stoll, H., Yariv, A., and Hunsperger, R.G. (1972). Optical waveguiding in proton-implanted GaAs. *Appl. Phys. Lett.* 21, 87-8.

Goell, J.E. (1969). A circular-harmonic computer analysis of rectangular dielectric waveguides. *Bell Syst. tech. J.* 48, 2133-60.

——— (1973). Electron-resist fabrication of bends and couplers for integrated optical circuits. *Appl. Opt.* 12, 729-36.

Hammer, J.M. and Philips, W. (1974). Low-loss single-mode optical waveguides and efficient high-speed modulators of $LiNb_xTa_{1-x}O_3$ on $LiTaO_3$. *Appl. Phys. Lett.* 24, 545-7.

——— , Channin, D.J., Duffy, M.T., and Wittke, J.P. (1972). Low-loss epitaxial ZnO optical waveguides. *Appl. Phys. Lett.* 21, 358-60.

Harrington, R.F. (1961). *Time harmonic electromagnetic fields.* McGraw-Hill, New York.

Izawa, T. and Nakagome, H. (1972). Optical waveguide formed by electrically induced migration of ions in glass plates. *Appl.Phys.Lett.* 21, 584-6.

Kaiser, P. and Ash, H.W. (1974). Low loss single-material fibres made from pure fused silica. *Bell Syst.tech.J.* 53, 1021-39.

Kaminow, I.P. and Carruthers, J.R. (1973). Optical waveguiding layers in $LiNbO_3$ and $LiTaO_3$. *Appl.Phys.Lett.* 22, 326-8.

Kogelnik, H. and Li, T. (1966). Laser beams and resonators. *Proc. IEEE* 54, 1312-29.

Miyazawa, S. (1973). Growth of $LiNbO_3$ single-crystal film for optical waveguides. *Appl.Phys.Lett.* **23**, 198-200.

Petermann, K. (1976a). Theory of single-mode single-material fibres. *Archiv Electron. & Übertragungstech.* 30, 147-53.

────── (1976b). *Theorie des Wulstleiters.* Dr.-Ing.-Dissertation an der Techn. Universität Braunschweig.

Rütze, U. (1977). Rigorous analysis of the optical rib guide using rectangular waveguide modes. *Archiv Electron. & Übertragungstech.* 31, 88-90.

Schlosser, W. (1964). Der rechteckige dielektrische Draht. *Arch.elekt.Übertr.* 18, 403-10.

────── and Unger, H.-G. (1966). Partially filled waveguides and surface waveguides of rectangular cross-section. In *Advances in Microwaves* 1. Academic Press Inc., New York, 319-92.

Schmidt, R.V. and Kaminow, I.P. (1974). Metal-diffused optical waveguides in $LiNbO_3$. *Appl.Phys.Lett.* 25, 458-60.

Smith, H.I. (1974). Fabrication techniques for surface-acoustic wave and thin-film optical devices. *Proc.IEEE* 62, 1361-87.

Somekh, S., Garmire, E., Yariv, A., Garvin, H.L., and Hunsperger, R.G. (1973). Channel optical waveguide directional couplers. *Appl.Phys.Lett.* 22, 46-7.

Standley, R.D., Gibson, W.M., and Rodgers, J.W. (1972). Properties of ion-bombarded fused quartz for integrated optics. *Appl.Opt.* 11, 1313-16.

Tien, P.K., Martin, R.J., Blank, S.L., Wemple, S.H. and Varnerin, L.J. (1972). Optical waveguides of single crystal garnet films. *Appl.Phys.Lett.* 21, 207-9.

Tien, P.K., Riva-Sanseverino, S. and Ballman, A.A. (1974). Light beam scanning and deflection in epitaxial $LiNbO_3$ electro-optic waveguides. *Appl.Phys.Lett.* 25, 563-5.

─── , ─── , Martin, R.J., Ballman, A.A., and Brown, H. (1974). Optical waveguide modes in single-crystalline $LiNbO_3$-$LiTaO_3$ solid-solution films. *Appl. Phys.Lett.* 24, 503-6.

Tomlinson, W.J., Weber, H.P., Pryde, C.A., and Chandross, E.A. (1975). Optical directional couplers and grating couplers using a new high-resolution photo-locking material. *Appl.Phys.Lett.* 26, 303-6.

Ulrich, R. and Weber, H.P. (1972). Solution-deposited thin films as passive and active light-guides. *Appl.Opt.* 11, 428-34.

Weber, H.P., Ulrich, R., Chandross, E.A., and Tomlinson, W.J. (1972). Light-guiding structures of photo-resist films. *Appl.Phys.Lett.* 20, 143-5.

Wei, D.T.Y., Lee, W.W. and Bloom, L.R. (1974). Large refractive index change induced by ion implantation in lithium niobate. *Appl.Phys.Lett.* 25, 329-31.

Zernike, F. (1975). Fabrication and measurement of passive components in integrated optics. In *Topics in Applied Physics*, Vol.7, Springer Verlag, Berlin, 201-41.

4
CLADDED-CORE FIBRES

Optical waveguides in the form of glass fibres find their most important application as the transmission medium in optical communication systems. With their potential for low transmission loss and low dispersion they can transmit wide-band signals on optical carrier waves over long distances. In these characteristics glass fibres differ greatly from the film and planar waveguides of previous sections. Film and strip guides as well as the other planar waveguide structures serve in optical components and as connecting lines in planar optical circuits. In these applications their length is only fractions of a millimetre up to a few millimetres. Guided-mode attenuation for such short waveguide sections need not be much lower than a few dB cm^{-1}, and in some component applications we may even tolerate much higher attenuation. As their transmission dispersion is in principle so low, it will not cause any noticeable signal distortion in such short sections and lines.

The requirements are entirely different for optical waveguides as transmission media. Such waveguides span the wide distances between communication centres and terminals, which are between a fraction of a kilometre and hundreds of kilometres apart. Depending on the actual distance within this range, the transmission loss should not be higher than 100 dB km^{-1}. For a guide length of several kilometres it should even remain below 10 dB km^{-1}. As a further requirement, the signals should not suffer too much distortion even through lengths of guide as high as 10 km or more. As soon as we have a transmission medium of such high standards at our disposal, we could also use it to span hundreds and even thousands of kilometres by placing repeaters every few kilometres or possibly only every few tens of kilometres. These repeaters boost the signals on the optical carriers by amplification and, if necessary, regenerate these signals to remove all or at least part of their distortion.

Such long-distance transmission requires optical waveguides to be as low in loss and dispersion as is possible by any reasonable means. They should therefore be made of materials which have the highest possible quality with respect to absorption, scattering, and dispersion. When we process these high-quality materials to fabricate low-loss and low-dispersion waveguides, we should introduce as little additional degradation as possible. The ultimate aim is to try to maintain the same low loss and low dispersion as in the bulk of the high-quality base materials. The main contenders for base materials in such waveguides are several highly transparent and homogeneous multicomponent glasses and pure fused silica. The waveguide structures that most closely maintain the high quality of their base materials are fibres. Even simple glass fibres of uniform refractive index could be used to trap light and guide it by total internal reflection at their walls. However, any foreign bodies that touch such a naked fibre, to support or accommodate it in a tube, would frustrate or perturb the total reflection and degrade the guidance and transmission of light.

Optical waveguides in the form of glass fibres therefore usually consist of a glass core which is surrounded by a glass cladding of somewhat lower refractive index. Any light waves which are to be utilized for signal transmission are now trapped by the core and guided by total reflection at the core-cladding boundary. Fig.4.1 shows a longitudinal cross-section of such a cladded-core fibre model. In addition to the core of refractive index n_1 and the coaxial cladding region of

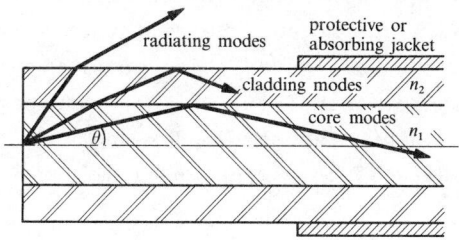

Fig.4.1. Cladded-core fibre with meridional light rays.

index n_2, a third region surrounds this model as a jacket. This jacket protects the fibre mechanically; it can also serve to absorb any light incident from inside or outside the fibre. For this purpose the jacket must be given a suitable refractive index and absorption constant; it will then prevent any crosstalk from other fibres in a bundle. Furthermore such a lossy jacket will also suppress any destructive interference with light rays out of the core; these light rays, instead of being totally reflected at the outer cladding boundary to go back to the core, are now absorbed by the jacket.

Fig.4.1 shows light rays which represent plane uniform wavelets from a point source on the fibre axis at its front face. Depending on the propagation angle θ of these light rays with the fibre axis, they will experience total reflection at the core-cladding boundary or at the outer cladding boundary; a third possibility is that for large propagation angles they will, at least partially, radiate into space. A light ray which is trapped by total reflection inside the core can form a guided mode of the core, or core mode. When trapped by the outer cladding boundary it can form a cladding mode. Light rays which radiate out into space form the radiating or space modes. Cladding modes and space modes have only secondary significance. When launching core modes under practical circumstances they can never be completely avoided. Fibre imperfections will also convert power from core modes into cladding and space modes. But these modes are mostly parasitic and undesired. They subtract power from core modes and degrade their transmission by interference. A suitable lossy jacket surrounding the cladding as in Fig.4.1 will absorb both the cladding and space modes. We will therefore focus attention first upon the core modes. The outer cladding boundary, and the lossy jacket around it, will be considered only as much as they influence the core modes and affect the fibre design for most suitable core-mode propagation.

4.1. RAY PICTURE OF CORE MODES

For some initial understanding of core modes, and to recognize their propagation limits and field distribution we continue with the ray picture of Fig.4.2. In it we disregard the

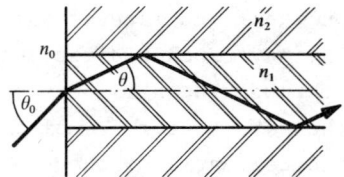

Fig.4.2. Longitudinal section at front end of cladded-core fibre model with meridional light ray.

lossy jacket and let the cladding of refractive index n_2 extend to infinity. A light ray incident onto the front face of the fibre at the core centre belongs to the class of meridional rays which have planes of incidence that all contain the fibre axis. Such a meridional ray incident from free space with refractive index $n_0 = 1$ at a propagation angle θ_0 with respect to the axis will be refracted upon entering the core and will continue with a propagation angle θ. From Snell's law equation (1.19) we have

$$\sin \theta_0 = (n_1/n_0)\sin \theta = n_1 \sin \theta . \qquad (4.1)$$

As long as θ remains smaller than the limiting angle θ_c of total reflection at the core-cladding interface, given by

$$\cos \theta_c = (n_2/n_1) , \qquad (4.2)$$

the core will trap this ray. Any meridional rays, therefore, which are incident on the front face of the fibre core at an angle θ_0 smaller than θ_{0c}, where

$$\sin \theta_{0c} = (n_1^2 - n_2^2)^{\frac{1}{2}}/n_0 = (n_1^2 - n_2^2)^{\frac{1}{2}} ,$$

will be trapped by total reflection inside the core. $\sin \theta_{0c}$ is the numerical aperture (N.A.) of the fibre core, and we have

$$\text{N.A.} = \sin \theta_{0c} = (n_1^2 - n_2^2)^{\frac{1}{2}} . \qquad (4.3)$$

So far we have considered only meridional rays. These rays are either incident on the front face at the core centre or, when incident off-axis, have such a direction that their zig-zag path always goes through the axis between reflections.

Fig.4.3 shows the wave vector $n_1 \vec{k}$ of a light ray of general orientation incident at a distance r from the core axis. With its propagation angle θ with respect to the axis,

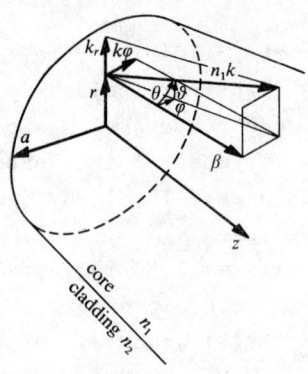

Fig.4.3. Wave vector of a skew light ray in a fibre core and its rectangular components in cylindrical coordinates.

its wave vector has the axial component

$$\beta = n_1 k \cos \theta \qquad (4.4)$$

while

$$k_r = n_1 k \sin \vartheta \qquad (4.5)$$

and

$$k_\varphi = n_1 k (\sin^2 \theta - \sin^2 \vartheta)^{\frac{1}{2}} \qquad (4.6)$$

are its radial and circumferential components, respectively, with the angle ϑ as indicated by Fig.4.3. Such rectangular components k_r, k_φ, and β for a plane wave solution of Maxwell's equation satisfy for the separation condition

$$k_r^2 + k_\varphi^2 + \beta^2 = n_1^2 k^2 . \tag{4.7}$$

What happens to such a skew ray when it is incident on the core wall depends not only on θ but also on its angles φ or ϑ at this point of incidence. Fig.4.4 shows the fibre with the projection of the light ray in the plane which is tangential to the core boundary at the point of incidence; the plane of incidence also appears in Fig.4.4. This latter plane

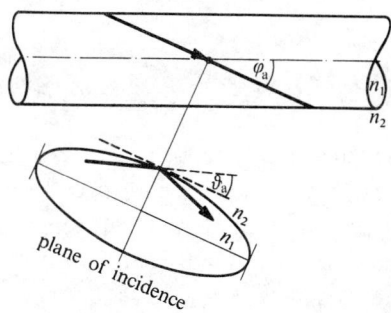

Fig.4.4. Projection of skew light ray on the plane tangential to its point of reflection and its plane of incidence.

contains the light ray and the normal to the tangential plane at the point of incidence. The plane of incidence is inclined at φ_a with respect to the core axis, where

$$\cos \varphi_a = \cos \theta / \cos \vartheta_a . \tag{4.8}$$

In this plane of incidence the ray has ϑ_a as the complement of its angle of incidence. The intersection of the core boundary and the plane of incidence forms an ellipse in this plane; the incident ray sees the boundary in its plane of incidence with the curvature of the ellipse. Curvature frustrates the total reflection and leads to radiation and loss of energy, which tunnels through the evanescent field region. It would seem therefore, that such non-meridional rays with $\varphi_a \neq 0$ could never form any guided modes of the core. In the present situation, however, the approximate analysis of reflection at curved boundaries which leads to this conclusion does not quite apply. In fact when such a ray forms a guided mode of the core, it drags its evanescent fields along in the axial direction at whatever phase velocity it travels. As long as this velocity is smaller than the intrinsic phase velocity $v_2 = \omega/(n_2 k)$ of the cladding medium, the evanescent field remains attached to the core and the mode and incurs no loss by leakage. The axial component β of the wave vector $n_1 \vec{k}$ must for this reason remain smaller than the wave number $n_2 k$ of the cladding medium, or its propagation angle θ must obey

$$\theta < \theta_c = \text{arc } \cos(n_2/n_1) \; .$$

Therefore, the same limiting angle θ_c from equation (4.2) or the numerical aperture (4.3) applies not only for meridional rays but for all rays which are completely trapped by the core.

Frustration of the total reflection due to the wall curvature in the plane of incidence becomes effective only when θ increases beyond θ_c. This frustrated total reflection will then prevail as long as the propagation angle ϑ_a in the plane of incidence remains below θ_c. $\vartheta_a = \theta_c$ represents the limit of total reflection even for its range of wall-curvature frustration. We designate the corresponding limiting value for θ by θ_l, and find θ_l as a function of φ_a, the inclination of the plane of incidence, from

$$\cos \theta_l = \cos \theta_c \cos \varphi_a \; . \qquad (4.9)$$

This limiting angle grows from $\theta_l = \theta_c$ at $\varphi_a = 0$ to $\theta_l = \pi/2$ at $\varphi_a = \pi/2$.

For nearly meridional rays with small inclinations φ_a, the wall curvature in the plane of incidence remains quite low. Light rays with θ near to θ_l for such small inclinations will suffer hardly any loss from the weak curvature frustration of total reflection. These light rays can form modes of propagation, which, strictly speaking, are leaky modes, but have so little attenuation that for most practical purposes, they still qualify as guided modes. With increasing inclination φ_a, the core boundary shows more and more curvature in the plane of incidence. Eventually, at $\varphi = \pi/2$, the plane of incidence coincides with the fibre cross-section and the core boundary curves with the core radius a. With so much curvature of the boundary those light rays with θ near θ_l reflect only partially and more power radiates out into the cladding. Leaky modes from such light rays suffer a correspondingly high leakage loss.

We restrict our attention now to the completely trapped rays with propagation angles in the range $\theta < \theta_c$. Fig.4.5 traces a non-meridional light ray on its skewed path down the

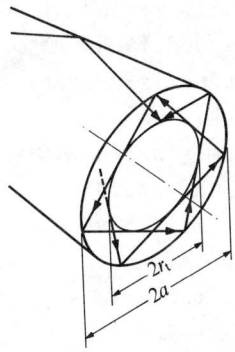

Fig.4.5. Annular region $r_t < r < a$ of core cross-section with traces of a non-meridional ray.

core by its projection onto the fibre cross-section. Its angles φ and ϑ depend on the distance r from the axis of the point of observation; but they obey

$$\cos \varphi = \cos \theta / \cos \vartheta \qquad (4.10)$$

everywhere. The angle ϑ has its largest value ϑ_a at the core wall, where the ray reflects back into the core. It decreases to zero at the radius

$$r_t = a[1 - (\sin \vartheta_a / \sin \theta)^2]^{\frac{1}{2}} , \qquad (4.11)$$

where the cross-sectional projection of the ray comes closest to the core axis and turns from decreasing to increasing radii. Repeated reflections of this ray down the core will let it move only between $r_t < r < a$. For any particular values of θ and ϑ_a, the circular centre region with $r < r_t$ remains free of any ray traces. The fields of the plane wavelets which this ray represents will only propagate in the annular region between r_t and a. Beyond a they show the evanescent character which is associated with total reflection; a similar evanescent behaviour of these fields also appears inside r_t. The cylinder with radius r_t forms a second boundary at which waves turn from propagating to evanescent fields. Just because rays only graze this boundary and never enter it, this cylinder seems to totally reflect these rays.

For any such trapped light rays to form a guided mode of the core, they need to satisfy phase conditions for self-consistent field solutions. In order for these rays to fit into the core they must form standing wave patterns by superimposing upon themselves. The standing wave pattern in the circumferential direction must have an even integer number $2l$ of nodes around the circumference. This requires the circumferential component k_φ of the wave vector to obey $k_\varphi = l/r$, and with equation (4.7) this leads to

$$k_r = [n^2 k^2 - \beta^2 - (l/r)^2] \qquad (4.12)$$

for the radial wave-vector component. Fig.4.6 illustrates this relation with the level of $n^2 k^2$ in the core and cladding

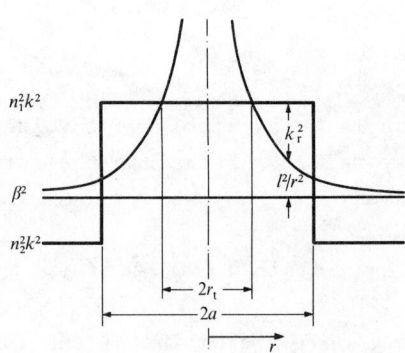

Fig.4.6. Profile of squared wave number and wave vector components in a cladded-core fibre.

and with a level of $\beta^2 > n_2^2 k^2$ in the range of guided core modes. Depending on the circumferential order l, the range of positive values for k_r^2 extends from $r = r_t$ to $r = a$. For $l = 0$ (corresponding to meridional rays) we obtain core modes with field distributions of axial symmetry. They have $r_t = 0$, and their fields extend over the full cross-section of the core including the core axis. With increasing l, corresponding to skew rays with larger and larger inclination φ_a in Fig.4.4, the inner boundary or caustic of these rays at $r = r_t$ moves to larger and larger radii. k_r^2 remains positive only in the annulus between r_t and a. The modal fields propagate in the radial direction and form a standing-wave pattern in this direction only in this annular region. Inside r_t and outside a, the radial wave vector component k_r turns imaginary and indicates evanescent fields in these ranges. Fig.4.7 illustrates the intensity distribution of

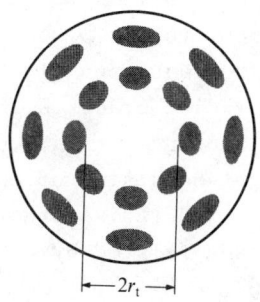

Fig.4.7. Transverse intensity distribution of a core mode with $l = 4$, $p = 2$.

this standing wave pattern; the particular example has transverse mode orders $l = 4$ in the circumferential direction, and $p = 2$ in the radial direction.

For the standing-wave pattern between r_t and a, the radial phase integral $\int_{r_t}^{a} k_r dr$, together with half the phase shift ϕ_a for total internal reflection at the core wall and half the phase shift $\pi/2$ for the apparent total reflection at the caustic with radius r_t, must equal an integer multiple p of π. With k_r from equation (4.12), this phase condition has the form

$$\int_{r_t}^{a} [n_1^2 k^2 - \beta^2 - (l/r)^2]^{\frac{1}{2}} dr + \phi_a/2 = (p - \frac{1}{4})\pi \qquad (4.13)$$

and represents the characteristic equation for guided modes of the core. The integer p counts the number of field nodes in the radial direction; it represents the radial order of the particular mode. When we evaluate the phase integral in equation (4.13), we obtain the more explicit form

$$a[n_1^2 k^2 - \beta^2 - (l/a)^2]^{\frac{1}{2}} - l \arccos \frac{l}{a(n_1^2 k^2 - \beta^2)^{\frac{1}{2}}} = (p - \frac{1}{4})\pi - \phi_a/2$$

$$(4.14)$$

of the characteristic equation. Any further evaluation of this equation would require the value of the phase shift ϕ_a of reflection at the core wall. Depending on the polarization of the field with respect to its plane of incidence, this phase shift follows either from equation (1.25) or equation (1.26). For sufficiently high radial orders p, however, ϕ_a will be small compared to $p\pi$, and under these conditions we can also neglect $\pi/4$ on the right-hand side of equation (4.14). The circumferential order l and radial order p of a core mode with phase constant β are then related by

$$a[n_1^2 k^2 - \beta^2 - (l/a)^2]^{\frac{1}{2}} - l \arccos \frac{l}{a(n_1^2 k^2 - \beta^2)^{\frac{1}{2}}} = p\pi \qquad (4.15)$$

Any such mode will go through cut-off when its rays have the propagation angle $\theta = \theta_c$. At this angle the phase constant β just equals the wave number $n_2 k$ of the cladding medium so that the level of β^2 in Fig.4.6 coincides with the level of $n_2^2 k^2$. The characteristic equation (4.15) reduces at cut-off to

$$p_c \pi = (V^2 - l^2)^{\frac{1}{2}} - l \arccos (l/V), \qquad (4.16)$$

where we have designated the radial mode order at cut-off by p_c and have introduced the fibre parameter

$$V = ka(n_1^2 - n_2^2)^{\frac{1}{2}} . \qquad (4.17)$$

This fibre parameter represents the corresponding characteristic quantity for the fibre to that introduced in equation (2.20) for film guides and in equation (3.53) and equation (3.67) for strip-loaded film guides and rib guides, respectively. According to

$$V = 2\pi \frac{a}{\lambda} (N.A.), \qquad (4.18)$$

the fibre parameter is proportional to the numerical aperture of the fibre core and to its radius a, relative to the light wavelength λ.

The characteristic equation (4.16) at cut-off yields the highest possible values of p and l for a given fibre parameter V, i.e. for a given fibre at a specific wavelength. In the (p,l)-plane, the curve p_c as a function of l as given by equation (4.16) limits the area within which we find the transverse orders l and p of all guided core modes. For a universal representation, we relate both p and l to V and plot p_c/V versus l/V in Fig.4.8 as a solid line. The broken lines

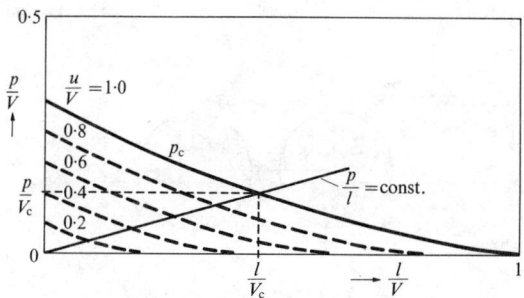

Fig.4.8. Radial mode order p versus circumferential order l for guided core modes. The solid line gives $p_c(l)$ at cut-off for $\theta = \theta_c$. The broken lines give $p(l)$ for specific propagation angles θ according to $u = n_1 ka \sin\theta$ (Gloge 1975).

in this diagram relate the transverse orders of core modes from rays which have specific values of propagation angles $\theta < \theta_c$. The parameter u for these lines is

$$u = a(n_1^2 k^2 - \beta^2)^{\frac{1}{2}} \qquad (4.19)$$

and is hence (from equation (4.15)) related to θ by

$$u = n_1 k \, a \, \sin \theta \, .\qquad(4.20)$$

The modes which are farthest from cut-off have the smallest propagation angle θ. With their small u-values and low transverse orders, their location is the lower left corner of the triangular area in Fig.4.8. The approximations of the present ray analysis restrict the application of equation (4.15) to sufficiently high radial orders p, so that depending on the magnitude of V, Fig.4.8 is not quite correct near its abscissa. For $V \gg 1$, however, any errors occur only very close to $p/V = 0$; we can then integrate over all mode orders p and l to determine the total number of guided core modes. In the (l,p)-plane any point with integers p and l below the limiting curve $p_c(l)$ represents a possible combination of transverse orders for a guided core mode. Any combination (l,p) of transverse orders can occur in two orientations, with two orthogonal polarizations for each orientation. Fig.4.9 illustrates these four possible combinations of orientation and polarization for the transverse orders $l = p = 1$. The shaded areas of the core

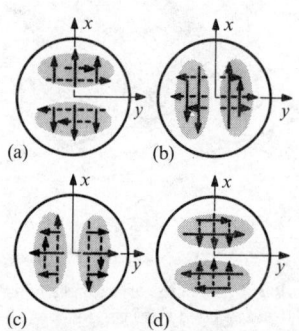

Fig.4.9. Guided core modes of equal transverse orders $l = p = 1$ with different orientation and polarization of their electric (solid arrow) and magnetic (broken arrow) fields.

cross-section indicate the intensity distribution, while the solid arrows with their direction and length indicate the polarization and strength, respectively, of the transverse electric field.

CLADDED-CORE FIBRES   303

With four possible combinations of field orientation and polarization for each pair of integers (l,p) we also have four core modes for each such pair of transverse orders. We therefore count the total number M of core modes by integrating p_c with respect to l from $l = 0$ to $l = V$ and then multiplying it by four:

$$M = 4 \int_0^V p_c(l)\, dl.$$

The result

$$M = V^2/2 \tag{4.21}$$

is the area underneath the limiting curve $p_c(l)$ in Fig.4.8 times $4V^2$. From equation (4.21) it would seem that the fibre core becomes monomode when the fibre parameter approaches $V = 2$, so that $M = 2$ and only the lowest-order mode in its two orthogonal polarizations exists. Actually, however, the previous approximations forbid the application of this expression to such low V-values and mode orders.

The approximate form (4.15) of the characteristic equation could also be used to evaluate the dispersion characteristics for any mode by solving it numerically for β as a function of k. Fig.4.8 offers a graphical solution of equation (4.15). To represent the phase constant β as a function of the fibre parameter V for modes of given order p and l from this graphical solution, we draw a straight line through the origin of Fig.4.8 with the slope p/l. The intersection of this line with the limiting line $p_c(l)$ yields the cutoff-value V_c for these transverse orders; its abscissa-value represents l/V_c and its ordinate-value p/V_c. When V increases beyond V_c we move along the straight line from this intersection towards the origin. Whenever we cross one of the broken lines in Fig.4.8, we have the corresponding value $u/V = (n_1^2 k^2 - \beta^2)^{\frac{1}{2}}/V$ for the respective abscissa-value l/V, from which follows β for the respective V. We defer graphical representations of β versus V for specific modes to a later section, where we obtain solutions of more accurate characteristic equations from a field analysis.

Let us now turn to the leaky modes of the cladded-core fibre and examine their characteristic equation as well as their range of radial and circumferential orders p and l. The skew rays which form these leaky modes propagate at angles θ with respect to the axis in the range

$$\theta_c < \theta < \theta_l .$$

Depending on the angle φ_a at which their plane of incidence on to the core cladding interface in Fig.4.4 is inclined with respect to the fibre axis, θ_l will grow from $\theta_l = \theta_c$ at $\varphi_a = 0$ to $\theta_l = \pi/2$ at $\varphi_a = \pi/2$. The axial phase constant of these modes depends on φ and ϑ as

$$\beta = n_1 k \cos\varphi \cos\vartheta . \tag{4.22}$$

Fig.4.10 gives a graphical presentation of equation (4.12) for

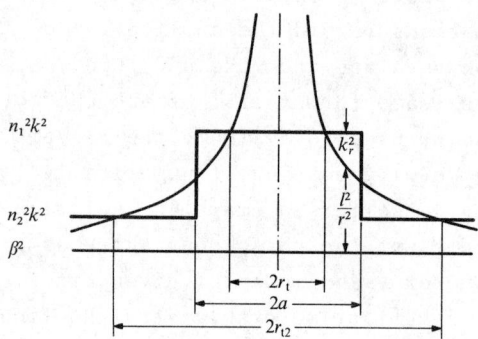

Fig.4.10. Profile of wave number squared and squared wave vector components for a leaky mode in a cladded-core fibre.

such leaky modes. With $\beta < n_2 k$ the square of the radial component k_r of the wave vector is positive not only in the annular region $r_t < r < a$, but also becomes positive again at a second turning point

$$r_{t2} = l/(n_2^2 k^2 - \beta^2) . \tag{4.23}$$

This turning point marks the transition of cladding fields

from evanescent character for $r < r_{t2}$ to a propagating wave for $r > r_{t2}$. The total reflection is frustrated for these modes and part of their energy tunnels through the evanescent field region to leak out into the radiating wave for $r > r_{t2}$. The power loss due to leakage increases when the evanescent field region narrows and when the purely imaginary radial component of the wave vector in this region becomes smaller. Hence with decreasing l and β the leaky modes become more strongly attenuated. If, on the other hand, l is very large and β only slightly smaller than $n_2 k$, the evanescent field region can be so wide and the transverse decay constant in part of this region so large that these leaky modes suffer hardly any attenuation.

The characteristic equation also maintains the same approximate form (4.15) for leaky modes, only now β is smaller than $n_2 k$. Since this equation was derived from the phase condition for standing waves inside the core, it takes no account of the power loss due to the frustration of total internal reflection. Therefore the phase constant but not the attenuation constant of leaky modes may be determined from it to a certain approximation.

The characteristic equation can also serve to relate radial to circumferential mode orders for leaky modes, to determine their limiting values, and count the total number of modes, including the leaky modes. At the limit of propagation of leaky modes for $\theta = \theta_l$, the complement ϑ_a of the angle of incidence is $\vartheta_a = \theta_c$, and we obtain for the phase constant along the boundary of the leaky mode region

$$\beta_l = n_2 k \cos \varphi_a . \qquad (4.24)$$

It ranges from $\beta_l = n_2 k$ for meridional rays with $\varphi_a = 0$ to $\beta_l = 0$ for rays which circulate in the transverse direction around the fibre core. With $\tan \varphi_a = l/\beta_l$ this limiting value for the propagation constant may be related to the circumferential mode order l according to

$$\beta_l^2 = n_2^2 k^2 - (l/a)^2 . \qquad (4.25)$$

The approximate characteristic equation (4.15) reduces in this limit to

$$p_l \pi = V - l \, \text{arc cot}(l/V) . \qquad (4.26)$$

The radial mode order p_l according to this equation is plotted in Fig.4.11 as a function of the circumferential mode order l. The line $p_l(l)$ marks the upper limit of the leaky mode

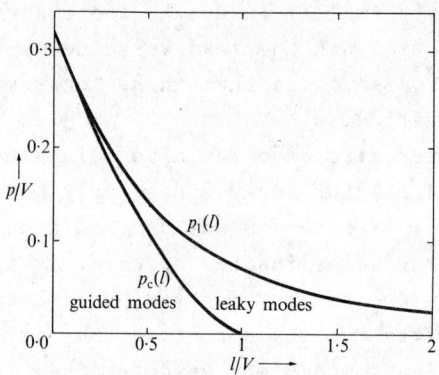

Fig.4.11. Leaky and guided mode regions in the (p/V)-(l/V) plane of relative radial and circumferential mode orders.

region in the (l,p)-plane. On its lower side the leaky mode region borders on the guided mode region along the line $p_c(l)$. The limit $p_l(l)$ appears to go out to infinitely large circumferential orders l. Actually, however, the highest possible circumferential mode order obtains when the angle φ approaches $\pi/2$ and the ray circulates in the transverse direction, with β equal to zero. Under these conditions we have

$$l \le n_1 k a ,$$

so that in the limit of $l_{max} = n_1 k a$ the integer multiple l_{max} of wavelengths $\lambda_1 = \lambda/n_1$ just fits around the circumference of the fibre core.

With p_l as a function of l according to equation (4.26), we can now count the total number of modes, guided as well as leaky, by integrating

$$M_t = 4 \int_0^{n_1 ka} p_l \, dl \, . \tag{4.27}$$

The factor 4 again takes account of the two orientations, with two orthogonal polarizations each, in which most of the modes can occur.

By substituting for p_l from equation (4.26) and transforming to the new variable of integration $x = l/V$ we obtain the total number of modes from the integral

$$M_t = \frac{4}{\pi} V^2 \int_0^{n_1/(n_1^2-n_2^2)^{\frac{1}{2}}} (1 - x \text{ arc cot } x) \, dx \, ,$$

which, upon integration, results in

$$M_t = V^2 \left[1 + \frac{2n_1}{\pi(n_1^2-n_2^2)^{\frac{1}{2}}} - \frac{2(2n_1^2-n_2^2)}{\pi(n_1^2-n_2^2)} \text{ arc cot } \frac{n_1}{(n_1^2-n_2^2)^{\frac{1}{2}}} \right]. \tag{4.28}$$

When the fibre cladding has only a slightly smaller refractive index n_2 than the core, the fibre parameter V according to equation (4.17) remains always much smaller than $n_1 ka$. We may then approximate arc cot $(n_1 ka/V) \simeq V/(n_1 ka)$, and the expression for the total number of modes reduces to nearly

$$M_t \simeq V^2 \left[1 - \frac{2}{\pi} (2(n_1-n_2)/n_1)^{\frac{1}{2}} \right] \, . \tag{4.29}$$

Out of this number, $V^2/2$ are guided modes, so that in addition to guided modes, we have

$$M_l = (V^2/2) \left[1 - \frac{4}{\pi} (2(n_1-n_2)/n_1)^{\frac{1}{2}} \right] \tag{4.30}$$

modes which are leaky. In the limit of $n_1 \simeq n_2$, and $V/(n_1 ka) \to 0$ as a consequence, the total mode number amounts to V^2, of which nearly half are leaky modes. The attenuation of some of these leaky modes can be so low, however, that for many practical purposes they appear equivalent to guided modes. We will determine the leaky mode attenuation in a subsequent section, when we have available the results of a more accurate and complete field analysis for guided and leaky modes.

The guided modes of the cladded-core fibre correspond to light rays with propagation angles $\theta < \theta_c$. All rays that are incident on the front face of the core from free space at angles $\theta_0 < \theta_{0c}$, with θ_{0c} from equation (4.3), are trapped by the core and excite guided modes. The condition $\theta_0 < \theta_{0c}$ for trapped rays is independent of the point of incidence of the ray on the front face of the core, nor does it depend on the azimuthal angle ϕ of the incident ray in Fig.4.12. In this sense any point on the front face of the fibre core has the same acceptance angle θ_{0c} or numerical aperture $\sin \theta_{0c}$ for trapped rays which excite guided modes. We may represent the

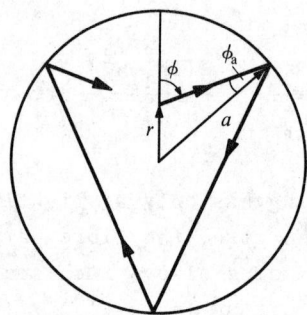

Fig.4.12. Cross-sectional projection of skew ray incident on the front face of a fibre at r with azimuthal angle ϕ.

cone of angle θ_{0c} by a circle of radius $\sin \theta_{0c} = (n_1^2 - n_2^2)^{\frac{1}{2}}$ on the fibre front face. Any point on the front face within the core radius has the same radius $\sin \theta_{0c}$ as this circle.

The situation turns out differently for leaky modes and incident rays that excite them. As long as rays are meridional, that is incident in a plane through the fibre axis, they have the same numerical aperture $\sin \theta_{0c}$ as trapped rays for guided modes. Thus there are no leaky modes with meridional rays. Skew rays, however, have a larger acceptance angle for leaky modes than they have for guided modes.

We let a skew ray be incident on the front face of the core at radius r and with an angle ϕ between its cross-sectional projection and the meridional plane in Fig.4.12. When this ray progresses to the core-cladding boundary, the angle ϕ decreases to ϕ_a according to

$$\sin \phi_a = (r/a) \sin \phi . \qquad (4.31)$$

The complement ϑ_a of its angle of incidence at the core cladding boundary is then given by

$$\sin \vartheta_a = \cos \phi_a \sin \theta . \qquad (4.32)$$

As long as ϑ_a remains smaller than the limiting angle θ_c for total internal reflection, a leaky or guided mode can trap this ray. We therefore find the limiting angle θ_l for such rays from

$$\sin \theta_l = \sin \theta_c / (1 - (r/a)^2 \sin^2 \phi)^{\frac{1}{2}}, \qquad (4.33)$$

and the numerical aperture for these rays is given by

$$N.A. = \sin \theta_{0l} = \sin \theta_{0c} / (1 - (r/a)^2 \sin^2 \phi)^{\frac{1}{2}}. \qquad (4.34)$$

It depends on r as well as on ϕ. For $r = 0$ as well as $\phi = 0, \pi$, corresponding to meridional rays, it reduces to $\sin \theta_{0c}$. In general, however, the N.A. forms an ellipse with its semi-minor axis for $\phi = 0, \pi$ equal to $\sin \theta_{0c}$ and its semi-major axis for $\phi = \pm \pi/2$ larger than $\sin \theta_{0c}$ by the factor $1/(1 - (r/a)^2)^{\frac{1}{2}}$. On the core boundary the semi-major axis widens beyond all bounds, which, of course, means only that no matter how small θ_{0c}, the limiting angle θ_{0l} will go as high as $\pi/2$ at $r = a$ for $\phi = \pm \pi/2$. Fig.4.13 illustrates the acceptance angle and numerical aperture for guided and leaky mode rays by showing corresponding circles and ellipses along a radius across the front face of the fibre.

4.2. FIELD SOLUTIONS FOR FIBRE MODES

Applications of glass fibres such as long-distance transmission media for wide-band signals in optical communication systems necessitate extreme specifications. To properly design fibres for these applications and to tolerate any defi-

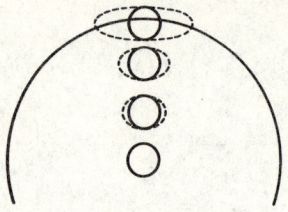

Fig.4.13. Circles and ellipses for the numerical aperture or acceptance angle of guided and leaky mode rays, respectively, across the front face of a cladded-core fibre.

ciencies only to a degree which these specifications allow, we need an accurate and general theory of wave-propagation in fibres. The ray approach of the previous section gives us some, but far from all the information which we desire for the characterization and design of optical fibre waveguides. Some of the results from the ray analysis also suffer from a lack of accuracy. We therefore devote this section to a solution of Maxwell's equations under the boundary conditions of optical fibres. As a fibre model for this solution we adopt the coaxially layered structure with the cross-section of Fig. 4.14. It has again a fibre core of refractive index n_1 and radius a surrounded by the cladding of refractive index n_2 extending to a radius b. The outer region beyond the radius b extends to infinity and has the refractive index n_3. Depending on the values which b, n_2, and n_3 assume, this model represents various kinds of fibres. If n_3 is complex, with an absorption constant $-n_3''$ as its imaginary component, and if the real component of this complex $\hat{n}_3$ differs not too much from n_2, we have a model for the cladded-core fibre with the absorbing jacket in Fig.4.1. Other specifications for n_2 and n_3 and b will be chosen once we have obtained the general solution and apply it to various fibre waveguides.

For a time-harmonic field solution, which oscillates with angular frequency ω, we start from Maxwell's equations (1.3). We adopt the coordinates r, φ, z of the circular cylinder of which the z-axis coincides with the fibre axis in Fig.4.14. By eliminating either $\vec{E}$ or $\vec{H}$ from Maxwell's equations, they reduce to the wave equations (1.5) or (1.6), but

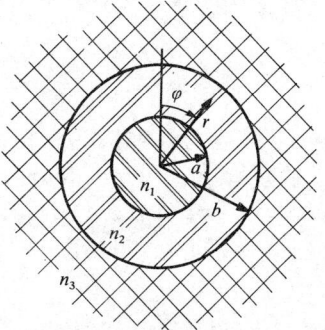

Fig.4.14. Cross-section of cladded-core fibre model and cylindrical coordinates for field analysis.

only for the Cartesian components of either $\vec{E}$ or $\vec{H}$. E_z and H_z are the only Cartesian components in the cylindrical coordinates of Fig.4.14. We therefore adopt these as the generating components of the total electromagnetic field. We aim for wave propagation in the z-direction and hence let all field components have the common z-dependence $\exp(-j\beta z)$; in particular we let the z-component of the electric field be $E_z \exp(-j\beta z)$ and the z-component of the magnetic field $H_z \exp(-j\beta z)$. The transverse distributions $E_z(r,\varphi)$ and $H_z(r,\varphi)$ must then satisfy the following transverse wave equation

$$(\nabla_t^2 + k_{rn}^2) \begin{Bmatrix} E_z \\ H_z \end{Bmatrix} = 0, \qquad (4.35)$$

where the Laplacian operator ∇_t^2 with respect to the transverse coordinates is in cylindrical coordinates:

$$\nabla_t^2 = \frac{1}{r} \frac{\partial}{\partial r} (r \frac{\partial}{\partial r}) + \frac{1}{r^2} \frac{\partial^2}{\partial \varphi^2} .$$

The separation constant

$$k_{rn}^2 = n^2 k^2 - \beta^2 \qquad (4.36)$$

corresponds to the transverse component of the wave vector in the ray picture. Depending on the cross-sectional region we have either

$$n = \begin{cases} n_1 \\ n_2 \\ n_3 \end{cases} \quad \text{and} \quad k_{rn} = \begin{cases} k_{r1} \\ k_{r2} \\ k_{r3} \end{cases} \quad \text{for} \quad \begin{cases} r<a \\ a<r<b \\ b<r \end{cases} \qquad (4.37)$$

in equation (4.35). Products of the form

$$Z_m(k_{rn}r) \begin{Bmatrix} \cos m\varphi \\ \sin m\varphi \end{Bmatrix}$$

solve equation (4.35), where $Z_m(k_{rn}r)$ represents a combination of two linearly independent cylinder functions of integer order m. More specifically we let

$$E_{z1} = A_E J_m(k_{r1}r) \cos m\varphi$$
$$H_{z1} = A_H J_m(k_{r1}r) \sin m\varphi \tag{4.38}$$

for $r \leq a$, and

$$E_{z2} = [C_E I_m(j\ k_{r2}r) + D_E K_m(j\ k_{r2}r)]\cos m\varphi$$
$$H_{z2} = [C_H I_m(j\ k_{r2}r) + D_H K_m(j\ k_{r2}r)]\sin m\varphi \tag{4.39}$$

for $a \leq r \leq b$, as well as

$$E_{z3} = F_E K_m(j\ k_{r3}r)\cos m\varphi$$
$$H_{z3} = F_H K_m(j\ k_{r3}r)\sin m\varphi \tag{4.40}$$

for $r \geq b$. The choice of Bessel functions $J_m(k_{r1}r)$ for the core region $r \leq a$ guarantees proper behaviour of the fields near the fibre axis. In the cladding region we need to start with a general combination of cylinder functions. We have chosen modified Bessel functions $I_m(jk_{r2}r)$ and modified Hankel functions $K_m(jk_{r2}r)$ here, because under normal conditions for core modes, we anticipate evanescent fields in the cladding region. These fields will then have imaginary components of wave vectors in radial direction. For the same reason we use the modified Hankel function in the outer cladding region beyond $r = b$; and only this function, because it will let the fields decay exponentially in positive r-direction. All longitudinal electric fields in equations (4.38) to (4.40) depend on φ as $\cos m\varphi$ while all longitudinal magnetic fields have $\sin m\varphi$ for their φ-dependence. This particular combination will enable us to satisfy boundary conditions for the other field components. We could also let E_z have $\sin m\varphi$ as φ-dependence and for H_z choose $\cos m\varphi$. Because of the axial symmetry of the fibre, the solution resulting from the latter φ-dependencies would differ from the first only by a rotation of fields with $\varphi = 2\pi/m$. Otherwise it would be identical.

The remaining components of the electromagnetic field follow from E_z and H_z by solving Maxwell's equations (1.3) for them in terms of E_z and H_z:

$$E_r = -\frac{j}{k_{rn}^2}[\beta\frac{\partial E_z}{\partial r} + \frac{\omega\mu_0}{r}\frac{\partial H_z}{\partial \varphi}]$$

$$E_\varphi = -\frac{j}{k_{rn}^2}[\frac{\beta}{r}\frac{\partial E_z}{\partial \varphi} - \omega\mu_0\frac{\partial H_z}{\partial r}]$$

$$H_r = \frac{j}{k_{rn}^2}[\frac{\omega\varepsilon_0 n^2}{r}\frac{\partial E_z}{\partial \varphi} - \beta\frac{\partial H_z}{\partial r}]$$

$$H_\varphi = -\frac{j}{k_{rn}^2}[\omega\varepsilon_0 n^2\frac{\partial E_z}{\partial r} + \frac{\beta}{r}\frac{\partial H_z}{\partial \varphi}].$$

(4.41)

The still unknown factors A_E and A_H to F_E and F_H, which we have allocated to the transverse functions for the longitudinal field components in equations (4.38), (4.39), and (4.40) give us enough freedom to satisfy boundary conditions at the two interfaces $r = a$ and $r = b$. We must match the tangential components E_φ and E_z, as well as H_φ and H_z at these boundaries, so that they remain continuous there. With four such tangential field components at two boundaries, these conditions provide us with a system of eight linear equations for the eight undetermined factors A_E to F_H. The system is homogeneous, hence for non-trivial solutions we must require its coefficient determinant to vanish. This constitutes the characteristic equation for modes of propagation. Any β-value that we obtain as a solution of it represents a propagation or phase constant of such a mode.

The system of 8 linear equations has an (8 × 8) determinant. To evaluate this bulky expression and obtain a relatively well-arranged characteristic equation we introduce a number of abbreviations (Kuhn 1974).

$$u = k_{r1}a = a(n_1^2 k^2 - \beta^2)^{\frac{1}{2}}$$

(4.42)

represents the radial phase parameter inside the core as in equation (4.19).

$$v = jk_{r2}a = a(\beta^2 - n_2^2 k^2)^{\frac{1}{2}}$$
(4.43)
$$w = jk_{r3}a = a(\beta^2 - n_3^2 k^2)^{\frac{1}{2}}$$

represent radial attenuation parameters for the evanescent fields in cladding and jacket. u, v, and w correspond to the transverse phase and attenuation parameters of the same designation in film guides. The fibre parameter V from equation (4.17) relates to u and v as

$$V^2 = u^2 + v^2 , \qquad (4.44)$$

just as it does in film guides. We will also find the longitudinal phase parameter

$$B = \frac{(\beta/k)^2 - n_2^2}{n_1^2 - n_2^2} \qquad (4.45)$$

useful to describe the dispersion characteristics of fibre modes. The relation

$$B = (v/V)^2 \qquad (4.46)$$

corresponds to equation (2.19) for film guides. A similar phase parameter

$$B_3 = \frac{(\beta/k)^2 - n_3^2}{n_2^2 - n_3^2} \qquad (4.47)$$

relates the effective index $N = \beta/k$ of the particular mode to the refractive index n_3 of the jacket and to the numerical aperture between cladding and jacket.

For the boundary conditions the cylinder functions must be evaluated at u and v as well as cv and cw, where

$$c = b/a \qquad (4.48)$$

is the ratio of cladding to core radius. We use the following abbreviations for combinations of cylinder functions

$$Y_m = \frac{J_m'(u)}{u\, J_m(u)} \quad ; \quad U_m = \frac{K_m'(wc)}{wc\, K_m(wc)}$$

$$W_0 = I_m(vc)\, K_m(v) - I_m(v)\, K_m(vc)$$

$$W_1 = I_m(v)\, K_m'(vc) - I_m'(vc)\, K_m(v)$$

$$W_2 = I_m(vc)\, K_m'(v) - I_m'(v)\, K_m(vc) \tag{4.49}$$

$$W_3 = I_m'(vc)\, K_m'(v) - I_m'(v)\, K_m'(vc).$$

The prime on any of the cylinder functions denotes differentiation with respect to its argument. By further abbreviating,

$$Z_1^{(H)} = Y_m(1 + vc\, U_m\, W_0/W_1)$$

$$Z_1^{(E)} = Y_m[1 + (n_3/n_2)^2\, vc\, U_m\, W_0/W_1]$$

$$Z_2^{(H)} = (c\, U_m\, W_2 - W_3/v)/W_1 \tag{4.50}$$

$$Z_2^{(E)} = [(n_3/n_2)^2\, c\, U_m\, W_2 - W_3/v]/W_1,$$

the characteristic equation eventually becomes

$$(Z_1^{(H)} + Z_2^{(H)})(n_1^2\, Z_1^{(E)} + n_2^2\, Z_2^{(E)}) = m^2\, N^2\, \Big\{ Z_1^{(H)}\, Z_1^{(E)}/(u^2\, B\, Y_m)^2 +$$

$$+ W_0^2/(n_2\, vc\, B_3\, W_1)^2 [(Y_m + \tfrac{W_2}{vW_0})(n_1^2\, Y_m + n_2^2\, \tfrac{W_2}{vW_0}) - (mN/(u^2 B))^2] -$$

$$- 2/[B\, B_3(u\, cv^2\, W_1)^2]\Big\}. \tag{4.51}$$

For a given fibre at a specific wavelength and considering all the abbreviations in equations (4.42) to (4.50), this characteristic equation has the axial phase constant β as the only unknown quantity. Values of β which solve this equation represent phase constants of fibre modes and allow us to determine, for example, all field factors A_H to F_H in equations (4.38), (4.39), and (4.40) relative to A_E. A_E would then represent the excitation coefficient of the particular mode. Any one of the other field factors from A_H to F_H could alternatively be taken as the excitation coefficient and the remaining field factors determined relative to this coefficient as the solution of the linear equation system. This linear system, because it is homogeneous, fixes its unknowns only up to a common factor, which remains arbitrary.

In the case of axially symmetric fields we have $m = 0$ and according to equation (4.41) only E_r and H_φ derive from E_z while H_z has only E_φ- and H_r-components associated with it. The transverse magnetic or E-field from E_z has only E_z and H_φ as tangential field components at each boundary, so that the boundary conditions result in 4 linear equations for the 4 field factors A_E, C_E, D_E, and F_E in equations (4.38) to (4.40). Hence, E_z and its components E_r and H_φ can satisfy all boundary conditions and form an axially symmetric mode solution. By the same token, H_z alone, and its components E_φ and H_r from equations (4.41), can satisfy all boundary conditions for an axially symmetric H-mode.

The characteristic equations for these axially symmetric H- and E-modes can be found from equation (4.51). With $m = 0$, this equation reduces to

$$(Z_1^{(H)} + Z_2^{(H)})(n_1^2 Z_1^{(E)} + n_2^2 Z_2^{(E)}) = 0, \qquad (4.52)$$

where

$$Z_1^{(H)} = -Z_2^{(H)} \qquad (4.53)$$

represents the characteristic equation for axially symmetric H-modes and

$$n_1^2 Z_1^{(E)} = -n_2^2 Z_2^{(E)} \qquad (4.54)$$

for axially symmetric E-modes. In the ray picture of modes these axially symmetric modes are formed by meridional rays.

Any non-meridional ray can only form an asymmetric mode with $m \neq 0$. In this case only a superposition of E_z- and H_z-fields can satisfy all boundary conditions. The modes lose their transverse character and become hybrid modes with longitudinal components of both their electric and magnetic fields

These hybrid modes are designated as HE_{mp}- and EH_{mp}-modes. The first index m corresponds to the integer m in the φ-dependence of fields for both groups of modes, HE and EH; it therefore designates the circumferential order of any mode. The second index p designates the radial order of a particular mode. p counts the order in which a particular HE_{mp}- or EH_{mp}-mode goes through cut-off with increasing V to become a guided mode of the fibre. The axially symmetric modes are designated in this nomenclature as H_{0p}- and E_{0p}-modes, respectively.

HE_{11} and EH_{11} are the lowest-order hybrid modes. One of them has a field configuration inside the core which resembles the field configuration of the H_{11}-mode in the round hollow-tube waveguide with perfectly conducting walls. The H_{11}-mode has the lowest cut-off of all modes in round metallic waveguides and is hence the fundamental mode in this type of round waveguide. The lowest-order hybrid mode of the cladded-core fibre which resembles the H_{11}-mode in field configuration has zero cut-off, as we shall see later, and constitutes the fundamental mode of the fibre. Because of this close correspondence to the H_{11}-waveguide mode, this particular hybrid mode is designated as an HE_{11}-mode while for the other lowest-order hybrid mode the designation EH_{11} remains.

This designation of the lowest-order hybrid modes also fixes the classification for all other modes. Just as the characteristic equation (4.52) for axially symmetric modes splits into equation (4.53) for H_{0p}-modes and into equation (4.54) for E_{0p}-modes, the more general form (4.51) can also be considered to contain two different characteristic equations, one for HE-modes and the other for EH-modes. The separation of the characteristic equation into two different forms will be discussed later for the simplified fibre model with the cladding extending to infinity. The HE_{11}-mode follows from a solution of one of these two equations and the EH_{11}-mode as a solution of the other characteristic equation. All modes which follow from the same characteristic equation as the HE_{11}-mode will then be designated as HE_{mp}-modes and all modes from solutions of the other equation as EH_{mp}-modes.

In order to solve the characteristic equation in its general form (4.51), we have to resort to numerical methods and need to specify the sizes of fibre core and cladding relative to the wavelength as well as the refractive index in core, cladding, and jacket. We will obtain such solutions later for specific cases of practical interest.

In many such cases, however, we do not actually need to solve the characteristic equation in the general and quite complicated form (4.51). It will rather suffice to consider certain limiting cases.

As one such case, we have core modes with fields so well confined to the core that their evanescent tails out into the cladding hardly extend to the outer cladding boundary. Also, if the cladding radius is large enough we have this same situation for most of the core modes. For both these cases, vc is sufficiently large compared to v that

$$I_m(vc) \gg \begin{cases} I'_m(v) \ K_m(vc)/K'_m(v) \\ I_m(v) \ K_m(vc)/K'_m(v) \end{cases}$$

as well as

$$I'_m(vc) \gg \begin{cases} I'_m(v) \ K'_m(vc)/K'_m(v) \\ I_m(v) \ K'_m(vc)/K_m(v) \ . \end{cases}$$

Abbreviating,

$$X_m = \frac{K'_m(v)}{v \ K_m(v)} \quad , \tag{4.55}$$

we can approximate

$$W_3/W_1 \simeq -v \ X_m$$

and

$$W_2/W_3 \simeq -W_0/W_1 \simeq I_m(vc)/I'_m(vc).$$

If we now divide equation (4.51) by

$$\left[1 - \frac{v}{w}\frac{K_m'(wv)I_m(vc)}{K_m(wv)I_m'(vc)}\right]\left[1 - \frac{n_3^2 v K_m'(wc)I_m(vc)}{n_2^2 w K_m(wc)I_m'(vc)}\right] \simeq \left(1+\frac{v}{w}\right)\left(1 + \frac{n_3^2 v}{n_2^2 w}\right),$$

we obtain the following approximate form of the characteristic equation:

$$(Y_m + X_m)(n_1^2 Y_m + n_2^2 X_m) = m^2 N^2/(u^2 B)^2 . \qquad (4.56)$$

This, of course, is the well-known characteristic equation for the round dielectric waveguide of refractive index n_1 embedded in a medium with refractive index n_2 that extends to infinity. It was first derived by Hondros and Debye (1910) who considered round dielectric wires in free space as waveguides for radio frequencies. We could also obtain this equation from equation (4.51) by letting $n_3 \to n_2$, or with $c = 1$, when we replace n_3 by n_2 everywhere.

The jacket which surrounds the cladded-core fibre should protect it mechanically but should also absorb any light rays that are not well guided by the core. To absorb such unwanted radiation and not reflect it back to the core, the jacked index n_3 should differ only little from the cladding index n_2 and the jacket should have a high enough absorption coefficient. However, if the jacket serves mainly to protect the fibre mechanically, its material should be chosen for this function and its refractive index might differ considerably from the cladding index. One possibility is a metallic jacket for mechanical protection. Such a jacket will also shield the fibre from any outside radiation. Metallic materials have a complex index of refraction at optical frequencies, with its imaginary component of the same order of magnitude as its real component. The absolute value of this complex index of refraction is large compared to unity, even at frequencies as high as the visible range of the optical spectrum. For a simplified model of a fibre with a metallic jacket or any other high-index material surrounding the cladding, we let n_3 go to infinity. In this limit the absolute value of w will also go to infinity, because β in equation (4.43) remains finite for any guided mode. The characteristic equation (4.51) reduces in this limit to

$$\left(Y_m - \frac{W_3}{v\,W_1}\right)\left(n_1^2\,Y_m + n_2^2\,\frac{W_2}{v\,W_0}\right) = m^2\,\frac{N^2}{u^4\,B^2} \quad . \qquad (4.57)$$

As the characteristic equation of the round metallic waveguide with a coaxial dielectric insert, it was first derived by Buchholz (1943) who also solved it for its modes of propagation and studied their dispersion and loss characteristics in detail.

Some simplification of the characteristic equation (4.51) also results when, with $n_3 = n_1$, the jacket assumes the same index of refraction as the core. We obtain in this case from equation (4.51)

$$(Z_1^{(H)} + Z_2^{(H)})(n_1^2\,Z_1^{(E)} + n_2^2\,Z_2^{(E)}) = \left\{m\,\frac{N}{B}\right\}^2 \left\{Z_1^{(H)}\,Z_1^{(E)}/(u^2\,Y_m)^2 + \right.$$
$$+ W_0^2/(n_2\,vc\,W_1)^2 \times \left[\left(Y_m + \frac{W_2}{v\,W_0}\right)\left(n_1^2\,Y_m + n_2^2\,\frac{W_2}{v\,W_0}\right) - (mN/(u^2B))^2\right] -$$
$$\left. - 2/(u\,cv^2\,W_1)^2\right\} \quad . \qquad (4.58)$$

A coaxially layered structure with $n_3 = n_1$ but $n_2 > n_1$ was studied earlier as a model for the dielectric tube waveguide to guide microwaves (Unger 1954). For microwave applications the dielectric tube presents an alternative to the dielectric rod waveguide which corresponds to the cladded-core fibre with $n_1 > n_2$ and the cladding extending to infinity. For transmitting microwaves in its fundamental mode of propagation, the dielectric tube has dispersion and loss characteristics which are in some respects superior to the characteristics of the dielectric rod.

As much as the cladded-core fibre represents the optical analogue to the dielectric rod waveguide, a doubly clad fibre with $n_2 > n_1 = n_3$ corresponds to the dielectric tube waveguide. The annular region of higher refractive index n_2 now serves to confine the fields by total internal reflection at its inner and outer boundaries. Guided mode fields are now evanescent under any conditions, not only in the outer cladding but also in the centre region. The main advantage of this annular fibre guide as compared to the simple cladded-core fibre lies in the large size of the annulus as compared to the core size. When the annular guide has the same difference in refractive index between the annulus and the centre region as well as the outer cladding, as the cladded core fibre has between core and cladding, the annulus can be larger in diameter than the core and still guide only the lowest-order fundamental mode. Also, under these conditions, the fundamental mode fields of the annular fibre distribute over a wider diameter than the dominant mode fields in the cladded core fibre. The larger cross-sectional sizes of the annulus and its fundamental mode makes it easier to launch this mode and to align such fibres for coupling and splicing.

4.3. GUIDED MODES FOR UNLIMITED CLADDING

Any fibre modes which we utilize for signal transmission in the cladded-core fibre should be guided by the core and have their fields and energy well confined inside or at least near the core. Their evanescent fields near the outer cladding boundary should in fact be so small that whatever jacket surrounds the cladding, it will not affect the core modes. In the first place it should not attenuate them too much by absorbing some of their energy.

We are aiming for fibres that have transmission loss as low as a few dB/km. For 1 dB/km the attenuation constant α as the real component of the propagation constant $\gamma = \alpha + j\beta$ would be $\alpha = 1 \cdot 15 \times 10^{-4}$ m^{-1}. At a wavelength $\lambda = 1$ μm and a refractive index of $n = 1 \cdot 5$, this compares to a phase constant $\beta = 0 \cdot 942 \times 10^{7}$ m^{-1}. Both differ by eleven orders of magnitude. To be sure that an absorbing jacket will not attenuate the core modes too much, its influence on core modes should cause a relative change of core mode characteristics of not much more than 10^{-11}. When we tolerate only such a minute perturbation, we need to extend the cladding to a large enough diameter for the intensity of evanescent fields to be just as low there relative to the intensity inside the core. Unless we want to specifically analyse these small effects due to the cladding limitation and the surrounding jacket, we can let the cladding extend to infinity and neglect the jacket altogether. Under present conditions this will not change any of the relevant core mode characteristics, except for the very small attenuation due to cladding absorption. The simplified fibre model with its cladding extending to infinity is therefore quite adequate to study many of the core-mode characteristics under practical conditions. Its characteristic equation (4.56), although of a transcendental form, is not too complicated to be solved graphically and many important mode characteristics may be derived from its solutions.

The field components in the core of this simplified fibre model have still the same general distribution as in the more general case. With equations (4.38) and (4.41) they are

$$E_{z1} = A_E\, J_m(ur/a)\cos m\varphi$$

$$H_{z1} = A_H\, J_m(ur/a)\sin m\varphi$$

$$E_{r1} = -j\frac{a}{u}[\beta\, A_E\, J_m'(ur/a) + \omega\mu_0\frac{ma}{ur} A_H\, J_m(ur/a)]\cos m\varphi$$

$$H_{r1} = -j\frac{a}{u}[\omega\varepsilon_0\, n_1^2\frac{ma}{ur} A_E\, J_m(ur/a) + \beta\, A_H\, J_m'(ur/a)]\sin m\varphi$$

$$E_{\varphi 1} = j\frac{a}{u}[\frac{m\beta a}{ur} A_E\, J_m(ur/a) + \omega\mu_0\, A_H\, J_m'(ur/a)]\sin m\varphi$$

$$H_{\varphi 1} = -j\frac{a}{u}[\omega\varepsilon_0\, n_1^2 A_E\, J_m'(ur/a) + \frac{m\beta a}{ur} A_H\, J_m(ur/a)]\cos m\varphi.$$

(4.59)

In the cladding, which now extends to infinity, the fields have the same evanescent or outward propagating character as in the jacket of the more general model. We therefore let $C_E = C_H = 0$ in equation (4.39) and are left with

$$E_{z2} = D_E\, K_m(vr/a)\cos m\varphi$$

$$H_{z2} = D_H\, K_m(vr/a)\sin m\varphi$$

$$E_{r2} = j\frac{a}{v}[\beta\, D_E\, K_m'(vr/a) + \omega\mu_0\frac{ma}{vr} D_H\, K_m(vr/a)]\cos m\varphi$$

$$H_{r2} = j\frac{a}{v}[\omega\varepsilon_0\, n_2^2\frac{ma}{vr} D_E\, K_m(vr/a) + \beta\, D_H\, K_m'(vr/a)]\sin m\varphi$$

(4.60)

$$E_{\varphi 2} = -j\frac{a}{v}[\frac{m\beta a}{vr} D_E K_m(vr/a) + \omega\mu_0 D_H K'_m(vr/a)]\sin m\varphi$$

$$H_{\varphi 2} = j\frac{a}{v}[\omega\varepsilon_0 n_2^2 D_E K'_m(vr/a) + \frac{m\beta a}{vr} D_H K_m(vr/a)]\cos m\varphi .$$

The conditions $E_{z1} = E_{z2}$ and $H_{z1} = H_{z2}$ at the core-cladding boundary relate the field factors A and D to each other by

$$D_E/A_E = D_H/A_H = J_m(u)/K_m(v) . \qquad (4.61)$$

With these relations the boundary condition $E\varphi_1 = E\varphi_2$ at $r = a$ leads to

$$A_E/A_H = -(Y_m + X_m)(\mu_0/\varepsilon_0)^{1/2} u^2 B/mN . \qquad (4.62)$$

Equations (4.61) and (4.62) allow us to specify one of the 4 field factors as the excitation coefficient of a particular mode and express the remaining 3 factors in terms of this excitation coefficient.

We will next define a quantity which by its magnitude measures the dominance of the longitudinal electric field E_z over the longitudinal magnetic field H_z. Equation (4.62) gives the ratio of both these field components at their respective maxima in their dependence on φ. It has the dimension of an impedance. In a homogeneous medium of refractive index $n = N = \beta/k$, corresponding to the effective index of a particular mode, a plane uniform wave has the wave impedance $(\mu_0/\varepsilon_0)^{1/2}/N$. The factor

$$\rho = -(Y_m + X_m) u^2 B/m \qquad (4.63)$$

then tells us to what extent the ratio A_E/A_H differs from the uniform plane wave impedance in a medium with index N. For $|\rho| > 1$, E_z dominates H_z compared to the ratio in a uniform plane wave.

An alternative expression for ρ follows from equation (4.53) when, instead of $E_{\varphi 1} = E_{\varphi 2}$ as for equation (4.52), we use the boundary condition $H_{\varphi 1} = H_{\varphi 2}$. Equating both expressions actually results in the characteristic equation (4.56).

This equation will therefore give us also the alternative formula for ρ. We replace $(Y_m + X_m)$ in equation (4.63) from equation (4.56) and obtain

$$\rho = -\frac{mN^2}{n_1^2(Y_m + (n_2^2/n_1^2)X_m)u^2 B} \,. \tag{4.64}$$

We now proceed to solve the characteristic equation and to classify its solutions as HE_{mp}- or EH_{mp}-modes. In terms of Y_m equation (4.56) represents a quadratic equation with the following two roots (Kapany and Burke 1972)

$$Y_m = -\frac{n_1^2 + n_2^2}{2n_1^2} X_m \mp \left[\left(\frac{n_1^2 - n_2^2}{2n_1^2} X_m\right)^2 + \frac{m^2 N^2}{n_1^2 (u^2 B)^2}\right]^{\frac{1}{2}} \,. \tag{4.65}$$

For a graphical representation and to examine limiting cases of its solution, we replace Y_m according to the identities

$$Y_m = \pm \frac{J_{m\mp 1}(u)}{u\, J_m(u)} \mp \frac{m}{u^2} \,. \tag{4.66}$$

We use the upper signs of this formula when we substitute into that root of equation (4.65) which has the minus sign on its radical. This particular part of the characteristic equation can then be written as

$$\frac{J_{m-1}(u)}{J_m(u)} = -\frac{n_1^2 + n_2^2}{2n_1^2} u X_m + \left\{\frac{m}{u} - u\left[m^2\left(\frac{1}{u^2} + \frac{1}{v^2}\right)\left(\frac{1}{u^2} + \frac{n_2^2}{n_1^2}\frac{1}{v^2}\right) + \left(\frac{n_1^2 - n_2^2}{2n_1^2} X_m\right)^2\right]^{\frac{1}{2}}\right\} \,. \tag{4.67}$$

In this equation we have also expressed the effective index N and phase parameter B by the transverse parameters u and v according to

$$N^2 = (n_2^2 u^2 - n_1^2 v^2)/(u^2 + v^2)$$
$$B = v^2/(u^2 + v^2) \,. \tag{4.68}$$

In the other part of the characteristic equation (4.65) with the positive sign on the radical we replace Y_m by equation

(4.66) with the lower signs of this identity. This part of equation (4.65) then follows as

$$\frac{J_{m+1}(u)}{J_m(u)} = \frac{n_1^2+n_2^2}{2n_1^2} uX_m + \left\{\frac{m}{u} - u\left[m^2\left(\frac{1}{u^2} + \frac{1}{v^2}\right)\left(\frac{1}{u^2} + \frac{n_2^2}{n_1^2}\frac{1}{v^2}\right) + \left(\frac{n_1^2-n_2^2}{2n_1^2}X_m\right)^2\right]^{\frac{1}{2}}\right\}. \quad (4.69)$$

Let us first decide which of these two parts of the characteristic equations has solutions that correspond to HE-modes. The other part will then yield the EH-solutions. We have stated earlier for the more general model of the cladded-core fibre with jacket that whatever mode resembles in its field distribution inside the core the H_{11}-mode of a round metallic waveguide will be designated as the HE_{11}-mode, so that it belongs to the HE_{mp}-mode family. Like the H_{11}-mode in a round waveguide, the HE_{11}-mode has the lowest cut-off and is therefore the fundamental mode of the fibre.

In order to examine equation (4.67) for such a low-cut-off solution we let $m = 1$. At cut-off the fields extend further and further out into the cladding, and their transverse attenuation parameter v goes to zero. With approximate expressions for modified Hankel-functions of small argument, we obtain

$$X_1 \simeq -v^{-2} + \ln(2/\gamma v), \quad (4.70)$$

where $\ln\gamma = 0.5772$ is Euler's constant. Using this approximation to X_1 for small arguments, equation (4.67) reduces to

$$\frac{J_0(u)}{uJ_1(u)} = \frac{2n_2^2}{n_1^2+n_2^2} \ln\frac{2}{\gamma v}. \quad (4.71)$$

With $v = 0$ at cut-off, the fibre parameter $V = (u^2 + v^2)^{\frac{1}{2}}$ equals the transverse phase parameter u. The lowest cut-off has the smallest $u = u_c$ at cut-off. Actually equation (4.71) has a solution with $u_c = 0$, because for $u \ll 1$ we have $J_0(u) \simeq 1$ and $J_1(u) = u/2$. Equation (4.71) then changes to

$$u^2 = (1+n_1^2/n_2^2)/\ln(2/\gamma v), \quad (4.72)$$

which lets u go to zero together with v, but at a different rate. Hence, this particular solution corresponds to a mode with zero cut-off and represents the fundamental HE_{11}-mode.

With the HE_{11}-mode resulting as a solution of equation (4.67), all other HE_{mp}-modes also follow as solutions of this part of the characteristic equation. Solutions of the other part (4.69) then correspond to EH_{mp}-modes. In tracing these correspondences back to the original form (4.65) of the characteristic equation, we note that the minus sign on the radical will yield HE-modes as solutions and the plus sign EH-modes.

This classification of modes suggests that the longitudinal electric field might dominate the longitudinal magnetic field in one of the two mode families, while the opposite dominance might prevail for the other mode family. We will therefore examine the magnitude of the factor ρ according to equations (4.63) and (4.64) which measures the dominance of E_z over H_z.

For HE-waves the form of equation (4.65) with the minus sign on the radical gives the following expression for ρ in equation (4.63):

$$Y_m + X_m = \frac{n_1^2 - n_2^2}{2n_1^2} X_m - \left[\left(\frac{n_1^2 - n_2^2}{2n_1^2} X_m \right)^2 + \frac{m^2 N^2}{n_1^2 (u^2 B)^2} \right]^{\frac{1}{2}}.$$

For real and positive values of v the term

$$X_m = - \frac{K_{m-1}(v)}{vK_m(v)} - \frac{m}{v^2} \tag{4.73}$$

is also real but always negative, so that the absolute value of ρ according to

$$|\rho| = \left\{ \frac{n_1^2 - n_2^2}{2n_1^2} |X_m| + \left[\left(\frac{n_1^2 - n_2^2}{2n_1^2} X_m \right)^2 + \frac{m^2 N^2}{n_1^2 (u^2 B)^2} \right]^{\frac{1}{2}} \right\} \frac{u^2 B}{m} \tag{4.74}$$

remains always larger than unity as long as

$$\left(\frac{n_1^2-n_2^2}{2n_1^2}X_m\right) + \frac{m^2N^2}{n_1^2(u^2B)^2} > \left(\frac{m}{u^2B} - \frac{n_1^2-n_2^2}{2n_1^2}|X_m|\right)^2.$$

When we replace X_m in this inequality by equation (4.73), it reduces to

$$K_{m-1}(v)/(v\,K_m(v)) > 0,$$

which is always satisfied for any real positive values of v. Solutions of the characteristic equation (4.65) with the minus sign on the radical which correspond to HE-modes therefore have $|\rho| > 1$. Their longitudinal electric field always dominates the longitudinal magnetic field.

For EH-modes the other part of equation (4.65) with the plus sign on the radical gives

$$Y_m + \frac{n_2^2}{n_1^2}X_m = -\frac{n_1^2-n_2^2}{2n_1^2}X_m + \left[\left(\frac{n_1^2-n_2^2}{2n_1^2}X_m\right)^2 + \frac{m^2N^2}{n_1^2(u^2B)^2}\right]^{\frac{1}{2}} \quad (4.75)$$

to be substituted into equation (4.64) for the evaluation of ρ. We obtain for its absolute value

$$\frac{1}{|\rho|} = \frac{u^2(N^2-n_2^2)}{2mN^2}|X_m| + \left\{\left[\frac{u^2(N^2-n_2^2)}{2mN^2}|X_m|\right]^2 + \frac{n_1^2}{N^2}\right\}^{\frac{1}{2}}. \quad (4.76)$$

Since the effective mode index $N = \beta/k$ is always smaller than the core index, we conclude from equation (4.76) that $|\rho|$ is always smaller than unity. Any solutions of the characteristic equation (4.65) with the plus sign on the radical have $|\rho| < 1$. For all EH-modes the longitudinal magnetic field dominates the longitudinal electric field.

The classification of hybrid modes and their designation as HE- and EH-modes derives from the resemblance of transverse field components for the fundamental HE_{11}-mode to the transverse field of the H_{11}-mode in a round metallic waveguide. It shows no such resemblance for the longitudinal fields. While the H_{11}-mode in a metallic waveguide is transverse electric with only a longitudinal magnetic field, the HE_{11}-fibre mode has a longitudinal electric field that may even dominates the longitudinal magnetic field. In general, it is only the second letter in this hybrid mode designation that indicates the dominating longitudinal field.

The axially symmetric modes with $m = 0$ are either transverse electric with only a magnetic field component in the longitudinal direction or transverse magnetic with only a longitudinal electric field. Their designation as H_{0p}-modes when they are transverse electric, and as E_{0p}-modes when they are transverse magnetic, follows the usual custom to designate a mode with transverse character according to its longitudinal field component. The characteristic equation (4.56) splits into

$$Y_m + X_m = 0 \qquad (4.77)$$

and

$$n_1^2 Y_m + n_2^2 X_m = 0, \qquad (4.78)$$

when we let $m = 0$ for axially symmetric modes. When we derive these equations from the boundary conditions for field expressions in equations (4.59) and (4.60) we obtain equation (4.77) for H-modes and equation (4.78) for E-modes. If, on the other hand, we start from equation (4.65) and let $m = 0$, the minus sign on the radical will lead to equation (4.77) and the plus sign to equation (4.78). Since for hybrid modes with $m \neq 0$ the minus sign corresponds to HE-modes and the plus sign to EH-modes, the H_{0p}-modes appear as the axially symmetric members of the EH-family, and the E_{0p}-modes as axially symmetric members of the HE-family. In accordance with this association the longitudinal field ratio ρ for HE-modes from equation (4.63) becomes infinitely large for E_{0p}-modes while ρ for EH-modes from equation (4.64) is zero with $m = 0$ for H_{0p}-modes.

We return now to the characteristic equations (4.67) for HE-modes and (4.69) for EH-modes and proceed to solve them graphically.

For a given circumferential order m of modes the left-hand sides of these equations depend only on the transverse phase parameter u. On the right-hand sides we specify the index ratio n_2/n_1 and the transverse attenuation parameter $v = (V^2-u^2)^{\frac{1}{2}}$. These right-hand sides then also depend only on u. Both equations may therefore be solved graphically by plotting both sides in the same diagram as a function of u. Values of u at which corresponding curves intersect are solutions of the respective equation.

Fig.4.15 illustrates this graphical solution for $m = 2$, $n_1/n_2 = 1\cdot 2$, and several values of the parameter v. The right-

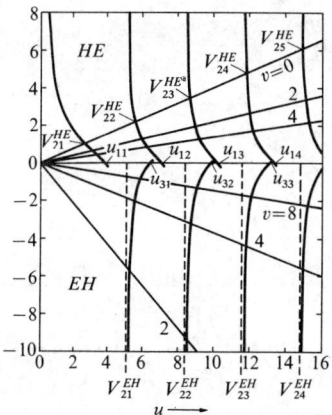

Fig.4.15. Graphical solution of the characteristic equations (4.67) and (4.69) for $m = 2$ and $n_1/n_2 = 1\cdot 2$.

hand sides of both equations are monotonic functions of u. Both start at zero for $u = 0$. In equation (4.67) for HE-modes the right-hand side increases monotonically with u, while equation (4.69) for EH-modes has a monotonically decreasing right-hand side. The left-hand sides have poles wherever $J_m(u) = 0$, and zero crossings at $J_{m\mp 1}(u) = 0$ between these poles. Fig.4.15 shows only those branches of the left-hand sides which intersect with the respective function of u on the right-hand sides. The upper half corresponds to HE-modes while the lower half represents the domain of EH-modes.

We consider first the limit of very large v values. With increasing v but constant u the right-hand sides of both equations (4.67) and (4.69) become smaller and smaller in absolute value and their curves in Fig.4.15 cling closer to the u-axis further and further out on this axis. In the limit of $v \to \infty$, they intersect the branches of the left-hand sides at the zero crossings of $J_{m-1}(u)$ for HE-modes, and of $J_{m+1}(u)$ for EH-modes. At such intersections the respective modes are very far from cut-off. Hence we obtain the far from cut-off limit u_∞^{HE} of the transverse phase parameter for the HE_{mp}-mode from the zero,

$$J_{m-1}(u_\infty^{HE}) = 0, \qquad (4.79)$$

of the pth-branch in the upper half of Fig.4.15. The limiting value u_∞^{EH} for the EH_{mp}-mode obtains likewise from the zero

$$J_{m+1}(u_\infty^{EH}) = 0 \qquad (4.80)$$

of the pth-branch in the lower half of Fig.4.15. By designating the first zero of $J_m(u)$ at a finite u-value with u_{m1} and the pth such zero as u_{mp}, we have the following limiting values very far above cut-off: A HE_{mp}-mode has $u_\infty = u_{m-1,p}$ and an EH_{mp}-mode has $u_\infty = u_{m+1,p}$.

We derive an even better approximation for these phase parameters far from cut-off when we start with $u_{m-1,p}$ from equation (4.79) or $u_{m+1,p}$ from equation (4.80) and evaluate the respective characteristic equations for large V-values. A perturbational analysis leads to

$$u = u_{m \mp 1,p} \left[1 - \frac{n_1^2 + n_2^2}{2n_1^2 V^2} \right], \qquad (4.81)$$

which is accurate to first order in V^{-1}. For HE_{mp}-modes $u_{m-1,p}$ from equation (4.79) should be substituted, and $u_{m+1,p}$ from equation (4.80) for EH_{mp}-modes. The relative amount

$$\Delta u / u_{m \mp 1,p} = - (n_1^2 + n_2^2)/(2n_1^2 V) ,$$

by which u remains below its far-from-cut-off limit $u_{m \mp 1,p}$ depends in this first-order approximation only on the index ratio n_2/n_1 and the fibre parameter V, but not on the particular mode. This approximation for u far from cut-off allows us now to calculate the transverse decay parameter $v = (V^2-u^2)^{\frac{1}{2}}$ as well as the longitudinal phase parameter $B = (v/V)^2$ under the same approximate conditions; from B in particular follows the effective mode index $N = \beta/k = B(n_1^2-n_2^2)+n_2^2$ and the phase constant β of the particular mode.

We examine next the other limit of the graphical solution in Fig.4.15 when the transverse decay parameter v decreases from large values far from cut-off to smaller values. As v decreases, the curves in Fig.4.15 which display the right-hand sides of equations (4.67) and (4.69) move away from the u-axis. In the HE-mode domain they attain larger and larger positive values, while for EH-modes they go lower into the negative region. Their intersections with the left-hand side branches of equations (4.67) and (4.69) move away from the Bessel function zeros on the u-axis to smaller values of u. In order for the modes to still be guided their effective index must stay above $N > n_2$. At $N = n_2$ the propagation angle θ in the ray picture reaches the limiting value $\theta_c = \arccos(n_2/n_1)$ for total internal reflection.

As we approach this cut-off limit the transverse attenuation parameter v goes to zero and u comes closer and closer to V. For small enough v we approximate X_m on the left-hand sides by

$$X_m \simeq -m/v^2 - 1/2(m-1) \qquad (4.82)$$

for $m \neq 1$ and by equation (4.70) for $m = 1$. Near cut-off the characteristic equation (4.67) for HE_{mp}-modes with $m \neq 1$ reduces to

$$\frac{J_{m-1}(u)}{J_m(u)} = \frac{n_2^2 u}{(n_1^2+n_2^2)(m-1)}, \qquad (4.83)$$

while for HE_{1p}-modes it has the form of equation (4.71). The straight line for $v = 0$ in Fig.4.15 represents the right-hand side of equation (4.83) for the respective values of m and

n_1/n_2. Its intersections with the positive branches of $J_{m-1}(u)/J_m(u)$ give the cut-off values $u_c = V_{mp}^{HE}$ for any of the HE_{mn}-modes. The recurrence formula

$$2(m-1) J_m(u) = u J_{m-1}(u) + u J_{m+1}(u)$$

leads to the following equivalent form of the near-cut-off approximation (4.83) for HE_{mp}-modes with $m > 1$:

$$\frac{J_{m-2}(u)}{J_m(u)} = -\frac{n_1^2 - n_2^2}{n_1^2 + n_2^2}. \tag{4.84}$$

HE_{1p}-modes of first-order circumferential dependence have the characteristic equation (4.71) near cut-off. Directly at cut-off, v goes to zero so that the corresponding cut-off value $u_c = V_{1p}^{HE}$ represents the pth zero of

$$J_1(u) = 0. \tag{4.85}$$

The first of these zeros is at $u_c = V_{11}^{HE} = 0$. We have already observed this cut-off at zero for the fundamental fibre mode when we identified it as the mode that corresponds to the fundamental H_{11}-mode in a round metallic waveguide, and when we designated it by HE_{11} as a consequence. Since the HE_{11}-mode is the fundamental mode, and because this mode has zero-cut-off, it deserves very special attention.

The transverse phase parameter u of this mode follows from equation (4.72) near cut-off and as long as it remains small compared to unity. The phase parameter $B = (v/V)^2$ of the HE_{11}-mode can, for this range, be expressed explicitly in terms of the fibre parameter V. We solve either of equations (4.71) or (4.72) for v and approximate $u = (V^2 - v^2)^{\frac{1}{2}} \simeq V$ to obtain, for example, with equation (4.71)

$$B = (1 \cdot 26/V^2) \exp\left\{-\frac{1 + (n_1/n_2)^2}{V} \frac{J_0(V)}{J_1(V)}\right\}. \tag{4.86}$$

The zero-point approximations for $J_0(V)$ and $J_1(V)$ simplify this formula with some loss of accuracy to

$$B = (1 \cdot 26/v^2) \exp\left\{-[1 + (n_1/n_2)^2]/v^2\right\} . \tag{4.87}$$

Let us now examine the cut-off characteristics of EH-modes. To approximate their characteristic equation (4.69) for small values of v we replace X_m on its right-hand side by equation (4.73). It then reduces to

$$\frac{J_{m+1}(u)}{u\,J_m(u)} = -\frac{n_1^2 + n_2^2}{n_1^2}\frac{m}{v^2} . \tag{4.88}$$

As v goes to zero the right-hand side of this equation grows negative beyond all limits. Eventually it intersects the branch curves of the left-hand sides at their poles. Hence

$$J_m(u_c) = 0 , \tag{4.89}$$

represents the cut-off condition of EH_{mp}-modes. The smallest value of u at which $J_m(u)$ goes to zero is $u = 0$. Equation (4.88) has, however, no solution for $v = u = 0$ because for $u \ll 1$ we obtain $J_{m+1}(u)/J_m(u) = u/2m$. Hence as u and v go to zero, the left-hand side of equation (4.88) approaches the constant value of $1/2m$, while the right-hand side goes to infinity in its absolute value. The lowest cut-off v_{m1}^{EH} of any EH_{m1}-mode occurs at the first zero crossing of $J_m(u)$ with a finite value of $u_c = v_{m1}^{EH}$.

To obtain approximations far from cut-off as well as near cut-off for axially symmetric H_{0p}- and E_{0p}-modes we examine their characteristic equations (4.77) and (4.78). As noted above, the transverse electric H_{0p}-modes belong to the EH-family. Their characteristic equation therefore also derives from equation (4.69) with $m = 0$:

$$\frac{J_1(u)}{u\,J_0(u)} = -\frac{K_1(v)}{v\,K_0(v)} . \tag{4.90}$$

Far from cut-off they behave like the other members of the EH_{mp}-family. The limiting value u_∞^H of their transverse phase parameter follows from equation (4.80) as the pth-zero

$u^H = u_{1p}$ of $J_1(u)$. For large values of the fibre parameter V, this phase parameter is approximated by equation (4.81) with $m = 0$ and the lower sign, that is $u_{1,p}$.

The axially symmetric members of the HE_{mp}-family are the transverse magnetic E_{0p}-modes. The characteristic equation of E_{0p}-modes therefore also derives from equation (4.67) with $m = 0$

$$\frac{J_1(u)}{u\,J_0(u)} = -\frac{n_2^2}{n_1^2}\frac{K_1(v)}{v\,K_0(v)}. \tag{4.91}$$

It differs from the corresponding equation for H_{0p}-modes only by the factor (n_2^2/n_1^2) on its right-hand side. Far above cut-off these axially symmetric modes also behave like the other members of the HE_{mp}-family. Their limiting value u_∞^E for very large V again follows from equation (4.80) as $u_\infty^E = u_{1p}$, the pth-zero of $J_1(u)$. In this limit far from cut-off any E_{0p}-mode therefore becomes degenerate with the H_{0p}-mode of the same order p. According to equation (4.81), for E_{0p}-modes now with its upper sign and with $u_{-1,p}=u_{1,p}$ the transverse phase parameter u of any E_{0p}-mode remains the same as for the corresponding H_{0p}-mode, up to first order in $1/V$. E_{0p}- and H_{0p}-modes of the same radial order p tend to degeneracy even when they are not very far above cut-off. As we recede more towards cut-off, however, their degeneracy splits due to the factor n_2^2/n_1^2 in equation (4.91).

Near cut-off, for $v \ll 1$, we introduce

$$\frac{K_1(v)}{v\,K_0(v)} = \frac{1}{v^2\,\ln(2/\gamma v)} \tag{4.92}$$

to approximate the right-hand sides of both characteristic equations (4.90) and (4.91). As v goes to zero at cut-off both equations require

$$J_0(u_c) = 0. \tag{4.93}$$

The H_{0p}-cut-off therefore coincides with the E_{0p}-cut-off, and we have

$$V^E_{0p} = V^H_{0p} \ . \tag{4.94}$$

At cut-off and not too far above it, any E_{0p}-mode is again degenerate with the H_{0p}-mode of the same transverse order p. E_{0p}- and H_{0p}-modes are thus degenerate near cut-off and far above cut-off; they differ in their dispersion characteristics only in an intermediate range. Furthermore this difference stems from the factor $(n_2/n_1)^2$ in the characteristic equation (4.91) for E_{0p}-modes; it diminishes when the cladding index n_2 is only slightly lower than the core index n_1.

We have now determined the near-cut-off and far-from-cut-off characteristics of all EH_{mp}- and HE_{mp}-modes of a fibre core in a cladding of infinite extent. The axially symmetric H_{0p}- and E_{0p}-modes have also been included in these considerations. From these results we find the range of values for the transverse phase parameter u to be rather limited for any particular mode. As the fibre parameter V increases, it reaches the cut-off value V^{EH}_{mp} at the pth-zero of $J_m(u)$. At cut-off, the transverse attenuation parameter v is zero, and the transverse phase parameter $u = u_c$ equals the fibre parameter V^{EH}_{mp}. Near cut-off, the fields extend very far out into the cladding. As the fibre parameter V increases beyond the respective cut-off value V^{EH}_{mp}, the particular EH_{mp}-mode moves above cut-off, v grows, and the fields confine themselves more and more to the core. High enough above cut-off for $V \gg V^{EH}_{mp}$, we have $v \simeq V$, and the phase parameter u approaches $u_{m+1,p}$ as its limiting value. $u_{m+1,p}$ according to equation (4.80) is the pth-zero of $J_{m+1}(u)$. Over the full range from cut-off to arbitrarily large values of V, the transverse phase parameter of an EH_{mp}-mode increases only from the pth-zero of $J_m(u)$ to the pth zero of $J_{m+1}(u)$. The asymptotic expression for large u,

$$J_m(u) = (2/\pi u)^{\frac{1}{2}} \cos(u - m\pi/2 - \pi/4) , \tag{4.95}$$

demonstrates that u very high above cut-off differs from u at cut-off only by about $\pi/2$.

For HE_{mp}-modes, u ranges over similar spans. For small V the fibre guides only the fundamental HE_{11}-mode. Its transverse phase parameter ranges between $u_c = V_{11}^{HE} = 0$ at cut-off and $u_{01} = 2\cdot 405$, the first zero of $J_0(u)$, very far above cut-off. The next modes to go above cut-off as V increases are the axially symmetric H_{01}- and E_{01}-modes; they have their cut-off at

$$V_{01} = 2\cdot 405, \tag{4.96}$$

the first zero of $J_0(u)$. When V increases beyond V_{01} it reaches $u_c = V_{21}^{HE}$ as the next cut-off value. V_{21}^{HE} follows from equation (4.84) with $m = 2$; for small index differences it is only slightly larger than V_{01}.

All other HE_{mp}-modes go through cut-off likewise at values $u_c = V_{mp}^{HE}$ which lie somewhat above the pth-zero of $J_{m-2}(u)$, depending on how much the core index differs from the cladding index. Far above cut-off for $V \gg V_{mp}^{HE}$ their transverse phase parameters according to equation (4.81) approach the pth-zero $u_{m-1,p}$ of $J_{m-1}(u)$. Far from cut-off an HE_{mp}-mode therefore becomes degenerate with the $EH_{m+2,p}$-mode. The total span over which u moves from cut-off at $u_c = V_{mp}^{HE}$ to $V \to \infty$ is now even smaller than the spacing between the pth-zero $J_{m-2}(u)$ and the pth-zero of $J_{m-1}(u)$.

Table 4.1 summarizes these results by listing the limiting forms of the characteristic equation at cut-off and very far above cut-off. Except for HE_{mp}-modes of circumferential order $m \geq 2$, these limiting forms depend only on the transverse phase parameter u and specify this parameter as a particular zero of the Bessel functions. The radial order p follows from the order of the respective zero crossing of the Bessel function in the cut-off condition. In the case of $J_1(u) = 0$ as the HE_{1p}-cut-off condition, $p = 1$ corresponds to $u = 0$. In all other cut-off conditions $p = 1$ corresponds to the first zero crossing of the respective Bessel function at a finite u-value.

4.4. WEAKLY GUIDING FIBRES

Optical waveguides in form of cladded-core fibres have usually a refractive index n_1 of the core, which is only slightly higher than the refractive index n_2 of the cladding.

TABLE 4.1

Limiting forms of characteristic equation for transverse phase parameter u

Circumferential order	HE_{mp}-modes (E_{0p} for $m = 0$) at cut-off far above	EH_{mp}-modes (H_{0p} for $m = 0$) at cut-off far above
$m = 0$	$J_0(u) = 0$	$J_1(u) = 0$	$J_0(u) = 0$	$J_1(u) = 0$
1	$J_1(u) = 0$	$J_0(u) = 0$	$J_1(u) = 0$ (excluding $u=0$)	$J_2(u) = 0$
≥ 2	$\dfrac{J_{m-2}(u)}{J_{m-1}(u)} = -\dfrac{n_1^2 - n_2^2}{n_1^2 + n_2^2}$	$J_{m-1}(u) = 0$	$J_m(u) = 0$	$J_{m+1}(u) = 0$

The reason for only a small index difference between core and cladding is to a certain extent dictated by the limited choice of suitable materials. But in some respects a small index difference also improves the transmission characteristics of the fibre. Optical fibres for long distance transmission are required to have low transmission loss and not to distort signals in transmission. Materials for such fibres should therefore have low absorption, little scattering, and low dispersion. To best fulfil these requirements limits the choice of materials to a few, notably fused silica and some compound silicate glasses. Core and cladding should preferably be made of the same base material to avoid too much internal tension when temperatures change as drastically as they do in a normal fabrication process. A suitable base material with low loss and low dispersion must then be modified differently for core and cladding in order to raise the index for the core or lower it for the cladding. The refractive index of fused silica, for example, can be raised or lowered by doping it with certain oxides. By choosing the proper doping method and avoiding any other contamination the doped material will have nearly the same low loss and dispersion as the base material. However, such modifications allow the refractive index to change only by small increments.

On the other hand often only a small index difference is desired to achieve suitable guidance and transmission characteristics. For a small index difference in a fibre with a certain core diameter $2a$, at a given wavelength λ, the fibre parameter $V = (2\pi a/\lambda)(n_1^2-n_2^2)^{\frac{1}{2}}$ will likewise be small. This limits the number of guided modes. If from equation (4.95) we have $V < 2\cdot 405$, the core will guide only the fundamental HE_{11}-mode. The fibre will transmit optical signals only in this one mode and distort these signals only because of the material dispersion and the delay dispersion of this one mode.

To remain in this single-mode range of the cladded core fibre for a given wavelength we need either to reduce the core to a very small diameter or have a low index difference between core and cladding. Fibres with such small cores are difficult to handle, and they pose serious problems in launching a wave into them or connecting such fibres for splicing. For launching and splicing, the core must be very precisely aligned with a suitable light source or with the connecting fibre, respectively. We therefore prefer to reduce the index difference in order to allow a larger core size and still remain with $(2\pi a/\lambda)(n_1^2-n_2^2)^{\frac{1}{2}}$ below 2·405 for single-mode guidance in the fibre core. To this end we aim for relative index differences of only a small fraction of a percent.

For larger values of the fibre parameter V, the core guides more than one mode; usually they will all participate in signal transmission. However, each mode suffers a different delay, so that in addition to signal distortion by material dispersion and the delay dispersion of the individual modes, the differences in delay between the modes will now also distort the signal. For small index differences the limiting angle θ_c of total reflection remains likewise small. All guided modes correspond to rays with propagation angles $\theta < \theta_c$. Such rays with small θ differ only little in path length from a ray on axis with $\theta = 0$. The delays which modes suffer on these different ray paths will spread only over a time interval which corresponds to this small difference in path length. A small index difference will therefore not only reduce the number of modes but, even more effectively, shorten the time interval of their delay spread.

These considerations show us that a small index difference between core and cladding is not only dictated by technological requirements and limitations but that it also allows larger and more convenient core sizes and that it leads to less signal distortion.

For low index differences the core guides only these modes with small propagation angles θ, and even these are only weakly guided, with their evanescent fields extending further out into the cladding than for a larger index step at the core-cladding boundary. Fibres with such small index differences of typically a few per cent or even less are called weakly guiding fibres (Gloge 1971). This designation corresponds to the designation of weakly guiding films, when the film index differs only little from the substrate index.

By assuming $n_1 \simeq n_2$, so that the relative index difference

$$\Delta = (n_1 - n_2)/n_1 \qquad (4.97)$$

is small compared to unity, the general theory of guided modes simplifies greatly, and many of their characteristics may be approximated by explicit formulas. First we note that with $\Delta \ll 1$ the numerical aperture of the cladded-core fibre is approximated by

$$\text{N.A.} = n_1 (2\Delta)^{\frac{1}{2}} \qquad (4.98)$$

and the fibre parameter by

$$V = n_1 k a (2\Delta)^{\frac{1}{2}} . \qquad (4.99)$$

The effective index $N = \beta/k$ of weakly guided core modes changes only between n_2 at cut-off and n_1 far above cut-off. Their phase parameter B from equation (4.45) may therefore be approximated by

$$B = (N - n_2)/(n_1 - n_2). \qquad (4.100)$$

We consider now a weakly guiding fibre with its cladding extending to infinity and apply the assumption $\Delta \ll 1$ to the results of the previous section. First we let $n_1 \simeq n_2$ in the characteristic equation (4.65) for guided modes. With $N \simeq n_1$ as a consequence of $n_1 \simeq n_2$, this equation reduces to

$$Y_m + X_m = \mp\, m/(u^2 B). \qquad (4.101)$$

From the considerations of the previous section, the minus sign on the right-hand side applies for HE-modes and the plus sign for EH-modes. We simplify equation (4.101) still further by replacing Y_m according to equation (4.66) and X_m according to

$$X_m = -\frac{K_{m \mp 1}}{v\, K_m} \mp \frac{m}{v^2} . \qquad (4.102)$$

In addition we express B by u and v as in equation (4.68). We use the upper sign of both equations (4.66) and (4.102) when we substitute for Y_m and X_m into equation (4.101) with its upper sign, and thus obtain the following approximate characteristic equation for HE-modes:

$$u \, J_m(u)/J_{m-1}(u) = v \, K_m(v)/K_{m-1}(v) \, . \tag{4.103}$$

With the lower signs of equations (4.66) and (4.102) together with the lower sign in equation (4.101), the characteristic equation for EH-modes is approximated by

$$u \, J_m(u)/J_{m+1}(u) = -v \, K_m(v)/K_{m+1}(v) \, . \tag{4.104}$$

With the recurrence relations

$$u \, J_m = 2(m+1) J_{m+1} - u \, J_{m+2}$$

and

$$v \, K_m = -2(m+1) K_{m+1} + v \, K_{m+2},$$

equation (4.104) may also be written as

$$u \, J_{m+2}(u)/J_{m+1}(u) = v \, K_{m+2}(v)/K_{m+1}(v) \, . \tag{4.105}$$

We note that for a given circumferential order m all these approximate forms of the characteristic equation depend only on the transverse parameters u and v, but not explicitly on any of the other fibre parameters. For any particular mode they render v a unique function of u, and the phase parameter $B = (v/V)^2$ also appears as a unique function of the fibre parameter $V = (u^2 + v^2)^{\frac{1}{2}}$ for any of the fibre modes.

The characteristic equation (4.103) for HE_{mp}-modes has the same form as equation (4.105), if this latter equation is regarded as the characteristic equation of $EH_{m-2,p}$-modes. Hence in this approximation for weakly guiding fibres any HE_{mp}-mode is degenerate with the $EH_{m-2,p}$-mode of the same radial order, but of a circumferential order that is lower by two than the circumferential order of the HE_{mp}-mode. The near-degeneracy of HE_{mp}-modes with $EH_{m-2,p}$-modes also occurs for a fibre with larger index differences between core and cladding, but then it is limited to a range of the fibre parameter V very large compared to the respective cut-off values V_{mp}^{HE} and $V_{m-2,p}^{EH}$. For large numerical apertures the HE_{mp}-modes and $EH_{m-2,p}$-modes must therefore be sufficiently far above cut-off to show this approximate degeneracy. In the limit of $V \to \infty$, they become completely degenerate, no matter how large the numerical aperture is. The limiting values for $V \to \infty$ of the transverse phase parameters $u_\infty^{HE_{m,p}} = u_{m-1,p}$ and $u_\infty^{EH_{m-2,p}} = u_{m-1,p}$ follow according to equations (4.79) and (4.80) as the very same zero of the Bessel function $J_{m-1}(u)$. But even at some distance from this limit the modes of this particular pairing are still very close to degeneracy. According to the far-above-cut-off approximation for the transverse phase parameter u in equation (4.81) and with $u_\infty^{HE_{mp}} = u_\infty^{EH_{m-2,p}}$ the difference in u between the HE_{mp}-mode and the $EH_{m-2,p}$-mode remains zero. For large values of the fibre parameter V, it is small at least to second order in $1/V$. Such a small difference in u will shift the phase constants of these nearly degenerate modes only very little; they propagate with nearly the same phase and group velocities and suffer nearly the same delay.

When we move closer to cut-off with such a pair of modes, their degeneracy splits rather more. According to equation (4.89), the $EH_{m-2,p}$-cut-off occurs at the pth-zero $u_c = V_{m-2,p}^{EH}$ of the Bessel function $J_{m-2}(u_c) = 0$. On the other hand, the cut-off for the HE_{mp}-mode occurs at values of $u_c = V_{mp}^{HE}$ that form the pth-solution, with increasing V, of equation (4.84). These solutions lie somewhat above the respective zeros of $J_{m-2}(u)$ and coincide with them only in the limit $n_1 \to n_2$. For a small but finite index difference between core and cladding the following expression for $\Delta V_{mp} = V_{m-2,p}^{EH} - V_{mp}^{HE}$ follows from a series expansion of equation (4.84) at $u_c = V_{m-2,p}^{EH}$:

$$\Delta V_{mp} = 2 \frac{n_1^2 - n_2^2}{n_1^2 + n_2^2} \frac{m-1}{V_{mp}^{HE}} . \qquad (4.106)$$

With this small difference in cut-off between an HE_{mp}-

mode and the corresponding $EH_{m-2,p}$-mode, but more and more degeneracy as we move far above cut-off, we expect that in the region of V-values between cut-off and infinity the modes of any particular such mode pair move closer and closer together in their propagation characteristics. To estimate how much they differ in these characteristics we let

$$u = u_w + \Delta u$$

and (4.107)

$$v = v_w + \Delta v ,$$

where u_w and v_w designate the transverse phase and attenuation parameters of an HE_{mp}-mode as they follow from the approximate characteristic equation (4.103) for a given value of the fibre parameter $V = (u_w^2 + v_w^2)^{\frac{1}{2}}$. The same values u_w and v_w also follow, of course, from the approximate characteristic equation (4.105) for the $EH_{m-2,p}$-mode of the nearly degenerate pair. For a weakly guiding fibre with a relative index difference $\Delta \ll 1$, the actual transverse parameters u and v differ only little from their respective approximate values u_w and v_w. Δu and Δv will then be small, of the order of Δ, and the more accurate characteristic equations (4.67) and (4.69) may be expanded about their weak guidance approximations. We obtain to first order in Δu, Δv, and Δ

$$\left[\frac{J_l(u)}{uJ_{l\pm 1}(u)}\right]' \Delta u \mp \left[\frac{K_l(v)}{vK_{l\pm 1}(v)}\right]' \Delta v = \mp \Delta \frac{K_l(v)}{vK_{l\pm 1}(v)} . \quad (4.108)$$

The upper signs apply for the $HE_{l+1,p}$-mode and the lower signs for the $EH_{l-1,p}$-mode of a nearly degenerate pair. With $m = l+1$ for the circumferential order of the HE-mode and $m-2 = l-1$ for the EH-mode, the two modes of the nearly degenerate pair differ by two in their order of circumferential dependence. u and v stand for u_w and v_w, but the index w has been omitted here for brevity. The prime on the Bessel function combinations denotes differentiation with respect to u and v, respectively, with evaluation of these derivatives at u_w and v_w.

For a given value of the fibre parameter V, the transverse parameters relate to each other as

$$v^2 = V^2 - u^2.$$

Their incremental changes are hence also related by

$$\Delta v = -(u_w/v_w) \Delta u. \tag{4.109}$$

We furthermore note that from the equivalent forms (4.103) and (4.105) of the characteristic equations for weak guidance we have

$$\left(\frac{K_l(v)}{vK_{l\pm1}(v)}\right)' = \pm \left(\frac{J_l(u)}{uJ_{l\pm1}(u)}\right)' \frac{du}{dv}, \tag{4.110}$$

where $(du/dv) = (du_w/dv_w)$ is the slope of the $u_w(v_w)$-characteristic for weak guidance. When we express Δv according to equation (4.109) and substitute for that derivative (which involves the modified Hankel functions) according to equation (4.110), we obtain from equation (4.108) the following formula:

$$\Delta u_{\mp} = \mp \Delta \frac{K_l(v)}{vK_{l\pm1}(v)} \bigg/ \left[\left(\frac{J_l(u)}{uJ_{l\pm1}(u)}\right)'\left(1 + \frac{u}{v}\frac{du}{dv}\right)\right] \tag{4.111}$$

for the change in transverse phase parameters due to a finite difference in refractive index between core and cladding. Δu_{-} designates the transverse phase parameter change of the $HE_{l+1,p}$-mode and follows from equation (4.111) with the upper signs on the right-hand side. Δu_{+}, on the other hand, is the corresponding change of the $EH_{l-1,p}$-mode and follows from equation (4.111) with the lower signs on the left-hand side. Due to a finite difference Δ of relative refractive indices, the transverse phase parameters of a nearly degenerate pair separate altogether by an amount

$$\Delta u_{lp} = \Delta u_{-} - \Delta u_{+}. \tag{4.112}$$

As a consequence the two modes of the pair have longitudinal phase parameters $B = 1 - (u/V)^2$ that now differ by

$$\Delta B = 2u_w \, \Delta u_{lp}/V^2 . \qquad (4.113)$$

For complete degeneracy between modes they propagate with exactly the same phase constant and maintain the same phase difference with respect to each other along any length of the fibre. A difference in phase parameter ΔB means

$$\Delta\beta = (n_1 - n_2) k \, \Delta B \qquad (4.114)$$

for the actual difference in phase constants. The finite index difference between core and cladding in a weakly guiding fibre splits the phase parameter of a nearly degenerate mode pair according to equation (4.113). The phase constants of these modes differ by

$$\Delta\beta = u_w \, \Delta u_{lp} / (n_1 k a^2) . \qquad (4.115)$$

As they propagate along the fibre they now shift in phase with respect to each other. When they are in phase at one point, it takes the distance

$$\Lambda = 2\pi/\Delta\beta \qquad (4.116)$$

for the modes to accumulate a total phase shift of 2π and again fall into phase. The distance Λ according to equation (4.116) is called the beat wavelength between any two modes of a waveguide. The nearly degenerate modes of a weakly guiding fibre have the beat wavelength

$$\Lambda = 2\pi n_1 k a^2 / (u_w \, \Delta u_{lp}) . \qquad (4.117)$$

Relative to the wavelength λ/n_1 in the core material, and multiplied by the square of the relative index difference Δ, this beat wavelength between any two nearly degenerate modes appears as

$$n_1 \Delta^2 \Lambda/\lambda = \tfrac{1}{2}(\dot{V}^2\Delta)/(u_w \Delta u_{lp}) ,$$

and for any pair of nearly degenerate modes depends only on the fibre parameter V. Fig.4.16 shows this normalized beat wavelength for a number of low-order pairs of nearly degenerate modes up to values of the fibre parameter of $V = 10$. The

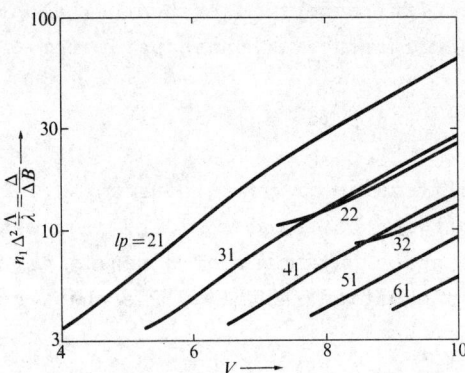

Fig.4.16. Beat wavelength between the nearly degenerate $HE_{l+1,p}$- and $EH_{l-1,p}$-modes in weakly guiding fibres with unlimited cladding.

beat wavelength is shortest near the respective cut-off. It increases with V but also, to some extent, with the transverse orders l and p. A value near cut-off of $n_1\Delta^2\Lambda/\lambda = 3\cdot4$ means for $n_1 = 1\cdot5$, $\lambda = 1$ μm, and $\Delta = 0\cdot01$ that the beat wavelength amounts to $\Lambda = 2\cdot3$ cm. Although this beat wavelength does not seem very long when we compare it to the length of fibre which we might use for signal transmission, it is still extremely long compared to the actual wavelength. Modes of the nearly degenerate pairs propagate near cut-off in phase synchronism along weakly guiding fibres for distances up to $\lambda/(n_1\Delta^2)$. Further above cut-off they maintain phase synchronism for much longer distances.

Under general conditions in a cladded core fibre, the fields of guided modes follow from equations (4.38) to (4.40) together with equation (4.41) for the transverse field components. The resulting expressions are quite complicated and of little use for discussing the field distribution. Letting the cladding extend to infinity affords some simplification. But even equations (4.59) and (4.60) are still too complicated for studying the nature of fields which are associated with guided modes. If, however, we assume the cladded-core fibre to have $\Delta \ll 1$ and consequently only weak guidance, the expressions for the transverse field components in equations (4.59) and (4.60) simplify considerably.

To evaluate these expressions for HE-waves we note that from equation (4.65) for $n_1 \simeq n_2$

$$Y_m + X_m \simeq -\frac{m}{u^2 B} .$$

The field factor ρ in equation (4.63) reduces under these circumstances to $\rho = 1$. For EH-waves we obtain from equation (4.75)

$$Y_m + \frac{n_2^2}{n_1^2} X_m = \frac{m}{u^2 B} ;$$

hence for these waves the field factor is $\rho = -1$. Therefore, for both classes of waves HE- as well as EH-waves, the ratio of longitudinal electric field maximum to longitudinal magnetic field maximum in the weakly guiding fibre equals nearly the intrinsic wave impedance of the fibre material.

With $A_E/A_H = \pm (\mu_0/\varepsilon_0)^{\frac{1}{2}}/N$ for HE- and EH-waves, respectively, and with $D_E/A_E = D_H/A_H$ according to equation (4.61), the field components inside the core of the weakly guiding fibre obtain from equation (4.59) as

$$E_{z1} = A_E J_m(ur/a)\cos m\varphi$$

$$H_{z1} = \pm A_E n_1 (\varepsilon_0/\mu_0)^{\frac{1}{2}} J_m(ur/a)\sin m\varphi$$

$$E_{r1} = \mp j A_E (\beta a/u) J_{m\mp 1}(ur/a)\cos m\varphi \qquad (4.118)$$

$$H_{r1} = -j A_E n_1 (\varepsilon_0/\mu_0)^{\frac{1}{2}} (\beta a/u) J_{m\mp 1}(ur/a)\sin m\varphi$$

$$E_{\varphi 1} = j A_E (\beta a/u) J_{m\mp 1}(ur/a)\sin m\varphi$$

$$H_{\varphi 1} = \mp j\, A_E\, n_1(\varepsilon_0/\mu_0)^{\frac{1}{2}}(\beta a/u) J_{m\mp 1}(ur/a)\cos m\varphi \qquad (4.118)$$

The field components in the cladding reduce from equation (4.60) to

$$E_{z2} = A_E [J_m(u)/K_m(v)] K_m(vr/a)\cos m\varphi$$

$$H_{z2} = \pm A_E\, n_2(\varepsilon_0/\mu_0)^{\frac{1}{2}} [J_m(u)/K_m(v)] K_m(vr/a)\sin m\varphi$$

$$E_{r2} = -j\, A_E(\beta a/v) [J_m(u)/K_m(v)] K_{m\mp 1}(vr/a)\cos m\varphi \qquad (4.119)$$

$$H_{r2} = \mp j\, A_E\, n_2(\varepsilon_0/\mu_0)^{\frac{1}{2}}(\beta a/v)[J_m(u)/K_m(v)] K_{m\mp 1}(vr/a)\sin m\varphi$$

$$E_{\varphi 2} = \pm j\, A_E(\beta a/v)[J_m(u)/K_m(v)] K_{m\mp 1}(vr/a)\sin m\varphi$$

$$H_{\varphi 2} = -j\, A_E\, n_2(\varepsilon_0/\mu_0)^{\frac{1}{2}}(\beta a/v)[J_m(u)/K_m(v)] K_{m\mp 1}(vr/a)\cos m\varphi$$

These expressions for field components in the core and cladding of modes in the weakly guiding fibre display the nearly transverse electromagnetic character of these modes. Any ray that forms a guided mode must travel at a propagation angle θ, which is smaller than the limiting angle $\theta_c = \arccos(n_2/n_1)$ of total reflection. With n_2 only slightly smaller than n_1, all $\theta < \theta_c$ remain quite small; the phase constant $\beta = n_1 k \cos\theta$ ranges only between $n_2 k$ at cut-off and $n_1 k$ far above cut-off. The transverse phase constant $(u/a) = n_1 k \sin\theta$ from equation (4.20), on the other hand, remains always small compared to $n_1 k$ for small propagation angles. We therefore have

$$(\beta a/u) \gg 1 \qquad (4.120)$$

for all modes of the weakly guiding fibre. Under these conditions the transverse attenuation constant in the cladding $(v/a) = (\beta^2 - n_2^2 k^2)^{\frac{1}{2}}$ also remains small compared to β so that

$$(\beta a/v) \gg 1. \qquad (4.121)$$

Due to equation (4.120) the longitudinal fields in the core

turn out to be weak compared to the transverse core fields, and due to equation (4.121), the longitudinal cladding fields are small compared to the transverse cladding fields. We infer this almost transverse character of the electric as well as the magnetic field also directly from the ray picture of the modes. Any ray represents a plane uniform wavelet with electromagnetic fields normal to its direction of propagation. If a ray propagates at small angles θ, its field components point in directions which, with respect to the fibre cross-section, form angles that are always smaller than θ. What angle they are actually directed with respect to the cross-section depends on their polarization. But since this angle remains always smaller than θ, the longitudinal components of their fields will always remain quite small compared to their transverse components.

The nearly transverse character for both the electric and magnetic fields of guided modes leads to a simple state of polarization of these modes. In the more general case of a cladded-core fibre with arbitrarily large index differences between core, cladding, and jacket, but no absorption in any of these regions, the transverse field components are in phase all over the fibre cross-section. Because A_E/A_H, for example, from equation (4.62) is a real quantity as well as D_E/D_H, the transverse field components in equations (4.59) and (4.60) appear indeed to be all in phase. These transverse fields are therefore linearly polarized all over the fibre cross-section. The longitudinal components in equations (4.59) and (4.60), however, have a phase shift of $\pi/2$ with respect to the transverse components. The total field of a fibre mode is therefore in a state of elliptical polarization that varies over the cross-section. In a weakly guiding fibre, however, the longitudinal field remains small compared to the linearly polarized transverse field. The nearly transverse character of the electromagnetic field hence makes these fields almost linearly polarized.

An even simpler state of polarization arises when we include the near degeneracy of HE_{mp}- and $EH_{m-2,p}$-modes in our considerations. We decompose the transverse field into its Cartesian components according to

$$E_x = E_r \cos\varphi - E_\varphi \sin\varphi$$

$$E_y = E_r \sin\varphi + E_\varphi \cos\varphi .$$

(4.122)

The transverse fields of equation (4.118) inside the core have the following Cartesian components:

$$E_{x1} = \mp j\, A_E (\beta a/u) J_{m\mp 1}(ur/a) \cos(m\mp 1)\varphi$$

$$E_{y1} = j\, A_E (\beta a/u) J_{m\mp 1}(ur/a) \sin(m\mp 1)\varphi$$

$$H_{x1} = -j\, A_E\, n_1 (\varepsilon_0/\mu_0)^{\frac{1}{2}} (\beta a/u) J_{m\mp 1}(ur/a) \sin(m\mp 1)\varphi$$

$$H_{y1} = \mp j\, A_E\, n_1 (\varepsilon_0/\mu_0)^{\frac{1}{2}} (\beta a/u) J_{m\mp 1}(ur/a) \cos(m\mp 1)\varphi,$$

(4.123)

where the upper signs apply for HE_{mp}-modes and the lower signs for EH_{mp}-modes. The cladding field of equation (4.119) has the following Cartesian components in the transverse direction:

$$E_{x2} = -j\, A_E (\beta a/v) [J_m(u)/K_m(v)] K_{m\mp 1}(vr/a) \cos(m\mp 1)\varphi$$

$$E_{y2} = \pm j\, A_E (\beta a/v) [J_m(u)/K_m(v)] K_{m\mp 1}(vr/a) \sin(m\mp 1)\varphi$$

$$H_{x2} = \mp j\, A_E\, n_2 (\varepsilon_0/\mu_0)^{\frac{1}{2}} (\beta a/v) [J_m(u)/K_m(v)] K_{m\mp 1}(vr/a) \sin(m\mp 1)\varphi$$

$$H_{y2} = -j\, A_E\, n_2 (\varepsilon_0/\mu_0)^{\frac{1}{2}} (\beta a/v) [J_m(u)/K_m(u)] K_{m\mp 1}(vr/a) \cos(m\mp 1)\varphi,$$

(4.124)

where again the upper signs apply for HE_{mp}-modes and the lower signs for EH_{mp}-modes. Note that these transverse Cartesian components differ in the order of their circumferential dependence from their small longitudinal components. An HE_{mp}-mode has transverse Cartesian field components with circumferential dependence of order $m - 1$ while the transverse Cartesian components of an EH_{mp}-mode have $m + 1$ as the order of their circumferential dependence.

An $HE_{l+1,p}$-mode is, however, nearly degenerate with the corresponding $EH_{l-1,p}$-mode. We therefore conclude that HE-modes and EH-modes of corresponding radial order p and equal circumferential order l for their Cartesian components form nearly degenerate pairs.

Let us assert now that this degeneracy is complete. Such an assertion will be valid, if only the indices of core and cladding differ by a small enough amount. Any combination of an $HE_{l+1,p}$-mode with an $EH_{l-1,p}$-mode will then likewise constitute a guided mode of the fibre because it has all the characteristics of such a mode: it propagates with one phase constant, so that both terms of the combination maintain the same phase difference all along the fibre and its transverse field distribution stays the same everywhere.

As one such combination with particularly unique polarization we superimpose an $HE_{l+1,p}$-mode with amplitude $A_E = 1$ and an $EH_{l-1,p}$-mode with amplitude $A_E = -1$. The HE-mode in this combination has the circumferential order $m = l + 1$ and the EH-mode has $m = l - 1$; their transverse Cartesian components are therefore both of circumferential order l. Inside the core we obtain the following Cartesian field components for this particular superposition of degenerate modes:

$$E_{z1} = J_{l+1}(ur/a)\cos(l+1)\varphi - J_{l-1}(ur/a)\cos(l-1)\varphi$$

$$H_{z1} = n_1(\varepsilon_0/\mu_0)^{\frac{1}{2}}[J_{l+1}(ur/a)\sin(l+1)\varphi + J_{l-1}(ur/a)\sin(l-1)\varphi]$$

$$E_{x1} = -j\ 2(\beta a/u)J_l(ur/a)\cos l\varphi \qquad (4.125)$$

$$H_{y1} = -j\ 2n_1(\varepsilon_0/\mu_0)^{\frac{1}{2}}(\beta a/u)J_l(ur/a)\cos l\varphi$$

$$E_{y1} = H_{x1} = 0.$$

When we superimpose the cladding fields of the $HE_{l+1,p}$-mode with $A_E = 1$ and the $EH_{l-1,p}$-mode with $A_E = -1$ we obtain for the longitudinal electric field

$$E_{z2} = [J_{l+1}(u)/K_{l+1}(v)]K_{l+1}(vr/a)\cos(l+1)\varphi -$$

$$- [J_{l-1}(u)/K_{l-1}(v)]K_{l-1}(vr/a)\cos(l-1)\varphi . \qquad (4.126)$$

To simplify this expression as well as the expressions for the other field components in the cladding we note that

$$J_{l+1}(u)/K_{l+1}(v) = -J_{l-1}(u)/K_{l-1}(v) \qquad (4.127)$$

is an alternative form of the approximate characteristic equation for modes of the weakly guiding fibre. It arises from equation (4.104) when we apply the appropriate recurrence relations for Bessel functions. With equation (4.127) we obtain as the Cartesian field components of our present superposition

$$E_{z2} = [J_{l+1}(u)/K_{l+1}(v)][K_{l+1}(vr/a)\cos(l+1)\varphi + K_{l-1}(vr/a)\cos(l-1)\varphi]$$

$$H_{z2} = n_1(\varepsilon_0/\mu_0)^{\frac{1}{2}}[J_{l+1}(u)/K_{l+1}(v)]$$
$$[K_{l+1}(vr/a)\cos(l+1)\varphi - K_{l-1}(vr/a)\sin(l-1)\varphi]$$

$$E_{x2} = -j2(\beta a/v)[J_{l+1}(u)/K_{l+1}(v)]K_l(vr/a)\cos l\varphi \qquad (4.128)$$

$$H_{y2} = -j2\, n_2(\varepsilon_0/\mu_0)^{\frac{1}{2}}(\beta a/v)[J_{l+1}(u)/K_{l+1}(v)]K_l(vr/a)\cos l\varphi$$

$$E_{y2} = H_{x2} = 0.$$

Because $\beta a \gg u, v$ the longitudinal fields in this superposition are still small compared to the transverse fields. These dominating transverse fields, however, now consist only of E_x and H_y in core and cladding; E_y and H_x are zero. The total field is therefore not only nearly linearly polarized like the fields of the constituent $HE_{l+1,p}$- and $EH_{l-1,p}$-modes; it is also nearly uniformly polarized all over the fibre cross-section. Because of this unique and simple state of polarization these particular superpositions of fibre modes are quite often taken as the alternative set of guided modes in weakly guiding fibres. They are designated as LP_{lp}-modes (Gloge 1971) for linearly polarized modes of circumferential order l for their transverse fields and radial order p. Actually a designation that expresses the uniform state of polarization over all the fibre cross-section would be more appropriate, because the constituent HE- and EH-modes already have the linear polarization in a weakly guiding fibre. We will nevertheless follow common practice and adopt the designation LP_{lp}-modes here.

Such a linearly and uniformly polarized mode with its predominantly transverse electromagnetic field has its transverse electric field perpendicular to its transverse magnetic field. The ratio of its transverse electric field strength to its transverse magnetic field strength is constant across each of the homogeneous regions, core as well as cladding, of the fibre. It is equal to the intrinsic wave impedance $(\mu_0/\varepsilon_0)^{\frac{1}{2}}/n$ of the particular cross-sectional region. In these respects the LP-modes behave like plane uniform waves, which also have this uniform polarization and the intrinsic impedance of the material as wave impedance.

Each of these LP_{lp}-modes occurs not only with $\cos l\varphi$ for the circumferential distribution of its transverse field components as in equations (4.125) and (4.128), but also with $\sin l\varphi$. Replacing $\cos l\varphi$ by $\sin l\varphi$ rotates the field distribution by $\varphi = \pi/(2l)$ but leaves the direction of polarization unchanged.

By superimposing an $HE_{l-1,p}$-mode with $A_E = 1$ and an $EH_{l+1,p}$-mode also with $A_E = 1$ from equations (4.123) and (4.124), we obtain an $LP_{l,p}$-mode with its transverse electric field in the y-direction and its transverse magnetic field in the x-direction. Both these transverse fields depend on φ as $\sin l\varphi$. Therefore, in its circumferential dependence this third $LP_{l,p}$-mode corresponds to the second $LP_{l,p}$-mode which resulted from equations (4.125) and (4.128) when we replaced $\cos l\varphi$ by $\sin l\varphi$. In its polarization, however, it differs from this second mode and is actually polarized orthogonal to it. A fourth $LP_{l,p}$-mode is finally obtained when we take the third mode and replace $\sin l\varphi$ by $\cos l\varphi$. This mode corresponds in its φ-dependence to the first LP_{lp}-mode as it appears in equations (4.125) and (4.128) but is polarized orthogonal to it.

Altogether then, each LP_{lp}-mode occurs in four different versions, two orthogonal polarizations, each with $\cos l\varphi$ and $\sin l\varphi$. Fig.4.9 illustrates these four different versions of transverse field polarization and distribution for the simple example of the LP_{11}-mode. The shaded area in this figure indicates the intensity distribution across the fibre core. The solid arrows give the direction of the electric field and the broken arrows the direction of the magnetic field. The relative length of both classes of arrows indicates the strength of the respective field.

The LP_{11}-mode of this particular example is a combination of the HE_{21}-mode with either the axially symmetric E_{01}-mode or the H_{01}-mode. Fig.4.17 illustrates these combinations of HE_{21} and E_{01} as well as of HE_{21} rotated by $\pi/4$ in its transverse field distribution and combined with H_{01}. The LP_{11}-mode in Fig.4.17(a) represents just the superposition of fields

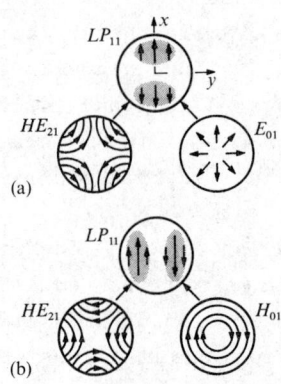

Fig.4.17. Transverse electric field and intensity distribution of two LP_{11}-modes and their composition from exact modes of the cladded-core fibre.

from equations (4.118) and (4.119) that leads to the LP_{11}-mode with the transverse fields of equations (4.125) and (4.128). With $m = 0$, equations (4.118) and (4.119) indeed describe the fields of E_{0p}-waves. The other polarization and distribution of transverse fields in Fig.4.17(b) also has its electric field polarized in the x-direction; it obtains from equations (4.118) and (4.119) when we replace $\cos m\varphi$ by $\sin m\varphi$ and vice versa. For $m = 0$, these new expressions describe the fields of H_{0p}-waves with $E_z = 0$.

The LP_{0p}-modes represent a special case. They cannot be combined from $HE_{l+1,p}$- and $EH_{l-1,p}$-modes, because with $l = 0$, the $EH_{l-1,p}$-mode would be of negative circumferential order which has no physical significance and hence such EH-modes do not exist. Actually any of the HE_{1p}-modes is by itself uniformly and linearly polarized, as may be readily verified by evaluating equations (4.123) and (4.124) with the upper signs and for $m = 1$. This leads to the LP_{0p}-field of one particular polarization

$$E_{z1} = J_1(ur/a)\cos \varphi$$

$$H_{z1} = n_1(\varepsilon_0/\mu_0)^{\frac{1}{2}} J_1(ur/a)\sin \varphi$$

$$E_{x1} = -j(\beta a/u) J_0(ur/a)$$

$$H_{y1} = -jn_1(\varepsilon_0/\mu_0)^{\frac{1}{2}}(\beta a/u) J_0(ur/a)$$

$$E_{y1} = H_{x1} = 0$$

$$E_{z2} = [J_1(u)/K_1(v)]K_1(vr/a)\cos \varphi$$

$$H_{z2} = n_2(\varepsilon_0/\mu_0)^{\frac{1}{2}}[J_1(u)/K_1(v)]/K_1(vr/a)\sin \varphi$$

$$E_{x2} = -j(\beta a/v)[J_1(u)/K_1(v)] K_0(vr/a)$$

$$H_{y2} = -jn_2(\varepsilon_0/\mu_0)^{\frac{1}{2}}(\beta a/v)[J_1(u) K_1(v)] K_0(vr/a)$$

$$E_{y2} = H_{x2} = 0.$$

(4.129)

The transverse fields of these LP_{0p}-modes have an axially symmetric distribution. The other polarization of LP_{0p}-modes with the electric field in the y-direction comes from equations (4.118) and (4.119) as well as (4.123) and (4.124) when all cos functions of φ are replaced by sine functions and vice versa.

The transverse field distribution of an LP_{lp}-mode has the circumferential dependence $\cos l\varphi$ or $\sin l\varphi$. With the radial coordinate it varies as $J_l(ur/a)$ inside the core, while in the cladding it decays according to $K_l(vr/a)$. The transverse phase parameter u for the radial dependence inside the core ranges between the pth-zero $u_{l-1,p}$ of $J_{l-1}(u)$ at cut-off and the pth-zero $u_{l,p}$ of $J_l(u)$ far above cut-off, discounting the zero of $J_l(u)$ at $u = 0$ for $l > 0$. Fig.4.18 shows the Bessel functions of zero and first order. Their zero crossings mark the u-ranges for LP_{0p}- and LP_{1p}-modes. They also indicate the radial field distribution inside the core for any u-value of such an LP-mode between cut-off and far above cut-off.

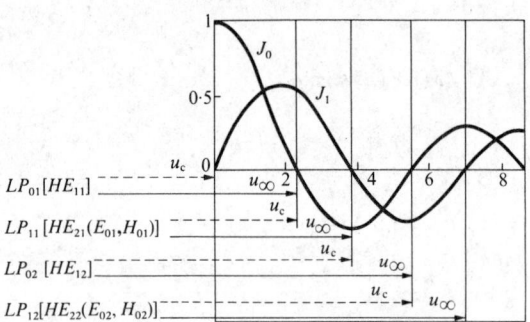

Fig.4.18. Range of transverse phase parameter u for LP_{0p}- and LP_{1p}-modes and their radial field distribution inside the fibre core.

The LP_{01}-mode is just the fundamental HE_{11}-mode of the fibre and has zero cut-off. Its transverse phase parameter ranges between $u_c = 0$ and $u_\infty = 2 \cdot 405$, the first zero crossing of $J_0(u)$. Near cut-off its fields are almost uniform across the core with only a slight decrease of $J_0(u)$ for small u. Far above cut-off its core fields fall off radially according to J_0 ($2 \cdot 405\ r/a$). The LP_{11}-fields depend on r as J_1 ($2 \cdot 405\ r/a$) at cut-off, far above cut-off their r-dependence follows from J_1 ($3 \cdot 83\ r/a$).

With their uniform and linear polarization in only one transverse direction all over the cross-section, the LP_{1p}-modes are well suited to describing field and intensity distribution as well as wave propagation in weakly guiding fibres. But there is one even more practical reason for expressing wave propagation in terms of these uniformly and linearly polarized modes. The sources which we use in optical communications preferably generate coherent or at least partially coherent light. We in fact prefer lasers as light generators

in optical communications. The laser output beam or wave
depends in its transverse field and intensity distribution on
the particular type of laser - gas, solid-state, or diode laser
- and on the form of its optical cavity. Usually, however,
these output beams or waves have predominantly transverse electromagnetic fields of a uniform and linear polarization. In
this respect they correspond directly to the LP_{lp}-modes of the
weakly guiding fibres. A similar state of polarization prevails for other types of optical waveguides. The various forms
of strip and rib guides used in planar optical circuits
have modes of propagation which under corresponding conditions
of weak guidance are likewise uniformly and linearly polarized.
When we launch coherent or partially coherent light from laser
sources into fibres, or when we connect fibres to the guiding
structures of planar optical circuits, we will primarily excite
the LP_{lp}-modes.

The dispersion characteristics of LP_{lp}-modes as well as of
their constituent $HE_{l+1,p}$- and $EH_{l-1,p}$-modes follow from any
of the approximate forms (4.103), (4.104), or (4.127) of the
characteristic equation. These approximate forms give u as an
implicit function of v for any of the modes. No other parameter appears in these equations. For any specific LP_{lp}-mode,
u is therefore a unique function of v and the phase parameter
$B = (v/V)^2$ follows as a unique function of the fibre parameter
$V = (u^2 + v^2)^{\frac{1}{2}}$:

$$B = 1 - u^2/v^2 .$$

From a numerical evaluation of the approximate characteristic
equation (Gloge 1971), Fig.4.19 shows B as a function of V for
a number of low-order LP_{lp}-modes. The general relation (4.45)
between phase constant β and phase parameter B reduces to the
approximate expression (4.100) for the small index difference
of a weakly guiding fibre. With B from Fig.4.19 for any value
of the fibre parameter, the phase constant is then approximately given by

$$\beta = [B(n_1 - n_2) + n_2] k . \qquad (4.130)$$

The LP_{lp}-modes of weakly guiding fibres propagate for some distance with phase constants from the numerical solution. Eventually, however, the small but finite difference between core and cladding index causes these modes to split into $HE_{l+1,p}$- and $EH_{l-1,p}$-modes. Only the LP_{0p}-modes, each of which is just the HE_{1p}-mode, maintain their transverse field distribution for any index difference and any fibre

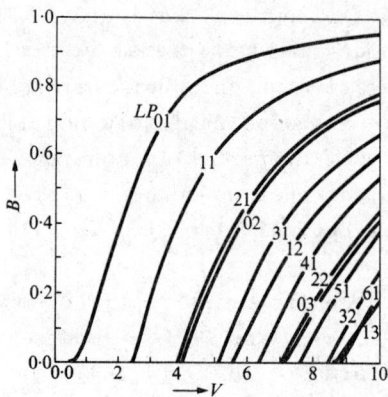

Fig.4.19. Phase parameter $B = (\beta^2/k^2 - n_2^2)/(n_1^2 - n_2^2)$ as a function of fibre parameter $V = ka(n_1^2 - n_2^2)^{\frac{1}{2}}$ for low-order LP_{lp}-modes of the weakly guiding fibre with unlimited cladding (Gloge 1971).

length. The breakup of all other LP_{lp}-modes begins whenever the distance along the fibre is no longer short compared to the beat wavelength between the constituent HE- and EH-modes of the nearly degenerate pair. We estimate this distance for an $LP_{l,p}$-mode at a given value of the fibre parameter V by taking the parameter values u and v from the numerical solution of the approximate characteristic equation, and then evaluate Δu_{lp} in equation (4.112) with the respective differences $\Delta u_{\mp}$ from equation (4.111). The phase parameter difference then follows from equation (4.113) and allows us to evaluate the phase difference and beat wavelength according to equations (4.114) and (4.116), respectively.

Instead of an exact numerical solution of the characteristic equation (4.104) for the weakly guiding fibre, one can also obtain approximate analytic solutions. For such analytic approximations we remember the relatively narrow range of u between cut-off at $V = u_c = u_{l-1,p}$, and $u_\infty = u_{lp}$ very far above cut-off for $V \to \infty$. We therefore base the analytic approximation on one of these limiting values as a starting point and obtain u as a function of V by integrating du/dV, for example, with cut-off as the starting point:

$$u = u_{l-1,p} + \int_{u_{l-1,p}}^{V} \frac{du}{dV} \, dV . \qquad (4.131)$$

For du/dV we differentiate the characteristic equation on both sides with respect to V and obtain

$$\frac{du}{dV} = \frac{u}{V} [1 - \kappa_l(v)] , \qquad (4.132)$$

where

$$\kappa_l(v) = \frac{K_l^2(v)}{K_{l-1}(v) \, K_{l+1}(v)} . \qquad (4.133)$$

If we now find a suitable analytic approximation for $\kappa_l(v)$ which, when substituted into equation (4.132), allows us to evaluate the integral in equation (4.131), we will obtain an even better analytic approximation for u as a function of V. Any errors which we introduce by approximating $\kappa_l(v)$ analytically will diminish through the integration of du/dV. Whatever remains as a small error in the integral in equation (4.131) will be of minor significance, in particular for higher order modes for which we start with large cut-off values $u_{l-1,p}$, and have only little to add all through the range of u up to the next root u_{lp} of the Bessel function $J_l(u)$ very far above cut-off.

As a suitable analytic approximation for equation (4.133) we introduce (Gloge 1971)

$$\kappa_l(v) \simeq 1 - (v^2 + l^2 + 1)^{-1/2} , \qquad (4.134)$$

which provides a reasonable fit for all values of v including the asymptotic approximation $\kappa_l \simeq 1 - 1/V$ for $v \to \infty$, where

$v \to V$. To evaluate the integral in equation (4.131) with $\kappa_l(v)$ according to equation (4.134) we replace v^2 by $v^2 - u^2$ and, with u varying only between $u_{l-1,p}$ and u_{lp}, let

$$v^2 \simeq V^2 - u^2_{l-1,p} \, .$$

We can then solve the differential equation (4.132) by separation of variables and obtain instead of equation (4.131)

$$u(V) = u_{l-1,p} \exp\{[\text{arc } \sin(s/u_{l-1,p}) - \text{arc } \sin(s/V)]/s\} \quad (4.135)$$

with

$$s = (u^2_{l-1,p} - l^2 - 1)^{1/2} \, . \tag{4.136}$$

This expression approximates the actual u-V characteristic of all modes of the weakly guiding fibre except the fundamental LP_{01}- or HE_{11}-mode. The HE_{11}-mode has zero cut-off and hence starts with $u = V = 0$. In the vicinity of the zero cut-off its transverse phase parameter follows according to equation (4.72) from

$$u^2 \simeq 2/\ln(2/\gamma v), \tag{4.137}$$

so that v remains quite small compared to u even for some distance from the zero cut-off. For small values of the fibre parameter V, u is therefore only slightly smaller than V. Far above cut-off, u approaches $u_{01} = 2 \cdot 405$, the first zero of $J_0(u)$. An approximation which not only yields these limiting values but provides an adequate fit anywhere in between has also been obtained by Gloge (1971) following a similar approach to the above. This approximation results in

$$u(V) = \frac{u_{01} V}{1 + [(u_{01}-1)^4 + V^4]^{1/4}} \tag{4.138}$$

for the HE_{11}-mode. Far enough above cut-off for $V \gg s$ both equations (4.135) and (4.138) reduce to

CLADDED-CORE FIBRES   361

$$u = u_{lp}\left(1 - \frac{1}{V}\right) . \tag{4.139}$$

When we use the analytical approximations in equations (4.135) or (4.138) to evaluate the phase parameter B, we find that this approximation for B deviates too little from the exact numerical solution to be noticed in the dispersion diagram (Fig.4.19).

When we employ fibres for the transmission of optical signals, we are not so much interested in the phase constants of individual modes but rather in the delay which these modes suffer during transmission. If the spectrum of the optical signals is confined to a narrow enough band, the group delay at the centre frequency of this band will represent the signal delay adequately. From equation (1.93), we must differentiate the phase constant β with respect to the angular frequency in order to obtain the group delay per unit axial distance. Equivalently, this group delay per unit distance is given by

$$\tau = \frac{1}{c}\frac{d\beta}{dk} \tag{4.140}$$

with c as the velocity of light in vacuum. When we differentiate

$$\beta = k[n_2^2 + (n_1^2 - n_2^2)B]^{1/2} , \tag{4.141}$$

we must take account of the k-dependence not only of B but also of n_1 and n_2 or Δ, respectively. Because of the material dispersion, as we discussed in section 1.3, the refractive index varies with frequency and hence with k. The derivative of β is therefore given by

$$\frac{d\beta}{dk} = \{n_2 n_2' + [B + \frac{1}{2} V \frac{dB}{dV}](n_1 n_1' - n_2 n_2')\}/[n_2^2 + (n_1^2 - n_2^2)B]^{1/2}, \tag{4.142}$$

where n_1' and n_2' are the group indices of the core and cladding material as defined by equation (1.99). Far above cut-off the phase parameter approaches unity ($B \simeq 1$) and becomes independent of the fibre parameter ($dB/dV \simeq 0$). The derivative of β then reduces to $d\beta/dk = n_1'$. Any mode which is sufficiently far above cut-off propagates with the group delay of a plane uniform wave in the core material.

In a weakly guiding fibre, n_1 and n_2 differ only very little. In this case equation (4.142) reduces to

$$\frac{d\beta}{dk} = n_2' + (n_1' - n_2') B + \frac{1}{2} [(n_1/n_2) n_1' - n_2'] V(dB/dV) . \quad (4.143)$$

Low-loss fibres are made of highly transparent materials with little absorption. Such materials have weak dispersion because both absorption and dispersion are caused by molecular or electronic resonances. In the case of weak dispersion, however, we have $k(dn/dk) \ll n$ in the expression $n' = n + k(dn/dk)$ for the group index. The group index is then not too different from the refractive index. For weakly guiding fibres moreover, core and cladding consist of similar materials, and not only the refractive indices, but also their dispersion characteristics resemble each other. In this case their relative differences are very small and differ only a little. We may then utilize the approximations

$$(n_1 - n_2)/n_2 \simeq (n_1' - n_2')/n_2' \ll 1 \quad (4.144)$$

with which equation (4.143) reduces to

$$d\beta/dk = n_2' + (n_1' - n_2') \, d(VB)/dV . \quad (4.145)$$

The group delay per unit distance is then given by

$$\tau = \frac{1}{c} [n_2' + (n_1' - n_2') d(VB)/dV] . \quad (4.146)$$

The first term within the brackets of this expression represents the delay contribution from material dispersion. In the present approximation material dispersion causes the same delay for all modes independent of their order or distance from cut-off. The second term in equation (4.146) gives the delay contribution from mode dispersion.

In the ray picture of guided modes, the zigzag path changes its direction when we shift the frequency or wavelength

of light. At cut-off the rays of any mode propagate at an
angle θ with respect to the fibre axis which is equal to the
limiting angle of total reflection at the core-cladding boundary. Far above cut-off the rays run more and more parallel
to the fibre axis. With decreasing θ the path length of rays
shortens and the phase shift in total reflection increases.
Both effects cause β to change non-linearly with frequency
and the delay to vary with frequency.

To evaluate the delay contribution from mode dispersion
we express the derivative $d(VB)/dV$ by

$$d(VB)/dV = 1 + (u/V)^2 - 2(u/V)(du/dV)$$

and introduce du/dV from equation (4.132)

$$d(VB)/dV = 1 - (u/V)^2 [1 - 2\kappa_l(v)] . \qquad (4.147)$$

With the transverse parameters u and v as unique functions of
V for any particular LP_{lp}-mode of the weakly guiding fibre,
the mode delay factor $d(VB)/dV$ also follows from equation
(4.147) as a unique function of V. This function may be
evaluated approximately by using the analytic approximations
for $u(V)$ in equations (4.135) and (4.138). Exact values of
$d(VB)/dV$ may be obtained from a computer solution of the characteristic equation (4.104).

Fig.4.20 shows the mode delay factor of a number of low-order modes of the weakly guiding fibre. Far above cut-off
for $V \simeq v \to \infty$ we have $\kappa_l \simeq 1$ and hence the mode delay factor
approaches unity in this limit. Within the approximation of
equation (4.139) this limiting value again yields the delay
of a plane uniform wave in the core material for any mode
sufficiently far above cut-off.

At cut-off we have $u = V$ and $v = 0$, and the mode delay
factor is given by

$$d(VB)/dV = 2\kappa_l(0) = \begin{cases} 0 & \text{for } l = 0,1 \\ 2(1-1/l) & \text{for } l \geq 2. \end{cases} \qquad (4.148)$$

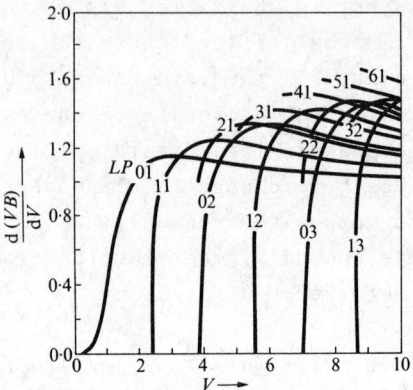

Fig.4.20. Mode delay factor of LP_{lp}-modes in weakly guiding fibres with unlimited cladding (Gloge 1971).

All LP_{0p}- and LP_{1p}-modes including the fundamental HE_{11}-mode start at their respective cut-off with the group delay of a plane uniform wave in the cladding material. We should expec such a group delay at cut-off for all modes that have meridio al rays in the ray picture. Such meridional rays see the core cladding interface as a plane boundary in their plane of incidence. Their propagation angle θ forms the complement of th angle of incidence and, at cut-off, it is equal to the limiti angle of total reflection. Their cut-off fields extend infinitely wide into the cladding without any exponential decay These cut-off fields therefore propagate as plane uniform waves in the cladding and also suffer the delay of such waves A similar extension of cut-off fields into the cladding occur for the LP_{1p}-modes so that at cut-off these modes also propagate as plane uniform waves in the cladding.

The situation at cut-off appears different for all LP_{lp} modes of circumferential order $l = 2$ and higher. Such modes have skew rays in the ray picture, which see the core-claddin boundary with curvature in their plane of incidence. At cut-off the complement of their angle of incidence remains smalle than the limiting angle $θ_c$ of total reflection. For a suffic iently large circumferential order and a correspondingly larg skew angle $φ_a$ in Fig.4.4 this angle in fact remains so small compared to θ_c that the phase shift in total reflection is practically independent of frequency. The delay which these modes of high circumferential order suffer at cut-off corresponds under such circumstances to the delay of a plane uniform wave which propagates on the spiralling zig-zag path of the skew ray within the core. This path is longer by the factor $1/\cos\theta_c = n_1/n_2$ than the direct path of a ray on axis so that with $\tau = n_1'/c$ for the delay of an on-axis ray, the high-order skew ray has a delay at cut-off according to $\tau = (n_1/n_2)(n_1'/c)$. For similar dispersion characteristics of core and cladding material, equation (4.144) holds, and we have $n_1/n_2 \simeq n_1'/n_2'$. The high-order skew ray at cut-off will then be delayed by

$$\tau = (n_1')^2/(n_2'c) \simeq (1/c)[n_2' + 2(n_1' - n_2')] . \quad (4.149)$$

From equations (4.146) and (4.148) we obtain the more exact expression

$$\tau = (1/c)[n_2' + 2(n_1'-n_2')(1-1/l)]$$

for the delay at cut-off of any LP_{lp}-mode with circumferential order $l \geq 2$. This expression approaches the simple formula (4.149) from the ray picture when $l \gg 1$. The cut-off delays are actually somewhat shorter than those given by the ray formula (4.149).

We note from Fig.4.20 that the cut-off delay represents also the maximum delay for any LP_{lp}-mode of circumferential order l large compared to its radial order p. If the fibre parameter V is large enough for the fibre to guide many modes, the mode closest to cut-off has the longest delay. The difference in delay per unit distance between this and the lowest order LP_{01}-mode will be

$$\Delta\tau = (n_1'-n_2')/c . \quad (4.150)$$

It corresponds again to the difference in path length between a skew ray at the propagation angle $\theta_c = \arccos(n_2/n_1)$ and a ray on axis.

The total range over which the delay of modes in a fibre with a large V-value spreads can actually be still larger, if we take the delay at cut-off of LP_{0p}- and LP_{1p}-modes into consideration. But this is only the case if the fibre parameter V is at, or only little above, the cut-off value of such a mode. Even then it will be only this one particular LP_{0p}- or LP_{1p}-mode that would widen the total delay spread to a maximum of

$$\Delta\tau = 2(n_1' - n_2')/c . \qquad (4.151)$$

For all practical purposes we can disregard such singular situations, where only one or the other LP_{0p}- or LP_{1p}-mode is at or very near cut-off, and take equation (4.150) as the total delay spread.

The power which a guided mode transmits along the fibre is partly confined to the core and propagates partly in the cladding. Near cut-off we expect a substantial fraction of the total power to be guided by the evanescent cladding fields while far above cut-off the fields concentrate inside the core and the total power will then likewise travel in the core.

To determine the total power and its portions inside the core and in the cladding we integrate the axial component of the Poynting vector

$$S_z = E_x H_y^* - E_y H_x^* \qquad (4.152)$$

over the respective part of the fibre cross-section. We take the x-polarized LP-modes to evaluate the power flow and designate the maximum value of the electric field at the core-cladding boundary by E_l. The field distribution in equation (4.125) has

$$E_l = 2(\beta a/u) J_l(u) . \qquad (4.153)$$

With E_l as a field factor the transverse field components are given by

$$E_x = \begin{Bmatrix} 1/n_1 \\ 1/n_2 \end{Bmatrix} \left(\frac{\mu_0}{\varepsilon_0}\right)^{\frac{1}{2}} H_y = E_l \begin{Bmatrix} J_l(ur/a)/J_l(u) \\ K_l(vr/a)/K_l(v) \end{Bmatrix} \cos l\varphi \quad (4.154)$$

in core and cladding, respectively.

We evaluate the integral

$$P_1 = \int_0^a \int_0^{2\pi} E_x H_y^* \, r \, dr \, d\varphi \quad (4.155)$$

with the transverse core fields in the upper line of equation (4.154) and obtain for the power which a mode guides inside the core

$$P_1 = (\mu_0/\varepsilon_0)^{\frac{1}{2}} (\pi a^2/n_1 \varepsilon_l) \, E_l^2 [1 - J_{l-1}(u) \, J_{l+1}(u)/J_l^2(u)] \, . \quad (4.156)$$

ε_l designates the Neumann number defined by

$$\varepsilon_l = \begin{cases} 1 & \text{for } l = 0 \\ 2 & \text{for } l \neq 0 \end{cases} \quad (4.157)$$

The approximate form (4.127) of the characteristic equation lets us express the Bessel functions of argument u by modified Hankel functions of argument v which leads to

$$P_1 = (\mu_0/\varepsilon_0)^{\frac{1}{2}} (\pi a^2/n_1 \varepsilon_l) \, E_l^2 [1 + v^2/(u^2 \kappa_l)] \quad (4.158)$$

with κ_l according to equation (4.133). Integrating the power flow density $S_z = E_x H_y^*$ over the cladding region results in

$$P_2 = (\mu_0/\varepsilon_0)^{\frac{1}{2}} (\pi a^2/n_2 \varepsilon_l) \, E_l^2 [1/\kappa_l - 1] \quad (4.159)$$

for the power which is guided in the cladding. The total power guided by a particular mode then becomes

$$P = P_1 + P_2 = (\mu_0/\varepsilon_0)^{\frac{1}{2}} (\pi a^2/n \varepsilon_l) \, E_l^2 \, v^2/(u^2 \kappa_l) \, , \quad (4.160)$$

where, with $n \simeq n_1 \simeq n_2$, we ignore the small difference in core and cladding index for the weakly guiding fibre. Of the total power in a mode the fraction

$$P_1/P = 1 - (u/V)^2 (1-\kappa_l) \qquad (4.161)$$

is guided by the core while the remaining fraction

$$P_2/P = (u/V)^2 (1-\kappa_l) \qquad (4.162)$$

travels in the cladding.

Very far above cut-off, when $u \ll V$, we obtain $P_1 = P$ and $P_2 = 0$. The fields confine themselves completely to the core and all the power also travels inside the core. Near cut-off, we have $u \simeq V$, and the transverse attenuation parameter v tends to zero. In this case, $\kappa_0 \simeq v^2 \log^2(2/\gamma v)$, $\kappa_1 = 1/[2 \log(2/\gamma v)]$ but $\kappa_l = 1 - 1/l$ for $l \geq 2$. At cut-off therefore, the LP_{0p}- and LP_{1p}-modes have all their power travelling in the cladding while all LP_{lp}-modes of circumferential order $l = 2$ or higher show at cut-off a ratio of core to cladding power of $l - 1$. The higher a mode is in circumferential order, the more of its power will travel inside the core even at cut-off. Fig.4.21 shows with its two scales on the vertical axis the power fractions in core and cladding for a number of low-order LP_{lp}-modes.

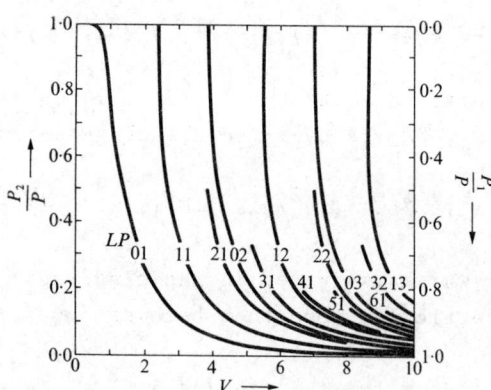

Fig.4.21. Power fractions in core and cladding of LP_{lp}-modes in weakly guiding fibres with unlimited cladding (Gloge 1971).

4.5. LEAKY MODES IN WEAKLY GUIDING FIBRES

In the previous three sections we have considered field solutions of cladded-core fibres, which correspond to guided modes. As long as core cladding and jacket are loss free and homogeneous, these guided modes propagate without attenuation, and a real phase constant β characterizes their field dependence in axial direction according to $\exp(-j\beta z)$. In radial and circumferential direction their fields inside the core, but not too close to the core centre, have a standing wave character as described by Bessel functions of the real argument (ur/a). In the cladding or jacket region the fields have in the radial direction the evanescent character of modified Hankel functions with the real argument vr/a or wr/a, respectively. In the asymptotic expansions of modified Hankel functions for large arguments the radial dependence is dominated by the exponential factor $\exp(-vr/a)$ or $\exp(-wr/a)$. This exponential decay guarantees proper behaviour of cladding or jacket fields out to infinity and lets the guided modes be source-free solutions of the field equations which have physical reality. We can excite these modes by a suitable arrangement of sources and utilize them for signal transmission.

In the ray picture of Section 4.1 for the cladded-core fibre with unlimited cladding, we found not only these guided modes but, in addition, a set of leaky modes. While the guided modes are formed by rays with propagation angles θ with respect to the fibre axis smaller than $\theta_c = \arccos(n_2/n_1)$, leaky modes are formed by rays which propagate at angles $\theta > \theta_c$. Their respective propagation angles must only be small enough so that in Fig.4.4 the complement ϑ_a to their angle of incidence on to the core-cladding interface remains smaller than θ_c. Depending at what angle φ_a their plane of incidence in Fig.4.4 is inclined to the fibre axis, their propagation angles θ range over

$$\cos\theta_c > \cos\theta > \cos\theta_c \cos\varphi_a . \qquad (4.163)$$

For meridional rays with $\varphi_a = 0$, this range shrinks to zero, while for circumferentially circulating rays with $\varphi_a = \pi/2$, this range extends from θ_c to $\pi/2$.

The guided modes propagate in loss-less fibres without

attenuation; the leaky modes, however, suffer loss from power which tunnels through an evanescent field region of the cladding near its inner boundary and then radiates out into the unlimited cladding. Their rays, although still incident on the core-cladding boundary in the range of total reflection, suffer frustration of this total reflection because the boundary is curved. The ray picture, as far as we completed it in section 4.1, did not give us any quantitative information on leaky-mode attenuation. The characteristic equation (4.15) which we derived from the ray picture holds for guided as well as leaky modes. But it involves only phase conditions, and does not account for any power loss due to energy tunnelling through the evanescent field region. Therefore this characteristic equation relates only the phase constant to the fibre parameter V for specific circumferential and radial mode orders l and p.

With attenuation due to radiation, the leaky modes will have a complex propagation constant $\alpha + j\beta$ and an axial dependence of their fields according to $\exp[(-\alpha-j\beta)z]$. As a consequence of this attenuation the radial phase parameter u in the core and the radial attenuation parameter v in the cladding will also assume complex values. By separating u and v into their real and imaginary components according to

$$u = u_r + j u_j \qquad v = v_r + j v_j ,$$

the separation conditions in equation (4.43) can also be split into their real and imaginary parts

$$u_r^2 - u_j^2 = (n_1^2 k^2 - \beta^2 + \alpha^2)a^2 \qquad u_r u_j = \alpha\beta a^2$$
$$v_j^2 - v_r^2 = (n_2^2 k^2 - \beta^2 + \alpha^2)a^2 \qquad -v_r v_j = \alpha\beta a^2 .$$

(4.164)

To determine these complex quantities, we note that leaky modes, just as guided modes, are source-free solutions of the field equations under the boundary conditions of the cladded-core fibre. Their complex transverse parameters u and v should therefore solve the respective characteristic equation, just as the real quantities u and v follow as solutions of the characteristic equation. In the ray picture leaky modes appear directly adjacent to guided modes whenever the propagation angle θ goes beyond the limit θ_c for guided modes. Also in the p,l-plane of radial versus circumferential mode orders in Fig.4.11, the leaky modes lie directly beyond the cut-off line $u = V$ for guided modes. In the ray picture therefore, leaky modes appear as the analytic continuation of guided modes into the domain below cut-off.

We also expect the same character of leaky modes when we derive them as solutions of field equations, and will therefore look for the analytic continuation of solutions of the characteristic equation below cut-off. As a sufficiently general but simple example, we take the weakly guiding fibre with unlimited cladding. Its characteristic equation (4.104) is

$$u J_{l-1}(u)/J_l(u) = -v K_{l-1}(v)/K_l(v) .$$

In order to find any analytic continuation below cut-off we expand this equation about the cut-off point, where we have $v = 0$ and $u = V_c$, with V_c as the respective value of the fibre parameter at cut-off. Expanding the left-hand side about V_c with $u = V_c + \Delta u$ leads to

$$u\, J_1(u)/J_0(u) = V_c \Delta u + 1/2\, (\Delta u)^2 \qquad \text{for} \quad l = 0$$

and (4.165)

$$u\, J_{l-1}(u)/J_l(u) = -V_c \Delta u \qquad \text{for} \quad l \geq 1.$$

For $l = 0$ we have included the quadratic term of the Taylor expansion, because in the case of the fundamental mode with $V_c = 0$, this is the lowest order term of the expansion.

The right-hand side of the characteristic equation reduces near cut-off for $|v| \ll 1$ to

$$v\, K_{l-1}(v)/K_l(v) = -1/\ln v \qquad \text{for} \quad l = 0$$

$$= -v^2 \ln(\gamma v/2) \qquad \text{for} \quad l = 1 \qquad (4.166)$$

$$= v^2/(2(l-1)) \qquad \text{for} \quad l \geq 2 .$$

Near cut-off the fibre parameter V will likewise deviate only little from its cut-off value V_c. With $V = V_c + \Delta V$ and $|v| \ll 1$, equation (4.44) reduces to

$$v^2 = 2V_c(\Delta V - \Delta u) \;. \tag{4.167}$$

When we substitute all these approximations into the characteristic equation, we obtain in the case of $l = 0$ and $V_c \ne 0$:

$$\Delta V - \Delta u = \frac{1}{2V_c} \exp\left[-\frac{2}{V_c \Delta u} + \frac{1}{V_c^2}\right] \;. \tag{4.168}$$

The solution of this equation for $\Delta V > 0$ moves from cut-off into the guided-mode region. At first we obtain $\Delta u \simeq \Delta V$, because the exponential on the right-hand side remains quite small.

For $\Delta V < 0$, however, equation (4.168) has no solution with $|\Delta u| \ll 1$. Hence, in the case of $l = 0$, no analytic continuation exists for the solution of the characteristic equation below cut-off. There are no leaky modes of circumferential order $l = 0$ with low radiation loss in the cladded-core fibre with unlimited cladding. This result agrees with the conclusion from the ray picture. Guided modes with $l = 0$ are formed by meridional rays. They see the core-cladding boundary with no curvature in their plane of incidence and are either totally reflected for $\theta < \theta_c$, or for $\theta > \theta_c$ are partially refracted; but they have no region of frustrated total reflection.

In the case of $l = 1$, the approximations in equations (4.165), (4.166), and (4.167) lead to the following characteristic equation near cut-off:

$$-\Delta u = (\Delta V - \Delta u) \ln(\Delta V - \Delta u) \;. \tag{4.169}$$

This equation has a solution $|\Delta u| \ll 1$ for small positive as well as negative ΔV. For $\Delta V > 0$ the solution for Δu is likewise positive and obeys $\Delta u < \Delta V$. It represents the guided mode approximation near cut-off. For $\Delta V < 0$, the Δu from equation (4.169) becomes complex with a negative real part. It represents the leaky mode approximation near cut-off and appears as an analytic continuation into the region of V below cut-off.

In the case of $l \geq 2$, the approximations in equations (4.165), (4.166) and (4.167) lead to

$$\Delta u = \Delta V / l \qquad (4.170)$$

as the characteristic equation near cut-off. This equation yields real values Δu of the same sign but smaller than ΔV by the factor $1/l$ for positive and negative ΔV. Hence, to this first approximation, the guided mode solution above cut-off continues analytically below cut-off, with real values of the transverse phase parameter u. This result shows us that, at least near cut-off, the leaky modes of circumferential order $l \geq 2$ will have nearly real transverse phase parameters and, from equation (4.164), only a small attenuation constant α.

To actually determine the leaky mode attenuation for $l \geq 2$, we utilize this result and assume $u_j \ll u_r$ and $\alpha \ll \beta$ for the complex phase parameter u and complex propagation constant of leaky modes. Equation (4.164) yields under these assumptions, and just below cut-off, nearly imaginary values for the transverse parameter v. But also further down from cut-off the relation $v^2 = V^2 - u^2$ together with $|u_j| \ll u_r$ will let $|v_j| \gg |v_r|$. To solve the characteristic equation for the weakly guiding fibre with unlimited cladding under these conditions, we use its alternative form equation (4.105), but replace the modified Hankel functions with nearly imaginary argument by normal Hankel functions with the nearly real argument $-jv = v_j - jv_r$. The characteristic equation may then be written as

$$D(u, -jv) \equiv u \frac{J_{l+1}(u)}{J_l(u)} + jv \frac{H_{l+1}^{(2)}(-jv)}{H_l^{(2)}(-jv)} = 0 . \qquad (4.171)$$

When we replace v by u according to $v^2 = V^2 - u^2$, the expression $D(u,-jv)$ depends for a given V only on the transverse parameter u. Because u is nearly real we can approximate the characteristic equation $D(u) = 0$ by its expansion about the real component u_r (Snyder and Mitchell 1974)

$$D(u_r) + ju_j \frac{dD}{du_r} = 0 \; .$$

Its imaginary part then appears as

$$\text{Im}[D(u_r)] + u_j \; \text{Re}[\frac{dD}{du_r}] = 0$$

and may be solved for u_j to give

$$u_j = -\frac{\text{Im}[D(u_r)]}{\text{Re}[\frac{dD}{du_r}]} \; . \tag{4.172}$$

To evaluate the imaginary component of $D(u_r)$ and the real component of dD/du_r in this expression, we neglect the small real component of v_r and let v be purely imaginary. We then obtain

$$\text{Im}[D(u_r)] = -v_j \; \text{Im}[H^{(2)}_{l+1}(v_j)/H^{(2)}_l(v_j)]$$

and, by utilizing the Wronskian expression

$$J_{l+1}(x) \, N_l(x) - N_{l+1}(x) \, J_l(x) = 2\pi/x \; ,$$

we get the simplified form

$$\text{Im}[D(u_r)] = -2/(\pi|H^{(2)}_l(v_j)|^2) \; . \tag{4.173}$$

To evaluate the real part of dD/du_r, we again let v be purely imaginary and have from equation (4.171)

$$\frac{dD}{du_r} = \frac{d}{du_r}\left[u_r \frac{J_{l+1}(u_r)}{J_l(u_r)}\right] - \frac{d}{dv_j}\left[v_j \frac{H^{(2)}_{l+1}(v_j)}{H^{(2)}_l(v_j)}\right] \frac{dv_j}{du_r} \; . \tag{4.174}$$

Under present conditions, u_r and v_j are related to V by $u_r^2 - v_j^2 = V^2$ so that for a fixed value of the fibre parameter V

$$dv_j/du_r = u_r/v_j \; . \tag{4.175}$$

With the appropriate recurrence relations for Bessel functions, the first term on the right-hand side of equation (4.174) evaluates as

$$\frac{d}{du_r}\left[u_r \frac{J_{l+1}(u_r)}{J_l(u_r)}\right] = u_r\left[1 - \frac{J_{l-1}(u_r) J_{l+1}(u_r)}{J_l^2(u_r)}\right] \quad (4.176)$$

and represents a real quantity. By virtue of the characteristic equation (4.171), the Bessel functions can be replaced by Hankel functions

$$\frac{d}{du_r}\left[u_r \frac{J_{l+1}(u_r)}{J_l(u_r)}\right] = u_r\left[1 - \frac{v_j^2}{u_r^2}\frac{H_{l-1}^{(2)}(v_j) H_{l+1}^{(2)}(v_j)}{[H_l^{(2)}(v_j)]^2}\right]. \quad (4.177)$$

But then the Hankel function combination $H_{l-1}^{(2)} H_{l+1}^{(2)}/(H_l^{(2)})^2$ must also be a real quantity. The second term on the right-hand side of equation (4.174) evaluates to

$$\frac{d}{dv_j}\left[v_j \frac{H_{l+1}^{(2)}(v_j)}{H_l^{(2)}(v_j)}\right]\frac{dv_j}{du_r} = u_r\left[1 - \frac{H_{l-1}^{(2)}(v_j) H_{l+1}^{(2)}(v_j)}{[H_l^{(2)}(v_j)]^2}\right], \quad (4.178)$$

which for the same reason as in equation (4.177) must be a real quantity. Hence, under present conditions, dD/du_r appears approximately as a real quantity by itself:

$$\text{Re }\frac{dD}{du_r} = \frac{v^2}{u_r}\frac{H_{l-1}^{(2)}(v_j) H_{l+1}^{(2)}(v_j)}{[H_l^{(2)}(v_j)]^2}. \quad (4.179)$$

From the two alternative forms

$$u \frac{J_{l\pm 1}(u)}{J_l(u)} = -jv \frac{H_{l\pm 1}^{(2)}(-jv)}{H_l^{(2)}(-jv)} \quad (4.180)$$

of the characteristic equation, we see that under present conditions, for $u \simeq u_r$ and $v \simeq jv_j$, both $H_{l+1}^{(2)}(v_j)/H_l^{(2)}(v_j)$ and

$H^{(2)}_{l-1}(v_j)/H^{(2)}_l(v_j)$ must be real quantities; hence $H^{(2)}_{l-1}(v_j)$, $H^{(2)}_{l+1}(v_j)$, and $H^{(2)}_l(v_j)$ have approximately all the same phase. The real part of dD/du_r according to equation (4.179) is therefore a positive quantity and by substituting from equation (4.173) and equation (4.179) into equation (4.172) the imaginary component of u according to

$$u_j = \frac{2}{\pi} \frac{u_r}{V^2} \frac{1}{|H^{(2)}_{l-1}(v_j) \, H^{(2)}_{l+1}(v_j)|} \qquad (4.181)$$

appears likewise as a positive quantity. We introduce this expression for u_j into equation (4.164) and solve it for the attenuation constant of leaky modes

$$\alpha = \frac{2}{\pi \beta a^2} \frac{u_r^2}{V^2} \frac{1}{|H^{(2)}_{l-1}(v_j) \, H^{(2)}_{l+1}(v_j)|} \, . \qquad (4.182)$$

Given a fibre parameter V and a pair of values u_r and v_j that solve the characteristic equation for a particular leaky mode and the respective value of V, we can evaluate β from equation (4.164) and then determine the attenuation constant of this mode.

We could also, without referring to any discrete mode, specify the angles θ and φ_a, or θ and ϑ_a of a ray in the leaky mode range. The transverse parameters u_r and v_j as well as the circumferential order l and axial phase constant β of the mode which belongs to such a ray are by virtue of equations (4.2), (4.4), (4.6), (4.17), and (4.20)

$$\begin{aligned} u_r &= V \sin\theta / \sin\theta_c \\ v_j &= V (\sin^2\theta - \sin^2\theta_c)^{\frac{1}{2}}/\sin\theta_c \\ l &= V (\sin^2\theta - \sin^2\vartheta_a)^{\frac{1}{2}}/\sin\theta_c \\ \beta a &= V \cos\theta / \sin\theta_c \, . \end{aligned} \qquad (4.183)$$

When we substitute these expressions into equation (4.182), the leaky-mode attenuation follows from

$$\alpha a = \frac{2\tan\theta \sin\theta}{\pi V \sin\theta_c} \frac{1}{|H^{(2)}_{l-1}(v_j) H^{(2)}_{l+1}(v_j)|} . \quad (4.184)$$

This formula allows us to determine the attenuation all through the leaky mode range

$$\theta_c < \theta < \arccos(\cos\varphi_a / \cos\theta_c) .$$

At the upper limit of this range, for $\vartheta_a = \theta_c$, we have $l = v_j$. For sufficiently large circumferential orders and arguments $l = v_j \gg 1$, the magnitude of the Hankel function product may be approximated by (Abramowitz and Stegun 1964)

$$|H^{(2)}_{l-1}(l) H^{(2)}_{l+1}(l)| \simeq 0.800 \, l^{-2/3} , \quad (4.185)$$

which leads to the following approximation of maximum leaky-mode attenuation at the upper limit $\vartheta_a = \theta_c$ of frustrated total reflection

$$\alpha a = 0.795 \frac{\tan\theta \, \sin\theta}{V^{1/3} \sin\theta_c} [(\sin\theta / \sin\theta_c)^2 - 1]^{1/3} . \quad (4.186)$$

Note that for extremely large fibre parameters V, the maximum leaky-mode attenuation remains quite low even at this upper limit.

Throughout the leaky-mode range we have $l > v_j$. If again we consider only leaky modes of large circumferential order ($l \gg 1$), and in addition assume

$$l - v_j \gg l^{1/3} , \quad (4.187)$$

we can use the approximate representation (Abramowitz and Stegun 1964)

$$|H^{(2)}_{l-1}(v_j) H^{(2)}_{l+1}(v_j)| = \frac{2}{\pi(l^2 - v_j^2)^{\frac{1}{2}}} \exp\{2l \, \text{arccosh} \frac{l}{v_j} - 2(l^2 - v_j^2)^{\frac{1}{2}}\} , \quad (4.188)$$

which holds throughout the leaky-mode range except, because of the restriction in equation (4.187), very near to its upper limit.

When we substitute this approximation for the magnitude of the product of Hankel functions into equation (4.184) and use equation (4.183) to express l and v_j by V, θ_c, and the ray angles, we obtain

$$\alpha a = \frac{\tan\theta \sin\theta}{\sin^2\theta_c} (\sin^2\theta - \sin^2\vartheta_a)^{\frac{1}{2}} \exp\left\{\frac{2V}{\sin\theta_c}\left[(\sin^2\theta_c - \sin^2\vartheta_a)^{\frac{1}{2}} - (\sin^2\theta - \sin^2\vartheta_a)^{\frac{1}{2}} \operatorname{arccosh}\left(\frac{\sin^2\theta - \sin^2\vartheta_a}{\sin^2\theta - \sin^2\theta_c}\right)^{\frac{1}{2}}\right]\right\} \quad (4.189)$$

This formula provides an excellent approximation for the attenuation over most of the leaky-mode range. When core and cladding differ only a little in their refractive index, we have $\theta_c \ll 1$ and because of $\vartheta_a \leq \theta_c$ also $\vartheta_a \ll 1$; we may then further simplify equation (4.189) by replacing $\sin\theta_c \simeq \theta_c$ and $\sin\vartheta_a \simeq \vartheta_a$. Except for the relatively few leaky modes which are of very high order both in circumferential as well as radial direction and therefore very close to their upper high-loss limit, all leaky modes have propagation angles θ not too much larger than θ_c. For $\theta_c \ll 1$ all these leaky modes have likewise $\theta \ll 1$, and we can replace $\tan\theta \simeq \sin\theta \simeq \theta$. The leaky mode attenuation is then given by

$$\alpha a = (\theta/\theta_c)^2 (\theta^2 - \vartheta_a^2)^{\frac{1}{2}} \exp\left\{\frac{2V}{\theta_c}\left[(\theta_c^2 - \vartheta_a^2)^{\frac{1}{2}} - (\theta^2 - \vartheta_a^2)^{\frac{1}{2}} \operatorname{arccosh}\left(\frac{\theta^2 - \vartheta_a^2}{\theta^2 - \theta_c^2}\right)^{\frac{1}{2}}\right]\right\}.$$

We have evaluated this expression for a representative value of the fibre parameter V and plotted $\alpha a/\theta_c$ versus θ/θ_c in Fig. 4.22 for several values of ϑ_a/θ_c in the range of practical interest. The leaky-mode attenuation in this diagram varies over many orders of magnitude for only moderate changes in θ and ϑ_a.

The leaky modes have their lowest attenuation just across the borderline between guided and leaky mode regions near cutoff, where $v_j \ll 1$. To evaluate leaky mode attenuation near this limit, we use the approximations for Hankel functions of

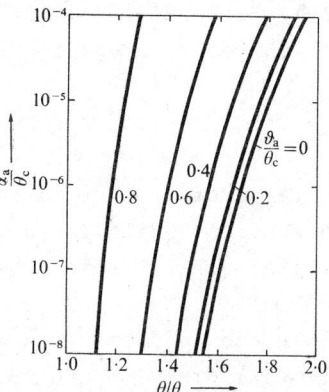

Fig.4.22. Leaky-mode attenuation in weakly guiding fibres with unlimited cladding as a function of propagation angle θ for different values of the complement ϑ_a to the angle of incidence of the constituent ray on to the core-cladding boundary and for $V = 50$.

small argument to obtain

$$\frac{1}{|H^{(2)}_{l-1}(v_j)\, H^{(2)}_{l+1}(v_j)|} = \left(\frac{\pi\, v_j^l}{2^l(l-1)!}\right)^2 \frac{l-1}{l}. \tag{4.190}$$

For $|v_j| \ll 1$, the phase parameter u_r differs only little from V, so that equation (4.184) reduces to the following formula for leaky-mode attenuation near cut-off

$$\alpha a = 2\pi(\sqrt{2\Delta}/V)\,\frac{(u_r^2 - V^2)^l}{[2^l(l-1)!]^2}\,\frac{l-1}{l}. \tag{4.191}$$

In this expression $\Delta = (n_1 - n_2)/n_2$ represents the relative index difference between core and cladding. This formula applies for any circumferential order $l \geq 2$. For large l, we use Stirling's formula (Abramowitz and Stegun 1964)

$$(l-1)! \simeq \sqrt{2\pi}\, e^{-l}\, l^{l-1/2} \tag{4.192}$$

and obtain

$$\alpha a = \sqrt{2\Delta}\,(l/V)\,[1.85(u_r^2 - V^2)/l^2]^l \tag{4.193}$$

for leaky modes of large circumferential order near cut-off. By solving this equation for u_r according to

$$u_r^2 = V^2 + 0.541 \; l^2 \; [\alpha a \; V/(l\sqrt{2\Delta})]^{1/l} \; , \qquad (4.194)$$

we can determine the range of transverse phase parameter values $u_r > V$ of leaky modes, which for a given fibre parameter and circumferential order l remain below a certain loss limit α.

The limiting value for u_r also sets a limit for the radial and circumferential orders p and l of leaky modes that propagate with this much loss, or less. The characteristic equation (4.15) from the ray picture relates p to l for guided and leaky modes. By introducing u from equation (4.20), this characteristic equation appears as

$$p\pi = (u^2 - l^2)^{\frac{1}{2}} - l \; \text{arc cos} \; (l/u).$$

If now we substitute for u its real part according to equation (4.194), we obtain the radial order p_α of leaky modes which have the attenuation α:

$$p_\alpha = \frac{1}{\pi} \left\{ [V^2 - l^2 \{1 - 0.541 \; (\alpha a V/(l\sqrt{2\Delta}))^{1/l}\}]^{\frac{1}{2}} - \right.$$

$$\left. - l \; \text{arc cos} \; l/[V^2 + 0.541 \; l^2 \; (\alpha a V/(l\sqrt{2\Delta}))^{1/l}]^{\frac{1}{2}} \right\} . \qquad (4.195)$$

Fig.4.23 shows the leaky mode region in the $p(l)$-diagram with two curves $p_\alpha(l)$ for different values of the leaky-mode loss parameter $\alpha a/\sqrt{2\Delta}$.

The total number of modes including leaky modes up to a certain leaky-mode attenuation α may now be determined by evaluating the integral

$$M_\alpha = 4 \int_0^{l_\alpha} p_\alpha \; dl \; . \qquad (4.196)$$

The upper limit extends to a circumferential order l_α which obeys the following condition

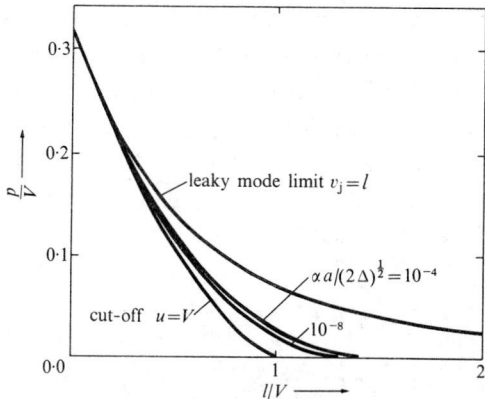

Fig.4.23. Radial versus circumferential orders of leaky modes with different values of the radiation loss parameter $\alpha a/(2\Delta)^{\frac{1}{2}}$ for $V = 50$.

$$(V/l_\alpha)^2 = 1 - 0.541 \, [\alpha a \, V/(l_\alpha \sqrt{2\Delta})]^{1/l_\alpha} . \qquad (4.197)$$

For l-values larger than this l_α, the radicand of the first term in equation (4.195) turns negative.

To facilitate an approximate analytic evaluation of the integral in equation (4.196) we note that for large l, the factor l^2 dominates the dependence of u_r on l in the second term of equation (4.194). The remaining factor $[\alpha a V/(l\sqrt{2\Delta})]^{1/l}$ changes only very little with l, because of the small exponent $1/l$. We therefore replace equation (4.194) by

$$u_r^2 = V^2 + 0.541 \, l^2 \, (\alpha a/\sqrt{2\Delta})^{1/V} . \qquad (4.198)$$

This approximation means that we still account for the correct value of u_r at and near $l = V$, but that we replace the actual dependence of u_r on l by the simple square law dependence

$$u_r^2 = V^2 + dl^2$$

with (4.199)

$$d = 0.541 \, (\alpha a/\sqrt{2\Delta})^{1/V} .$$

When we substitute this simple square-law dependence of $u_r(l)$

into the characteristic equation and compare the resulting $p_\alpha(l)$ with $p_\alpha(l)$ from equation (4.195) we notice very little difference. During integration of equation (4.196), this difference will diminish even further, so that the square-law approximation is well justified when we determine the number of modes. With u_r according to equation (4.199) the integration extends between the limits

$$l = 0 \ldots V/(1-d)^{\frac{1}{2}} .$$

To integrate the first term of p_α

$$\frac{4}{\pi} \int_0^{V/(1-d)^{\frac{1}{2}}} (u^2 - l)^{\frac{1}{2}} dl = \frac{4}{\pi} \int_0^{V/(1-d)^{\frac{1}{2}}} (V^2 - l^2(1-d))^{\frac{1}{2}} dl ,$$

we transform to $L = l(1-d)^{\frac{1}{2}}$ as the new variable of integration and obtain

$$\frac{4}{\pi} \int_0^{V/(1-d)^{\frac{1}{2}}} (u^2 - l^2)^{\frac{1}{2}} dl = \frac{4}{\pi\sqrt{1-d}} \int_0^{V} (V^2 - L^2)^{\frac{1}{2}} dL = V^2/(1-d)^{\frac{1}{2}} . \quad (4.200)$$

To integrate the second term of p_α we transform to $x = l/V$ as the new variable of integration and, with $d \ll 1$, expand the inverse cosine into a Taylor series about $(l/V) = x$

$$\arccos \frac{l}{u} = \arccos x + \frac{d}{2} x^3/(1-x^2)^{\frac{1}{2}} . \quad (4.201)$$

Integrating the second term with this expansion between $l = 0 \ldots V$ leads to

$$\frac{4}{\pi} \int_0^V l \arccos (l/u) dl = \frac{4}{\pi} V^2 \int_0^1 \left(x \arccos x + \frac{d}{2} \frac{x^4}{(1-x^2)^{\frac{1}{2}}} \right) dx$$

$$= (V^2/2)(1 + 3d/4) . \quad (4.202)$$

The upper limit of integration $l = V$ is dictated in this approximation by the range in which the two Taylor series terms remain real. The error which this shift of integration limit introduces, as well as the error which we accumulate when, near $x = 1$, the first-order term of the Taylor series expansion is no longer small, lend to the results the character of an estimate rather than an analytic approximation. By collecting both contributions from equation (4.200) and equation (4.202), the total number of modes up to a leaky mode loss α is

$$M_\alpha = \frac{V^2}{2} [1 + 0.135 \, (\alpha a/\sqrt{2\Delta})^{1/V}]. \qquad (4.203)$$

The relative increment of the number of leaky modes up to a loss of α then amounts to

$$\Delta M_\alpha / M = 0.135 \, (\alpha a/\sqrt{2\Delta})^{1/V} . \qquad (4.204)$$

Fig.4.24 shows this number of leaky modes relative to the number of guided modes as a function of the fibre parameter V

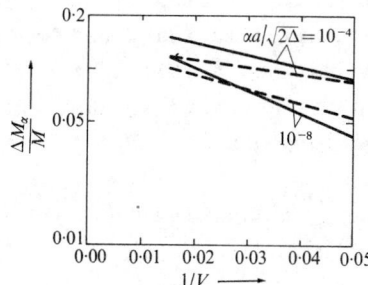

Fig.4.24. Number of leaky modes relative to number of guided modes as a function of the fibre parameter V for different values of the attenuation parameter $\alpha a/(2\Delta)^{\frac{1}{2}}$. Broken lines according to equation (4.204); solid lines according to equation (4.196) with equation (4.195).

for different values of the attenuation parameter $\alpha a/\sqrt{2\Delta}$. The broken lines represent the approximation in equation (4.204), while the solid lines give more accurate results from a numerical integration of equation (4.196) with $p_\alpha(l)$ from equation (4.195). But even this more accurate numerical integration can give reliable results only in the range of V-values over which the lines of Fig.4.24 extend.

So far we have treated leaky modes only as a result arising from the ray picture and as formal solutions of the field equations under the boundary condition of cladded core fibres with unlimited cladding. As the latter field solutions, we found leaky modes to be analytic continuations of guided modes into the domain below their respective cut-offs. We have not yet looked for their physical significance nor have we examined if they behave properly as physically real solutions when we go out to infinity in the radial direction.

For guided modes we ascertained the proper behaviour of their evanescent cladding fields for larger and larger radii and consequently acknowledged them as physically real solutions, each of which can exist by itself, independently from other fields. Leaky modes, in contrast to guided modes, have not only a complex propagation constant with the real part accounting for the attenuation due to power leakage away from the guide; they also have complex parameters for their transverse field dependence. The radial cladding parameter v in particular is predominantly imaginary but has also a small real component v_r. From equation (4.183) of the ray picture the imaginary component v_j is positive so that according to equation (4.164) the real component

$$v_r = -\alpha\beta a^2/v_j$$

must be negative.

To examine the cladding fields for such a field parameter at large radii, we use the asymptotic expression for Hankel functions and find

$$H_l^{(2)}(-jv\frac{r}{a})\exp-(\alpha+j\beta)z = \left(\frac{2}{\pi\frac{vr}{a}}\right)^{\frac{1}{2}} j^{l-1} \exp\left\{(\alpha\beta a^2/v_j - jv_j)\frac{r}{a} - (\alpha+j\beta)z\right\} .$$

Hence at large enough radii far out in the cladding, the leaky wave propagates almost as a plane wave with a phase constant

$$\beta_2 = [(v_j/a)^2 + \beta^2]^{\frac{1}{2}}$$

in a direction which makes the angle

$$\theta_2 = \arctan(v_j/\beta a)$$

with the fibre axis. When we introduce v_j and β from equation (4.183) of the ray picture, we find $\beta_2 = n_2 k$, the wave number of the cladding region and

$$n_2 \cos\theta_2 = n_1 \cos\theta.$$

This θ_2 is the propagation angle that results from the propagation angle θ inside the core by refraction at an interface of core and cladding index.

The negative real part v_r together with α lends to the cladding field a long way out the character of a transversely attenuated plane wave. It propagates in the direction θ_2 but decays aperiodically transverse to this direction at an angle $\theta_2 - \pi/2$ with respect to the axis. In the radial direction, therefore, this wave increases exponentially with the increment $v_r/a = \alpha\beta a/v_j$.

Far out in the cladding the leaky mode corresponds in its field distribution and direction of propagation to the transversely attenuated plane wave of the kind which is also associated with leaky modes out of prism or grating couplers for planar waveguides. With its cladding field increasing exponentially in the radial direction, it has physical reality and significance only over a limited part of the fibre cross-section out to a certain radius in the cladding. This radius, within which leaky modes offer a meaningful description, depends on the propagation angle θ_2 of their radiating cladding fields and on the axial distance of the point of observation from the fibre excitation. Fig.4.25(a) illustrates the excitation and propagation of a leaky mode. In general a source distribution at the front end of the fibre excites not only guided modes, but also a radiating field. Part of this radiating field in the vicinity of the fibre and not too far from the source of excitation may conveniently be described

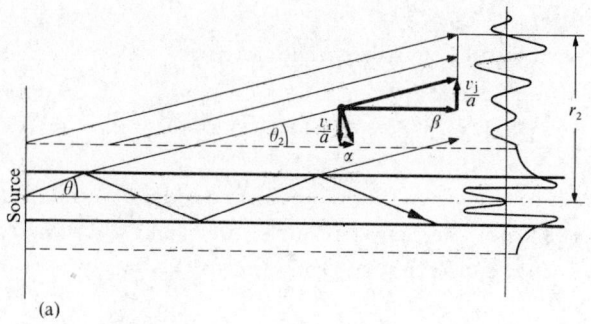

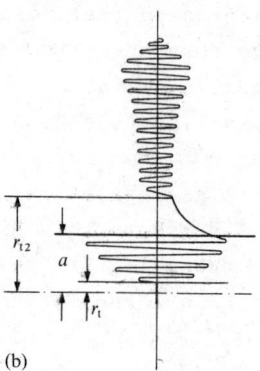

Fig.4.25. a) Propagation direction θ_2 and attenuation components α and v_r/a of the radiating cladding fields near their source of excitation.

b) Transverse field distribution of a leaky mode in the core as well as in the evanescent and radiating field region of the cladding.

by leaky modes.

Fig.4.25(b) depicts the transverse field distribution of one such leaky mode at a certain distance from the source. Inside the core the field has the radial standing wave pattern of a Bessel function. The cladding field near the core has evanescent character first, but further out changes to the oscillating character of a radiating field. Its amplitude in the range of oscillation increases nearly exponentially in the radial direction. But at a radius $r_2 \simeq z/\tan\theta_2$, depending on the distance z from the source and the propagation angle θ_2, the amplitude decreases again and ceases to exist much beyond r_2. We observe a larger field amplitude of the leaky mode further out rather than nearer to the fibre, because the radiating field at r_2 originates from a fibre section closer to the source, where the leaky mode has not suffered so much radiation loss than further away from the source.

Beyond the radius r_2 in Fig.4.25(a) the particular leaky mode with θ_2 no longer represents by itself the radiation field. Other components of the radiation field must also be considered here in order to account for the decline and disappearance of fields beyond r_2.

The intensity of the leaky mode radiation field in the cladding relative to the intensity of its fields in the core depends very much on the radial extend of the evanescent field region in the cladding. This region begins at the core-cladding boundary at $r = a$ and reaches the turning point r_{t2} according to equation (4.23). Throughout this region the radial component of the wave number

$$k_r = (n_2^2 k^2 - \beta^2 - l^2/r^2)^{\frac{1}{2}}$$

is purely imaginary, so that the field decays altogether by the exponential factor

$$\exp\left\{-\int_a^{r_{t2}} (l^2/r^2 + \beta^2 - n_2^2 k^2)^{\frac{1}{2}} \, dr\right\}$$

between $r = a$ and $r = r_{t2}$. If only this factor becomes sufficiently small, the radiation fields of the leaky mode will be extremely weak and the leaky mode attenuation rather low.

The special significance of leaky modes in signal transmission on fibres lies in the fact that a certain fraction of them, as described by equation (4.204), can have extremely low radiation loss. Their leaky-mode attenuation can actually be so low that they suffer not much more total loss than guided modes. For all practical purposes, such leaky modes then qualify as guided modes and may take part in signal transmission just as guided modes do.

4.6. WEAKLY GUIDING FIBRE WITH JACKET

The cladding which surrounds the core of a glass fibre has only a finite diameter. Actual fibres very often have another layer around this cladding which serves various purposes; it can give additional strength to the fibre and protect it mechanically. We have also mentioned already that such an additional layer may serve to improve the transmission characteristics of the fibre. If made of an absorbing material, it reduces any optical interference from outside, and in particular the crosstalk between different fibres in a bundle. If this layer has a refractive index which is nearly equal to or only little larger than the cladding index, it strips off any light rays that, due to whatever fibre imperfection, might radiate from the core modes of the fibre. In this function, the outer layer prevents such spurious radiation reflecting to the core, and reduces the interference of core modes with such outer cladding reflections.

For many of these and other purposes, but in particular the last one, all three indices, n_1 of the core, n_2 of the cladding, and n_3 of the jacket, differ only slightly from each other. The whole cladded-core fibre, including the outer jacket, is then a weakly guiding structure, for which all guided modes have particularly simple field distributions and propagation characteristics, similar to those we encountered in the weakly guiding fibre with a cladding of infinite extent.

We could find these field distributions and propagation characteristics by assuming for the general field expressions in section 4.2 that the refractive indices differ only little. The general form of the characteristic equation would, under these conditions, also reduce to a simpler approximate form. It would reveal pairs of nearly degenerate modes which, when suitably combined, result in modes of uniform and linear polarization for their transverse fields. These LP-modes, their field distributions, and their characteristic equation can, however, also, and more directly, be found by starting with such a uniformly and linearly polarized field, which satisfies the field equations, and subject this field to the boundary conditions of the weakly guiding fibre with a limited cladding but a jacket of infinite extent around it.

Since the index differences are small, the limit θ_c for total reflection also remains small, and any ray belonging to a guided mode of the structure propagates under a small angle θ with respect to the axis. Its transverse fields will then be large compared to the longitudinal fields and have almost the character of uniform plane waves. The transverse electric field will, in particular, be perpendicular to the transverse magnetic field, and the ratio of their amplitudes will be $(\mu_0/\epsilon_0)^{\frac{1}{2}}/n$, the intrinsic wave impedance of the material with refractive index n. Starting with the Cartesian components E_x and H_y as transverse fields, they must solve the wave equations (1.5), (1.6) and, in the coordinates of the circular cylinder, can therefore be written in terms of Bessel functions and harmonic functions. Inside the core we let

$$E_{x1} = H_{y1}(\mu_0/\epsilon_0)^{\frac{1}{2}}/n_1 = A\, J_l(ur/a)\, \cos l\varphi \qquad (4.205)$$

with only the Bessel function $J_l(ur/a)$, to avoid any singularities on the axis. In the cladding the more general solution

$$E_{x2} = H_{y2}(\mu_0/\epsilon_0)^{\frac{1}{2}}/n_2 = [C\, I_l(vr/a) + D\, K_l(vr/a)]\, \cos l\varphi \qquad (4.206)$$

is required in order that the boundary conditions at both interfaces $r = a$ and $r = b$ of this intermediate region can be satisfied. The modified Bessel and Hankel functions describe radially evanescent fields as long as the transverse attenuation parameter v remains real.

The jacket with refractive index n_3 is assumed to extend from $r = b$ to infinity. Its fields can therefore be described by just the modified Hankel function

$$E_{x3} = H_{y3}(\mu_0/\epsilon_0)^{\frac{1}{2}}/n_3 = F\, K_l(wr/a)\, \cos l\varphi . \qquad (4.207)$$

The longitudinal field components derive from these transverse fields according to

$$E_{zi} = -j((\mu_0/\varepsilon_0)^{\frac{1}{2}}/(n_i^2 k)) \partial H_{yi}/\partial x$$

$$H_{zi} = -j((\varepsilon_0/\mu_0)^{\frac{1}{2}}/k) \partial E_{xi}/\partial y ,$$
(4.208)

where the index i denotes the respective cross-sectional region, 1 for the core, 2 for the cladding, or 3 for the jacket. We evaluate the partial derivatives in equations (4.208) by utilizing recurrence relations and obtain E_{zi} and H_{zi} in the following form, which lends itself well to satisfying the boundary conditions,

$$E_{z1} = j\frac{A}{n_1 ka}[u\,J_{l+1}(ur/a)\cos\varphi\cos l\varphi - l(a/r)J_l(ur/a)\cos(l-1)\varphi]$$

$$H_{z1} = j\left(\frac{\varepsilon_0}{\mu_0}\right)^{\frac{1}{2}}\frac{A}{ka}[u\,J_{l+1}(ur/a)\sin\varphi\cos l\varphi + l(a/r)J_l(ur/a)\sin(l-1)\varphi]$$

$$E_{z2} = j\frac{1}{n_2 ka}\{v[-CI_{l+1}(vr/a)+DK_{l+1}(vr/a)]\cos\varphi\cos l\varphi -$$

$$- l\frac{a}{r}[CI_l(vr/a)+DK_l(vr/a)]\cos(l-1)\varphi\}$$

$$H_{z2} = \frac{j}{ka}\{v[-CI_{l+1}(vr/a)+DK_{l+1}(vr/a)]\sin\varphi\cos l\varphi +$$

$$+ l\frac{a}{r}[CI_l(vr/a)+DK_l(vr/a)]\sin(l-1)\varphi\}$$

$$E_{z3} = j\frac{F}{n_3 ka}[wK_{l+1}(wr/a)\cos\varphi\cos l\varphi - l\frac{a}{r}K_l(wr/a)\cos(l-1)\varphi]$$

$$H_{z3} = j\left(\frac{\varepsilon_0}{\mu_0}\right)^{\frac{1}{2}}\frac{F}{ka}[wK_{l+1}(wr/a)\sin\varphi\cos l\varphi + l\frac{a}{r}K_l(wr/a)\sin(l-1)\varphi].$$
(4.209)

The boundary conditions at the interfaces between core, cladding, and jacket require all tangential components to be continuous. Since all refractive indices are assumed to differ only little, we satisfy these boundary conditions for transverse field components adequately by just matching the Cartesian components E_{xi} and H_{yi} at the interfaces. This leads to the relations

$$A = C \, I_l(v)/J_l(u) + D \, K_l(v)/J_l(u) \tag{4.210}$$

$$F = C \, I_l(cv)/K_l(cw) + D \, K_l(cv)/K_l(cw)$$

for the field factors A, C, D, and F in equations (4.205), (4.206), and (4.207) with $c=b/a$. These relations also match the last terms of the longitudinal fields in equation (4.209) at the boundaries.

For continuity of the complete longitudinal fields we need only to match the first terms in equation (4.209) at the boundary. By substituting for A and F from equation (4.210), continuity of these first terms of E_{zi} and H_{zi} leads to the following system of two linear equations for the field factors C and D

$$\begin{bmatrix} u \dfrac{J_{l+1}(u)}{J_l(u)} I_l(v) + v \, I_{l+1}(v) & u \dfrac{J_{l+1}(u)}{J_l(u)} K_l(v) - v \, K_{l+1}(v) \\[6pt] w \dfrac{K_{l+1}(cw)}{K_l(cw)} I_l(cv) + v \, I_{l+1}(cv) & w \dfrac{K_{l+1}(cw)}{K_l(cw)} K_l(cv) - v \, K_{l+1}(cv) \end{bmatrix} \begin{bmatrix} C \\ D \end{bmatrix} = 0. \tag{4.211}$$

A non-trivial solution for C and D requires the coefficient determinant of this system to be zero. This condition represents the characteristic equation and from equation (4.211) may be written as

$$\hat{J}_l(u) - \hat{K}_l(v) = \frac{I_{l+1}(v) \, K_{l+1}(cv)}{I_{l+1}(cv) K_{l+1}(v)} \frac{(\hat{J}_l(u) + \hat{I}_l(v))(\hat{K}_l(cw) - \hat{K}_l(cv))}{\hat{I}_l(cv) + \hat{K}_l(cw)}, \tag{4.212}$$

where we use the following abbreviations

$$\hat{J}_l(u) = J_l(u)/[u \, J_{l+1}(u)] \; ; \quad \hat{I}_l(v) = I_l(v)/[v \, I_{l+1}(v)] \; ;$$

$$\hat{K}_l(v) = K_l(v)/[v \, K_{l+1}(v)]. \tag{4.213}$$

For $c = b/a = 1$ the cladding shrinks to zero, and we are left with the jacket of index n_3 surrounding the fibre. The characteristic equation (4.212) reduces for $c = 1$ to

$$\hat{J}_l(u) = \hat{K}_l(w),$$

which corresponds to the approximate form (4.103) for a weakly guiding fibre with unlimited cladding. For $c \to \infty$, as well as for $n_3 = n_2$ and consequently $w = v$, equation (4.212) reduces directly to the approximate form (4.103). In both limiting cases we return again to the weakly guiding fibre with a single unlimited cladding.

For a perturbation solution of the characteristic equation (4.212), we assume that the jacket modifies the core modes of the fibre only slightly (Kawakami and Nishida 1975). We will then take the fibre with the same core radius a and the same core and cladding indices n_1 and n_2, but an unlimited cladding, as a reference structure. Its guided modes have transverse phase and attenuation parameters u and v which depend on the fibre parameter $V = (u^2 + v^2)^{\frac{1}{2}}$ according to the solutions of

$$\hat{J}_l(u) - \hat{K}_l(v) = 0. \qquad (4.214)$$

As long as the cladding of the actual fibre is thick enough and the particular mode far enough from cut-off so that

$$K_{l+1}(cv)/K_{l+1}(v) \simeq \frac{1}{(c)^{\frac{1}{2}}} \exp[-(c-1)v] \ll 1, \qquad (4.215)$$

the right-hand side of equation (4.212) will remain small. The factor $I_{l+1}(v)K_{l+1}(cv)/[I_{l+1}(cv)K_{l+1}(v)]$ on the right-hand side of this equation is only of the order of $\exp[-2(c-1)v]$. Under these conditions, we can base a perturbational analysis on the solution of equation (4.214) for the weakly guiding reference fibre and let the actual transverse parameters be

$$u + \Delta u, \qquad v + \Delta v$$

with small perturbations Δu and Δv. The transverse attenuation parameter for the jacket fields

$$w = a\,(\beta^2 - n_3^2 k^2)^{\frac{1}{2}}, \qquad (4.216)$$

can be expressed in terms of u and v according to

$$w^2 = v^2(n_1^2-n_3^2)/(n_1^2-n_2^2) - u^2(n_3^2-n_2^2)/(n_1^2-n_2^2) \qquad (4.217)$$

and can likewise be considered to be perturbed by the jacket according to

$$w + \Delta w .$$

Assuming small perturbations $|\Delta u| \ll |u|$, $|\Delta v| \ll |v|$, and $|\Delta w| \ll |w|$, we now expand the characteristic equation about its unperturbed solution for the reference fibre and obtain to first order in these perturbations

$$\hat{J}'_l(u)\,\Delta u - \hat{K}'_l(v)\,\Delta v = R(u,v,w), \qquad (4.218)$$

where $R(u,v,w)$ represents the right-hand side of equation (4.212) according to

$$R = \frac{I_{l+1}(v) K_{l+1}(cv)}{I_{l+1}(cv) K_{l+1}(v)}\; \frac{(\hat{J}_l(u)+\hat{I}_l(v))(\hat{K}_l(cw)-\hat{K}_l(cv))}{\hat{I}_l(cv) + \hat{K}_l(cw)}, \qquad (4.219)$$

and the prime on $\hat{J}_l(u)$ and $\hat{K}_l(v)$ denotes differentiation with respect to u and v, respectively. From $V^2 = u^2 + v^2$, we obtain for a fixed value of the fibre parameter

$$u\,\Delta u + v\,\Delta v = 0, \qquad (4.220)$$

with which we solve equation (4.218) for Δu

$$u\,\Delta u = \frac{-R(u,v,w)}{(1/u^2 + 1/v^2)(1 - 2l\,\hat{K}_l(v))} . \qquad (4.221)$$

The axial phase constant from $\beta^2 = (n_1^2 k^2 - u^2)/a^2$ is likewise perturbed by the jacket and with $\beta\Delta\beta = -u\Delta u/a^2$ follows from

$$\beta \, \Delta\beta = \frac{R(u,v,w)}{a^2(1/u^2+1/v^2)(1-2l\,\hat{k}_l(v))} \, . \qquad (4.222)$$

We consider the effect of the jacket on core modes, first for a lossless structure without absorption in the jacket, and hence a real refractive index n_3. As long as w^2 from equation (4.217) remains positive, we obtain a real perturbation for the phase constant from equation (4.222). This shift in phase constant will be the only change in propagation characteristics caused by the jacket. If, however,

$$v^2 < u^2 \frac{n_3^2 - n_2^2}{n_1^2 - n_3^2} \simeq u^2 \frac{n_3 - n_2}{n_1 - n_3} \, , \qquad (4.223)$$

we obtain imaginary values for the unperturbed parameter w from equation (4.217), and in equation (4.222), $R(u,v,w)$ as well as $\Delta\beta$ become complex. The imaginary component of $\Delta\beta$ represents an attenuation constant

$$\alpha = -\mathrm{Im}(\Delta\beta) = -\mathrm{Im}\left[\frac{R(u,v,w)}{\beta a^2(1/u^2+1/v^2)(1-2l\,\hat{k}_l(v))}\right]. \qquad (4.224)$$

It accounts for the loss which under these conditions the particular mode suffers due to leakage into the jacket. This leakage occurs whenever the jacket has a large enough index n_3 to frustrate the total reflection at the core-cladding boundary. In Fig.4.10 for the wave vector components of the ray picture, the asymptotic level β^2 must be higher than $n_3^2 k^2$ to avoid this frustration of total reflection and any leakage loss into the cladding. Since the unperturbed β of any core mode is always smaller than $n_1 k$, the jacket must have $n_3 < n_1$, if all guided modes are not to turn into leaky waves.

To survey the range of u and v for which a particular mode remains lossless and guided, we plot v as a function of u according to the characteristic equation (4.214) for the weakly guiding fibre with unlimited cladding. The straight line

$$v = u[(n_3^2-n_2^2)/(n_1^2-n_3^2)]^{\frac{1}{2}} \qquad (4.225)$$

divides the v-u plane in Fig.4.26 into a guided mode and a leaky mode range. Any particular mode remains leaky along

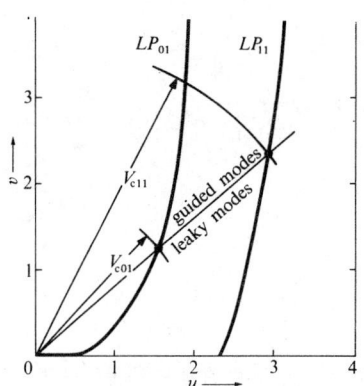

Fig.4.26. Transverse phase and attenuation parameter of weakly guiding fibre modes for infinite cladding and guided and leaky mode ranges as well as first-order cut-off values for a jacket with $(n_3^2-n_2^2)/(n_1^2-n_3^2) = 0.7$.

its curve $v(u)$ below this straight line; it becomes a truly guided mode only after its curve has crossed this line. The intersection defines, in a first-order approximation, a new cut-off for the respective mode. The distance of this intersection to the origin of the v-u plane gives the cut-off value

$$V_c = (u_c^2 + v_c^2)^{\frac{1}{2}} \qquad (4.226)$$

for the fibre parameter $V = ka(n_1^2-n_2^2)^{\frac{1}{2}}$. We note from Fig. 4.26 that as soon as the straight line has a finite slope due to $n_3 > n_2$, all modes move up in cut-off to larger values of V. The dominant HE_{11}- or LP_{01}-mode assumes a finite cut-off value V_{c01} in this first-order approximation.

All cut-off values V_c according to this approximation appear independent of the ratio $c = b/a$ of cladding to core radius. Actually, of course, this cut-off also depends on the thickness of the cladding, and hence on c. A more accurate approximation is needed to determine the cut-off as a function of c. For a second-order approximation of V_c we let $u + \delta u$, and $v + \delta v$ be the perturbed transverse parameters at cut-off. With $w = 0$ at cut-off, the first-order perturbation of equation (4.212) is

$$\hat{J}'_l(u) \, \delta u + \hat{K}'_l(v) \delta v = R(u,v,0) \, , \qquad (4.227)$$

while from $V_c^2 = u^2(n_1^2-n_2^2)/(n_1^2-n_3^2)$, we obtain the perturbation relation

$$V_c \, \delta V_c = u \, \delta u \, (n_1^2-n_2^2)/(n_1^2-n_3^2)$$

$$= (V_c^2/u) \, \delta u \, . \qquad (4.228)$$

Equation (4.225) relates the perturbations δu and δv according to

$$\delta v = \delta u \, [(n_3^2-n_2^2)/(n_1^2-n_3^2)]^{\frac{1}{2}}$$

$$= (v/u) \, \delta u \, . \qquad (4.229)$$

When we substitute for δv from equation (4.229) into equation (4.227) and for δu in turn from equation (4.227) into equation (4.228), we obtain the following perturbational expression for the cut-off value of V_c

$$\delta V_c = V_c \, R(u,v,0)/[u \, \hat{I}'_l(u) + v \, \hat{K}'_l(v)] \, . \qquad (4.230)$$

With $R(u,v,0)$ from equation (4.219) this expression may be evaluated to give

$$\delta V_c = - \frac{v^2}{V_c} \frac{I_{l+1}(v) \, K_{l+1}(cv) \, K_{l+1}(v)}{I_{l+1}(cv) \, K_l^2(v)} [\hat{K}_l(v) + \hat{I}_l(v)] . (4.231)$$

The modified Bessel and Hankel functions of this expression

are always positive, so that this correction to the first-order approximation of the fibre parameter at cut-off is always negative, and shifts the cut-off value V_c somewhat down from the intersection in Fig.4.26.

Above cut-off, when a particular mode has changed its leaky mode character into that of a truly guided mode, the lossless jacket around the cladding causes only a shift in phase constant from that of the reference guide. This phase shift may to first order be determined from the perturbational expression (4.222). Close to cut-off, this expression is, however, not very accurate because in this range, the jacket changes the fields and the mode character substantially from that of the reference fibre.

We obtain more accurate expressions for propagation constants near cut-off, when we start with exact cut-off values and calculate the fibre parameter as well as the transverse phase and attenuation parameters as first-order perturbations of their cut-off values. The cut-off values are solutions of the following set of equations:

$$\hat{J}_l(u) - \hat{K}_l(v) = R(u,v,0) \tag{4.232}$$

$$u^2 + v^2 = V_c^2 \tag{4.233}$$

$$u = [(n_1^2 - n_3^2)/(n_1^2 - n_2^2)]^{\frac{1}{2}} V_c . \tag{4.234}$$

With $u + \Delta u$, $v + \Delta v$, w, and $V_c + \Delta V$ as mode and fibre parameters near cut-off, we have

$$\hat{J}_l(u+\Delta u) - \hat{K}_l(v+\Delta v) = R(u+\Delta u, v+\Delta v, w) ,$$

which we expand about their cut-off values to obtain the first-order perturbation

$$[\hat{J}'_l(u) - \frac{\partial R}{\partial u}] \Delta u - [\hat{K}'_l(v) + \frac{\partial R}{\partial v}] \Delta v = R(u,v,w) - R(u,v,0) . \tag{4.235}$$

We maintain the difference form on the right-hand side, because there are cases in which the partial derivative $\partial R/\partial w$ does not exist in the limit of $w \to 0$. On the left-hand side of equation (4.235) we may neglect $\partial R/\partial u$ compared to $\hat{J}'_l(u)$ as well as $\partial R/\partial v$ compared to $\hat{K}'_l(v)$ since R as well as its derivatives are small (of the order $\exp[-2(c-1)v]$).

Depending on the circumferential order l of the modes, the difference form on the right-hand side leads to different expressions, and equation (4.235) appears as follows:

$$J'_l(u)\Delta u - K'_l(v)\Delta v =$$

$$= \frac{I_{l+1}(v)K_{l+1}(cv)}{I_{l+1}(cv)K_{l+1}(v)}[\hat{J}_l(u)+\hat{I}_l(v)][\hat{K}_l(cv)+\hat{I}_l(cv)] \begin{cases} 1/\ln(\gamma cw/2) & l=0 \\ \left(\frac{cw}{2}\right)^2 \frac{\ln(\gamma cw/2)}{(\frac{1}{2}+\hat{I}_l(cv))^2} & l=1 \\ \left(-\frac{c^2w^2}{8l^2(l-1)(\frac{l}{2}+\hat{I}_l(cw))^2}\right) & l\geq 2 \end{cases}$$

(4.236)

In addition we obtain from $u^2 + v^2 = V^2$ as the first-order perturbation

$$u\,\Delta u + v\,\Delta v = V_c\,\Delta V \qquad (4.237)$$

and from $u^2 + w^2 = [(n_1^2-n_3^2)/(n_1^2-n_2^2)]V^2$ the first-order perturbation

$$2u\,\Delta u + w^2 = 2\,[(n_1^2-n_3^2)/(n_1^2-n_2^2)]V_c\Delta V\,. \qquad (4.238)$$

For $l = 0$ and $l = 1$ the logarithmic singularity at $w = 0$ on the right-hand side of equation (4.236) lets $w^2/\Delta u$ and $w^2/\Delta v$ approach zero in the cut-off limit of $\Delta V \to 0$. Near cut-off we may therefore neglect w^2 in equation (4.238) so that

$$\Delta u = [(n_1^2-n_3^2)/(n_1^2-n_2^2)]^{\frac{1}{2}}\,\Delta V \qquad (4.239)$$

as well as

$$\Delta v = [(n_3^2-n_2^2)/(n_1^2-n_2^2)]^{\frac{1}{2}}\,\Delta V\,. \qquad (4.240)$$

Introducing both these perturbations into equation (4.236), we obtain w as a function of ΔV, for $l = 0$ according to

$$\ln(2/\gamma cw) = \frac{V_c(\hat{\mathcal{J}}_0(u)+\hat{R}_0(v))\;(\hat{I}_0(cv)+\hat{R}_0(cv))\;K_1(cv)\;I_1(v)}{\Delta V(u^2\;\hat{\mathcal{J}}_0^2(u)\;+\;v^2\;\hat{R}_0^2(v))K_1(v)\;I_1(cv)}, \tag{4.241}$$

and for $l = 1$ according to

$$c^2 w^2 \ln(2/\gamma cw) = \frac{\Delta V\;I_2(cv)\;K_2(v)\;[1+2\hat{I}_1(cv)]^2\;[u^2\;\hat{\mathcal{J}}_1^2(u)+v^2\;\hat{R}_1^2(v)]}{V_c\;I_2(v)\;K_2(cv)\;[\hat{\mathcal{J}}_1(u)\;+\;\hat{R}_1(v)][\hat{I}_1(cv)\;+\;\hat{R}_1(cv)]}. \tag{4.242}$$

For $l \geq 2$ the right-hand side of equation (4.236) depends on w as w^2 and has no singularity in the cut-off limit of $w \to 0$. Furthermore the right-hand side is proportional to nearly $\exp[-2(c-1)v]$ and remains rather small. If, for these reasons, we let it be zero, the same dependence between u and v obtains near cut-off as in the reference fibre. Any modes of circumferential order $l = 2$ and higher have therefore nearly the same characteristics near their cut-off from equation (4.225), as in the reference fibre with unlimited cladding.

Below cut-off the guided modes turn into leaky modes with little attenuation at first due to power leakage into the jacket, but more and more attenuation as we recede from cut-off into the leaky-mode domain. From the first-order perturbation, the attenuation constant of these leaky modes is given by equation (4.224) where w has now turned into an imaginary quantity.

If a particular mode is far enough above the cut-off of the reference fibre with infinite cladding to have $v \gg 1$ in equation (4.224), and if at the same time the corresponding leaky mode is far enough below cut-off of the actual fibre to have $|w| \gg 1$, the attenuation constant is approximately given by

$$\alpha \simeq \frac{2}{\beta a^2}\;\frac{(2/v)\exp[-2(c-1)v]}{(1/u^2\;+\;1/v^2)}\;\frac{v\;|w|}{v^2\;+\;|w|^2}. \tag{4.243}$$

The dependence on v and c in this approximation is dominated by the exponential factor. It accounts for the exponential decay of the evanescent cladding fields in radial direction and for the cladding thickness. Under the assumption $v \gg 1$, for which equation (4.243) approximates the attenuation constant, we are far enough above cut-off to replace

$$v = (V^2 - u^2)^{\frac{1}{2}}$$

by

$$v \simeq V - u_{lp}^2/2V,$$

where u_{lp} as the pth-zero of $J_l(u)$ is the limiting value for u very far above cut-off. With this approximation for v the attenuation constant according to equation (4.243) depends nearly exponentially also on the fibre parameter V. Its logarithm then decreases almost linearly with V over the range for which both v and $|w|$ are large compared to unity. Fig. 4.27 illustrates this linear dependence of $\log \alpha$ on V for a

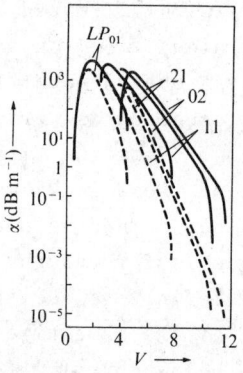

Fig.4.27. Attenuation constant of low-order leaky modes in a cladded core fibre with lossless jacket and $a = 15$ μm, $n_1 = 1\cdot 5$, $(n_1-n_2)/n_1 = 0\cdot 0018$, $(n_1-n_3)/n_1 = 0\cdot 00033$. Solid lines, $c = 1\cdot 5$. Broken lines, $c = 1\cdot 9$ (Kawakami and Nishida 1975).

representative example of a weakly guiding fibre with lossless jacket. It shows the attenuation constant of a few low-order modes in the leaky-mode region. These curves have been taken from a numerical evaluation of leaky modes in weakly guiding fibres by Kawakami and Nishida (1975). The range of fibre parameter values V for which $\log \alpha$ decreases nearly linearly with V is limited with increasing V by the cut-off of the particular mode in the jacketed fibre at the transition of this mode from a leaky mode to a guided mode. At this cut-off, α goes to zero, and remains so in the guided-mode region. With decreasing values of V the range of linear dependence of $\log \alpha$ on V is limited when we approach the cut-off of the respective mode in the reference fibre with unlimited cladding. Near this cut-off, the transverse attenuation parameter v is small compared to unity and the first-order perturbation of equation (4.224) no longer applies. A numerical evaluation of the characteristic equation (4.212) shows that the leaky mode attenuation of LP_{0p}- and LP_{1p}-modes decreases again as we approach the cut-off of the respective mode in the reference fibre with unlimited cladding.

4.7. GUIDED MODE ATTENUATION

The previous sections on cladded-core fibres have assumed the core and cladding as well as the surrounding jacket to consist of perfectly uniform materials and the interfaces between core and cladding and cladding and jacket to be coaxial cylinders of perfect circular geometry. Core and cladding material were assumed to be without absorption, and also the effects of a lossy jacket have not yet been considered quantitatively. Under these conditions of absorptionless and homogeneous fibre materials and perfect cylindrical geometry, the guided modes propagate independently of each other and without any loss due to absorption or scattering and radiation. Of course, the leaky modes also propagate independently of each other under these perfect conditions, but they suffer loss due to the frustration of their total reflection at the core-cladding interface and the ensuing tunnelling of power through the evanescent field region into cladding and jacket. The leaky modes are thus attenuated by radiation of power out of the guiding structure. We have found explicit though approximate expressions for the attenuation constant of leaky modes in the preceding sections, where for a weakly guiding structure the total internal reflection at the core-cladding interface was frustrated by the curvature of this interface, or by a surrounding jacket with a refractive index higher than the cladding index but lower than the core index.

In this section we will discuss the attenuation of guided modes. Guided modes lose power due to absorption of the material and due to scattering at its inhomogeneities. Inhomogeneities of the material as well as deviations of the fibre geometry from its perfect form of circular and coaxial cylinders for its interfaces cause guided modes to interact with one another and interchange power. In addition they lose power to radiation. In this section we will not consider the effects of geometrical imperfections for now and will assume material inhomogeneities to be random in nature with uniform statistical characteristics for any of the three cross-sectional regions. We will also let the correlation of these random fluctuations reach only over distances that are short compared to the light wavelength. Under these conditions the Rayleigh scattering law applies and the scattering loss is accounted for by an effective absorption coefficient according to equation (1.4) as imaginary component n'' of the now complex refractive index $\hat{n}_i = n_i - jn_i''$. This procedure neglects any scattering of power from one guided mode into another and is therefore accurate only for weakly guiding fibres where guided-mode interaction due to Rayleigh scattering remains quite small. Core and cladding of glass fibres usually differ very little in refractive index and only the jacket may have a substantially different index as well as the surrounding free space in the case of a fibre without any jacket. In these cases the outer cladding diameter is large enough, however, to consider all power that scatters into the many cladding modes as being lost by radiation.

We also include material absorption in the absorption coefficient n'' and instead of real values for the refractive indices now have complex values with different real and imaginary components depending on absorption and Rayleigh scattering in core, cladding, and jacket region. Instead of real wave numbers $n_i k$ we have complex wave numbers $n_i k - j\alpha_i$ now, where α_i represents the attenuation constant of a plane uniform wave in region i.

The general form of the characteristic equation (4.51) as well as its reduced forms, equation (4.56) for a fibre with unlimited cladding, equation (4.58) for a fibre with an annular core region, and the approximate forms, equation (4.57) for a metallic jacket and equations (4.103) and (4.104) for weakly guiding fibres - all these forms of the characteristic equation now turn into complex equations due to the complex character of any one or all of the refractive indices $\hat{n}_i$. The transverse parameters u, v, and w which solve these equations for guided modes will likewise turn complex, and the guided modes will assume complex values $\gamma = \alpha + j\beta$ for their propagation constants with α as the guided mode attenuation. Large absorption constants as, for example, a lossy jacket might have for efficient cladding-mode absorption will require a numerical solution of the respective characteristic equation in its complex form.

Usually, however, we can count on material absorption and scattering to be low enough to render $\alpha_i \ll n_i k$. This is the case in particular for high-quality low-loss glasses, which are used for the core and cladding of low-loss fibres. A situation will then prevail quite similar to low-loss films in section 2.4, where the respective characteristic equations were solved by a perturbational approach. Under practical conditions low-loss fibres have much less absorption and scattering than even the best of films. The same perturbational approach for the attenuation analysis is therefore even more justified for low-loss fibres. Due to such low absorption and scattering, the transverse field distribution of guided modes changes only little from its loss-free distribution and the transverse wave parameters u, v, and w in core, cladding, and jacket deviate only slightly from their real and positive values, u_0, v_0, w_0. As in equation (2.100) we let $j\Delta u$, $j\Delta v$, $j\Delta w$ be the perturbations of u_0, v_0, w_0, which equations (2.103) relate to the α_i of core, cladding, and jacket as well as to the guided-mode attenuation α. Here again the absorption and scattering effect only a purely imaginary perturbation of u_0, v_0, and w_0. We now write any of the characteristic equations of the cladded-core fibre without absorption and scattering in the form of equation (2.104) and obtain equation (2.107) as a first-order perturbational expression for guided-mode attenuation. Only the film thickness d in equation (2.107) needs to be interpreted as the core radius a, and n_0, α_0, and D_0 replaced by n_3, α_3, and D_3, respectively, as jacket index, jacket loss, and partial derivative of D with respect to n_3. The guided mode attenuation α of a cladded-core fibre surrounded by an unlimited jacket with low absorption and scattering in all three cross-sectional regions thus follows from

$$\alpha k \beta a^2 = \frac{(vw\, D_u\, n_1 \alpha_1 - uw\, D_v\, n_2 \alpha_2 - uv\, D_w\, n_3 \alpha_3) k^2 a^2 - uvw\, (D_1 \alpha_1 + D_2 \alpha_2 + D_3 \alpha_3)}{vw\, D_u - uw\, D_v - uv\, D_w} . \quad (4.244)$$

β, u, v, w, in this perturbational expression are phase constant as well as transverse wave parameters of the particular mode in the lossless structure, and as such solve the lossless characteristic equation

$$D(u,v,w,n_1,n_2,n_3) = 0 . \quad (4.245)$$

As the first of a number of special cases, we consider the cladded-core fibre without any jacket in free space. For this case we have $n_3 = 1$ and $\alpha_3 = 0$, and the perturbational expression for guided mode attenuation reduces to

$$\alpha k \beta a^2 = \frac{(v\, D_u\, n_1 \alpha_1 - u\, D_v\, n_2 \alpha_2)\, k^2 a^2 - uv\, (D_1 \alpha_1 + D_2 \alpha_2)}{v\, D_u - u\, D_v - (uv/w)\, D_w} . \quad (4.246)$$

Its characteristic equation (4.51) must be written in the form of equation (4.245) in order to evaluate its partial derivatives according to equation (2.106). Equation (4.246) corresponds to equation (2.108) for the guided-mode attenuation in a film guide with empty surrounding space.

Next we consider the cladded-core fibre with unlimited cladding. Its characteristic equation (4.56) contains of course, neither n_3 nor w so that D_3 and D_w vanish, and we obtain the following attenuation formula:

$$\alpha k\beta a^2 = \frac{(v\,D_u\,n_1\alpha_1 - u\,D_v\,n_2\alpha_2)k^2 a^2 - uv\,(D_1\alpha_1 + D_2\alpha_2)}{vD_u - uD_v}. \quad (4.247)$$

All three expressions (4.244), (4.246), and (4.247) approach the same limiting value if the particular mode goes very far above cut-off. In this limit the fields confine themselves completely to the core, and the transverse attenuation parameters v and w as well as ka become large compared to u. As a consequence $\alpha = \alpha_1$; any mode far enough above cut-off suffers the loss α_1 of a plane uniform wave in the core material.

We turn our attention now to weakly guiding fibres which differ only little in their refractive indices of core cladding and jacket. Their characteristic equation (4.212) depends only on the transverse wave parameters u, v, w, but shows no explicit dependence on n_1, n_2 or n_3. When written in the form

$$D \equiv \hat{J}_l(u) - \hat{K}_l(v) - R(u,v,w) = 0, \quad (4.248)$$

its partial derivatives with respect to u, v, and w appear as

$$D_u = \hat{J}'_l(u) - R_u, \quad D_v = \hat{K}'_l(v) - R_v, \quad D_w = R_w, \quad (4.249)$$

where the index u, v, or w on R denotes partial differentiation with respect to the particular variable. With these partial derivatives we obtain the following first-order perturbational expression for guided-mode attenuation:

$$\alpha = \frac{vw(\hat{J}'_l - R_u)\,\alpha_1 + uw(\hat{K}'_l + R_v)\,\alpha_2 + uv\,R_w\,\alpha_3}{vw(\hat{J}'_l - R_u) + uw(\hat{K}'_l + R_v) + uv\,R_w}. \quad (4.250)$$

Any guided mode that is not too close to cut-off in a fibre with not too small a ratio $c = b/a$ of cladding to core radius will only be slightly modified by the jacket. Its evanescent cladding fields decay to low values at the boundary between cladding and jacket. The right-hand side $R(u,v,w)$ as well as its partial derivatives with respect to u, v, and w are under these conditions small (of the order

$\exp[-2(c-1)v])$ and may be neglected in equation (4.250). The guided mode attenuation is then simply given by

$$\alpha = \frac{vw\,\hat{J}'_l\,\alpha_1 + uw\,\hat{K}'_l\,\alpha_2 + uv\,R_w\,\alpha_3}{w\,[v\,\hat{J}'_l + u\,\hat{K}'_l]}. \qquad (4.251)$$

We retain the third term in the numerator on the right-hand side despite the small factor R_w which it contains. This term accounts for guided-mode attenuation due to absorption and Rayleigh scattering in the jacket. There are two practical cases in which this loss contribution from the jacket is of special significance. One is the case of small values w, which brings us back to near cut-off for the particular mode. The other case concerns a jacket with intentionally high jacket loss for cladding mode absorption and the suppression of crosstalk between parallel fibres in a bundle. Such a lossy jacket has an absorption coefficient $n''_3 = \alpha_3/k$ which may be so much larger than the core and cladding absorption that the term with R_w in equation (4.251) may not only be significant in guided-mode attenuation but may even dominate the contributions from core and cladding loss. In the latter situation the guided-mode attenuation from the absorption in a lossy jacket alone follows from

$$\alpha = \frac{uv\,R_w\,\alpha_3}{w\,[v\hat{J}'_l + u\hat{K}'_l]}.$$

The last case for which we consider guided-mode attenuation in detail here is the weakly guiding fibre with unlimited cladding. For this fibre model we have $R_w = 0$, and equation (4.251) reduces to

$$\alpha = \frac{v\,\hat{J}'_l\,\alpha_1 + u\,\hat{K}'_l\,\alpha_2}{v\,\hat{J}'_l + u\,\hat{K}'_l}. \qquad (4.252)$$

The derivatives of the combinations $\hat{J}_l(u)$ and $\hat{K}_l(v)$ of Bessel and modified Hankel functions are

CLADDED-CORE FIBRES   407

$$\hat{J}'_l(u) = \frac{J_{l-1}(u)}{u\, J_{l+1}(u)} \left(1 - \frac{J_l^2(u)}{J_{l-1}(u)\, J_{l+1}(u)}\right)$$

$$\hat{K}'_l(v) = \frac{K_{l-1}(v)}{v\, K_{l+1}(v)} \left[\frac{K_l^2(v)}{K_{l-1}(v)\, K_{l+1}(v)} - 1\right]. \qquad (4.253)$$

If we introduce κ_l from equation (4.133) and use the characteristic equation (4.103) to express Bessel functions by modified Hankel functions we obtain

$$\hat{J}'_l = -\frac{K_{l-1}(v)}{u\, K_{l+1}(v)} \left(1 + \frac{u^2}{v^2}\kappa_l\right)$$

$$\hat{K}'_l = \frac{K_{l-1}(v)}{v\, K_{l+1}(v)} (\kappa_l - 1). \qquad (4.254)$$

With these expressions for $\hat{J}'_l$ and $\hat{K}'_l$ the contribution to guided-mode attenuation from absorption and scattering loss α_1 of the core material follows as

$$\frac{v\, \hat{J}'_l\, \alpha_1}{v\, \hat{J}'_l + u\, \hat{K}'_l} = [1 + (u/V)^2 (\kappa_l - 1)]\, \alpha_1 \qquad (4.255)$$

while the cladding loss α_2 contributes to guided-mode attenuation according to

$$\frac{u\, \hat{K}'_l\, \alpha_2}{v\, \hat{J}'_l + u\, \hat{K}'_l} = (u/V)^2 (1-\kappa_l)\, \alpha_2. \qquad (4.256)$$

When we compare these results with equations (4.161) and (4.162) we note that the core loss α_1 contributes to the guided-mode attenuation with $(P_1/P)\,\alpha_1$ and cladding loss with $(P_2/P)\,\alpha_2$, where P is the total guided mode power and P_1 and P_2 are the fractions of P which a particular mode guides in core and cladding, respectively. In terms of these power fractions, the guided mode attenuation is given by

$$\alpha = (P_1/P)\,\alpha_1 + (P_2/P)\,\alpha_2 \,. \tag{4.257}$$

Fig.4.21 displays the power fractions in core and cladding for a number of low order LP_{lp}-modes in the weakly guiding fibre with unlimited cladding. These power fractions can now be substituted into equation (4.257) to determine the guided-mode attenuation for a given loss in core and cladding. If the fibre has only core or cladding loss or if one is so much dominated by the other to be neglected, then the curves in Fig.4.21 give directly the factors $\alpha/\alpha_1 = P_1/P$ and $\alpha/\alpha_2 = P_2/P$ by which the guided mode loss is smaller than the bulk loss in core or cladding. When both core and cladding have losses, a superposition of both leads to the guided-mode loss. As an example, the ratio α_1/α_2 has been assumed to be either 0·1 or 10, and equation (4.257) evaluated by taking the power fractions from Fig.4.21. The results are shown in Fig.4.28.

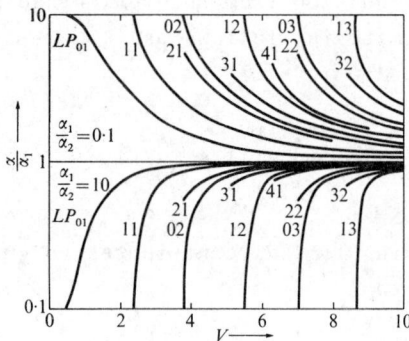

Fig.4.28. Guided-mode attenuation of LP_{lp}-modes in weakly guiding fibres with unlimited cladding and core loss α_1 and cladding loss α_2.

Attenuation formulas such as equation (4.257), which express the contribution from a particular region of the guide cross-section by the power fraction which the mode guides in this region, also follow directly from the conservation of energy. They hold for weakly guided modes which have a nearly transverse electromagnetic field. The power which a wave with an effective electric field strength E loses in a volume element dV due to a dielectric loss factor ε'' is

$$-dP = \omega \varepsilon'' \varepsilon_0 |E|^2 \, dV .$$

The dielectric loss factor is related to the refractive index n and the plane wave attenuation constant α_n by

$$\varepsilon'' = 2n \, \alpha_n/k .$$

Along an element of length dz of a waveguide, a wave then loses the power

$$-dP = 2\left(\frac{\varepsilon_0}{\mu_0}\right)^{\frac{1}{2}} \int_A n \, \alpha_n \, |E|^2 \, dA \, dz$$

due to whatever absorption and scattering causes α_n. The integration extends over the total cross-section of the waveguide and takes account of different values for n and α_n in different cross-sectional regions. The weakly guided mode with its nearly transverse electromagnetic field has a power flow density in axial direction according to the Poynting vector

$$\vec{S} = \vec{E} \times \vec{H}^* .$$

Since its fields have locally the character of uniform plane waves, their electric and magnetic fields are orthogonal to each other and their ratio is equal to the intrinsic wave impedance

$$E/H = (\mu_0/\varepsilon_0)^{\frac{1}{2}}/n .$$

In terms of power flow density, their power loss per unit distance then becomes

$$-dP/dz = 2 \int \alpha_n \, S \, dA . \qquad (4.258)$$

For such a uniform rate of power loss the total power,

$$P = \int S \, dA ,$$

which the wave transmits decays exponentially along z according to

$$P = P_0 \exp(-2\alpha z) . \qquad (4.259)$$

Note that the exponent contains the decay constant 2α for power. This power is proportional to the square of the field amplitudes, and 2α therefore corresponds to an attenuation constant α for field amplitudes. Conservation of energy requires the power change $dP/dz = -2\alpha P$ from equation (4.259) to equal the power loss in equation (4.258). When solved for α this leads to the following attenuation formula for weakly guided modes:

$$\alpha = \int \alpha_n \, S \, dA/P . \qquad (4.260)$$

For a waveguide with cross-sectional regions of different but homogeneous α_n the integration over each such region gives $\alpha_n \, P_n/P$ with P_n/P as the power fraction which the mode guides in region n. Summing over all regions results in a guided-mode attenuation according to

$$\alpha = \sum_n (\alpha_n \, P_n)/P , \qquad (4.261)$$

of which equation (4.257) is the special form for a weakly guiding fibre with only two different cross-sectional regions.

We return now to the weakly guiding fibre with unlimited cladding and examine its guided-mode attenuation near and far above cut-off. In general we obtain from equation (4.252) together with equations (4.255) and (4.256)

$$\alpha = \alpha_1 + (\alpha_2 - \alpha_1)(1 - \kappa_l) \, u^2/v^2 . \qquad (4.262)$$

Near cut-off the approximations for κ_l as well as for the characteristic equation lead to $\alpha \simeq \alpha_2$ for $l = 0$ and $l = 1$. LP_{lp}-modes of zero- and first-order circumferential dependence guide nearly all of their power in the cladding when they are close to cut-off, and therefore suffer the cladding loss as attenuation. For LP_{lp}-modes of circumferential order $l \geq 2$, the

CLADDED-CORE FIBRES 411

approximations near cut-off for κ_l as well as $u \simeq V$ lead to

$$\alpha = \alpha_1 + (\alpha_2 - \alpha_1)/l . \qquad (4.263)$$

Depending on their circumferential order l, the fraction $1/l$ of total power is guided in the cladding near cut-off. For large circumferential orders, the core loss α_1 therefore dominates in guided-mode attenuation even near cut-off.

Far above cut-off we approximate $1 - \kappa_l \simeq 1/v \simeq 1/V$ and obtain

$$\alpha = \alpha_1 + (\alpha_2 - \alpha_1) \, u_{lp}/V^3 , \qquad (4.264)$$

where the asymptotic value u_{lp} for the transverse phase parameter represents the pth-zero of the Bessel function $J_l(u)$. With increasing values of the fibre parameter V any difference $(\alpha_2 - \alpha_1)$ between core and cladding loss contributes to the guided-mode attenuation only with a factor that diminishes as u_{lp}/V^3.

An approximation for the attenuation of all but the fundamental HE_{11}-mode results when we introduce κ_l according to equation (4.134) into equation (4.262). The attenuation formula which we thus obtain reads

$$\alpha = \alpha_1 + (\alpha_2 - \alpha_1) \, \frac{2}{V^2(v^2 + l^2 + 1)^{\frac{1}{2}}} . \qquad (4.265)$$

it holds not only near and far above cut-off but provides a simple and useful approximation throughout.

4.8. SINGLE-MODE FIBRES

The previous sections on cladded-core fibres provide much of the information on guided modes including their field distribution and propagation characteristics. They form a broad and firm enough base to discuss the application of such fibres for signal transmission, and they guide us in the design of fibres for these applications.

Fibre design depends very much on which guided modes we intend to utilize for signal transmission. Off-hand it seems desirable to have the signal propagate only in one mode. The signal will then experience the delay and attenuation of only this mode and any interference from signal power in other modes will be avoided. To make absolutely sure that the signal really travels in one mode only, the waveguide should guide only this particular mode and attenuate all other possible modes by absorption or power leakage. It should be a single-mode guide.

The cladded-core fibre with a cladding of infinite extent guides only the dominant HE_{11}-mode if the fibre parameter $V = ka(n_1^2 - n_2^2)^{\frac{1}{2}}$ remains smaller than $u_{01} = 2 \cdot 405$, the first zero of the Bessel function $J_0(u)$. This mode has zero cut-off and therefore exists for any value of the fibre parameter V. At small V-values, however, its evanescent fields extend very far into the cladding and with so little field confinement, the mode is only loosely bound to the fibre core.

In actual fibres the cladding can only be of limited extent and we need to surround the fibre by a jacket, if only to protect it mechanically. We should, however, adjust the cladding diameter and choose the jacket so that all desirable features of the dominant mode for signal transmission are maintained and no degrading effects are introduced by the limitation of the cladding and the surrounding jacket.

Before we discuss the jacket design we will first list the dominant mode characteristics which are relevant and of specific significance for single-mode fibres. For many of these characteristics the weakly guiding fibre with an unlimited cladding can serve as an adequate model. Fig.4.29 presents a number of important parameters for this fibre model, which characterize the fundamental mode. The fibre parameter V as the independent variable in this presentation has been limited between $V = 0 \cdot 6$ and $V = 2 \cdot 405$. At the lower limit of this range the core begins to effectively guide the dominant mode by confining its fields and power inside or near to it. At the upper limit the LP_{11}-mode goes beyond cut-off, and the fibre loses its single-mode character. The vertical scale in Fig.4.29 has been chosen linear for parameter values larger than $0 \cdot 1$ and logarithmic for values below $0 \cdot 1$. This choice of scale displays all parameters adequately over the full V-range of Fig.4.29 and allows accurate reading everywhere

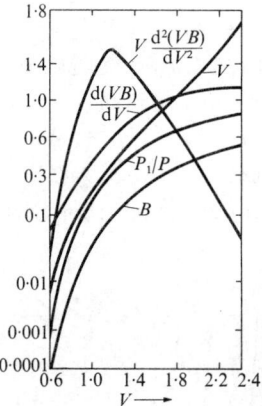

Fig.4.29. Field and propagation parameters of the fundamental mode in a weakly guiding fibre with unlimited cladding (Gloge 1975).

(Gloge 1975).

The transverse phase parameter u varies between $0 \cdot 6$ and $1 \cdot 69$ while the transverse attenuation parameter v increases from $0 \cdot 006$ at $V = 0 \cdot 6$ to $1 \cdot 71$ at the LP_{11}-cut-off. The transverse attenuation parameter defines an effective radius of the particular mode according to

$$r_e = a \, (1 + 1/v) \, . \qquad (4.266)$$

For large enough v, when the modified Hankel function approaches its asymptotic expression,

$$K_l(v) \simeq (\pi/2v)^{\frac{1}{2}} \exp(-v) \, ,$$

the evanescent cladding fields at the effective mode radius r_e have decayed to less than $1/e$ of their amplitudes at the core-cladding interface. The shape of the curves in Fig.4.30 for modified Hankel functions of low order allows a similar interpretation of r_e as effective mode radius also for small and intermediate values of v. With this interpretation the effective mode radius of the fundamental mode decreases from as much as $168a$ at $V = 0 \cdot 6$ to only $1 \cdot 58a$ at the LP_{11}-cut-off. Together with this contraction of fields near to and inside the core, we observe the corresponding confinement of power inside the core. At $V = 0\cdot 6$, the core guides only $0\cdot 1$ per cent of the total power in the fundamental mode. At the LP_{11}-cut-off, however, it guides nearly 90 per cent of the total power. The effective refractive index $N = \beta/k$ differs only little from the cladding index n_2 at $V = 0\cdot 6$ and even at the LP_{11}-cut-off, it has approached the core index just about halfway.

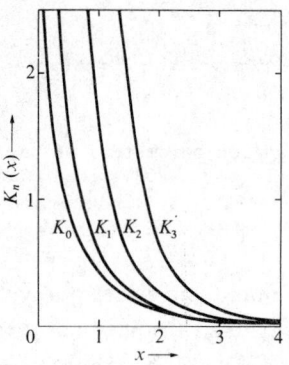

Fig.4.30. Modified Hankel functions of low integral order.

The delay factor $d(VB)/dV$ which according to equation (4.146) determines the group delay due to mode dispersion is also plotted in Fig.4.29. It is small at $V = 0\cdot 6$, but increases quite rapidly with V and levels off at the LP_{11}-cut-off where it approaches its maximum value. The maximum value of $d(VB)/dV$ occurs, however, beyond this cut-off at $V = 2\cdot 90$.

Optical signals which we transmit in fibre guides consist of spectral components which spread over a certain wavelength or frequency band. This frequency band is not only determined by the bandwidth of the signals which modulate the optical carrier, it even spreads over a much wider band because the optical carriers lack complete coherence even when they originate from nominally coherent laser sources.

For a transmission medium in which the group delay varies with frequency, the different spectral components of an optical signal travel with different delays, the signal disperses in time and is distorted. In section 1.3, we have treated pulse widening of partially coherent light pulses of a plane uniform wave in a dispersive medium. For a narrow-band signal we found the frequency derivative $d\tau/df$ of the group delay τ to account adequately for pulse widening and signal distortion. To characterize the transmission medium with respect to dispersion we introduced the dispersion coefficient

$$cf \; d\tau/df = k d^2\beta/dk^2$$

as a dimensionless quantity. Fig.1.17 shows the dispersion factor for plane uniform waves due to material dispersion in several glasses of high optical quality.

This dispersion coefficient for guided modes in weakly guiding fibres with unlimited cladding is found when we differentiate $d\beta/dk$ according to equation (4.145) with respect to k:

$$k(d^2\beta/dk^2) = k(dn_2'/dk) + k[d(n_1'-n_2')/dk] \; d(VB)/dV +$$
$$+ [(n_1'-n_2')(n_1 n_1' - n_2 n_2')/(n_1^2-n_2^2)] \; Vd^2(VB)/dV^2. \tag{4.267}$$

Equation (4.145) assumes that not only the refractive indices of core and cladding but also their dispersion characteristics are very similar and that the approximation in equation (4.144) holds. Under these conditions, we may also approximate

$$n_1 n_1' - n_2 n_2' \simeq (n_1'+n_2')(n_1-n_2) \tag{4.268}$$

to obtain the following expression for the dispersion coefficient:

$$k(d^2\beta/dk^2) = k(dn_2'/dk) + k[d(n_1'-n_2')/dk] \; d(VB)/dV +$$
$$+ [(n_1'^2-n_1'^2)/(n_1+n_2)] \; Vd^2(VB)/dV^2 \; . \tag{4.269}$$

If, in addition, we have $kdn/dk \ll 1$, as in most glasses of good optical quality, equation (4.269) simplifies still further to

$$k(d^2\beta/dk^2) = k(dn_2'/dk) + (n_1'-n_2') V d^2(VB)/dV^2 . \quad (4.270)$$

The first term of this dispersion coefficient accounts for material dispersion, and the second term represents the dispersion of the waveguide mode. Fig.4.29 shows the dispersion coefficient of the fundamental mode of the weakly guiding fibre with unlimited cladding relative to the difference $n_1'- n_2'$ of group indices in core and cladding. The fundamental mode dispersion is low at small values of V, it first increases and goes through a maximum of

$$V [d^2(VB)/dV^2] = 1\cdot 53$$

at $V = 1\cdot 18$ where the group delay τ rises quite steeply with frequency. At the border of the single-mode range where the LP_{11}-mode goes through cut-off, we have

$$V [d^2(VB)/dV^2] = 0\cdot 045. \quad (4.271)$$

Beyond the LP_{11} cut-off the fundamental mode dispersion decreases still further and goes through zero at $V = 2\cdot 90$. This is a point of inflection for the phase constant as a function of frequency. Beyond this point the dispersion coefficient of the fundamental mode remains negative for all larger values of the fibre parameter V. It passes another extremum at $V = 4\cdot 57$ where it assumes its largest negative value of $-0\cdot 12$ and then approaches zero asymptotically.

The condition for single-mode operation restricts the fibre design to $V < 2\cdot 405$. Actually a value of V quite close to this LP_{11} cut-off offers the most desirable characteristics in the single-mode range. At this limit the fields confine themselves best to the core, and the core guides as large a fraction of the total power as is possible for single-mode operation. Furthermore the mode dispersion is as weak as it can be in the single-mode range. It can only be weaker for $V < 0\cdot 65$ where the fields extend much too far into the cladding and the fundamental mode is only loosely guided by the core.

With V fixed near $2\cdot 4$, and the single-mode fibre to be designed for operation at a specific wavelength, we still need to decide on what index difference or core radius we should choose to meet

$$a(n_1^2 - n_2^2)^{\frac{1}{2}} = 0\cdot 38\ \lambda. \qquad (4.272)$$

Large core sizes require a small index difference but offer the advantage of a larger effective radius of the fundamental mode. Such large core and mode sizes facilitate launching of the fundamental mode and allow wider tolerances in connecting and splicing of single-mode fibres. The small index difference for large core sizes renders the guidance of the fundamental mode quite weak, however, and the fibre becomes very sensitive to imperfections. Imperfections, such as curvature of the fibre or any other deviation of the core from the perfect geometry of a circular cylinder, will couple the fundamental mode to the radiation field and cause it to lose power by radiation into the unlimited cladding of our present fibre model.

A large index difference, on the other hand, offers more effective guidance of the fundamental mode. But the small core size which follows from equation (4.272) for a large index difference and the small mode diameter in such a core makes it more and more difficult to launch the fundamental mode with adequate efficiency and imposes strict tolerances for connecting and splicing of such fibres.

Actual single-mode fibres for operation at wavelengths of GaAlAs-lasers or luminescent diodes between 0·8 and 0·9 μm have relative index differences in the range

$$\Delta = 0\cdot 001\ \text{to}\ 0\cdot 01$$

with $n \simeq 1\cdot 5$. For the lower limit of this range of index differences, the fibre can have 12 μm core diameter. At the upper limit, for $\Delta = 0\cdot 01$, the core diameter must be smaller than 4 μm to maintain single-mode operation.

So far we have let the cladding of the single-mode fibre extend to infinity. Actual fibres have, of course, a cladding of only a limited size and a jacket which surrounds this fibre for mechanical protection and possibly also for other purposes. Before we consider this jacket and its proper design for best fundamental mode propagation we will first discuss the effect of a finite cladding diameter $2b$ for a fibre in free space. With $n_3 = 1$ for $r > b$ and $n_2 \simeq n_1$ but substantially larger than unity, the outer cladding boundary has a rather large limiting angle $\theta_{c3} = \arccos(n_3/n_2)$ for total reflection. Furthermore its diameter needs to be much larger than the core diameter, so that the evanescent cladding fields of the fundamental core mode have decayed to small enough values and do not feel the cladding boundary. The cladding diameter should also be sufficiently large to give mechanical strength to the fibre and ease its handling. Altogether the cladding parameter

$$V_2 = kb(n_2^2 - n_3^2)^{\frac{1}{2}}$$

for a fibre in free space will be so much larger than 2·4 that this fibre guides many cladding modes by total reflection at its outer cladding boundary. Typically we have $2b \simeq 100$ μm so that for $n_2 \simeq 1\cdot5$ and $n_3 = 1$ we obtain at a wavelength of $\lambda \simeq 1$ μm a cladding parameter $V_2 = 350$. According to equation (4.21) such a fibre guides $M = V_2^2/2 \simeq 6\cdot1 \times 10^4$ cladding modes. Fig.4.31 shows the phase parameter and the effective index of refraction of a number of low-order HE_{2p}- and EH_{2p}- modes as a function of the fibre parameter V or the core radius a relative to the wavelength λ. These curves were obtained from a numerical solution of the characteristic equation (4.51). The dispersion characteristic of any particular cladding mode has two distinct ranges. The transition between these ranges occurs near the cut-off value $V = V_c = u_{l-1,p}$ of the corresponding mode in the core with unlimited cladding. For V larger than this cut-off value, the respective mode confines its fields more and more to the core, and in field distribution as well as dispersion characteristic approaches the corresponding core mode for unlimited cladding. For V smaller than the respective cut-off value, the fields expand to the outer core boundary and remain evanescent only beyond this boundary. They are now trapped and guided by this outer boun-

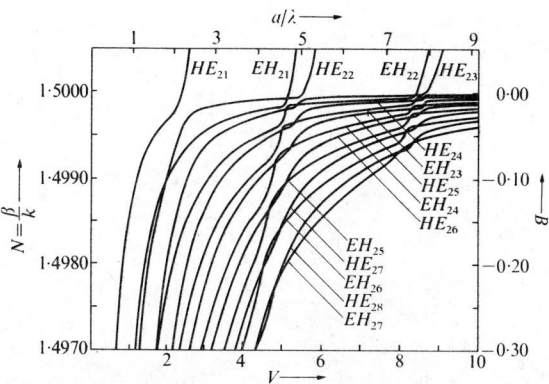

Fig.4.31. Dispersion diagram for HE_{2p}- and EH_{2p}-modes of cladded-core fibres in free space with $n_1 = 1.51$, $n_2 = 1.50$, $n_3 = 1$, and $c = b/a = 10$ (Kuhn 1975a).

dary and correspond in field distribution and dispersion to cladding modes, which the core modifies but little. Fig.4.31 depicts that part of the dispersion characteristics in which the modes change from cladding to core modes and where the phase constants of these modes are close to the phase constant of the fundamental mode.

Next in cut-off and phase constant to the fundamental mode is the LP_{11}-mode of the weakly guiding fibre or its constituent H_{01}-, E_{01}-, and HE_{21}-modes. Fig.4.32 compares phase parameter and effective index of these next higher-order modes with the fundamental-mode phase parameter. H_{01}-, E_{01}-, and HE_{21}-modes differ so little in phase constant that it does not show in Fig.4.32. The transition from cladding to core mode with increasing fibre parameter V depends on the ratio c of cladding to core diameter. This transition and the cladding and core mode character become more pronounced the larger this ratio c. For $c > 5$ and $V > 1.5$, the HE_{11}-mode shows no difference from the fundamental core mode for infinite cladding. The same is true for the LP_{11}-mode, if only $V > 2.5$.

We consider now an unlimited jacket surrounding the cladding and at first let it be loss free. For its refractive index n_3 we distinguish between the three cases for which Fig.4.33 illustrates the index profiles. The double-step pro-

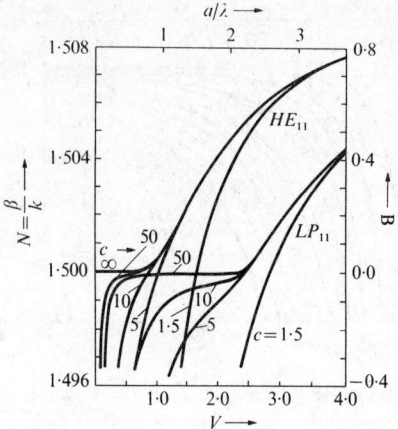

Fig.4.32. Dispersion diagram for the HE_{11}-, E_{01}-, H_{01}-, and H_{21}-modes of cladded-core fibres in free space with n_1 = 1·51, n_2 = 1·50, n_3 = 1 (Kuhn 1975a).

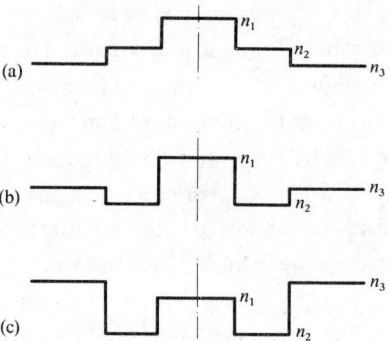

Fig.4.33. Cladded-core fibres with different index profiles. (a) Double-step fibre; (b) W-fibre; (c) leaky-mode fibre.

file with $n_1 > n_2 > n_3$ in Fig.4.33(a) and the W-profile with $n_1 > n_3 > n_2$ can guide modes without any loss. The third profile, however, with $n_3 > n_1 > n_2$ supports only leaky waves, because any ray inside the core which would be totally reflected at the core-cladding boundary suffers frustration of its total reflection at the outer cladding boundary. Some of its energy tunnels through the cladding and leaks out into the jacket.

We assume the refractive indices n_1, n_2, and n_3 to be close enough to each other for the weakly guiding fibre approximation of section 4.6 to apply. Fig.4.34 presents dispersion

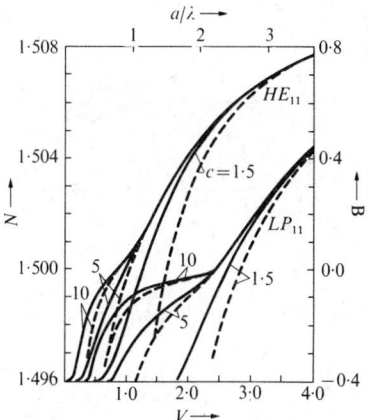

Fig.4.34. Dispersion diagram for HE_{11}- and LP_{11}-modes in double step fibres with $n_1 = 1 \cdot 51$, $n_2 = 1 \cdot 50$, $n_3 = 1 \cdot 496$ (Kuhn 1975a). Broken lines for comparison with Fig.4.32 for $n_3 = 1$.

curves for the fundamental and next higher-order mode for a typical case of a double-step fibre. In contrast to the large index step between cladding and jacket in Fig.4.32, the small index step reduces the variation of the effective N to between n_3 and n_2 in the range where the modes have the character of cladding modes. The range of core-mode character remains much the same as in case of the fibre in free space.

The situation is quite different for the W-fibre. The effective index N of its guided modes ranges over $n_3 < N < n_1$. At cut-off of any particular mode it is $N = n_3$. By evaluating the characteristic equation (4.212) under the cut-off condition $w = 0$ we get the fibre parameter value V_c at cut-off for any particular mode as a function of the diameter ratio $c = b/a$ and of the cladding parameter $V_3 = ka(n_2^2-n_3^2)^{\frac{1}{2}}$ at cut-off. Fig.4.35(a) shows V_3 versus $(V_c/V_3)^2 = (n_1^2 - n_2^2)/(n_1^2 - n_3^2)$

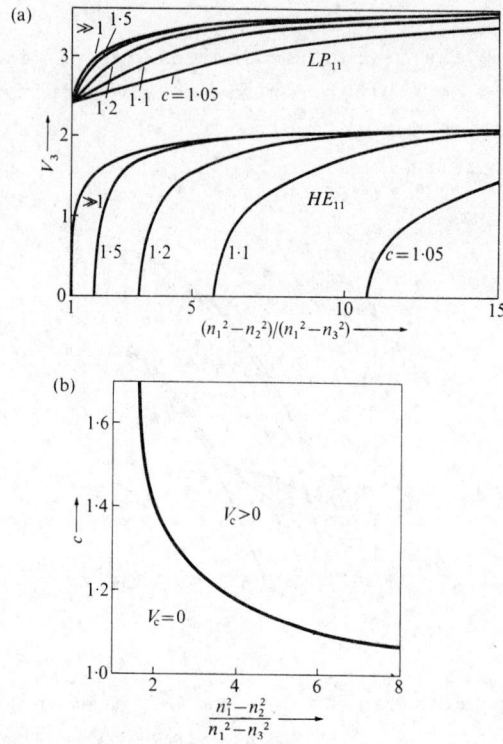

Fig.4.35. (a) HE_{11}- and LP_{11} cut-off values for $V_3 = ka(n_3^2-n_1^2)^{1/2}$ in weakly guiding W-fibres. (b) Zero cut-off limit of $c = b/a$ for the fundamental mode in W-fibres.

with c as a parameter for the fundamental (HE_{11}-)mode cut-off and the cut-off of the next higher (LP_{11}-)modes. The region between any pair of curves that belongs to the same parameter value c represents the single-mode range. For ratios of the permittivity differences $(n_1^2 - n_2^2)/(n_1^2 - n_3^2)$ which are not too much above unity, and for moderate values of the diameter ratio c, the fundamental mode still has zero cut-off, as in the fibre with a single cladding. However, as c and $(n_1^2 - n_2^2)/(n_1^2 - n_3^2)$ become larger, the fundamental mode assumes finite cut-off values V_3 and V_c. Fig.4.35(b) shows the zero cut-off limit for the fundamental mode by plotting the c-values versus the values of $(n_1^2 - n_2^2)/(n_1^2 - n_3^2)$ at which the HE_{11}-cut-off curves in Fig.4.35(a) separate from the abscissa.

The outer cladding of the W-fibre with large enough n_3 and c raises not only the HE_{11} cut-off to finite values of V it also shifts the LP_{11} cut-off as well as the cut-off of all other higher order modes to substantially higher V-values. The shift in LP_{11} cut-off from $V_c = 2\cdot 405$ to higher V-values occurs for any $(n_1^2 - n_2^2) > (n_1^2 - n_3^2)$ and $c > 1$, but is larger when both $(n_1^2 - n_2^2)/(n_1^2 - n_3^2)$ and c differ more from unity. It lends to the W-fibre the advantage of a larger core and fundamental mode size for single-mode operation.

We will now take absorption of the jacket into consideration. The jacket needs to be more or less lossy in order to prevent crosstalk from other fibres in a bundle and also to absorb the power that converts from the fundamental mode into cladding or leaky modes at fibre imperfections. We examine first what loss such absorption in the jacket causes for the fundamental mode. Fig.4.36 shows the HE_{11}-loss due to absorption in the jacket from a numerical solution of the general characteristic equation for complex values $n_3(1-j\tan\delta)^{\frac{1}{2}}$ of the jacket index. The attenuation constant is plotted as a

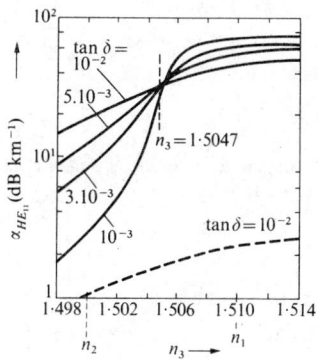

Fig.4.36. Fundamental mode attenuation in a cladded-core fibre with a lossy jacket for different loss tangents with $n_1 = 1\cdot 51$, $n_2 = 1\cdot 50$, $V = 2\cdot 18$, $\lambda = 1$ μm (Kuhn 1975). Solid lines for $c = b/a = 6$; broken line for $c = 7$.

function of the jacket index n_3 with its loss tangent as a parameter. For the particular core and cladding indices $n_1 = 1\cdot 51$, $n_2 = 1\cdot 50$, and $c = 6$, as well as the fibre parameter

$V = 2\cdot 18$, the transition from guided to leaky wave in the case of the zero-loss tangent occurs at $n_3 = 1\cdot 5047$. The jacket absorption attenuates the HE_{11}-mode also in the guided-mode range. But in this guided mode range the HE_{11}-attenuation decreases with decreasing loss tangent. Near cut-off, where the HE_{11}-mode in the loss-free structure changes from a guided to a leaky mode, the attenuation curves cross over and reverse their dependence on the loss tangent. In the leaky-mode range the attenuation due to jacket absorption decreases when the loss tangent is increased.

The LP_{11}-mode in Fig.4.37 has similar loss characteristics due to jacket absorption. In the case of a loss-free

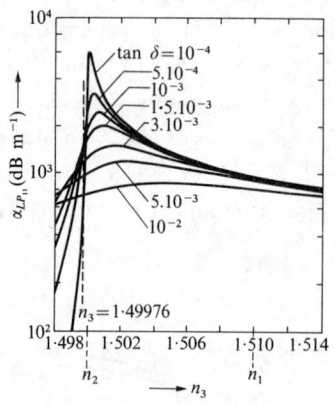

Fig.4.37. LP_{11}-attenuation in a cladded-core fibre with a lossy jacket for different loss tangents $n_1 = 1\cdot 51$, $n_2 = 1\cdot 50$, $V = 2\cdot 18$, $\lambda = 1$ μm, $c = 10$ (Kuhn 1975b).

jacket and the particular set of parameter values of Fig.4.37 it changes from a guided to a leaky mode at $n_3 = 1\cdot 49976$. Near this transition we observe again the cross-over of attenuation curves.

If a cladded core fibre with a lossy jacket is to be designed for single-mode operation, we should choose the outer cladding diameter large enough so that the cladding absorption causes not too much HE_{11}-attenuation. But we should also maintain as much LP_{11}-loss as possible, to reduce any undesired interference from this mode. For high LP_{11}-loss the outer cladding diameter should be kept as low as possible. Figs. 4.38 and 4.39 demonstrate how HE_{11}- and LP_{11}-loss depend on the ratio c of core to cladding diameter. For the HE_{11}-loss

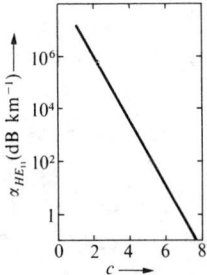

Fig.4.38. Fundamental mode attenuation in a cladded-core fibre with a lossy jacket versus ratio of cladding to core diameter (Kuhn 1975b). $n_1 = 1\cdot 51$, $n_2 = 1\cdot 50$, $n_3 = 1\cdot 5015$, $\tan \delta = 3 \times 10^{-3}$, $V = 2\cdot 18$, $\lambda = 1\mu$ m.

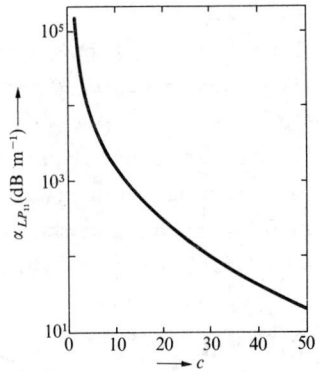

Fig.4.39. LP_{11}-attenuation in cladded-core fibre with lossy jacket versus ratio of cladding to core diameter (Kuhn 1975b). $n_1 = 1\cdot 51$, $n_2 = 1\cdot 50$, $n_3 = 1\cdot 5015$, $\tan \delta = 3 \times 10^{-3}$, $V = 2\cdot 18$, $\lambda = 1$ μm.

in Fig.4.38 we observe a nearly linear descent of $\log \alpha$ with c. It reflects the exponential decay of evanescent cladding fields. To have less than 1 dB km^{-1} of extra HE_{11}-loss from jacket absorption in the typical example of Fig.4.38, the ratio of cladding to core diameter should be larger than $c = 6\cdot 8$. The LP_{11}-loss due to jacket absorption amounts still to more than

3 dB mm^{-1} at this particular ratio of core to cladding diameter.

Transmission loss is one deciding factor that degrades signal transmission and therefore needs to be considered in fibre design. The other main effect for the degradation of transmission is signal distortion due to delay dispersion. We will therefore examine now what possibilities we have of reducing delay dispersion by a suitable choice of jacket. Material dispersion as well as mode dispersion contribute to signal distortion in a single-mode fibre. In the case of a weakly guiding fibre without jacket and its cladding extending to infinity, the dispersion coefficient is given by equation (4.270). This expression for the dispersion coefficient also assumes that core and cladding materials have similar dispersion characteristics. Under these conditions, the dispersion coefficient in equation (4.270) consists of two separate terms, the first for material dispersion and the second for mode dispersion. The dispersion coefficient for material dispersion of a number of glasses of good optical quality is plotted in Fig.1.17. This quantity is normally positive except towards the long-wavelength limit of the transparent range, where the effects of vibrational resonances on the material dispersion might dominate those of electronic resonances. Fused silica, for example, has a dispersion coefficient that goes through zero at $\lambda = 1\cdot 28$ μm and becomes negative at longer wavelengths.

The dispersion coefficient for mode dispersion of the fundamental mode in a weakly guiding fibre with unlimited cladding is plotted in Fig.4.29 as a function of the fibre parameter V. It also remains positive in the single-mode region of V, just as the material dispersion coefficient normally does. Hence in this region both contributions from material and mode dispersion add and there is no chance to compensate one against the other.

For the most interesting part of the single-mode region between $V = 2$ and $2\cdot 4$, the factor $Vd^2(VB)/dV^2$ decreases from $0\cdot 3$ to $0\cdot 045$. Usually in single-mode fibres, the core and cladding index and their respective group indices differ by less than 1 per cent. In the spectral range $\lambda < 1$ μm, the dispersion coefficient for material dispersion is larger than $0\cdot 01$

for all glasses in Fig.1.17. Under normal conditions therefore, material dispersion dominates mode dispersion by a rather wide margin.

Beyond the single-mode region at $V = 2\cdot 90$, the mode dispersion goes through zero and becomes negative. It reaches a negative maximum at $V = 4\cdot 57$. If we want to utilize this negative mode dispersion to compensate the positive material dispersion, we must suppress the LP_{11}- and other higher-order modes or shift their cut-offs to higher V-values. In addition the negative mode dispersion must be made large enough to be of the same absolute magnitude as the positive material dispersion. A weakly guiding fibre with its small index difference has too little mode dispersion to accomplish this. We need to consider larger index differences for which the weakly guiding fibre approximations no longer hold.

The mode dispersion coefficient for the fundamental HE_{11}-mode from a numerical evaluation (Kawakami and Nishida 1974) of the rigorous characteristic equation (4.56) of the cladded core fibre with unlimited cladding is plotted in Fig.4.40(a) for the range of negative mode dispersion as a function of the fibre parameter V for one particular ratio of cladding to core index. The negative peak at $V = 4\cdot 57$ has been transferred to Fig.4.40(b)

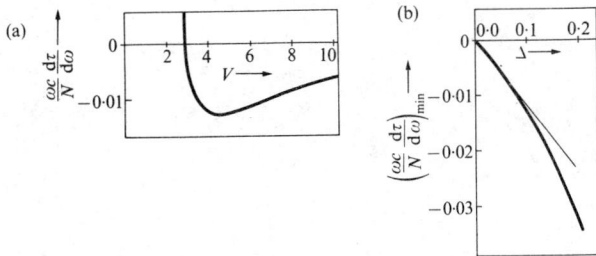

Fig.4.40. (a) HE_{11}-dispersion factor in a cladded-core fibre with unlimited cladding for $\Delta = 0\cdot 1$.

(b) Peak of negative HE_{11}-dispersion versus relative index difference $\Delta = (n_1-n_2)/n_1$ (Kawakami and Nishida 1974).

and complemented by the peak values for other index ratios n_2/n_1. Fig.4.40(b) thus gives the maximum negative dispersion coefficient as a function of relative index difference. The straight line which is the tangent to this curve at the origin represents the approximation for weakly guiding fibres. By comparing Fig.1.17 with Fig.4.40(b) we note that we require a relative index difference of the order of $\Delta = 0.1$ to compensate material dispersion by mode dispersion.

The peak of negative mode dispersion occurs at a V-value where the core guides not only the fundamental mode, but where higher-order core modes are also above cut-off. To shift the cut-off of these higher-order modes beyond the peak of negative HE_{11}-dispersion, we limit the cladding to a finite diameter and surround it by a second cladding of a suitable refractive index n_3 in the range

$$n_1 > n_3 > n_2 .$$

This second cladding should not have the character of a jacket for absorption and mechanical protection. It should rather be of the same high quality with low loss and low material dispersion as the core and cladding materials.

The H_{01}-mode is the first higher-order mode in a cladded core fibre with unlimited cladding. Its effective index $N = \beta/k$ at the V-value, where the negative HE_{11}-dispersion has its peak is plotted in Fig.4.41 as a function of relative index difference between core and cladding. For a given index

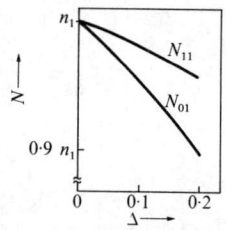

Fig.4.41. Effective indices N_{11} and N_{01} of HE_{11}- and H_{01}-modes at the peak of negative HE_{11}-dispersion in cladded-core fibres with unlimited cladding versus relative index difference $\Delta = (n_1-n_2)/n_1$ (Kawakam and Nishida 1974).

difference we read the effective H_{01}-index from this diagram as N_{01}. If we now choose $n_3 > N_{01}$, the H_{01}-mode turns into a leaky mode and we have shifted its cut-off beyond this particular V-value. For $\Delta = 0.1$ and $V = 4.57$, we have

$N_{01} = 0.9554\ n_1$ and n_3 must be larger than this value for N_{01}. To obtain sufficient leaky mode attenuation for the H_{01}-mode under these conditions, we should raise n_3 above N_{01} by a certain margin. An upper limit to n_3 is set by the change which the second cladding effects for the fundamental HE_{11}-mode. Fig.4.41 also shows the effective HE_{11}-index N_{11} at the negative peak of HE_{11}-dispersion as a function of Δ. If n_3 is raised above N_{11}, the HE_{11}-wave will also turn into a leaky wave. The choice of suitable index for the outer cladding is therefore limited by

$$N_{11} > n_3 > N_{01}.$$

Any W-fibre with n_3 in this range will guide only the HE_{11}-mode at the negative peak of its mode dispersion coefficient.

When we limit the cladding to a finite diameter and surround it with a second cladding of raised index n_3, we not only shift the HE_{11}- and H_{01}-cut-offs to finite and higher V-values, respectively. We also change their dispersion characteristic. Fig.4.42 demonstrates the change in HE_{11} disper-

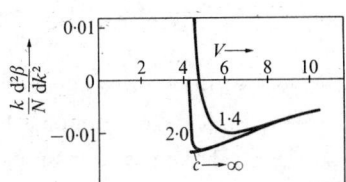

Fig.4.42. Normalized dispersion coefficient of the fundamental HE_{11}-mode in W-fibres versus the fibre parameter V for different ratios $c = b/a$ of cladding to core diameter and $n_3 = 0.98\ n_1$ (Kawakami and Nishida 1974).

sion by comparing the dispersion coefficient for the infinite cladding of Fig.4.40(a) with that of the W-fibre with inner cladding to core diameter $c = 2$ and 1.4. As long as $c > 2$ in this representative numerical example, the HE_{11}-dispersion near its negative peak and further above cut-off changes only little from the HE_{11}-dispersion in a cladded-core fibre with unlimited cladding.

4.9. MULTIMODE FIBRES

Cladded-core fibres for single-mode operation must have a fibre parameter V which is smaller than 2.4. The outer cladding or jacket of a W-fibre with an index n_3 between core and cladding index extends the single-mode region to somewhat higher V-values. But at optical wavelengths even for W-fibres the fibre cores must be extremely thin or the index difference quite small to remain in the single-mode region. For fibre cores of such microscopic dimensions and fundamental modes with equally narrow transverse field distributions, it is difficult to launch the fundamental mode and achieve sufficient launching efficiency. Connectors and splices for such small core and mode sizes must keep extremely tight tolerances in order to transmit the fundamental mode with low enough loss. If we reduce the index difference between core and cladding or, in case of the W-fibre, apply an outer cladding which differs less in index from the core than the inner cladding, we can extend the single mode region to larger core and mode sizes. This alleviates launching and splicing problems but it weakens the guiding capabilities of the fibre. Fundamental mode transmission in such weakly guiding fibres becomes very sensitive to bending of the fibre and may be seriously degraded due to fibre imperfections.

As another drawback, single-mode fibres can only be excited from single-mode sources. To launch the fundamental mode efficiently, we need to match the transverse field distribution of the source to that of the fundamental fibre mode. Any mismatch leads to loss of power into leaky modes or to the radiation field.

For all these disadvantages of single-mode fibres and despite their potentially excellent transmission characteristics, there is considerable practical interest in cladded-core fibres with much larger core sizes and sometimes also much larger index differences between core and cladding. Such fibres have correspondingly high values of their fibre parameter V and, according to equation (4.21), guide as many as

$$M = V^2/2$$

modes in their core. A large index difference between core and cladding means a large numerical aperture

$$\text{N.A.} = \sin\theta_{0c} = (n_1^2 - n_2^2)^{\frac{1}{2}}$$

with substantial values for the limiting angle θ_{0c} of total reflection in Fig.4.2. The core captures all light rays which are incident on its front face at angles smaller than θ_{0c} and guides them in one or other of its many modes of propagation.

Into such multimode fibres, light signals cannot only be launched from single-mode laser sources, but also, even more efficiently, from lasers which oscillate in many modes of different transverse orders. Even multimode sources of only partially coherent or incoherent light, such as superluminescent or light emitting diodes, can be connected to multimode fibres and still excite them with acceptable launching efficiencies.

Compared to single-mode fibres, therefore, multimode fibres offer the following advantages. They cannot only be connected to single-mode lasers but also to multimode lasers and even incoherent sources. In all these cases, light signals can be launched into them with reasonable efficiency. Much wider tolerances for the alignment of these sources with the multimode fibre may be allowed than for a single-mode laser with a single-mode fibre. To connect and splice multimode fibres requires also much less precision than for single-mode fibres.

The price which we pay for so much freedom in the choice of sources and so much convenience in connection and splicing lies in inferior transmission characteristics. Since multimode or incoherent sources excite many or even most of the guided modes as well as leaky modes, light signals will propagate in all these modes and suffer the delay and dispersion of these modes as well as the respective mode attenuation. In general all these propagation characteristics differ from mode to mode. The signal components in different modes will change their amplitude and form during transmission, depending on the different dispersion and attenuation characteristics, and arrive at different times corresponding to the delay of the various modes. Their superposition at the fibre end may then deviate considerably from the optical signal at the fibre input.

If we use a laser which oscillates in only one or a few low transverse-order modes, and excite with it only the fundamental mode of the multimode fibre or a few of its low-order modes, the signal will remain only in these modes of excitation as long as the fibre maintains its straight and circular geometry. Any geometrical imperfections as well as index variations will couple the modes with one another. The modes interact and signal power also converts from the low-order mode of excitation to modes of higher order. Further along the multimode fibre, the signal will then not only travel in its original modes of excitation but also in some or many of the other propagating modes.

To assess how multimode effects influence signal transmission, we need to examine the propagation characteristics of all these modes and their dependence on fibre and jacket parameters. Many of these modes and in particular those of low order are far above cut-off in a multimode fibre. All the other modes are then either high in circumferential order l or radial order p or of high order in both l and p. Under these conditions, which prevail quite generally in multimode fibres, we can make a number of approximations.

With the total number of guided modes given approximately by $M = V^2/2$ and the values of the transverse phase parameter u for a particular mode equal to V at cut-off, $u_c = V_c$, there are $\nu = V_c^2/2 = u_c^2/2$ modes with cut-off values below $u_c = V_c$. We can therefore label these modes with ν in the sequence of their cut-off values. The νth-mode will then have its cut-off occurring approximately at

$$u_c = V_c = (2\nu)^{\frac{1}{2}} . \qquad (4.273)$$

With the near degeneracy of the weakly guiding fibre there will, of course, always be four LP_{lp}-modes, and two for the lowest circumferential order $l = 0$, that go beyond cut-off at the same value u_c.

We will first examine the group delay of the different

modes, because the delay differences between modes degrade transmission quite seriously in multimode fibres. In a cladded-core fibre with jacket under multimode conditions most guided modes are far enough above cut-off, and their fields confined so much to the core, that the jacket has no influence on the dispersion characteristics of these modes. The weakly guiding fibre with an unlimited cladding may then serve as an adequate model. The group delay of its modes is given by equation (4.146) with $d(VB)/dV$ following from equation (4.147). If we approximate $\kappa_l(v)$ according to equation (4.134) and use $u^2 + v^2 = V^2$ to express v by u and V, we obtain

$$\frac{d(VB)}{dV} = 1 + \frac{u^2}{V^2}\left[1 - \frac{2}{(V^2 - u^2 + l^2 + 1)^{\frac{1}{2}}}\right].$$

Of the many guided modes in the multimode fibre, we ignore those few which are close to cut-off, and with $\kappa_l(v) \simeq 1$ obtain approximately

$$\frac{d(VB)}{dV} = 1 + \frac{u^2}{V^2}. \qquad (4.274)$$

The transverse phase parameter u in this expression ranges only between its cut-off value $u_c = u_{l-1,p}$ and u_{lp}, the next higher zero crossing of the Bessel function $J_l(u)$ of next higher order l. Except for the lowest-order modes, this range is relatively narrow so that in equation (4.274) we may replace u by u_c. The lowest-order modes have such small u-values that for them $(u/V)^2$ in equation (4.274) is quite insignificant to start with. The delay factor is now given by

$$\frac{d(VB)}{dV} = 1 + \frac{u_c^2}{V^2} \qquad (4.275)$$

and with equations (4.21) and (4.273) may be expressed by the total number of modes M and the index ν of the particular modes in the sequence of their cut-off:

$$\frac{d(VB)}{dV} = 1 + \frac{\nu}{M}. \qquad (4.276)$$

The group delay per unit distance in equation (4.146) may now also be written in terms of the mode number M and mode index ν according to

$$\tau = \frac{1}{c} [n'_2 + n'_1 \Delta (1 + \frac{\nu}{M})] \ . \qquad (4.278)$$

It shows a delay difference

$$\delta\tau = \frac{n'_1 \Delta}{cM} \qquad (4.279)$$

between modes that go through cut-off next to each other. This delay difference between neighbouring modes is nominally independent of the mode index ν. Actually, of course, for all $l > 0$ we have always 4 nearly degenerate LP_{lp}-modes going through cut-off at the same V_c with practically no delay difference among each other. Nevertheless equations (4.278) and (4.279) indicate to us, that on the average all modes of the cladded-core fibre are nearly equally spaced in group delay. When signal components travel in all these modes along a fibre of length L, they will follow each other at the fibre end in equal time intervals $\delta\tau L$. Their total delay spread $\Delta\tau L$ follows in the present approximation from the relative delay spread

$$\Delta\tau/\tau = \Delta \ . \qquad (4.280)$$

With signal components potentially travelling in so many modes, we need to examine next the attenuation which these components suffer in different modes. We assume here a multi-mode fibre of perfect geometry and without any material defects. Its mode attenuation then arises from bulk loss in the core, cladding, and jacket and in addition from radiation in the case of leaky modes. When for modes not too close to cut-off, we consider the attenuation from bulk loss in the core and cladding alone, the weakly guiding fibre with unlimited cladding appears again as an adequate model. Except for the lowest-order fundamental mode, its guided-mode attenuation is to a good approximation given by equation (4.265). For moderate and large circumferential orders l we may further simplify this expression and obtain, when we replace v^2 by $V^2 - u^2$,

$$\alpha = \alpha_1 + (\alpha_2 - \alpha_1) \frac{u^2}{V^2(V^2 - u^2 + l^2)^{\frac{1}{2}}} \quad . \tag{4.281}$$

According to this formula any excess bulk loss $\Delta\alpha = \alpha_2 - \alpha_1$ of the cladding over the bulk loss α_1 in the core raises the mode attenuation by

$$\alpha_c = (\alpha_2 - \alpha_1) \frac{u^2}{V^2(V^2 - u^2 + l^2)^{\frac{1}{2}}} \quad . \tag{4.282}$$

If, for a given fibre parameter V, we restrict this additional attenuation to

$$\alpha_c = (\alpha_2 - \alpha_1) \, d/V \tag{4.283}$$

and specify

$$d = \frac{\alpha_c}{\alpha_2 - \alpha_1} V \tag{4.284}$$

so that this restriction is obeyed, we obtain a limiting value for u from equation (4.282), which must not be exceeded in order to stay below α_c. By solving

$$\frac{u^2}{V^2(V^2 - u^2 + l^2)^{\frac{1}{2}}} = \frac{d}{V}$$

for u, this limiting value appears in terms of V, d, and l:

$$u_{cl}^2 = dV[(V^2 + l^2 + d^2V^2/4)^{\frac{1}{2}} - \frac{dV}{2}] \quad . \tag{4.285}$$

The approximate form of the characteristic equation

$$p\pi = (u^2 - l^2)^{\frac{1}{2}} - l \arccos (l/u) \tag{4.286}$$

from equation (4.15) of the ray picture is adequate for all modes of the multimode fibre except for those few of very low order p. When we introduce the limiting value u_{cl} for u from equation (4.285) into this characteristic equation, we obtain p/V as a function of l/V with the loss factor d as the only other parameter. The diagram of relative mode orders p/V versus l/V in Fig.4.43 shows curves for specific loss factors d. The area below any such curve represents the number of modes which suffer less additional attenuation due to the excess

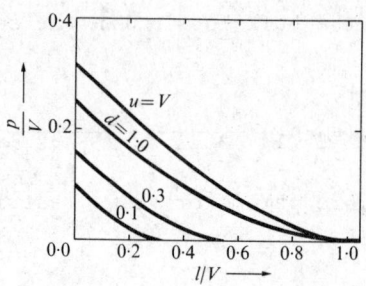

Fig.4.43. Diagram of relative mode orders p/V versus l/V with limiting curves for excess cladding attenuation $\alpha_c = (\alpha_2 - \alpha_1) \, d/V$.

cladding loss than the loss factor d specifies from equation (4.284). This area depends only on d so that the number M_{cl} of modes with less attenuation than

$$\alpha = \alpha_1 + (\alpha_2 - \alpha_1) \, d/V$$

is also a function of d and V alone according to

$$M_{cl} = f(d) V^2 / 2 \ . \qquad (4.287)$$

Fig.4.44 displays $f(d)$ from a numerical integration of

$$M_{cl} = 4 \int p_{cl}(l) \, dl$$

with $p_{cl}(l)$ according to equation (4.286) and $u = u_{cl}$.

The range of d for which $f(d)$ in equation (4.287) yields meaningful mode numbers is limited by the cut-off condition $u = V$ for guided modes. In order to stay in the guided mode domain of Fig.4.43 the limiting value u_{cl} in equation (4.285) must remain smaller than V. As a function of l this limiting value grows largest for $l = V$ in the lower right-hand corner

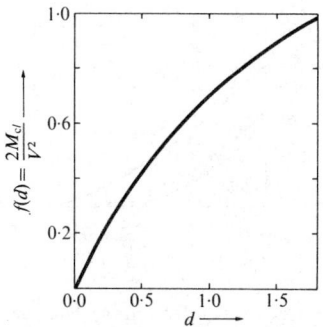

Fig.4.44. Relative number of modes with excess cladding loss less than $\alpha_c = (\alpha_2-\alpha_1)\,d/V$ versus loss factor d.

of the guided mode domain. For u_{cl} to remain smaller than V in this corner requires the loss factor d to be specified below the value which solves

$$d[(2 + d^2/4)^{\frac{1}{2}} - \frac{d}{2}] = 1.$$

This loss factor is $d = 1$. If d exceeds this value, we obtain $u_{cl} > V$ and move partially into the domain of leaky modes in Fig.4.43. The attenuation which then results from (4.283) is, of course, still due to bulk loss in the core and cladding, but it adds to the leaky mode attenuation from radiation. For $d = 1$, the limiting curve $p_{cl}(l)$ just crosses over into the leaky mode domain at $l = V$.

In equation (4.278) we have used the mode index ν to determine the group delay of specific modes. This mode index from equation (4.273) as well as the mode number M may also be introduced into equation (4.282) to relate the attenuation to ν and M:

$$\alpha = \alpha_1 + (\alpha_2-\alpha_1)\,\frac{\nu}{M(2M + l^2 - 2\nu)^{\frac{1}{2}}}. \qquad (4.288)$$

A group of modes that ranges in cut-off value between u_c and $u_c + du_c$ has mode indices between $u_c^2/2$ and $u_c^2/2 + u_c\,du_c$. It occupies a strip in the p,l-diagram as shown in Fig.4.45. Within this strip circumferential orders occur between $l = 0$

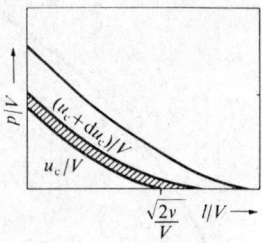

Fig.4.45. Diagram of relative mode orders p/V versus l/V with incremental area for modes with index values between $\nu = u_c^2/2$ and $\nu+d\nu = u_c^2/2 + u_c\, du_c$.

and $l = u_c = \sqrt{2\nu}$. For a given mode index ν, therefore, the attenuation α can have values within a range that is bound by

$$\frac{\nu}{M(2M)^{\frac{1}{2}}} < \frac{\alpha - \alpha_1}{\alpha_2 - \alpha_1} < \frac{\nu}{M(2M - 2\nu)^{\frac{1}{2}}}. \qquad (4.289)$$

For low-order modes with $\nu \ll M$, this range remains rather narrow. But for higher orders with ν approaching M near cut-off, the upper limit of this range can extend as high up as $\nu/M \simeq 1$, and accounts for the large fraction of power which modes of low circumferential order l guide in the cladding near cut-off.

Equation (4.278) tells us that modes which are close to each other in their index values ν travel with nearly the same group delay. If we consider a large enough increment from ν to $\nu + \Delta\nu$ so that it contains all circumferential orders from $l = 0$ to $l = \sqrt{2\nu}$, this group of modes suffers on the average the relative excess attenuation

$$\left(\frac{\alpha-\alpha_1}{\alpha_2-\alpha_1}\right)_{av.} = \frac{1}{M}\left(\frac{\nu}{2}\right)^{\frac{1}{2}} \int_0^{\sqrt{2\nu}} \frac{dl}{(2(M-\nu) + l^2)^{\frac{1}{2}}} = \frac{1}{V}\left(\frac{\nu}{M}\right)^{\frac{1}{2}} \operatorname{arc\,tanh} \left(\frac{\nu}{M}\right)^{\frac{1}{2}}.$$
$$(4.290)$$

Fig.4.46 shows this average excess attenuation as a function of the relative mode index ν/M. For low mode indices, this average attenuation remains quite low, but for ν approaching M, it can be as high as

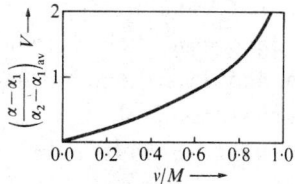

Fig.4.46. Average excess attenuation due to excess cladding loss in a fibre with unlimited cladding versus relative mode index.

$$\left(\frac{\alpha - \alpha_1}{\alpha_2 - \alpha_1}\right)_{\text{av.}} \simeq \frac{1}{V} \ln V \quad \text{for} \quad M - \nu = 2. \quad (4.291)$$

If a source excites all circumferential orders l of equal group delay with the same power, this group of modes will suffer the average excess attenuation according to equation (4.290). A high bulk loss α_2 in the cladding will then reduce modes of high index ν by strong attenuation but cause only little excess attenuation for low-index modes. A lossy cladding may thus effectively reduce the mode volume, the delay spread of signals, and improve the transmission characteristic of multimode fibres.

More effective suppression of high-index modes with still less excess attenuation of low-index modes may be provided by a lossy jacket around the cladding. In this case, the cladding is made of a high quality material with the same low bulk loss as the core and only the jacket is furnished with absorption. All modes with fields well confined to the core will suffer little attenuation but those closer to cut-off with cladding fields reaching the lossy jacket will be attenuated much more strongly.

To determine mode attenuation for a cladded-core fibre with a lossy jacket requires the solution of the rather complicated characteristic equation (4.51) for complex values of n_3. Even the weakly guiding fibre with an unlimited lossy jacket still leaves us with the characteristic equation (4.212) to be solved, again for complex n_3. Perturbational solutions for a small absorption constant of the jacket and for a jacket that modifies the guided modes of the weakly guiding fibre with unlimited cladding only slightly are still quite involved. We therefore refer to a simplified procedure used by Kuhn (1975c) that determines the attenuation for only a few of the many modes and finds the attenuation of the remaining modes by interpolation.

This procedure starts by asserting that guided modes extend their fields further into the cladding towards the jacket the higher the order ν with which they pass through cut-off. It then concludes that the mode attenuation must also increase monotonically with the mode index ν. To check this assertion, the guided mode attenuation of a multimode fibre with an unlimited lossy jacket has been determined by solving the characteristic equation (4.51) for a complex jacket index $n_3(1-j\tan\delta)^{1/2}$ with representative values for core and cladding index, and size. The jacket index and its loss tangent for this numerical solution were chosen close to a W-profile for efficient high-order mode suppression. The attenuation constants are plotted in Fig.4.47 versus mode index ν with one index number for each H_{0p}- and E_{0p}-mode, but two such

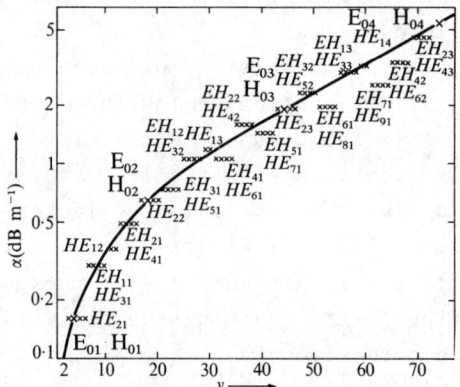

Fig.4.47. Attenuation of the first 76 modes in a cladded-core fibre with an unlimited lossy jacket $n_1 = 1\cdot51$, $n_2 = 1\cdot50$, $n_3 = 1\cdot5015$, $\tan\delta = 2 \times 10^{-4}$, $V = 21\cdot8$, $\lambda = 1$ μm, $c = b/a = 1\cdot1$ (Kuhn 1975c).

numbers for each HE_{mp}- and EH_{mp}-mode to account for their two orthogonal polarizations. The two modes of a nearly degenerate pair which form the LP_{lp}-modes have the same attenuation constants up to at least four decimals. The line in Fig. 4.47 connects all attenuation constants of axially symmetric modes. The attenuation constants of all other modes cluster quite near to this line. For large circumferential orders, they lie below this line. The same tendency for α to decrease with l near a given mode index ν also appears in equation (4.288) for the weakly guiding fibre with bulk loss in the unlimited cladding. For high numbers of the mode index ν, a curve such as in Fig.4.47 represents an upper bound for the mode attenuation. We will nevertheless use it to interpolate the attenuation of all asymmetric modes from the attenuation of axially symmetric modes.

For this purpose, the H_{0p}-attenuation of the cladded core fibre with lossy jacket is plotted in Fig.4.48 versus

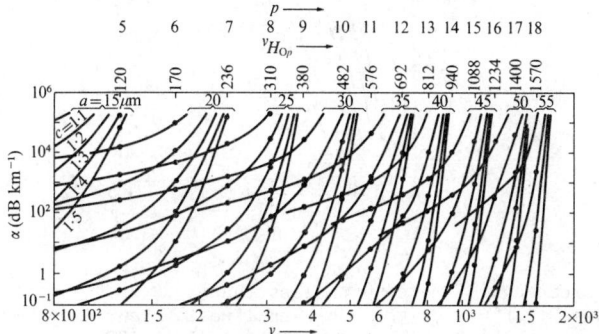

Fig.4.48. H_{0p}-attenuation versus mode index for cladded-core fibres with lossy jacket (Kuhn 1975c). $n_1 = 1\cdot 51$, $n_2 = 1\cdot 50$, $n_3 = 1\cdot 5015$, $\tan \delta = 2 \times 10^{-4}$, $\lambda = 1$ μm. The lines which connect calculated points of H_{0p}-attenuation for specific fibre parameters represent upper bounds for all other modes.

mode index ν for the same representative values of core, cladding, and jacket index as in Fig.4.47, but different core sizes and different ratios of cladding to core diameter. The range of attenuation constants in Fig.4.48 is limited between $10^{-1} \ldots 10^5$ dB km^{-1}, which includes all values of practical interest. With the radial order p of the H_{0p}-modes going as high as 18, as many as $\nu = 1600$ modes are guided by the fibre.

Fig.4.48 allows us to estimate the fraction of guided modes below a specified attenuation constant. We draw a horizontal line at this attenuation and for given core radius a and diameter ratio $c = b/a$ read the mode index ν at which the respective attenuation curve crosses this horizontal line.

As a result of this evaluation, the percentage of modes below two specified values of attenuation is shown in Fig.4.49 versus the core radius for different ratios of core and cladding diameter. This diagram can be used in the design of fibres for different objectives.

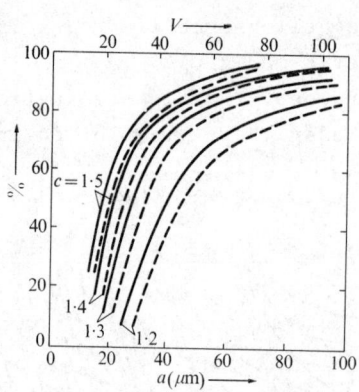

Fig.4.49. Percentage of modes with less than 1 dB km^{-1} (solid lines) and less than 0·1 dB km^{-1} (broken lines) attenuation due to a lossy jacket around cladded-core fibres (Kuhn 1975c). $n_1 = 1·51$, $n_2 = 1·50$, $n_3 = 1·5015$, $\tan\delta = 2·10^{-4}$, $\lambda = 1$ μm.

As one objective the multimode fibre might be designed to guide a certain percentage of its modes with less than a specified excess loss. If in the numerical example of Fig. 4.49 up to 90 per cent of the modes in a fibre of 40 μm core radius are required to have less than 1 dB km^{-1} excess attenuation, the cladding radius should be 1·47 times the core radius.

Another design objective might require the lossy jacket to suppress a certain percentage of modes in order to reduce the delay spread of signals in transmission. All modes in the sequence of their index ν are on the average uniformly spaced in delay. When we suppress the fraction of modes from $\nu = \nu_s$ to M, we reduce the relative delay spread from $\Delta\tau/\tau = \Delta$ to

$\Delta\tau/\tau = (\nu_s/M)\Delta$. In the numerical example of Fig.4.49 for a fibre of 40 µm core radius, nearly 50 per cent of the modes have more than 1 dB km^{-1} excess loss when the cladding is 20 per cent wider than the core.

REFERENCES

Abramowitz, H. and Stegun, I.A. (1964). *Handbook of mathematical functions*. US Government Printing Office, Washington, D.C.

Buchholz, H. (1943). Der Hohlleiter von kreisförmigen Querschnitt mit geschichtetem dielektrischen Einsatz. *Ann.Phys*. 43, 313-68.

Gloge, D. (1971). Weakly guiding fibres. *Appl. Opt.* 10, 2252-58.

Gloge, D. (1975). Propagation effects in optical fibres. *IEEE Trans. MTT* 23, 106-120.

Hondros, D. and Debye, P. (1910). Elektromagnetische Wellen an dielektrischen Drähten. *Ann. Phys*. 32, 465-76.

Kapany, N.S. and Burke, J.J. (1972). *Optical waveguides*. Academic Press, New York.

Kawakami, S. and Nishida, S. (1974). Characteristics of a doubly clad optical fibre with a low index inner cladding. *IEEE J. Quant. Electron.* 10, 879-87.

———— and ———— (1975). Perturbation theory of a doubly clad optical fibre with a low-index inner cladding. *IEEE J. Quant. Electron.* 11, 130-38.

Kuhn, M.H. (1974). The influence of the refractive index step due to the finite cladding of homogeneous fibres on the hybrid properties of modes. *Archiv Elektron. & Übertragungstech*. 28, 393-401.

———— (1975a). *Wellenausbreitung in Kern-Mantel-Fasern mit absorbierender Hülle*. Dr.-Ing.-Dissertation an der Technischen Universität Braunschweig.

———— (1975b). Optimum attenuation of cladding modes in homogeneous single-mode fibres. *Archiv Elektron. & Übertragungstech*. 20, 201-04.

———— (1975c). Lossy-jacket design for multimode cladded-core fibres. *Archiv Elektron. & Übertragungstech*. 29, 353-5.

Snyder, A.W. and Mitchell, D.J. (1974). Leaky rays on circular optical fibres. *J. Opt. Soc. Amer.* 64, 599-607.

Unger, H.-G. (1954). Dielektrische Rohre als Wellenleiter. *Archiv Elektron. & Übertragungstech*. 8, 241-52.

5
GRADED-INDEX FIBRES

The cladded-core fibre with core, cladding, and jacket regions of uniform and constant refractive index, and abrupt transitions of index values at interfaces between these regions, represents, in most cases, only a mathematical model for actual fibres. In reality, fibres will not have sharp index steps but a more or less smooth transition in index from one cross-sectional region to the other. Only the liquid-core fibre (Payne and Gambling 1972) makes a distinct exception. It consists of a glass capillary which is filled by a highly transparent liquid of a higher refractive index than the glass. At the interface between core liquid and the capillary glass, the refractive index indeed changes abruptly, while the liquid core is of uniform and constant index throughout. Most types of solid core fibres, however, have graded index transitions even if they were meant to have a stepped index profile. This smoothing of nominally abrupt transitions arises from diffusion processes during manufacturing. Such graded transitions are not actually detrimental to the transmission of optical signals and may therefore be easily tolerated. If the index transition is suitably graded or if we have a graded index profile of proper index distribution throughout the fibre core, we may even improve the transmission characteristics considerably.

For the cladded-core fibre we have noted in equation (4.150) the large delay differences between rays on axis which correspond to modes very far above cut-off, and rays with propagation angles θ near θ_c corresponding to modes near cut-off. This delay difference is mainly due to the difference in optical path lengths of these rays. If the index is graded according to Fig.5.1, the meridional ray with propagation angle θ_1 at the axis will curve towards the axis according to the ray equation (2.243) and be trapped by the profile as long as θ_1 is not too large. Its geometrical path will again be longer than for an on-axis ray; the optical path length, however, is not so long because in the regions off-axis with

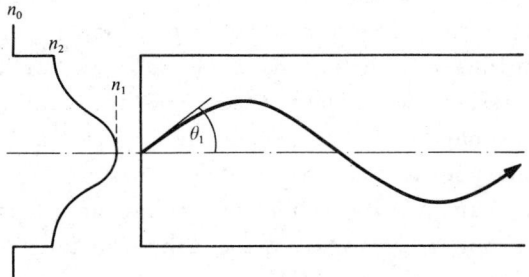

Fig.5.1. Refractive index profile and meridional ray in a graded-index fibre.

lower index, the light travels faster than on axis. The profile may thus serve to equalize the delay for all rays which the profile traps.

As long as the total change of index from one homogeneous region to the other occurs in a distance which is short compared to the wavelength, the cladded-core fibre model with sharp index steps represents an adequate model for the analysis of wave propagation. If, however, the transition in index extends over distances in the order of the light wavelength or if all of the core region has a graded index profile, we need to take this index distribution into account for the analysis of wave propagation.

Graded-index fibres, just like cladded-core fibres, will under practical conditions not be naked as Fig.5.1 might suggest. They will tend to be surrounded by a jacket for mechanical strength and protection and to absorb any unwanted radiation. This jacket thus serves the same purposes as in cladded-core fibres, and it affects the modes which propagate in the graded-index core in a similar way to modes in homogeneous cores. We will therefore not consider this jacket here and concentrate on wave propagation in the graded-index core. The uniform cladding of refractive index n_2 will then be assumed to extend to infinity.

5.1. RAY ANALYSIS

We will start the analysis of wave propagation in

graded-index fibres with the ray picture of modes. In this picture we will not only learn to understand how a graded-index core guides a mode but also be able to determine at least approximately with what phase and attenuation constants these modes propagate in a graded-index fibre and what the limits of mode propagation are. For a simple and transparent treatment, we want to apply the ray equation (2.243) of geometric optics and therefore assume that the refractive index changes only gradually over the distance of a wavelength. Under this assumption wave propagation may be represented by the superposition of plane uniform wavelets, each of which in turn is described by a ray that follows a path which solves the ray equation (2.243).

Under perfect conditions in a graded-index fibre, the index has an axially symmetric distribution which does not change along the fibre. In the coordinates r, φ, z of a circular cylinder with the fibre axis as the z-axis, the index distribution depends only on r but not on φ and z. With $n = n(r)$ therefore, the ray equation (2.243) has in these cylindrical coordinates the r-component

$$\frac{d}{ds}\left(n\frac{dr}{ds}\right) - nr\left(\frac{d\varphi}{ds}\right)^2 = \frac{dn}{dr}, \qquad (5.1)$$

while its φ- and z-components are

$$n\frac{dr}{ds}\frac{d\varphi}{ds} + \frac{d}{ds}\left(nr\frac{d\varphi}{ds}\right) = 0 \qquad (5.2)$$

and

$$\frac{d}{ds}\left(n\frac{dz}{ds}\right) = 0, \qquad (5.3)$$

respectively. Note that in these equations r represents the radial coordinate of the cylindrical system while s still measures the geometrical path length along the ray.

To solve these component equations for a typical ray, we need to specify initial conditions. Without loss of generality we let a light ray be incident from free space with refractive index n_0 on to the front face of the fibre at

$z = 0$, off axis by $r = r_0$ at $\varphi = 0$. We let its angle of incidence be θ_0 and its plane of incidence have the orientation $\varphi = \phi_0$ in the cylindrical coordinates of the fibre. The index step from $n_0 = 1$ to $n(r_0)$ refracts this ray, and it continues to propagate inside the fibre with an initial propagation angle θ_n with respect to the axis that follows from Snell's law equation (1.19) according to

$$n(r_0) \sin \theta_n = \sin \theta_0 \ .$$

Refraction at the front face leaves the angle ϕ_0 unchanged, of course, so that incident and refractive rays have a projection on to the fibre cross-section under this angle ϕ_0. Fig.5.2

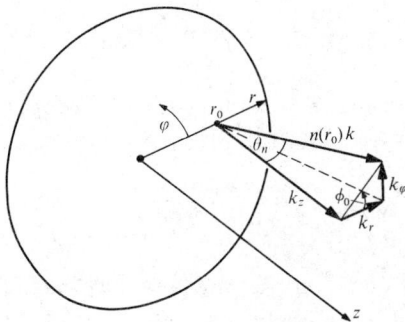

Fig.5.2. Wave vector components of refracted ray incident on fibre front face at r_0 and $\varphi = z = 0$ with propagation angle θ_n and its plane of incidence inclined at ϕ_0.

illustrates the refracted ray at the fibre front face by depicting the corresponding wave vector of magnitude $n(r_0)k$ and its cylindrical components:

$$k_r = n(r_0)k \sin \theta_n \cos \phi_0$$
$$k_\varphi = n(r_0)k \sin \theta_n \sin \phi_0 \qquad (5.4)$$
$$k_z = n(r_0)k \cos \theta_n$$

For the solution of the ray-equation system the refracted ray has the following initial conditions:

$$r = r_0 \qquad \varphi = z = 0$$

$$\frac{dr}{ds} = \sin\theta_n \cos\phi_0$$

$$r_0 \frac{d\varphi}{ds} = \sin\theta_n \sin\phi_0 \qquad (5.5)$$

$$\frac{dz}{ds} = \cos\theta_n \;.$$

Equation (5.3) may be integrated directly along s and, under these initial conditions, leads to

$$n \frac{dz}{ds} = n(r_0) \cos\theta_n \;. \qquad (5.6)$$

The axial component of the wave vector along the ray path is $k_z = nk(dz/ds)$. From equation (5.6) it follows as

$$k_z = n(r_0)k \cos\theta_n \qquad (5.7)$$

and hence remains constant along any ray path of the graded index profile. We turn next to equation (5.2), multiply it by r and then integrate it along s from r_0 to r:

$$nr^2 \frac{d\varphi}{ds} = n(r_0) \, r_0 \sin\theta_n \sin\phi_0 \;. \qquad (5.8)$$

The circumferential component $k_\varphi = nkr(d\varphi/ds)$ of the wave vector along the ray path follows from this equation as

$$k_\varphi = (r_0/r)n(r_0)k \sin\theta_n \sin\phi_0 \;. \qquad (5.9)$$

Along the ray path, it changes by the factor (r_0/r) from its initial value in equation (5.4).

In equation (5.1) of the system of ray-equation components we may now utilize equation (5.6) to replace the derivatives with respect to s by those with respect to z. To this end we multiply equation (5.1) by n and obtain

$$n^2(r_0) \cos^2\theta_n \left[\frac{d^2r}{dz^2} - r\left(\frac{d\varphi}{dz}\right)^2\right] = \frac{1}{2}\frac{d}{dr}(n^2) . \qquad (5.10)$$

When we divide equation (5.8) by equation (5.6), the following expression for $d\varphi/dz$ results:

$$\frac{d\varphi}{dz} = \frac{r_0}{r^2} \tan\theta_n \sin\phi_0 . \qquad (5.11)$$

We substitute this expression for $d\varphi/dz$ into equation (5.10) and arrive at the following differential equation for $r(z)$:

$$\frac{d^2r}{dz^2} - \frac{r_0^2}{r^3}\tan^2\theta_n \sin^2\phi_0 - \frac{1}{2n^2(r_0)\cos^2\theta_n}\frac{d}{dr}(n^2) = 0. \qquad (5.12)$$

The initial conditions for the integration of this equation are from equation (5.5)

$$r(0) = r_0 \quad \text{and} \quad \left.\frac{dr}{dz}\right|_{z=0} = \tan\theta_n \cos\phi_0 .$$

Upon multiplying equation (5.12) by dr/dz, it can be integrated along z and, with the latter form of initial conditions, leads to

$$\frac{dr}{dz} = \left[\frac{n^2(r)}{n^2(r_0)\cos^2\theta_n} - 1 - \tan^2\theta_n \sin^2\phi_0 \left(\frac{r_0}{r}\right)^2\right]^{\frac{1}{2}} . \qquad (5.13)$$

This expression for the rate of change of the radial distance r of the ray with axial distance z, in terms of the index profile $n(r)$ and the initial conditions, allows a ray to be traced and, for $dr/dz = 0$, determines the extreme values of r under specific initial conditions. The radial component $k_r = nk(dr/ds)$ of the wave vector along the ray follows also from equation (5.13), when with $dr/ds = (dr/dz)(dz/ds) = (dr/dz)k_z/nk$, we introduce the axial component k_z from equation (5.7) to obtain k_r in terms of the index profile and the initial conditions r_0, θ_n, ϕ_0 as

$$k_r = \left[n^2k^2 - n^2(r_0)k^2\cos^2\theta_n - n^2(r_0)k^2\sin^2\theta_n \sin^2\phi_0 \left(\frac{r_0}{r}\right)^2\right]^{\frac{1}{2}} \qquad (5.14)$$

We arrive at the same expression for k_r more directly by invoking the separation condition

$$k_r^2 + k_\varphi^2 + k_z^2 = n^2 k^2 \qquad (5.15)$$

for the present plane-wave solution, and by substituting for k_φ and k_z the solutions of the ray equation in equations (5.7) and (5.9), respectively. In terms of the local propagation angle θ and the angle ϕ at which the projection of the local wave vector $n(r)k$ on to the fibre cross-section is inclined with respect to the local radius vector $\vec{r}$, the components of the local wave vector are given by

$$k_r = n(r)\, k\, \sin\theta\, \cos\phi,$$

$$k_\varphi = n(r)\, k\, \sin\theta\, \sin\phi, \qquad (5.16)$$

$$k_z = n(r)\, k\, \cos\theta.$$

In this form they, of course, also satisfy the separation condition (5.15).

In order to understand the field distribution of waves which propagate along graded-index fibres and to determine the limits within which an index profile guides waves, we will now trace a ray by plotting its projection on to the fibre cross-section. We let this ray be incident at $r = r_0$, $\varphi = 0$, and $z = 0$ with θ_0 and ϕ_0. For $\phi_0 < \pi/2$, as in Fig.5.3(a), this ray first moves outward, but the index gradient lets it curve towards the fibre axis. dr/dz decreases and, if $dr/dz = 0$, the ray turns from increasing to decreasing r-values. Equation (5.13) for dr/dz yields the following condition for this turning point r_2:

$$n^2(r)k^2 - k_z^2 - k_\varphi^2(r_0)\left(\frac{r_0}{r}\right)^2 = 0.$$

Here we have introduced k_z and $k_\varphi(r_0)$ from equations (5.7) and (5.9), respectively. To evaluate this turning point condition, we plot $n^2 k^2 - k_z^2$ and $k_\varphi^2(r_0)\left(\frac{r_0}{r}\right)^2$ versus r in

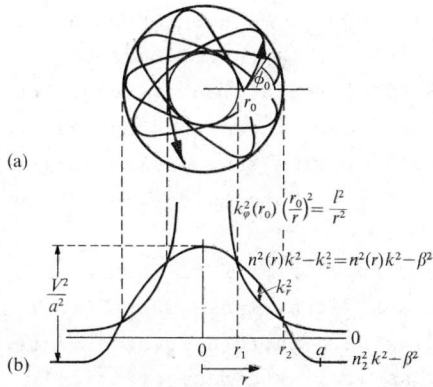

Fig.5.3. Cross-sectional projection of a ray in a graded-index fibre (a) with turning points at r_1 and r_2 and graphical presentation (b) of the square of its radial wave vector component.

Fig.5.3(b). The first curve represents the square of the index profile $n(r)$ while the second curve in form of the second-order hyperbola just shows the r^{-2}-dependence of the square of k_φ. At any intersection of these two curves, we have $dr/dz = 0$, and the ray turns from increasing to decreasing r-values or vice versa. Fig.5.3 shows two such turning points. At the outer turning point r_2, the ray turns from increasing to decreasing r-values. At the inner turning point r_1, it turns from decreasing to increasing r-values. Whether there is an outer turning point depends on the index profile and the initial conditions. For a given inclination ϕ_0 of the plane of incidence of the ray, both r_0 and θ_n must be small enough for this outer turning point to occur. We will defer the discussion of these conditions for the outer turning point to a later part of this section, and assume at present that we have such an outer turning point. The ray will then be trapped by the profile and never leave it. The inner turning point r_1 will always be different from zero as long as we start with a finite value of

$$k_\varphi(r_0)r_0 = n(r_0)k\ r_0\ \sin\theta_n\ \sin\phi_0 .$$

Hence, as long as r_0, θ_0, and ϕ_0 are all different from zero, we have an inner turning point at a finite value r_1. The turning points r_1 and r_2 represent boundaries of an annular region beyond which the ray never moves. It seems that the ray experiences total reflection at these boundaries, so that beyond them we have only the evanescent fields which are associated with total reflection. Essentially, however, the cross-sectional regions outside and inside this annulus remain field free.

If such a trapped ray is to form a mode of the graded-index fibre, it must superimpose upon itself in such a way as to form a standing wave pattern in both cross-sectional directions, radially as well as circumferentially. k_φ is the wave vector component in circumferential direction. To fit around any circumference $2\pi r$ with an integer number l of periods 2π, k_φ must satisfy

$$k_\varphi = l/r . \tag{5.17}$$

This condition fixes the initial values for the incident ray at

$$n(r_0) \, k \, r_0 \sin\theta_n \sin\phi_0 = l \tag{5.18}$$

in order that this ray forms a mode of circumferential order l.

The wave vector component in the radial direction k_r, is given by equation (5.14). If we introduce k_φ according to equation (5.17) for a mode of circumferential order l and let $k_z = \beta$ be the axial phase constant of this mode, its radial phase constant appears as

$$k_r = [n^2(r) \, k^2 - \beta^2 - l^2/r^2]^{\frac{1}{2}} . \tag{5.19}$$

This is the same expression as equation (4.12) for the cladded-core fibre, only now the index $n(r)$ depends on the radial distance from the axis. The phase condition for a standing wave pattern in the radial direction with an integer number p of nodes may again be formulated as in the case of the cladded-core fibre and leads to the following characteristic equation for modes of transverse orders p and l in a graded-index fibre:

$$p\pi = \int_{r_1}^{r_2} [n^2(r)k^2 - \beta^2 - l^2/r^2]^{\frac{1}{2}} \, dr \, . \qquad (5.20)$$

In this approximate form the characteristic equation neglects the phase shift $\pi/2$ of the ray at the turning points r_1 and r_2 and is therefore valid only for sufficiently large radial orders p. The limits of integration in the radial phase integral of the characteristic equation are the turning points of Fig.5.3, where the integrand of the phase integral vanishes. In the notation of equation (5.20), they follow as solutions of

$$n^2(r)k^2 - \beta^2 = l^2/r^2 \, . \qquad (5.21)$$

Given a profile $n(r)$ of the graded-index fibre, the characteristic equation may be solved for specific values of k, β, and l by numerically evaluating its phase integral to obtain its radial order p under these conditions.

Graded-index fibres which correspond to the cladded-core fibre with a single unlimited cladding have an index profile as shown in Fig.5.1 with a maximum index value n_1 in the fibre axis, a profile diameter $2a$, and a cladding region of uniform index n_2. We may use the same relative index difference

$$\Delta = \frac{n_1^2 - n_2^2}{2n_1^2} \simeq \frac{n_1 - n_2}{n_1} \qquad (5.22)$$

as for the cladded-core fibre to describe the index profile by

$$n^2 = n_1^2[1 - 2\Delta g(r/a)] \, , \qquad (5.23)$$

where the profile function $g(r/a)$ ranges from $g(0) = 0$ to $g(1) = 1$ and remains at unity for $r > a$ in the cladding region. In correspondence with the cladded-core fibre, we may furthermore use the fibre parameter

$$V = ka(n_1^2 - n_2^2)^{\frac{1}{2}} = n_1 ka \sqrt{2\Delta} \qquad (5.24)$$

and the phase parameter

$$B = \frac{(\beta/k)^2 - n_2^2}{n_1^2 - n_2^2}$$

to write the characteristic equation (5.20) in the form

$$p\pi = \frac{1}{a} \int_{r_1}^{r_2} [V^2(1 - g(r/a) - B) - (la/r)^2]^{\frac{1}{2}} \, dr \, . \qquad (5.25)$$

At cut-off, for any of the guided modes, the phase constant β equals the wave number of the cladding

$$\beta_c = n_2 k$$

and the phase parameter B vanishes. The characteristic equation reduces under these cut-off conditions to

$$p_c \pi = \frac{1}{a} \int_{r_1}^{r_2} [V_c^2(1 - g(r/a)) - (la/r)^2]^{\frac{1}{2}} \, dr \, . \qquad (5.26)$$

It relates the radial order p_c of that mode which is at cut-off at V_c to its circumferential order l. By evaluating equation (5.26) for a given profile function $g(r/a)$, we may plot p_c/V as a function of l/V just as in Fig.4.8, but now for the graded-index fibre with this particular profile function.

As a simple example for illustration, we pick an index profile with a parabolic profile function according to

$$g(r/a) = \begin{cases} (r/a)^2 & \text{for} \quad r < a \\ 1 & \text{for} \quad r > a \, . \end{cases} \qquad (5.27)$$

This parabolic index profile serves not only as a model for similar but not quite parabolic index distributions. Its guided modes also have some unique propagation characteristics,

GRADED-INDEX FIBRES

as will follow from the solution of the characteristic equation. Of greatest practical significance are the small delay differences between guided modes of the parabolic profile.

We introduce the parabolic profile function according to equation (5.27) into the characteristic equation (5.25) and evaluate the integral with the following result:

$$p = (1-B) \, V/4 - l/2 \, . \qquad (5.28)$$

At cut-off for $B = 0$, this characteristic equation reduces to the following relation between the fibre parameter $V = V_c$ at cut-off and the radial order p_c of modes at cut-off:

$$p_c = V_c/4 - l/2 \, .$$

Fig.5.4 illustrates these equations graphically with straight lines $p(l)$ for different values of the phase parameter B.

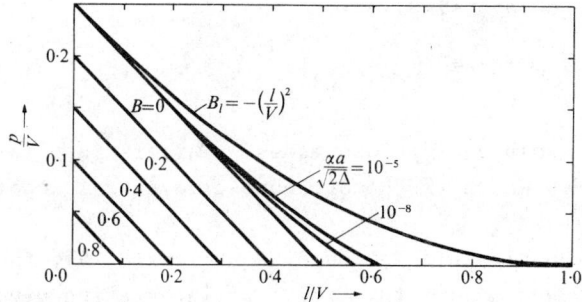

Fig.5.4. Radial order p versus circumferential order l of modes in a graded-index fibre with parabolic index profile.

For the total number of guided modes, we need to sum all numbers p_c from equation (5.26) for each particular integer of l. Under multimode conditions when the graded-index fibre guides a large number of modes, the maximum circumferential order is likewise a very large number and the summation for the total mode count may be replaced by an integration over l. The total number of guided modes is then given by

$$M = \frac{4}{\pi a} \int_0^{l_{max}} \int_{r_1}^{r_2} [V^2(1 - g(r/a)) - (la/r)^2]^{\frac{1}{2}} \, dr \, dl \, . \quad (5.29)$$

The factor 4 in front of the integral accounts for the two orthogonal polarizations with two different field distributions each in which modes of given orders p and l occur. These four degenerate modes of the same orders p and l correspond to the four nearly degenerate LP_{lp}-modes of the cladded-core fibre with a homogeneous core. Only for the few modes with axially symmetric field distributions and $l = 0$ is there again just one field distribution for each polarization and hence only two modes for each pair of numbers $l = 0$ and p. Under multimode conditions, these few axially symmetric modes remain insignificant, and the mode count in equation (5.29) approximates the total number of guided modes quite well.

The integral in equation (5.29) represents the area underneath the curve $p_c(l)$. In the case of the parabolic profile given by equation (5.27) this area in Fig.5.4 is triangular, and we obtain simply

$$M = V^2/4 \quad (5.30)$$

for the total number of guided modes. This is just half the number of modes which the homogeneous core guides according to equation (4.21).

In the double integration of equation (5.29), the turning points r_1 and r_2 are the limits of integration over r; they depend on the circumferential order l. For $l = 0$, the integration over r extends from $r = 0$ to $r = a$, where we have $g = 1$. As l increases, the turning points approach each other until at $l = l_{max}$, they coincide to form a double root of

$$V^2[1 - g(r/a)] = (la/r)^2 \, . \quad (5.31)$$

Fig.5.5 illustrates these limits of integration in the $l(r)$ plane. A change in the order of integration in equation

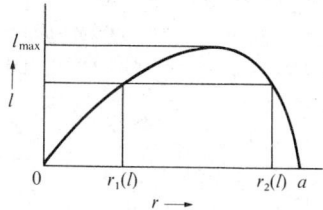

Fig.5.5. Limits of integration for the total number of guided modes in equation (5.29).

(5.29) leads to

$$M = \frac{4}{\pi a}\int_0^a \int_0^{rV[1-g(r/a)]^{\frac{1}{2}}/a} [V^2(1-g(r/a)) - (la/r)^2]^{\frac{1}{2}}\, dl\, dr, \tag{5.32}$$

which can be readily integrated over l to yield

$$M = \frac{V^2}{a^2}\int_0^a [1 - g(r/a)]\, r\, dr. \tag{5.33}$$

The integral in this expression corresponds to the volume of the body of revolution that the profile represents as a function of φ and r. For given Δ and a, this volume is largest in the homogeneous core, which hence has the largest number of guided modes.

Guided modes in graded-index fibres with unlimited cladding of index n_2 have their cut-off at $\beta = n_2 k$. For this value of the phase constant, the radial component k_r of the wave vector and the integrand in equations (5.20) and (5.25) remain imaginary but become smaller and smaller for large r far out in the cladding, reaching zero at infinity. The turning points r_1 and r_2 remain at finite values at cut-off, so that the cut-off fields are evanescent beyond r_2 and throughout the cladding. They decay to extremely small values further out in the cladding.

If for any such mode, the wave number k or the fibre parameter V decrease below their values at cut-off, k_r and the integrands in equations (5.20) and (5.25) will go through zero a third time in addition to their zero crossings at r_1 and r_2. This third zero crossing or turning point r_3 is shown in Fig.5.6; it occurs at first for very large values of r. Beyond r_3 the wave vector component k_r turns real again, and

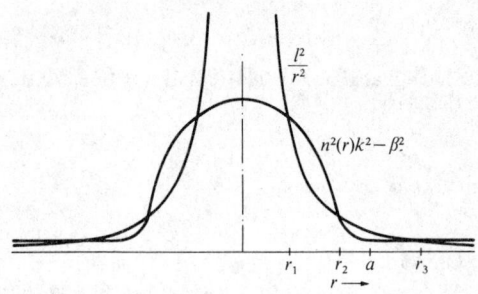

Fig.5.6. Graphical representation of the square of the radial wave vector component of a leaky mode with a third turning point in the cladding.

the fields change from their evanescent character between r_2 and r_3 back again to a propagating wave in radial direction. They form a radiating field beyond r_3 that extracts power from the field inside the core. Below cut-off therefore all guided modes of finite order l turn into leaky modes. The rays which form such leaky modes still move between the two turning points r_1 and r_2 but their apparent total reflection at the outer turning point is now frustrated. Some of their energy tunnels through the evanescent field region and, beyond r_3, radiates out into the cladding. The attenuation of these leaky waves remains very low at first and grows larger only further down from cut-off.

The leaky mode ceases to exist when the profile curve $n^2(r)k^2 - \beta^2$ in Fig.5.6 rises so high that the turning points r_2 and r_3 coincide. For $r_2 = r_3$ and beyond there is no evanescent field region left, the rays are no longer trapped by the index profile and move right out into the cladding. Such rays form cladding modes or radiating modes of the fibre.

For a sufficiently sharp transition from the graded-index profile to the homogeneous cladding at $r = a$, the turning points r_2 and r_3 will coincide at $r_2 = r_3 = a$ when we reach the high-loss limit of the leaky mode region. The phase constant β_l at this limit follows from equation (5.19) under the condition $k_r(a) = 0$

$$\beta_l = (n_2^2 k^2 - (l/a)^2)^{\frac{1}{2}} . \qquad (5.34)$$

With increasing circumferential order l, this limiting value for the leaky-mode phase constant decreases more and more from its cut-off value $n_2 k$ for guided modes. For $\beta < n_2 k$, the phase parameter B turns negative and at the leaky-mode limit $\beta = \beta_l$ is given by $B_l = -(l/V)^2$. When we introduce this limiting value B_l into the characteristic equation and extend the upper limit of integration to the upper limit $r_2 = a$ of the turning point at the boundary of the leaky-mode region, we obtain

$$p_l \pi = \frac{1}{a} \int_{r_1}^{a} \{V^2[1 - g(r/a)] - l^2[(a/r)^2 - 1]\}^{\frac{1}{2}} dr \qquad (5.35)$$

as the limiting form of the characteristic equation for leaky modes. The lower turning point r_1 follows in this case as the first zero of the integrand in the phase integral from

$$V^2[1 - g(r_1/a)] = l^2[(a/r_1)^2 - 1] . \qquad (5.36)$$

Given a specific profile function $g(r/a)$ within the constraints of equation (5.23), the phase integral in equation (5.35) may be evaluated and p_l/V plotted as a function of l/V. The area underneath this curve times $4V^2$ represents the total number of modes, guided as well as leaky. The number of leaky modes alone is then given by the area between the curves p_l/V and p_c/V both as a function of l/V.

For illustration we return to the example of a parabolic index profile with a profile function $g(r/a)$ according to equation (5.27). The characteristic equation (5.28) also holds for leaky waves, but now with their negative values for the phase parameter B. The limit of the leaky-mode range occurs at $B_l = -(l/V)^2$, and the leaky modes at this limit have the radial order

$$p_l = [1 + (l/V)^2] V/4 - l/2 . \qquad (5.37)$$

Fig.5.4 shows this limit in the (p/V)-(l/V) diagram. To count the total number of modes, guided as well as leaky, we integrate $4p_l(l)$ between the limits $l = 0$ and V and obtain

$$M_t = V^2/3 . \qquad (5.38)$$

Of this number

$$M_l = M_t - M = V^2/12 \qquad (5.39)$$

are leaky modes.

The ray analysis of leaky modes in graded-index fibres allows us also to estimate the attenuation which these modes suffer when they lose power to their radiation fields. For this estimate we take a ray which forms a leaky mode of the fibre and let the power P be associated with this ray. P may also be interpreted as a measure of the total power which the leaky mode transmits in axial direction. When this ray reaches the outer turning point at $r = r_2$, it will gradually turn around and move back to the fibre axis. Of the power P, which the ray carried before it turned around, not all will be deflected back, but some will tunnel through the evanescent field region between r_2 and r_3 and feed the radiating field beyond r_3. Between r_2 and r_3 the field decreases exponentially with the magnitude of the purely imaginary radial component

$$k_r = -j(\beta^2 + l^2/r^2 - n^2(r)k^2)^{\frac{1}{2}}$$

of the wave vector as its exponential decay constant. Altogether through the evanescent field region the field intensity decays by the factor

$$\exp\left[-2 \int_{r_2}^{r_3} \{\beta^2 + l^2/r^2 - n^2(r)k^2\}^{\frac{1}{2}} \, dr\right] .$$

To a certain approximation, we can state that of the power P

which is incident in the ray at r_2, the fraction

$$\Delta P = P \exp\left[-2 \int_{r_2}^{r_3} \{\beta^2 + l^2/r^2 - n^2(r)\,k^2\}^{\frac{1}{2}}\, dr\right] \quad (5.40)$$

will tunnel through the evanescent field region and radiate out into the cladding. This approximation should hold well as long as $\Delta P \ll P$, because then the detailed reflection processes at r_2 and r_3 are of little significance for this power ratio. We can furthermore expect this expression to be a valid approximation, since in the graded-index fibre the index changes only gradually in the radial direction and renders the transitions from propagating to evanescent field regions rather smooth.

Equation (5.40) represents the power fraction which the ray loses each time it touches the outer turning point and turns back from the caustic into the annulus. Before it returns again to this outer caustic, it moves a distance along the fibre that is given by

$$\Delta z = 2 \int_{r_1}^{r_2} \left(\frac{dz}{dr}\right) dr \,. \quad (5.41)$$

The derivative (dz/dr) in this integral represents the rate at which the ray proceeds in axial direction per change of radial position. We may substitute for this derivative from equation (5.13) and obtain

$$\Delta z = 2\beta \int_{r_1}^{r_2} (n^2(r)k^2 - \beta^2 - l^2/r^2)^{-1/2}\, dr \,. \quad (5.42)$$

With the leaky ray losing power ΔP once every axial distance Δz, the leaky mode which it forms will decay exponentially in power according to $P \sim \exp(-2\alpha z)$; its attenuation constant is related to the axial power flow P and its rate of decay dP/dz by $\alpha = -(1/2P)(dP/dz)$. The ratio $-(1/P)(dP/dz)$ corresponds to $-(1/P)(\Delta P/\Delta z)$ in the leaky ray picture; we therefore obtain the following formula for the leaky-mode attenuation:

$$\alpha = \frac{\exp\left[-2 \int_{r_2}^{r_3} (\beta^2 + l^2/r^2 - n^2(r)k^2)^{1/2} \, dr\right]}{4\beta \int_{r_1}^{r_2} (n^2(r)k^2 - \beta^2 - l^2/r^2)^{-1/2} \, dr} . \quad (5.43)$$

In terms of the fibre parameter V and the negative phase parameter B of the particular leaky mode, and with the profile function $g(r/a)$ of the fibre, this leaky-mode attenuation can also be written as

$$\alpha a = \frac{\exp\left[-\frac{2}{a} \int_{r_2}^{r_3} \{V^2[g(r/a) - 1 + B] + (la/r)^2\}^{1/2} \, dr\right]}{4\beta \int_{r_1}^{r_2} \{V^2[1 - g(r/a) - B] - (la/r)^2\}^{-1/2} \, dr} . \quad (5.44)$$

This result agrees with a more reliable field solution for leaky modes in graded-index fibres (Petermann 1975).

To illustrate this leaky-mode attenuation, we return again to the example of a parabolic profile function according to equation (5.27). Evaluation of the denominator integral in equation (5.44) for the distance between successive reflections results in this case in the following formula for leaky-mode attenuation:

$$\frac{\alpha a}{\sqrt{2\Delta}} = \frac{1}{2\pi} \exp\left[-\frac{2}{a} \int_{r_2}^{r_3} \{V^2[g(r/a) - 1 + B] + (la/r)^2\}^{1/2} \, dr\right]. \quad (5.45)$$

The remaining integral in this expression may also be evaluated for the parabolic profile and $\alpha a/\sqrt{2\Delta}$ represented in terms of V, p, and l. From this evaluation we can obtain the relative orders p/V and l/V for given values of V and $\alpha a/\sqrt{2\Delta}$ and plot them in Fig.5.4. The two lines in the leaky-mode region of Fig.5.4 represent the transverse mode orders for specific values of the attenuation parameter $\alpha a/\sqrt{2\Delta}$ and the typical fibre parameter value $V = 47$.

The cut-off for guided modes and the limit for leaky modes correspond also to limiting values for the angle θ_0 within which the rays for such modes must be incident on to the front face of the fibre. For guided modes, the cut-off occurs when the axial phase constant β according to equation (5.7) decreases to $\beta = n_2 k$, the wave number of the homogeneous cladding. For the initial propagation angle θ_n, we therefore obtain the cut-off value θ_{nc} from

$$\cos \theta_{nc} = n_2/n(r_0). \tag{5.46}$$

Snell's law then leads to a cut-off value θ_{0c} for the angle of incidence that follows from

$$\sin \theta_{0c} = (n^2(r_0) - n_2^2)^{\frac{1}{2}}. \tag{5.47}$$

In terms of n_1 and n_2 and the profile function $g(r/a)$, this formula may also be written as

$$\sin \theta_{0c} = (n_1^2 - n_2^2)^{\frac{1}{2}} (1 - g(r_0/a))^{\frac{1}{2}}. \tag{5.48}$$

According to these expressions the graded-index fibre has a numerical aperture that depends on the radial distance r_0 from the fibre centre at which the ray is incident on its front face. For $r_0 = 0$, this numerical aperture is largest and corresponds to the index difference between profile centre and cladding. Away from the fibre axis, the numerical aperture decreases, always corresponding to the difference in index between the point of incidence and the cladding. At the boundary of the profile for $r = a$, the numerical aperture diminishes to zero. Fig.5.7 illustrates this change of numerical aperture for a parabolic index profile by showing a number of circles for different radial distances which with their radii represent the numerical aperture.

Leaky modes have larger propagation angles θ_n than the respective cut-off values θ_{nc} for guided modes. Their limiting values at the high-loss boundary of the leaky-mode range follow from the condition $k_r(a) = 0$ which in terms of the initial conditions for a ray incident on the front face of

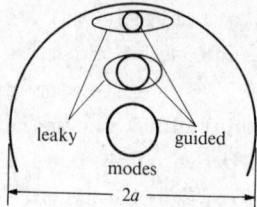

Fig.5.7. Circles and ellipses representing the numerical aperture for guided and leaky modes, respectively, in a graded-index fibre.

the fibre appears as

$$n_2^2 - n^2(r_0)\cos^2\theta_{nl} - n^2(r_0)\sin^2\theta_{nl}\sin^2\phi_0 (r_0/a)^2 = 0 . \quad (5.49)$$

If we solve this equation for $\sin\theta_{nl}$ and, with Snell's law replace $\sin\theta_{nl}$ by $\sin\theta_{0l}$, we obtain for the limiting value θ_{0l} of the angle of incidence at the boundary of the leaky-mode range

$$\sin\theta_{0l} = \left[\frac{n^2(r_0) - n_2^2}{1 - (r_0/a)^2 \sin^2\phi_0}\right]^{\frac{1}{2}} . \quad (5.50)$$

In terms of n_1 and n_2 and the profile function $g(r_0/a)$, this limiting value is given by

$$\sin\theta_{0l} = (n_1^2 - n_2^2)^{\frac{1}{2}}\left[\frac{1 - g(r_0/a)}{1 - (r_0/a)^2 \sin^2\phi_0}\right]^{\frac{1}{2}} . \quad (5.51)$$

If the ray is incident in a plane of incidence with $\phi_0 = 0$, it forms a meridional ray in the fibre and hence, a mode with $l = 0$ and axially symmetric field distributions. Just as in case of the cladded-core fibre with unlimited cladding, there are no axially symmetric leaky modes. For $\phi_0 = 0$ and hence $l = 0$, the limiting value θ_{0l} coincides with the cut-off value θ_{0c}. If, however, the ray is incident at a finite distance r_0 from the fibre axis and if its plane of incidence is inclined at ϕ_0 with respect to the radius vector to the point of incidence, then the limiting angle θ_{0l} is larger than θ_{0c}. For a given distance r_0, the numerical aperture $\sin\theta_{0l}$ as a function of ϕ_0 forms an ellipse with its semi-minor axis for $\phi_0 = 0$, π equal to $\sin\theta_{0c}$ and its semi-major axis for $\phi_0 = \pm\pi/2$ larger than $\sin\theta_{0c}$ by the factor $[1-(r_0/a)^2]^{-1/2}$. In these respects the numerical aperture for leaky modes in graded-index fibres corresponds to that of cladded-core fibres, except that for a core of uniform refractive index θ_{0c} is constant over the core cross-section while in the graded-index fibre it depends on r_0 through the profile function. Fig.5.7 for a parabolic profile function shows in addition to the circles for $\sin\theta_{0c}$ the ellipses for $\sin\theta_{0l}$ along a radius across the front face of the fibre.

5.2. FIELD SOLUTIONS FOR MODES IN A PARABOLIC INDEX PROFILE

The ray analysis of modes in graded-index fibres involves a number of approximations which limit the accuracy of its results and restrict its application to modes of high radial order in multimode fibres. It also yields no detailed information on the distribution and polarization of its modal fields. We will therefore derive more accurate field solutions for guided modes in graded-index fibres. As a representative example of considerable practical interest, we take the parabolic index profile with a profile function according to equation (5.27). For this simple index distribution we will be able to express the modal fields in terms of known functions and also derive expressions in a closed form for their propagation characteristics. By comparing the results from these more accurate field solutions for the parabolic index profile with the results of the ray analysis for the same index profile, the range can be estimated over which the ray analysis remains valid. In a subsequent section we will derive approximate field solutions for more general index profiles. The errors which these approximations involve can also be estimated when we compare the more accurate field analysis with the approximate field solutions, again for the parabolic index profile.

We analyse the fields of the graded-index fibre in the

coordinates r, φ, z of the circular cylinder in which the refractive index $n(r)$ changes only in the radial direction. By eliminating from Maxwell's equations the magnetic field vector $\vec{H}$, we arrive at the following system of second-order partial differential equations for the components of the electric field vector $\vec{E}$:

$$\nabla^2 \vec{E} + n^2(r) k^2 \vec{E} = -\nabla \{\vec{E} \nabla_r [\log_e n^2(r)]\} . \quad (5.52)$$

The operator ∇_r on the right-hand side denotes the gradient in radial direction. Since $n(r)$ changes only in this direction, $\nabla(n^2)$ has only the radial component $\nabla_r(n^2) = \vec{u}_r \, \partial n^2/\partial r$. The operator ∇^2 on the left-hand side of equation (5.52) is an abbreviation for

$$\nabla^2 = \nabla \nabla - \nabla \times \nabla \times ,$$

but for Cartesian components of $\vec{E}$, it corresponds to the Laplacian.

We assume the refractive index to change only gradually in the radial direction, so that the term on the right-hand of equation (5.52) remains small enough to be neglected. The parabolic index profile will obey this assumption if only it extends over a large enough fibre cross-section with a small enough total difference Δ in relative refractive index. Under practical conditions such a smooth profile with only small total differences in refractive index is dictated by technological limitations. We will therefore neglect the right-hand side of equation (5.52) now, but later on, when we have solutions without this term available, we will examine its range of validity by including this term in a perturbational analysis.

With the right-hand side of equation (5.52) equal to zero, this vector equation may be separated into three scalar wave equations, one for each Cartesian component of $\vec{E}$. Its x-component, for example, is the following scalar wave equation for the x-component E_x of $\vec{E}$:

$$(\nabla^2 + n^2(r)k^2)E_x = 0 . \quad (5.53)$$

The non-uniform refractive index of the index profile appears in this equation only as the factor $n^2(r)$ by which k^2 is multiplied.

To solve this scalar wave equation for guided modes of the graded-index fibre, we let

$$E_x = \psi(r) \begin{Bmatrix} \cos l\varphi \\ \sin l\varphi \end{Bmatrix} e^{-j\beta z} . \tag{5.54}$$

The z-dependence according to $\exp(-j\beta z)$ in this product postulates a wave that propagates with a phase constant β along the fibre. The periodic φ-dependence either according to $\cos l\varphi$ or according to $\sin l\varphi$ with integer values l lets this wave have the circumferential order l. When we introduce this trial product into the wave equation we can separate the variables and obtain the following ordinary differential equation for the function $\psi(r)$ that describes the radial distribution of E_x:

$$\frac{d^2\psi}{dr^2} + \frac{1}{r}\frac{d\psi}{dr} + k_r^2 \psi = 0 . \tag{5.55}$$

The separation constant k_r is given by

$$k_r^2 = n^2(r)k^2 - \beta^2 - l^2/r^2 \tag{5.56}$$

and corresponds to the radial component of the wave vector in equation (5.19) of the ray analysis. It appears here as the wave number in that part of the scalar wave equation that by separation of variables splits off as the r-dependence.

The equations which we have derived so far hold for any axially symmetric index profile with only gradual index changes in the radial direction but no index change in the longitudinal direction. We now let $n^2(r)$ have the parabolic distribution

$$n^2 = n_1^2 [1 - 2\Delta (r/a)^2] , \tag{5.57}$$

but at first (in contrast to equations (5.23) and (5.27)) let this parabolic distribution extend without limit in the radial direction and not be truncated at $r = a$. In this unlimited parabolic profile the refractive index vanishes at $r = a/\sqrt{2\Delta}$ and turns imaginary for larger radii. We are, however, not too much concerned with this region since the solutions which interest us do not extend their fields as far out as this radius. With the radial wave number k_r of the unlimited parabolic profile given by

$$k_r^2 = n_1^2 k^2 [1 - 2\Delta(r/a)^2] - \beta^2 - l^2/r^2 , \qquad (5.58)$$

the differential equation (5.55) assumes a form that corresponds to the equation of the parabolic cylinder (Abramowitz and Stegun 1964). Solutions of this equation which decay exponentially at large radii are

$$\psi(r) = \left(\sqrt{2}\,\frac{r}{w}\right)^l L_q^{(l)}\left(2\,\frac{r^2}{w^2}\right) e^{-r^2/w^2} . \qquad (5.59)$$

$L_q^{(l)}$ in this expression represents the generalized Laguerre polynomial of order l and degree q, where l as well as q assume only integer values. $L_q^{(l)}(u)$ is given by the polynomial form

$$L_q^{(l)}(u) = \sum_{\nu=0}^{q} \binom{q+l}{q-\nu} \frac{(-u)^\nu}{\nu!} . \qquad (5.60)$$

Up to second degree these polynomials are

$$L_0^{(l)} = 1$$

$$L_1^{(l)} = 1 + l - u \qquad (5.61)$$

$$L_2^{(l)} = (1/2)(l+1)(l+2) - (l+2)u + u^2/2 .$$

The parameter w is related to the reference radius a of the profile, to the maximum index n_1 on axis, and to the relative index difference Δ by

$$w = (2a/(n_1 k \sqrt{2\Delta}))^{\frac{1}{2}} . \qquad (5.62)$$

This parameter dominates in determining how far the fields extend in radial direction. In the radial field distribution of equation (5.59) the particular Laguerre polynomial is multiplied by the Gaussian function $\exp(-r^2/w^2)$. This function drops at $r = w$ to $1/e$ of its maximum value at $r = 0$. Beyond $r = w$ it decays quite rapidly to extremely small values at radii which are only little larger than w. The parameter w for the parabolic index fibre corresponds to the parameter of the same notation for the graded-index slab with a parabolic index profile. The slab parameter w is given by equation (2.273) and depends on the profile parameters and on wavelength just as w in equation (5.62) depends on the corresponding quantities of the fibre. We also used the parameter w in equation (3.7) to characterize the transverse field distribution of beam waves in beam waveguides. For all these transverse field distributions we have polynomials which are multiplied by the Gaussian function. w is in all these cases the distance from the guide centre where the Gaussian function has decayed to $1/e$ of its maximum value at the guide centre. For the lowest-order mode in all these guiding structures the polynomial reduces to unity and the Gaussian function alone describes the transverse field distribution. For these fundamental modes therefore, the parameter w directly measures the transverse field extension and as such is called mode radius or spot radius of the fundamental mode.

When we introduce the solutions (5.59) for the radial field distribution into equation (5.54), the x-component of the electric field is completely determined as a function of all three coordinates. For the other components of the electromagnetic field of guided modes, we refer to the ray analysis and note that the propagation angle θ always remains smaller than the limiting angle θ_{nc} as follows from the maximum index difference. In the unlimited parabolic profile this index difference can be rather large, but if we restrict our attention to modes with fields extending not too far from the axis, they will be guided by the small index difference in the narrow region through which their fields extend radially, and their propagation angles will remain correspondingly small. Small θ means that the longitudinal field is small compared to the transverse field so that the total field under these conditions is nearly transverse electromagnetic. In addition, it has locally very nearly the character of a uniform plane wave. We can therefore conclude that the E_x in the above solution represents the dominating electric field component and that the associated magnetic field has

$$H_y = n(\varepsilon_0/\mu_0)^{\frac{1}{2}} E_x \qquad (5.63)$$

as the dominating component. The polarization of this field is nearly linear and uniform over any fibre cross-section, which leads us to designate these guided modes as LP_{lp}-modes, just as in case of the weakly guiding cladded-core fibre, where a suitable superposition of nearly degenerate modes resulted in modes with the same simple state of polarization. The two index values l and p of this nomenclature again designate the circumferential and transverse order of modes. The circumferential order l appears in $\cos l\varphi$ and $\sin l\varphi$ for the circumferential field distribution and counts half the number of field nodes around any circumference; it also gives the order of the Laguerre polynomial for the radial field dependence. The radial order p is related to the degree q of the Laguerre polynomial of the radial field distribution by $p = q + 1$, and q counts the number of field nodes along any radius, including the field node at $r = 0$ for all modes with non-zero q. Just as in the cladded-core fibre we also have in the graded-index fibre four LP_{lp}-modes for each set of integers l and p. Two of these modes are polarized in the x-direction and the other two in the y-direction. Each of these pairs has two different circumferential orientations of the field distribution depending on whether $\cos l\varphi$ or $\sin l\varphi$ is used in equation (5.54). Only the axially symmetric LP_{0p}-modes occur only in pairs of orthogonal polarization.

The phase constant β with which an LP_{lp}-mode propagates along the fibre follows from equation (5.55) when we substitute its solution from equation (5.59). For this expression to solve equation (5.55), the phase constant must be

$$\beta = n_1 k \left[1 - \frac{\sqrt{2\Delta}}{n_1 k a} (4q+2l+2)\right]^{1/2} . \qquad (5.64)$$

The LP_{01}-mode with $q = 0$ is the lowest order and hence the fundamental mode. From equation (5.64) its phase constant is

$$\beta = n_1 k \left[1 - 2\sqrt{2\Delta}/(n_1 k a)\right]^{1/2} . \qquad (5.65)$$

As we expect for the fundamental mode, it has the largest phase constant of all guided modes. Since $\Delta \ll 1$ but $n_1 k a \gg 1$, the phase constant of the fundamental mode is only a little smaller than the intrinsic wave number $n_1 k$ in the centre of the profile. Higher-order modes have somewhat smaller phase constants decreasing with their transverse orders q and l; but these phase constants, too, differ only a little from $n_1 k$.

Still smaller are the differences which the guided modes of the parabolic index profile show in delay. The group delay per unit distance for each mode is given by $\tau = d\beta/d\omega$. In order to see what delay differences the guiding structure alone causes without material dispersion, we let the relative index difference Δ be independent of frequency. The different glasses which we combine to form the index gradient are here assumed to have equal dispersion characteristics. Under these conditions, the group delay per fibre length follows from equation (5.64) as

$$\tau = (n_1'/c)\left[1 + \Delta(n_1 k a)^{-2} (2q+l+1)^2\right] , \qquad (5.66)$$

where $n_1' = d(n_1 k)/dk$ is the group index of the material in the centre of the profile. Since $\Delta \ll 1$, but $n_1 k_1 a \gg 1$, delay differences between modes are indeed quite small.

To estimate the total delay spread as the difference between the group delay of the slowest and fastest mode, we need to determine first which modes under specific conditions are above cut-off and guided by the fibre. In the present model of a parabolic index profile that extends radially to infinity, the different modes have their cut-off at

$$n_1 ka = (2\Delta)^{\frac{1}{2}}(4q+2l+2) . \tag{5.67}$$

Under this condition, the phase constant goes to zero and for $n_1 ka$ smaller than the cut-off value of equation (5.67), β turns purely imaginary. The mode then no longer propagates in the axial direction but decays exponentially. This transition at cut-off from a propagating wave to aperiodically decaying fields is typical for waveguides of finite cross-section bound by perfectly reflecting sidewalls, and is also observed for metallic waveguides. In the present model of an unlimited parabolic index profile, the index decreasing beyond zero to ever larger imaginary values renders the cross-sectional boundaries of this profile perfect reflectors.

In an actual graded-index fibre with parabolic index profile, the index will decrease according to equation (5.23) only out to a finite radius a and from thereon will stay constant at the index n_2 of the cladding. For our present purposes, we may assume again that this uniform cladding extends without limit in the radial direction. From the ray analysis, we have learned that such an index profile guides a mode without any radiation into the cladding only as long as its phase constant remains larger than the wave number $n_2 k$ of the cladding. An exact field analysis would lead to the same conclusion. In such a field analysis the cladding fields would be described by modified Hankel functions $K_l(vr/a)$ of argument $vr/a = r(\beta^2 - n_2^2 k^2)^{\frac{1}{2}}$, just as in the uniform cladding of a cladded-core fibre. These modified Hankel functions represent evanescent fields for large radii only as long as $\beta > n_2 k$, so that their argument remains real. For $\beta < n_2 k$ this argument turns purely imaginary and the far-out fields represent radiating waves.

An exact field analysis of the truncated parabolic profile requires a more general solution than equation (5.59) for the radial field distribution. The tangential field components of this solution at $r = a$ must be matched to the field solution in the cladding, which is in terms of modified Hankel functions. We have derived this solution for the corresponding planar problem of the graded-index slab with a truncated parabolic profile. We will, however, not pursue it here, because it would involve us in quite extensive numerical work (Kirchhoff 1973a). Instead we will accept equation (5.59) as an accurate enough solution for guided modes in the truncated profile and even close to their cut-off. Far enough above cut-off, their outer turning point r_2 remains at a safe distance below the profile radius a, so that the evanescent fields beyond r_2 have decayed to small enough values to not really feel the truncation of the profile. But even at cut-off, r_2 usually stays at some distance from a, as we shall see presently.

Let us therefore continue with the assumption that equation (5.64) for the phase constant of the unlimited parabolic index profile also holds for guided modes of the truncated profile, even at cut-off. The cut-off condition $\beta = n_2 k$ then leads to the relation

$$4p_c = V - 2l + 2 \qquad (5.68)$$

for the radial order of an LP_{lp}-mode at cut-off in terms of its circumferential order l and the fibre parameter V from equation (5.24). This relation indicates a finite cut-off at $V_c = 2$ even for the fundamental LP_{01}-mode. Actually, however, the fundamental mode of the truncated parabolic profile has zero cut-off just as the fundamental HE_{11}-mode of the cladded-core fibre with unlimited cladding. As far as this fundamental mode or other modes of low radial order are concerned, the errors which we make by keeping (5.59) and (5.64) for the truncated profile are just too large to reveal the proper cut-off characteristics of these modes.

Since equation (5.68), as we have now seen, represents a valid approximation only for sufficiently large radial orders p_c, we may further simplify it to

$$4p_c = V - 2l .$$

This relation agrees with the corresponding cut-off relation from the ray analysis for guided modes in a parabolic profile. In retrospect, we note that for $p = q + 1 \gg 1$, the phase constant from equation (5.64) is equivalent to the phase constant from the characteristic equation (5.28) of the ray analysis.

We will now determine the turning points r_1 and r_2 at cut-off and check what distance r_2 keeps from the profile radius a. When we introduce the cut-off condition $\beta = n_2 k$ into equation (5.56) and subject it to $k_r = 0$ as the condition for a caustic, we obtain

$$(r/a)^4 - (r/a)^2 = (l/V)^2 .$$

This quadratic equation for $(r/a)^2$ has the two solutions

$$(r_{1,2}/a)^2 = \tfrac{1}{2} \mp \left\{ \tfrac{1}{4} - (l/V)^2 \right\}^{\tfrac{1}{2}} . \qquad (5.69)$$

The minus sign on the square root corresponds to the inner turning point r_1 and the plus sign to the outer turning point r_2. According to this latter solution, the cut-off value for the outer turning point shifts from $r_2 = a/\sqrt{2}$ for circumferential orders close to their upper limit $l = V/2$ to $r_2 = a$ for low circumferential orders. Except for LP_{lp}-modes of low order l near cut-off, the fields assume evanescent character at some distance from the boundary of the profile, and we may safely use the field solutions for the unlimited parabolic index profile.

For the group delay per fibre length according to equation (5.66), we have noted the very small difference from mode to mode. With all guided modes for a specific fibre parameter V known to us, we can now determine the total delay spread. According to equation (5.66) the lowest order fundamental mode travels fastest and has the shortest delay per fibre length

$$\tau = (n_1'/c)[1 + \Delta\,(n_1 k a)^{-2}] . \qquad (5.70)$$

The delay increases with $(2q+l+1)^2$ so that the modes closest to cut-off travel slowest and suffer most delay. From the cut off condition (5.68), we have

$$4q + 2l + 2 = V \qquad (5.71)$$

for the modes near or at cut-off. These modes have the delay per fibre length

$$\tau = (n_1'/c)[1 + \Delta\ (n_1 k_1 a)^{-2}\ v^2/4] \ . \tag{5.72}$$

The maximum delay spread between slowest and fastest modes, relative to the group delay per unit distance

$$\tau_1 = n_1'/c$$

of a uniform plane wave in a homogeneous medium of group index n_1' follows from equations (5.70) and (5.72) as

$$\frac{\Delta\tau}{\tau_1} = \frac{\Delta^2}{2} (1 - \frac{4}{v^2}) \ . \tag{5.73}$$

For multimode fibres with a large value for the fibre parameter, this expression is well approximated by

$$\frac{\Delta\tau}{\tau_1} = \frac{\Delta^2}{2} \ . \tag{5.74}$$

The relative delay spread for guided modes thus appears to be smaller by the factor $\Delta/2$ than the relative delay spread in equation (4.280) of guided modes in a cladded-core fibre. For weakly guiding fibres with small index differences between core and cladding, or across the index profile, the factor $\Delta/2$ means a substantial reduction in delay spread.

The reason why the group delay in the parabolic index profile varies so little between all guided modes as compared to the relatively large delay spread of modes in the cladded-core fibre is to be seen from the respective ray picture of modes. In the cladded-core fibre, the large difference in path-length between an on-axis ray for modes far from cut-off and the zigzag ray at θ near θ_c for modes near cut-off causes the relative delay to differ maximally by nearly Δ. In the parabolic index profile, the light travels just so much faster at some distance away from the axis, due to the smaller index there, that all rays which form guided modes have nearly the same effective optical length. With so much reduction in delay spread, the parabolic profile seems to be the optimum profile with respect to delay equalization, or at least very near to it. To find out if there are profiles with even less delay spread and to determine how much deviation from this near-optimum profile we can tolerate, we will consider more general profiles in subsequent sections and also take more detailed account of material dispersion then.

Before we proceed to derive field solutions for modes in graded-index fibres with more general profiles, we will establish first the conditions under which we may neglect the right-hand side of equation (5.52) and describe wave propagation by a scalar wave equation such as equation (5.53). We will use the solution of the scalar wave equation for an unlimited parabolic index profile as an example, but the results will be general enough to hold also for other profiles.

In general, the electric field $\vec{E}$ consists of a transverse vector $\vec{E}_t$ and an axial component E_z. We may also split the vector wave equation (5.52) into its transverse and longitudinal components. For an axially symmetric distribution with a refractive index that does not change in the axial direction, we find the following transverse part of the vector wave equation

$$\nabla^2 \vec{E}_t + n^2 k^2 \vec{E}_t = - \nabla_t \left(E_r \frac{1}{n^2} \frac{dn^2}{dr} \right) , \qquad (5.75)$$

in which $n = n(r)$ depends only on r; ∇_t denotes the transverse gradient. We let $\vec{E}_{t0}$ be a solution from the scalar wave equation for a guided mode with a phase constant β. $\vec{E}_{t0}$ then obeys the following transverse wave equation

$$\left(\nabla^2 - \frac{\partial^2}{\partial z^2} \right) \vec{E}_{t0} + (n^2 k^2 - \beta^2) \vec{E}_{t0} = 0 . \qquad (5.76)$$

To estimate what change the right-hand side of equation (5.75) effects for the phase constant β of this mode, we consider equation (5.75) to be a perturbed form of equation (5.76) with its right-hand side as the perturbation. We then take the inner product of equation (5.75) with $\vec{E}_{t0}$, and integrate over a full fibre cross-section S. The perturbation which the right-hand side of equation (5.75) adds to the unperturbed quantity $(n^2 k^2 - \beta^2)$ is then given by

$$\delta = \frac{\int_S \vec{E}_{t0} \nabla_t \left(E_r \frac{1}{n^2} \frac{dn^2}{dr} \right) dS}{\int_S \vec{E}_{t0}^2 \, dS} . \qquad (5.77)$$

As long as

$$|\delta| \ll n_1^2 k^2 - \beta^2 , \qquad (5.78)$$

the right-hand sides of equations (5.52) and (5.75) may be neglected altogether.

We estimate δ by evaluating the integrand of its numerator integral for the $\vec{E}_{t0}$ of an x-polarized LP_{lp}-mode of the parabolic index profile. With $E_r = E_x \cos\varphi$ and $E_\varphi = -E_x \sin\varphi$, we obtain

$$\vec{E}_{t0} \nabla_t \left(E_r \frac{1}{n^2} \frac{dn^2}{dr} \right) =$$

$$= E_x \cos^2\varphi \frac{\partial}{\partial r} \left(E_x \frac{1}{n^2} \frac{dn^2}{dr} \right) - E_x \sin\varphi \frac{1}{n^2 r} \frac{dn^2}{dr} \frac{d}{d\varphi} (E_x \cos\varphi) .$$

For the parabolic index profile, the first and second derivatives of n^2 with respect to r may be approximated by

$$\frac{1}{n^2} \frac{dn^2}{dr} \simeq -4\Delta \frac{r}{a^2} , \qquad \frac{d}{dr}\left(\frac{1}{n^2} \frac{dn^2}{dr} \right) \simeq -\frac{4\Delta}{a^2} ,$$

while for the transverse derivatives of E_x we use the following, somewhat drastic, approximations

$$\frac{\partial E_x}{\partial r} \simeq k_r E_x , \qquad \frac{\partial}{\partial \varphi}(E_x \cos\varphi) \simeq -l E_x \tan l\varphi \cos\varphi .$$

With these approximations and with $k_r r < k_r a$ of the order of $2q$, we find the following upper bound for δ:

$$\delta < \frac{2\Delta}{a^2}(4q + 2l + 2) .$$

The quantity $n_1^2 k^2 - \beta^2$, on the other hand, follows from equation (5.64) for an LP_{lp}-mode as

$$n_1^2 k^2 - \beta^2 = (2\Delta)^{\frac{1}{2}}(n_1 k/a)(4q + 2l + 2) .$$

We therefore satisfy the condition (5.78), if only we have

$$(2\Delta)^{\frac{1}{2}} \ll n_1 ka \ . \tag{5.79}$$

In terms of the maximum numerical aperture N.A.$_{max} = (n_1^2 - n_2^2)^{\frac{1}{2}}$ of the graded-index fibre and its maximum index n_1, this condition may also be formulated as

$$\text{N.A.}_{max} \ll n_1^2 \, ka \ . \tag{5.80}$$

Although it has been derived for the parabolic index profile, we may also apply it to other profiles as long as the index does not change too abruptly.

5.3. FIELD SOLUTIONS FOR MULTIMODE GRADED-INDEX FIBRES

The ray analysis for guided modes in graded-index fibres led to the approximate form (5.25) of the characteristic equation which implicitly relates the phase parameter B to the transverse mode orders p and l and the fibre parameter V. We will now attempt to find a more accurate version of this form of characteristic equation by starting from Maxwell's equations and assuming at first only that the index varies gradually enough for equation (5.79) to hold. We can then base the analysis on the scalar wave equation (5.53), and with the product solution of equation (5.54) for an x-polarized electric field, find its radial distribution $\psi(r)$ by solving the differential equation (5.55). We let the radial wave number k_r have only two zero crossings between $r = 0$ and ∞ so that we have only the two turning points r_1 and r_2 in Fig. 5.8. The method which we use here to solve equation (5.55) can also be applied when there are more than two turning points, but it is most easily explained for just two turning points.

Fig.5.8 indicates the two turning points as intersections of $n^2(r)k^2$ and $\beta^2 + l^2/r^2$; it also anticipates the character of the solution $\psi(r)$ in the different regions. Between r_1 and r_2 for positive k_r^2, the solution has the oscillating character of a standing wave; below r_1 and beyond r_2, it decays aperiodically.

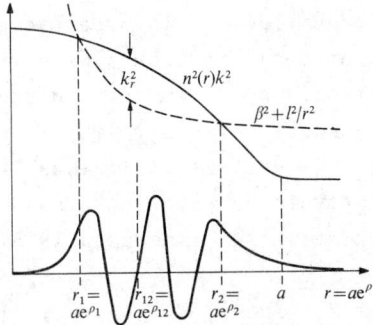

Fig.5.8. Graphical representation of k_r^2 with two turning points r_1 and r_2 and solutions $\psi_1(r)$ and $\psi_2(r)$ below and above r_{12}.

For the solution of equation (5.55) we transform to the new independent variable ρ according to

$$r = a\, e^\rho, \qquad (5.81)$$

which leads to the following differential equation for $\psi(\rho)$:

$$\frac{d^2\psi}{d\rho^2} + k_\rho^2 \psi = 0, \qquad (5.82)$$

where

$$k_\rho^2 = r^2 k_r^2. \qquad (5.83)$$

With ρ ranging from $-\infty$ to $+\infty$ when r goes from 0 to ∞, the solution $\psi(\rho)$ must be finite in the limit $\rho \to -\infty$ and go to zero in the limit $\rho \to +\infty$. Between $\rho_1 = \ln(r_1/a)$ and $\rho_2 = \ln(r_2/a)$, we have $k_\rho^2 > 0$ and $\psi(\rho)$ has oscillating character; below ρ_1 and above ρ_2, the solution is evanescent.

In the vicinity of a turning point ρ_i, k_ρ^2 as a function of ρ might be approximated by

$$k_\rho^2 \simeq \left.\frac{dk_\rho^2}{d\rho}\right|_{\rho=\rho_i}(\rho-\rho_i) \qquad (5.84)$$

in equation (5.82). For such a linear ρ-dependence of k_ρ^2 near its zero crossing, equation (5.82) is solved by Airy functions (Abramowitz and Stegun 1964). In order to use Airy functions as approximate solutions throughout, we divide the whole range of ρ into two regions, one around ρ_1 and the other around ρ_2 with ρ_{12} as the boundary between region 1 and region 2. We distinguish between quantities in regions 1 and 2 with the index 1 or 2, respectively. Approximate solutions in each region may then be written as (Timmermann 1975)

$$\psi_{1,2} = a_{1,2} \, k_\rho^{-1/2} \, \mu_{1,2}^{1/6} \, \text{Ai}(-z_{1,2}), \tag{5.85}$$

where $\text{Ai}(-z_{1,2})$ denotes the Airy function of argument $-z_1$ or $-z_2$, and $z_{1,2}$ follows from

$$z_{1,2} = \left(\frac{3}{2} \mu_{1,2}\right)^{2/3} \tag{5.86}$$

with

$$\mu_1 = \int_{\rho_1}^{\rho} k_\rho \, d\rho \quad \text{and} \quad \mu_2 = \int_{\rho}^{\rho_2} k_\rho \, d\rho. \tag{5.87}$$

The particular choice of Airy function in equation (5.85) lets ψ_1 ($\rho \to -\infty$) remain finite and ψ_2 ($\rho \to \infty$) go to zero, as we require for our guided-mode solution. Since we apply these approximations for the full range between ρ_1 and ρ_2, we need to match them at an intermediate point ρ_{12}, so that they not only equal each other in value,

$$\psi_1(\rho_{12}) = \psi_2(\rho_{12}), \tag{5.88}$$

but have also the same slope at ρ_{12},

$$\left.\frac{d\psi_1}{d\rho}\right|_{\rho=\rho_{12}} = \left.\frac{d\psi_2}{d\rho}\right|_{\rho=\rho_{12}} \tag{5.89}$$

These two conditions fix the ratio a_1/a_2 of field factors in equation (5.85) and also contain the characteristic equation for the eigenvalues or phase constants of guided modes. We satisfy them both most conveniently by choosing for the boundary ρ_{12} between region 1 and region 2 suitable zero crossings of ψ_1 and ψ_2, just as shown in Fig.5.8. The ith-zero of the Airy function $Ai(a_i) = 0$, counting from its turning point, is given by (Abramowitz and Stegun 1964)

$$a_i = - [I(1 + \frac{5}{32I^2} + \ldots)]^{2/3} , \qquad (5.90)$$

where

$$I = (3\pi/2)(i-1/4) . \qquad (5.91)$$

We now let

$$a_m = - \left(\frac{3}{2} \mu_m\right)^{2/3} \qquad (5.92)$$

be the mth-zero of $Ai(-z_1)$ with

$$\mu_m = \int_{\rho_1}^{\rho_{12}} k_\rho \, d\rho , \qquad (5.93)$$

and let

$$a_n = - \left(\frac{3}{2} \mu_n\right)^{2/3} \qquad$$

be the nth-zero of $Ai(-z_2)$ with

$$\mu_n = \int_{\rho_{12}}^{\rho_2} k_\rho \, d\rho . \qquad (5.94)$$

Since, according to equations (5.90) and (5.91), the zero crossing a_i depends only on i, the sum

$$\mu_{12} = \mu_m + \mu_n = \frac{2}{3}[(-a_m)^{3/2} + (-a_n)^{3/2}] \qquad (5.95)$$

depends likewise only on m and n. The radial order $p = q + 1$ of a particular mode represents the total number q of zero crossings of $\psi(\rho)$ between ρ_1 and ρ_2 plus one. With m zero crossings in region 1 and n in region 2, and the mth-zero of ψ_1 coinciding with the nth-zero of ψ_2, we have

$$q = m + n - 1 . \tag{5.96}$$

With these relations we can now satisfy the condition (5.95) by

$$\int_{\rho_1}^{\rho_2} k_\rho \, d\rho = \mu_{12}(m,n) . \tag{5.97}$$

This actually represents the characteristic equation while the other condition (5.89) fixes the ratio a_1/a_2 of field factors in equation (5.85). From equations (5.81) and (5.83) it follows that $k_\rho \, d\rho = k_r \, dr$. The integral in the characteristic equation (5.97) can therefore be replaced by the radial phase integral over k_r and equation (5.97) written as

$$\int_{r_1}^{r_2} k_r \, dr = \mu_{12}(q,m) , \tag{5.98}$$

where, in addition, we have used equation (5.96) to express n by m and q, so that μ_{12} appears as a function of q and m. With this dependence on m, the characteristic equation and with it the phase constant of guided modes still depend on the particular zero crossing at which we place the boundary between regions 1 and 2. Since both ψ_1 and ψ_2 are approximations to the exact solution, that lose accuracy with the distance from their respective turning points, we minimize this distance by placing the boundary between regions 1 and 2 as close to the centre between ρ_1 and ρ_2 as is possible with this method. We achieve this best possible approximation under present circumstances by fixing

$$m = \begin{cases} q/2 & \text{for } q \text{ even} \\ (q+1)/2 & \text{for } q \text{ odd} . \end{cases} \tag{5.99}$$

The higher the radial order of modes, the larger will be m and the closer ρ_{12} to the centre point between ρ_1 and ρ_2. With this choice of m, the characteristic equation becomes independent of m but separates into two forms depending on whether q is even or odd:

$$\int_{r_1}^{r_2} k_r \, dr = \begin{cases} \mu_{12}^{(e)}(q) & \text{for } q \text{ even} \\ \\ \mu_{12}^{(o)}(q) & \text{for } q \text{ odd}. \end{cases} \quad (5.100)$$

The right-hand sides may be replaced by equation (5.95) and the zero crossings of Airy functions in turn expressed by their series representation. Including all terms of the series representation up to the order $1/q$, we obtain the characteristic equations in the form

$$\int_{r_1}^{r_2} k_r \, dr = (q+\tfrac{1}{2})\pi \begin{cases} 1 + \dfrac{10}{9\pi^2} \dfrac{1}{(2q-1)(2q+3)} + \ldots & \text{for } q \text{ even} \\ \\ 1 + \dfrac{10}{9\pi^2} \dfrac{1}{(2q+1)^2} + \ldots & \text{for } q \text{ odd}. \end{cases} \quad (5.101)$$

For large $q \simeq p$, the right-hand sides of this equation may be approximated by $p\pi$. Equation (5.101) then corresponds in all details to the characteristic equation (5.20) from the ray analysis for guided modes. The additional terms on the right-hand side of equation (5.101) improve the accuracy as compared to the ray analysis. From values of the phase parameter B which solve this characteristic equation for a given fibre parameter V and transverse orders l and q, the dispersion characteristics may be evaluated as well as the field distribution for the respective modes.

Before we evaluate the characteristic equation (5.101) any further, let us examine under what conditions we can expect it to give accurate solutions. With the Airy function in equation (5.85), we have actually not solved equation (5.82) but the differential equation

$$\frac{d^2\psi}{d\rho^2} + (k_\rho^2 + \delta_\rho^2)\psi = 0, \qquad (5.102)$$

which differs in its coefficient from k_ρ^2 by

$$\delta_\rho^2 = \frac{5}{36}\left(\frac{k_\rho}{\mu}\right)^2 - \frac{5}{16}\left[\frac{1}{k_\rho^2}\frac{dk_\rho^2}{d\rho}\right]^2 + \frac{1}{4k_\rho^2}\frac{d^2 k_\rho^2}{d\rho^2}. \qquad (5.103)$$

In this expression μ designates either μ_1 or μ_2 according to equation (5.87), depending on whether we consider ψ_1 or ψ_2. We take δ_ρ^2 as a perturbation of k_ρ^2; then, if for a particular mode the exact phase constant, as eigenvalue of equation (5.82), is β, and the eigenvalue of equation (5.102), as the approximate phase constant from the characteristic equation, is β_0, we find for the difference of their squares

$$\delta\beta^2 = \beta^2 - \beta_0^2 \qquad (5.104)$$

the following perturbational expression (Timmermann 1975)

$$\delta\beta^2 = \frac{\int_{-\infty}^{\infty} \delta_\rho^2 \psi^2 d\rho}{a^2 \int_{-\infty}^{\infty} e^{2\rho}\psi^2 d\rho}. \qquad (5.105)$$

ψ in this expression represents the respective modal solution of equation (5.102), and is hence given by equation (5.85).

The perturbation δ_ρ^2 has been evaluated by Timmermann (1975) for the parabolic index profile according to equation (5.57). Fig.5.9 compares δ_ρ^2 to k_ρ^2 for modes with the specified value for the ratio of $l/(2q+l+1)$. The curves are plotted versus the radial coordinate r, which is normalized in Fig.5.9 with respect to

$$r_{20} = (a/\sqrt{2\Delta})\left[1 - [\beta/(n_1 k)]^2\right]^{\frac{1}{2}}; \qquad (5.106)$$

r_{20} is the outer, and only, turning point for meridional rays in the parabolic profile.

For the Airy-function solution ψ_1 that starts from the lower turning point, the perturbation is $\delta_{\rho 1}^2$, for ψ_2 it is $\delta_{\rho 2}^2$.

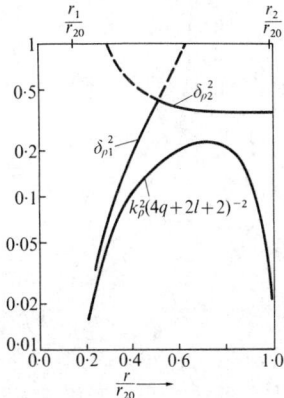

Fig.5.9. Perturbation δ_ρ^2 of the square k_ρ^2 of wave number in the Airy equation (5.102) for a parabolic index profile and modes with $l/(2q+l+1) = 0\cdot 3$. The turning points r_1 and r_2 are normalized with respect to the turning point $r_{20} = (a/\sqrt{2\Delta})\{1 - [\beta/(n_1 k)]^2\}^{\frac{1}{2}}$ for meridional rays (Timmermann 1975).

To keep the perturbation small, the Airy function zeros for matching ψ_1 to ψ_2 should best be chosen so that r_{12} is near the cross-over of $\delta_{\rho 1}^2$ and $\delta_{\rho 2}^2$. The quantity $k_\rho^2(4q+2l+2)^{-2}$ remains of the same order of magnitude as δ_ρ^2 over most of the region between r_1 and r_2. For modes of high enough order so that $(4q+2l+2)$ is large, the perturbation δ_ρ^2 is insignificant compared to k_ρ^2. The change in phase constant which this perturbation effects has also been evaluated by Timmermann (1975) for the parabolic index profile and is plotted in Fig.5.10 for modes of two different circumferential orders as a function of the radial order q. The difference $\delta\beta^2$ increases quite rapidly towards low radial orders q. But even at q as low as 2, we read $\delta\beta^2 < 0\cdot 2 n_1 k \sqrt{2\Delta}/a$ which, under normal conditions in graded-index fibres, is quite insignificant. Except for the modes of very lowest order and in particular excluding the fundamental mode, the Airy function solution provides formulae for mode dispersion and field distribution of adequate accuracy for most practical purposes.

To solve the characteristic equation (5.101), we need to

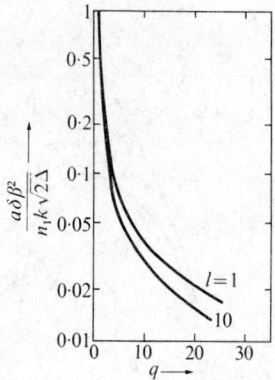

Fig.5.10. Difference in the square of phase constants from the exact solution and the Airy function approximation for a parabolic index profile (Timmermann 1975).

specify the index distribution $n(r)$ and then evaluate the phase integral. One particular index distribution for which an analytic solution of the phase integral was found in equation (5.28) of the ray analysis is the parabolic profile. With $n(r)$ according to equation (5.57) we therefore obtain as solutions of equation (5.101)

$$\beta = n_1 k \left[1 - \frac{\sqrt{2\Delta}}{n_1 k a} (4\mu_{12}^{(e,o)}/\pi + 2l) \right]^{1/2} . \qquad (5.107)$$

This equation differs from equation (5.28) only because we have $\mu_{12}^{(e,o)}$ on the right-hand side of the characteristic equation instead of $p\pi$ as in the characteristic equation (5.20) of the ray analysis. For large enough radial orders $p = q+1$, this right-hand side is approximated by $\mu_{12}^{(e,o)} = (q+1/2)\pi$ for both even and odd integers q. The phase constant from equation (5.107) is then identical to the result in equation (5.64) of the exact field solution for the unlimited parabolic profile. For this particular profile, therefore, all additional terms in the series representations of equation (5.101) for μ_{12} do not improve the accuracy of the characteristic equation. The errors which the Airy-function approximation introduces into the field solution are obviously too large for these higher-order terms in equation (5.101) to render the characteristic

GRADED-INDEX FIBRES   487

equation any more accurate. Although we can draw this conclusion only for the parabolic profile, it will in all likelihood also hold for other index profiles. We will therefore henceforth let

$$\mu_{12}^{(e,o)} = (p - \tfrac{1}{2})\pi \tag{5.108}$$

and accept

$$\int_{r_1}^{r_2} (n^2(r)k^2 - \beta^2 - l^2/r^2)^{\frac{1}{2}} dr = (p - \tfrac{1}{2})\pi \tag{5.109}$$

as the best possible approximation for the characteristic equation from the Airy-function solution of the scalar wave equation.

5.4. VARIATIONAL ANALYSIS AND POWER-SERIES ANALYSIS.

The ray analysis as well as the Airy-function approximations fail for low-order modes of graded-index fibres. Even the field solutions for the unlimited parabolic index profile cannot be applied when the profile is truncated at such small values of the fibre parameter, V, that it guides only a few low-order modes. We have in particular no method yet with which to treat the graded-index fibre for the transmission of its fundamental mode under single-mode conditions.

Such a method is provided by a variational analysis, for which we first set up a stationary expression for the modal field distribution, and then solve the variational problem by the Rayleigh-Ritz method (Okoshi and Okamoto 1974). We intend to analyse with the variational procedure not only graded-index fibres with only gradual index changes but also the truncated profiles of Fig.5.11 with sharp index steps between graded-index core and homogeneous cladding. In particular the quasi-W profile of Fig.5.11(b) might have favourable characteristics for single-mode operation.

We let the homogeneous cladding in Fig.5.11 extend to infinity. For an x-polarized LP_{lp}-mode in such a fibre, the electric field can be set up as in equation (4.124) with the modified Hankel function describing the radial field distribution in the cladding according to

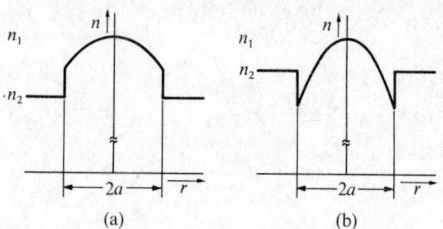

Fig.5.11. Truncated index profiles with index steps at core-cladding boundary.

$$\psi_2(r) = K_l(vr/a) ,\qquad(5.110)$$

where $v = a(\beta^2 - n_2^2 k^2)^{\frac{1}{2}}$ represents the transverse attenuation parameter for the evanescent cladding fields. If there were no index step we would match the radial core field distribution $\psi(r)$ to the cladding field $\psi_2(r)$ by requiring both

$$\psi(a) = \psi_2(a) \quad\text{and}\quad \left.\frac{d\psi}{dr}\right|_a = \left.\frac{d\psi_2}{dr}\right|_a .\qquad(5.111)$$

Dividing one of these equations through the other eliminates the field factors of ψ and ψ_2 and leads to the boundary condition

$$\left.\frac{1}{\psi}\frac{d\psi}{dr}\right|_{r=a} = \frac{1}{a} X_l(v) \qquad(5.112)$$

with

$$X_l(v) = v\, K'_l(v)/K_l(v) .\qquad(5.113)$$

Although this boundary condition is exact only for core cladding boundaries without an index step, we may, to a good approximation, apply it also when a small index step occurs at the boundary. Note that this boundary condition corresponds to the weakly-guiding-fibre approximation, and, for the cladded-core fibre with $\psi = J_l(ur/a)$ results directly in the characteristic equation (4.103) for the weakly guiding case.

Let us also formulate the boundary condition for the

graded-index core in Fig.5.12 that is surrounded by a first inner cladding of refractive index n_2 and outer radius b and an outer cladding of refractive index n_3 that extends to infinity. The outer cladding might either represent the jacket for mechanical protection or absorption of unwanted radiation. It might also, for single-mode operation, serve to shift the cut-off for the next higher order modes to larger values of the fibre parameter V. In the latter respect, this structure corresponds to the W-fibre with homogeneous core.

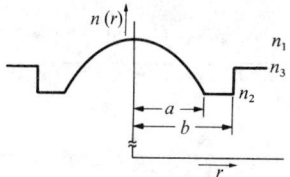

Fig.5.12. Refractive index profile of graded-index core with inner cladding of index n_2 and outer unlimited cladding of index n_3.

The radial field distribution of the inner cladding must now, for completeness, be represented by a combination of modified Bessel and Hankel functions

$$\psi_2(r) = A[I_l(vr/a) + D\, K_l(vr/a)] \qquad \text{for } a < r < b \qquad (5.114)$$

with the transverse attenuation parameter again given by $v = a(\beta^2 - n_2^2 k^2)^{\frac{1}{2}}$ and A and D as field factors to be determined by boundary conditions. In the unlimited outer cladding the evanescent field distribution is properly described by the modified Hankel function

$$\psi_3(r) = C\, K_l(wr/a) \qquad \text{for } r > b. \qquad (5.115)$$

For this region the transverse attenuation parameter is $w = a(\beta^2 - n_3^2 k^2)^{\frac{1}{2}}$, and the field factor C will also be determined by the boundary conditions.

In order to match fields at the boundary $r = b$ between

inner and outer cladding we assume the index n_3 to differ only little from n_2 and, within the weakly-guiding-fibre approximation, require

$$\psi_2(b) = \psi_3(b) \quad \text{and} \quad \left.\frac{d\psi_2}{dr}\right|_b = \left.\frac{d\psi_3}{dr}\right|_b . \tag{5.116}$$

When we subject the field distributions in equations (5.114) and (5.115) to these boundary conditions, we can eliminate A and C by dividing one condition by the other and solve the resulting equation for D:

$$D = \frac{w\, I_l(cv)\, K'_l(cw) - v\, I'_l(cv)\, K_l(cw)}{v\, K'_l(cv)\, K_l(cw) - w\, K_l(cv)\, K'_l(cw)} . \tag{5.117}$$

$c = b/a$ denotes the ratio of cladding to core radius. We can now match the fields at the boundary between core and inner cladding by subjecting $\psi_2(a)$ to equations (5.111) and obtain

$$\left.\frac{1}{\psi}\frac{d\psi}{dr}\right|_{r=a} = \frac{1}{a} K \tag{5.118}$$

with

$$K = v\, \frac{I'_l(v) + D\, K'_l(v)}{I_l(v) + D\, K_l(v)} . \tag{5.119}$$

For the doubly cladded fibre we hence have the quantity K according to equation (5.119) instead of $X_l(v)$ according to equation (5.113) to substitute into the boundary condition (5.118). Let us therefore designate either one by just K and substitute for K either $K = X_l(v)$ from equation (5.113), when we have only a single unlimited cladding, or from equation (5.119) when the graded core has the double cladding of Fig.5.12. In the limiting cases when $n_2 = n_3$, and $v = w$ as a consequence, or when b grows beyond all bounds, we have $D \to \infty$ and $K = X_l(v) = v\, K'_l(v)/K_l(v)$. For the other limiting case of $c \to 1$, when the inner cladding shrinks to zero thickness, we obtain from equations (5.117) and (5.119) the limiting value $X_l(w) = w\, K'_l(w)/K_l(w)$.

A stationary form in terms of $\psi(r)$ can now be set up as follows:

$$S(\psi) = K\,\psi^2(a) - \int_0^a \left(\frac{d\psi}{dr}\right)^2 r\,dr + \int_0^a \left[n^2(r)k^2 - \beta^2 - \frac{l^2}{r^2}\right]\psi^2(r)\,r\,dr. \tag{5.120}$$

That this form is indeed stationary with respect to variations of $\psi(r)$ from the distribution $\psi_0(r)$ that solves the differential equation (5.55) under the boundary condition (5.118) may be seen when we let $\psi(r) = \psi_0(r) + \delta\eta(r)$ and then examine the extremum condition for $\delta = 0$:

$$\frac{\partial}{\partial \delta} S(\psi)\bigg|_{\delta=0} = 2\psi_0(a)\,K\eta(a) - 2\int_0^a \frac{d\psi_0}{dr}\frac{d\eta}{dr}\,r\,dr +$$

$$+ 2\int_0^a \left[n^2(r)k^2 - \beta^2 - \frac{l^2}{r^2}\right]\psi_0(r)\,\eta(r)\,r\,dr = 0.$$

Integrating the second term by parts, we obtain

$$\left\{K - a\left[\frac{1}{\psi_0}\frac{d\psi_0}{dr}\right]_{r=a}\right\}\eta(a)\,\psi_0(a) +$$

$$+ \int_0^a \left\{\frac{d}{dr}\left(r\,\frac{d\psi_0}{dr}\right) + \left[n^2(r)k^2 - \beta^2 - \frac{l^2}{r^2}\right]r\,\psi_0\right\}\eta(r)\,dr = 0. \tag{5.121}$$

For this extremum condition to be satisfied for an arbitrary distribution $\eta(r)$, both expressions in parenthesis must vanish. When the first of these expressions vanishes the boundary condition is satisfied, while for the second expression to vanish satisfies the wave equation (5.55). This shows that equation (5.120) has indeed the stationary character with respect to ψ_0.

We now describe the index profile according to equation (5.23) by the profile function $g(r/a)$ and the relative index difference Δ, so that henceforth we deal with the following stationary form

$$S(\psi) = K\psi^2(a) + (n_1^2 k^2 - \beta^2)\int_0^a \psi^2\,r\,dr - \int_0^a \left[\left(\frac{d\psi}{dr}\right)^2 + \frac{l^2}{r^2}\psi^2\right]r\,dr -$$

$$- 2n_1^2 k^2 \Delta \int_0^a g(r/a)\,\psi^2\,r\,dr. \tag{5.122}$$

In order to solve the variational problem with this stationary form by the Rayleigh-Ritz method, we need to express $\psi(r)$ by a set of functions $F_{lk}(r)$ which should be orthogonal with respect to the domain $r = 0 \ldots a$. If, in addition, this set is complete in the sense that any solution of the homogeneous differential equation can be described by this set, the Rayleigh-Ritz method should in principle lead to an exact solution. Such a set of functions which is orthogonal with respect to the domain $0 \le r \le a$ as well as complete with respect to equation (5.55) is available in terms of the Bessel functions

$$F_{lk}(r) = \frac{\sqrt{2}}{a} \frac{J_l(u_k r/a)}{J_l(u_k)} \qquad (5.123)$$

with

$$u_k = \begin{cases} u_{1,k-1} & \text{for } l = 0 \\ u_{l-1,k} & \text{for } l \ne 0 , \end{cases}$$

where $u_{n,k}$ denotes the kth root of $J_n(u) = 0$. We now express $\psi(r)$ in terms of this set of functions according to

$$\psi(r) = \sum_{k=1}^{\infty} a_k F_{lk}(r) \qquad (5.124)$$

and introduce this series expression for $\psi(r)$ into the stationary form (5.122).

The individual functions $F_{lk}(r)$ obey the condition

$$\frac{1}{F_{lk}} \frac{dF_{lk}}{dr} = -\frac{l}{r}$$

at the core-cladding boundary $r = a$. The series expansion (5.124) in terms of these functions can therefore not, in general, satisfy the boundary condition (5.118). It is able to describe the actual field distribution adequately only in the open interval $0 \le r \le a$, excluding the boundary proper. However, this boundary condition involves the first derivative of ψ. When $d\psi/dr$ is obtained from the series expansion

(5.124) through term by term differentiation, it will still converge, though not uniformly. The subsequent variational analysis uses no higher-order derivatives, and takes account of the boundary condition (5.118) by requiring the parenthesis in the first term of equation (5.121) to vanish. For the Rayleigh-Ritz procedure, the boundary condition (5.118) is hence not an essential one, but only a residual condition, which need not be satisfied by the individual terms of the trial series (5.124).

The orthonormality relation

$$\int_0^a F_{lk}(r) \, F_{lj}(r) \, r \, dr = \delta_{kj}$$

facilitates the evaluation of the stationary form, and we obtain the following expressions for the first two integrals on the right-hand side of equation (5.122):

$$\int_0^a \psi^2(r) \, r \, dr = \sum_k a_k^2 \tag{5.125}$$

$$\int_0^a \left[\left(\frac{d\psi}{dr}\right)^2 + \frac{l^2}{r^2} \psi^2 \right] r \, dr = \sum_k \sum_j a_k \, a_j \, X_{kj} \tag{5.126}$$

where

$$X_{kj} = (u_k/a)^2 \, \delta_{kj} - 2l/a^2 \, .$$

The last integral on the right-hand side of equation (5.122) contains the profile function; it may be expressed as

$$\int_0^a g(r) \, \psi^2(r) \, dr = \sum_k \sum_j a_k \, a_j \, C_{kj} \tag{5.127}$$

with

$$C_{kj} = \frac{2}{a^2 \, J_l(u_k) \, J_l(u_j)} \int_0^a g\left(\frac{r}{a}\right) J_l(u_k \, r/a) \, J_l(u_j \, r/a) \, r \, dr . \tag{5.128}$$

When these expressions, together with

$$\psi^2(a) = \frac{2}{a^2} \sum_k \sum_j a_k \, a_j \, , \tag{5.129}$$

are substituted into the stationary form, it appears as the following function of the expansion coefficients a_k:

$$S(a_1, a_2 \ldots) = \frac{2K}{a^2} \sum_k \sum_j a_k a_j + (n_1^2 k^2 - \beta^2) \sum_k a_k^2 - \sum_k \sum_j a_k a_j X_{kj} - 2n_1^2 k^2 \Delta \sum_k \sum_j a_k a_j C_{kj}. \quad (5.130)$$

Following the Rayleigh-Ritz procedure, we make equation (5.124) the best possible approximation to a solution of equations (5.55) and (5.118) by minimizing $S(a_1, a_2 \ldots)$. For this minimization, we require each partial derivative of $S(a_1, a_2 \ldots)$ with respect to all the a_i to be zero:

$$\frac{\partial S}{\partial a_i} = \frac{4K}{a^2} \sum_k a_k + 2(n_1^2 k^2 - \beta^2) a_i - 2 \sum_k a_k X_{ik} - 4n_1^2 k^2 \Delta \sum_k a_k C_{ik} = 0.$$

These minimum conditions form a homogeneous system of linear equations

$$\sum_k a_k S_{ik} = 0 \qquad i = 1, 2 \ldots \quad (5.131)$$

with coefficients

$$S_{ik} = 2K + l + (u^2 - u_i^2) \delta_{ik} - 2V^2 C_{ik} \quad (5.132)$$

where

$$u = a(n_1^2 k^2 - \beta^2)^{1/2} \quad (5.133)$$

and

$$V = n_1 k a (2\Delta)^{1/2}$$

are a transverse phase parameter as in equation (4.42) and the fibre parameter as in equation (5.24), respectively.

For non-trivial solutions of equation (5.131), its coefficient determinant must vanish:

$$\det(S_{ik}) = 0.$$

This condition constitutes the characteristic equation of the graded-index fibre.

When we abbreviate the terms of each element in the coefficient determinant by

$$\Omega = 2K + l, \qquad U_i = u^2 - u_i^2, \qquad \text{and} \qquad Y_{ik} = 2V^2 C_{ik},$$

the characteristic equation assumes the form

$$\begin{vmatrix} \Omega + U_1 - Y_{11} & \Omega - Y_{12} & \cdots \\ \Omega - Y_{21} & \Omega + U_2 - Y_{22} & \cdots \\ \vdots & \vdots & \end{vmatrix} = 0.$$

With

$$D = \begin{vmatrix} U_1 - Y_{11} & -Y_{12} & \cdots \\ -Y_{21} & U_2 - Y_{22} & \cdots \\ \vdots & \vdots & \end{vmatrix} \qquad (5.134)$$

and Z_{kj} as the cofactor of the element $D_{kj} = U_k \delta_{kj} - Y_{kj}$ of the matrix of D, the characteristic equation may be written as

$$D + \Omega \sum_k \sum_j Z_{kj} = 0$$

or alternatively as

$$-\frac{1}{\Omega} = \frac{1}{D} \sum_k \sum_j Z_{kj}. \qquad (5.135)$$

In the latter form of the characteristic equation, the cladding characteristics appear on its left-hand side, while the right-hand side lumps together all elements which depend on the core profile.

The step-index fibre with homogeneous core and cladding regions represents a case for which this variational solution reduces to a closed form of the characteristic equation that proves this procedure to be valid at least for this and related profiles. With $g(r/a) = 0$ in equation (5.128), we obtain $C_{ik} = Y_{ik} = 0$ for all i and k. Under these conditions, equation (5.134) reduces to

$$D = \prod_k U_k$$

while the cofactors Z_{kj} are now given by

$$Z_{kj} = \delta_{kj} \prod_{i \neq j} U_i .$$

The latter product consists of all U_i as factors except U_j. When these expressions are substituted into the characteristic equation, it simplifies to

$$-\frac{1}{\Omega} = \sum_k \frac{1}{U_k} . \qquad (5.136)$$

The summation on its right-hand side may now be extended to infinity and, according to

$$\sum_{k=1}^{\infty} \frac{1}{U_k} = \sum_{k=1}^{\infty} \frac{1}{u^2 - u_k^2} = -\frac{J_l(u)}{2u\, J_{l-1}(u)}$$

(Watson 1922) may be expressed by those Bessel functions for which the u_k represent the zero crossings in equation (5.123). When the core is embedded in only a single unlimited cladding, the cladding parameter K is given by equation (5.113), and the characteristic equation appears as

$$-v\, K_{l-1}(v)/K_l(v) = u\, J_{l-1}(u)/J_l(u) .$$

This corresponds exactly to equation (4.104) for the weakly guiding fibre with a homogeneous core and an unlimited cladding.

For graded-index fibres with a given profile function $g(r/a)$, the integrals in equation (5.128) must be evaluated and the characteristic equation solved numerically. Okoshi and Okamoto (1974) have obtained numerical results for low-order modes in a number of different profiles including the truncated parabolic profile. They have found that in order to obtain accurate dispersion curves for these lower-order modes, the series expansion (5.124) needs to employ up to but not more than ten terms.

Index profiles of graded-index fibres may conveniently be described by a power series of radius r for their profile function $g(r/a)$. In most practical cases, in particular those, where the index gradient arises from diffusion of material components, such a power-series expansion contains only even-order terms, so that the profile function is given by

$$g(r/a) = g_2(r/a)^2 + g_4(r/a)^4 + \ldots g_N(r/a)^N . \qquad (5.137)$$

For a profile with no index step at the core-cladding boundary, the coefficients of this expansion obey the condition

$$\sum_k g_k = 1 . \qquad (5.138)$$

For such power-series profiles, the differential equation (5.55) for the radial field distribution appears likewise with power series for its coefficients. Under these conditions, a power-series expansion for $\psi(r)$ lends itself readily to solve equation (5.55) and find the dispersion characteristics and field distribution of low-order guided modes (Kirchhoff 1973b).

When equation (5.55) is transformed to $\rho = r/a$ as the new independent variable and when the transverse phase parameter u from equation (5.133) as well as the fibre parameter V from equation (5.24) are introduced, we obtain

$$\frac{d^2\psi}{d\rho^2} + \frac{1}{\rho}\frac{d\psi}{d\rho} + [u^2 - V^2 g(\rho) - (l/\rho)^2]\psi = 0 . \qquad (5.139)$$

This equation is of the form

$$\frac{d^2\psi}{d\rho^2} + f_1(\rho)\frac{d\psi}{d\rho} + f_2(\rho)\psi = 0 \qquad (5.140)$$

with

$$f_1(\rho) = 1/\rho$$

and

$$f_2(\rho) = q_0/\rho^2 + q_2 + q_4 \rho^2 + \ldots q_{N+2} \rho^N ,$$

where

$$q_0 = -l^2$$

$$q_2 = u^2$$

$$q_4 = -v^2 g_2$$

$$q_{2n} = -v^2 g_{2(n-1)}$$

$$\vdots$$

from the power-series representation (5.137) of the profile function. For a power-series expansion of its solution, we examine first the range of convergence. Any point ρ, where either $f_1(\rho)$ or $f_2(\rho)$ or both are singular, is called a singular point of the differential equation. If ρ_0 represents a point near which $f_1(\rho)$ and $f_2(\rho)$ are analytical, a solution as a power series in $\rho-\rho_0$ converges within a circle whose centre is ρ_0 and which extends to the singular point of the differential equation nearest to ρ_0. Equation (5.140) has two regular singular points, one at $\rho_0 = 0$ and the other at $\rho_0 \to \infty$. It is therefore possible to expand its solutions into power series relative to the singularity at the origin with convergence for all $|\rho| < \infty$. For the series expansion of its solution, we write equation (5.139) as

$$\rho^2 \frac{d\psi}{d\rho^2} + \rho \frac{d\psi}{d\rho} + Q(\rho) \psi = 0 , \qquad (5.141)$$

where

$$Q(\rho) = \sum_{n=0}^{(N/2)+1} q_{2n} \rho^{2n} .$$

We set up the series

$$\psi(\rho) = \rho^s \sum_{k=0}^{\infty} c_k \rho^k, \qquad (5.142)$$

in which the power s and the coefficients c_k are so determined that this series solves equation (5.141). By comparing coefficients of equal power of ρ, the following set of recurrence relations is found for the even-order coefficients c_{2n}:

$$c_0^2(s^2+q_0) = 0$$

$$c_2[(s+2)^2 + q_0] + c_0 q_2 = 0$$

$$\vdots \qquad (5.143)$$

$$c_{2n}[(s+2n)^2 + q_0] + c_{2(n-1)} q_2 + \cdots + c_0 q_{2n} = 0.$$

All odd-order coefficients c_{2n+1} vanish due to the even-order character of the power-series profile. The first of these equations requires that

$$s = \pm l$$

for $c_0 \neq 0$. Only the plus sign yields solutions which are regular at $\rho = 0$. With $s = l$, the remaining recurrence relations lead to the following formula for the expansion coefficients:

$$c_{2n} = -\frac{1}{4n(n+l)}[c_{2(n-1)}u^2 - c_{2(n-2)}v^2 g_2 - \cdots - c_{2(n-N+1)}v^2 g_N]. \qquad (5.144)$$

With these coefficients and with $s = l$, equation (5.141) is hence solved by a power series according to

$$\psi(\rho) = \rho^l \sum_{n=0}^{\infty} c_{2n} \rho^{2n}. \qquad (5.145)$$

This series converges for all finite values of ρ. However, for large values of the fibre parameter V, the convergence becomes rather slow. The power-series solution is therefore in general not too well suited for low-order modes very far above cut-off or for high-order modes, as is actually the case only under multimode conditions in graded-index fibres. Under single-mode conditions, on the other hand, or when the fibre guides only a few low-order modes, the power-series solution with its fast convergence under these conditions presents a universal and efficient method of mode analysis.

Substitution of ψ and $d\psi/dr$ from equation (5.145) into equation (5.118) results in the characteristic equation for the graded-core fibre model with a single or double cladding, depending on whether we use K from equation (5.113) or from equation (5.119) for its evaluation.

Kirchhoff (1973b) has tested this power-series method for a number of profiles of which the cosine profile with

$$g(r/a) = \frac{1}{2}[1 - \cos(r/a)] \qquad (5.146)$$

proved particularly critical. Any such profile that gradually trails out into the cladding over a substantial fraction of its radius requires many terms of the power series for its field representation in order to describe the evanescent fields in the outer core region adequately. To adapt the cosine function to the power-series analysis, it was first expanded into a power series itself up to the term containing ρ^8. Fig.5.13 displays the dispersion characteristics of this profile by plotting the phase parameter B versus the fibre parameter V for the fundamental HE_{11}-mode and the next higher-order LP_{11}-modes. For comparison, the dispersion characteristics of these modes in the step-index fibre and in the truncated parabolic profile with $g(r/a) = (r/a)^2$ for $r < a$ and $g(r/a) = 1$ for $r > a$ also appear in this diagram. The truncated parabolic profile needs 20 terms of the power-series solution (5.145) for fibre parameter values up to $V = 20$, in order to reduce the relative error in B to below 10^{-6}. The cosine profile on the other hand needs up to 90 terms to achieve the same degree of accuracy under equivalent conditions.

5.5. SINGLE-MODE OPERATION

The preceding sections on the analysis of guided and

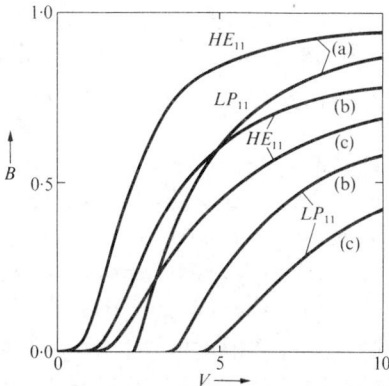

Fig.5.13. Phase parameter as a function of fibre parameter for the fundamental (HE_{11}-) mode and the next higher-order (LP_{11}-) modes in (a) the step-index fibre; (b) the truncated parabolic profile; (c) the cosine profile according to equation (5.146).

leaky modes provide a broad and firm enough theoretical base to investigate the propagation characteristics which are relevant for optical signal transmission. The results of this mode analysis will also enable us to design graded-index fibres for different modes of operation.

For single-mode operation we do not actually need a graded-index core. The cladded-core fibre with a homogeneous core has excellent transmission characteristics under single-mode condition. Its core radius must be small enough to satisfy equation (4.96). But for a low enough difference between core and cladding index, we can have sizable cores which still do not yet guide any of the higher-order modes. Furthermore, if we operate near the cut-off of the next higher order (H_{01}-) mode, the HE_{11}-fields are quite well confined to the core to provide fairly good guidance, and the dispersion coefficient in Fig.4.29 is low enough for the mode dispersion to be insignificant compared to material dispersion, even for fibre materials of low dispersion. Only if we tune the wavelength of the optical carrier for signal transmission to a value where the material dispersion has a minimum will the mode dispersion of a single-mode fibre remain as the dominant source for signal distortion. The cladded-core fibre with a W-index profile even offers the possibility of compensating normal material dispersion with a negative dispersion coefficient of mode dispersion. In all these respects therefore, the single-mode fibre with a homogeneous core has all the desired characteristics of a single-mode optical waveguide.

We will, nevertheless, consider single-mode operation of graded-index fibres because single-mode fibres, even if they are meant to have a homogeneous core, in most practical cases actually have a varying index across their core region. When preforms for fibres are fabricated and the fibres are pulled from them or when fibres are pulled out of molten glasses, the high temperatures of these processes will always cause diffusion of material components. Even if we start with a homogeneous core region in a preform or have a homogeneous glass melt in a crucible for the core, the heating during pulling will let components of the core and cladding materials diffuse into each other. In the final product, the index will then not step abruptly from its core value down to the cladding value but will decrease gradually; the index profile in the core will be more or less graded. We therefore need to investigate what changes such a graded index effects on the transmission characteristics. Possibly we might gain some benefit, even for single-mode operation, when the index is properly graded.

For the step-index fibre with a homogeneous core and an unlimited cladding, we have found the fibre parameter value near cut-off of the next higher-order mode at $V_{c1} = u_{01} = 2 \cdot 405$ to yield the most favourable fundamental mode characteristics in the single-mode range. This point of single-mode operation below the cut-off for the next higher-order modes will in general, and for the same reasons also be the most favourable in graded-index fibres.

In the truncated parabolic profile with $g(r/a) = (r/a)^2$ for $r < a$, but $g(r/a) = 1$ for $r > a$ as in equation (5.27), the fibre parameter at cut-off of the LP_{11}^- (H_{01}^-, E_{01}^-, H_{21}^-) modes has the value $V_{c1} = 3 \cdot 53$. Compared to the limiting value $V_{c1} = 2 \cdot 405$ for the step-index fibre, this represents an increase by a factor of $1 \cdot 47$ for the single-mode limit of this particular index profile. However, this higher single-

mode limit does not necessarily mean that the fundamental
mode fields extend over a wider fibre cross-section. Due to
the graded index, the fields actually confine themselves more
to the centre of the profile, and although the relative core
radius is larger in the truncated parabolic profile at the
single-mode limit, the mode radius remains nearly the same
as in the step-index fibre. Fig.5.13 shows that the dominant
HE_{11}-mode of the step-index fibre has nearly the same phase-
parameter value B at the edge of the single-mode range as this
mode has it in the truncated parabolic profile at the border
of its single-mode range. Field confinement to the high-index
region of the core is hence as effective in the truncated
parabolic profile at the single-mode limit as it is in the
step-index fibre. There is also not much difference in dispersion characteristics. Fig.5.14 compares the group delay

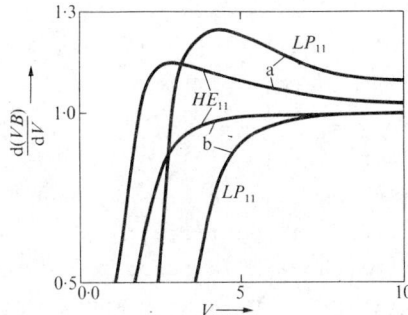

Fig.5.14. Delay factor as a function of fibre parameter for the fundamental (HE_{11}-) mode and the next higher-order (LP_{11}-) modes in
(a) the step-index fibre; (b) the truncated parabolic profile.

factor of the fundamental mode in the truncated parabolic
profile with that of the step-index fibre. Near their respective single-mode limits, both curves have such slopes that
the dispersion factor $Vd^2(VB)/dV^2$ is nearly equal for the fundamental mode at this limit. Altogether the examination of
the fundamental mode characteristics in the single-mode range
of the truncated parabolic profile reveals no significant
differences when compared with those of homogeneous-core
fibres. We therefore find no reason to prefer graded-index

fibres to fibres with homogeneous cores for single-mode operation. Should the fabrication process lead to gradual transitions between core and cladding index or even to index profiles which are graded throughout the core, then these profiles are quite acceptable for single-mode operation, if only they meet the other requirements for optical signal transmission.

A typical process that leads to a gradual index transition between core and cladding is diffusion between core and cladding glasses when the fibre preform is produced or the fibre proper pulled from it or from molten glasses at elevated temperatures (Chan et al. 1970). Initially, before any diffusion takes place, the core glass of the preform or melt or of the fibre may be assumed to have the homogeneous concentration N_0 of a second species that raises its index by $\Delta n = (n_1 - n_2)$ above that of the cladding. During diffusion this concentration changes to $N(r,t)$ which, as a function of radius r and time t, follows from the diffusion equation

$$\frac{\partial^2 N}{\partial r^2} + \frac{1}{r}\frac{\partial N}{\partial r} - \frac{1}{D}\frac{\partial N}{\partial t} = 0 \ . \tag{5.147}$$

This differential equation assumes a small enough concentration N so that the diffusion coefficient D remains independent of concentration. With the initial conditions $N = N_0$ for $0 < r < a$ and $N = 0$ for $r > a$ of a step-index profile at time $t = 0$, the solution of equation (5.147) is

$$\frac{N}{N_0} = \int_0^\infty \exp\left(-\frac{Dt}{a^2} u^2\right) J_0\left(\frac{ur}{a}\right) J_1(u)\, du \ . \tag{5.148}$$

Fig.5.15 compares this concentration for two different values of the relative diffusion time Dt/a^2 with the concentration of the step-index profile. For small concentrations N of a second species, the refractive index changes nearly linearly with N and is hence given by

$$n = (n_1 - n_2) N/N_0 + n_2 \ . \tag{5.149}$$

Under these conditions, the curves in Fig.5.15 also represent

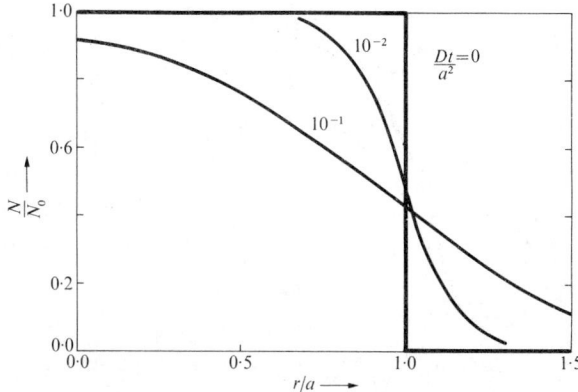

Fig.5.15. Relative concentration of a second species in the core glass of a cladded-core fibre before and after two different relative diffusion times Dt/a^2 (Chan et al. 1970).

the index distributions of the profile before diffusion and after the two different relative diffusion times. With more diffusion, the core region of raised index spreads to larger radii but with less index difference. Eventually, even the maximum index difference between core centre and cladding decreases substantially, so that the field confinement and guidance of modes become much weaker.

The dispersion characteristics of the dominant HE_{11}-mode in the cladded-core fibres with the diffused core-cladding boundary of Fig.5.15 have been computed by Chan et al. (1970). These authors approximate the continuous variation in refractive index by a staircase function and solve the wave equation for this cylindrically stratified medium. They find that five to nine steps in the stratification are sufficient to approximate the continuously varying index profile adequately for the dispersion analysis. From their numerical results, the phase parameter B of the HE_{11}-mode has been plotted as a function of the frequency parameter V in Fig.5.16. This frequency parameter is still defined by $V = ka(n_1^2 - n_2^2)^{\frac{1}{2}}$, where n_1 and n_2 designate the refractive indices in core and cladding before any diffusion takes place. From the dispersion curve of the step-index fibre, the phase parameter decreases stead-

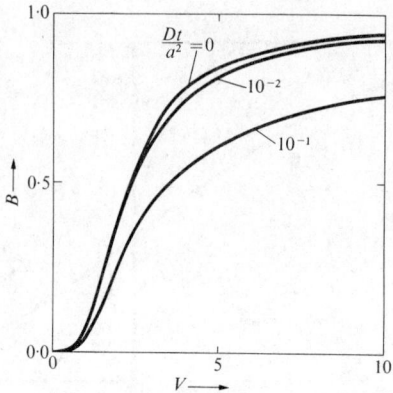

Fig.5.16. Phase parameter of the fundamental mode in a cladded-core fibre with a diffused core-cladding boundary for different values of the relative diffusion time Dt/a^2 (Chan et al. 1970).

ily with more diffusion. But the diffusion has only a slight effect on the dispersion characteristics, provided it is weak enough to let the relative diffusion time remain below $Dt/a^2 < 10^{-2}$. For more diffusion, the effect becomes more pronounced and even the limiting value which B approaches at large V-values is smaller than unity. It corresponds to the relative concentration which the second species of the core glass has in the fibre centre in Fig.5.15. To see more clearly the change in dispersion of the fundamental mode when the core-cladding boundary is more diffused, we examine the delay factor of this mode. Fig.5.17 compares this delay factor again for the two relative diffusion times of Figs.5.15 and 5.16 with that of the corresponding step-index fibre. The dispersion factor $V d^2(VB)/dV^2$ which is responsible for signal distortion follows from the slope of these curves. The positive slope which the delay factor has as a function of V in the step-index fibre near its single-mode limit at $V_{cl} = 2 \cdot 405$ increases when we operate at the same value of V in a fibre with a diffused core-cladding boundary. To maintain the dispersion factor at the same low level as in the step-index fibre at $V = 2 \cdot 4$, we therefore need to shift to larger values of V. For operation under single-mode conditions, however, V may be shifted to larger V-values only as much as the single-mode limit shifts up due to the diffusion of the core-cladding

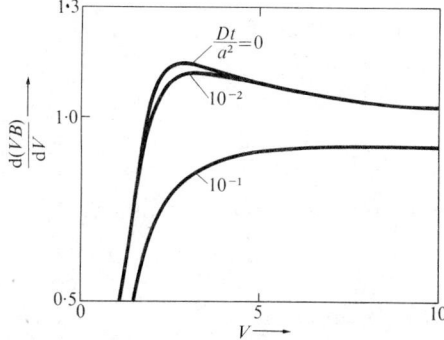

Fig.5.17. Delay factor of the fundamental mode in a cladded-core fibre with a diffused core-cladding boundary for different values of the relative diffusion time Dt/a^2 (Chan et al. 1970).

boundary and the ensuing change in index profile.

Under normal conditions, though, material dispersion is so much stronger than this mode dispersion that these effects remain of a secondary character and we need not worry too much about their minor consequences for signal transmission.

Of much greater significance is the larger effective core size and the smaller index difference to which a strongly diffused core-cladding boundary leads. Due to these changes, the fundamental mode fields extend over a wider fibre cross-section and the mode is not bound as tightly to the fibre as with the step-index profile. On one hand, the larger mode size eases launching and splicing problems, but on the other hand the weaker guidance makes the fundamental mode more sensitive to any imperfections of the fibre.

5.6. DELAY DIFFERENCES

In multimode operation graded-index fibres have much better transmission characteristics than fibres with homogeneous cores. The guided modes in homogeneous cores differ as much in their group delay as their respective paths in the ray picture differ in length. Signals that propagate in several or all of the guided modes suffer these different delays and spread in time, with much signal distortion as a consequence.

When instead of a homogeneous core the core index is graded to a near parabolic profile, the ray paths of different guided modes, although they still differ in geometrical length, nearly all have the same optical length. A properly graded core in a graded-index fibre thus equalizes the delay of guided modes. Only little delay spread remains and signal distortion due to such delay spread reduces by orders of magnitude.

The group delay τ_g of an individual mode follows in the ray picture from the integration

$$\tau_g = \int \tau(r) \, ds \qquad (5.150)$$

of the delay $\tau(r)$ per unit length along the ray path s. With $\tau = n'/c$ and n' as the local group index in the index profile, we can also write

$$\tau_g = \frac{1}{c} \int n'(r) \, ds \ . \qquad (5.151)$$

From this integral expression, we infer that the group delay depends not only on the geometrical form of the ray path and its refractive index distribution but rather on the distribution of group index along the ray path. In addition, a more accurate evaluation of group delay should consider that not only the refractive index changes with frequency, so that the group index also appears in equations (5.151) for the delay. It should also take account of the whole ray path changing with frequency. We include all these effects when we start from the characteristic equation of the individual modes, which, in the implicit form $D(\beta,\omega) = 0$, gives β as a function of frequency. By differentiating this dispersion relation according to

$$\tau = \frac{d\beta}{d\omega} = -\frac{\partial D}{\partial \omega} \bigg/ \frac{\partial D}{\partial \beta} \ , \qquad (5.152)$$

we obtain the group delay τ per unit fibre length to whatever accuracy the characteristic equation offers.

For multimode operation the characteristic equation (5.109) from the field solution in terms of Airy functions appears as a sufficiently accurate approximation. It gives the dispersion relation accurately for all guided modes except those of very low radial order. In multimode operation where usually a very large number of modes are guided by the core and participate in signal transmission, the few modes of low radial order are of minor significance and we accept that their dispersion is only approximately accounted for. Under these conditions, it might even suffice to further simplify this characteristic equation and use its form (5.20) from the ray analysis as the dispersion relation. From one or the other form of characteristic equation, the dispersion relation $D(\beta,\omega) = 0$ appears as

$$D(\beta,\omega) \equiv \int_{r_1}^{r_2} k_r(r,\beta,\omega)\,dr - \mu(p) = 0 \;, \qquad (5.153)$$

where $\mu(p)$, with $p = q + 1$, represents the right-hand side of equation (5.109), if we choose this more accurate characteristic equation. When we use the approximate form (5.20) from the ray picture, we have simply $\mu(p) = p\pi$.

To calculate the group delay according to equation (5.152), we need to differentiate $D(\beta,\omega)$ partially with respect to ω. $D(\beta,\omega)$ depends on frequency due to the frequency-dependent refractive index $n(r,\omega)$ and to $k = \omega/c$ in

$$k_r = [n^2(r,\omega)k^2 - \beta^2 - l^2/r^2]^{\frac{1}{2}} \;.$$

In a graded-index fibre, the index profile in the core is formed by mixing glasses of different refractive indices or by doping a base glass with certain additives to change its refractive index. In general, for such glass mixtures or for a base glass with additives, not only the refractive index but also its dispersion depend on the component ratio or mole fraction in the mixture, or on the amount of additives. In an inhomogeneous material, the mole fraction of the mixture or the amount of additives changes locally. The refractive index will then be a function of position and frequency. For the index distribution in equation (5.23) with circular symmetry, not only n_1 depends on frequency but also

Δ and even the profile function g.

We simplify the situation somewhat by restricting our considerations to a combination of only two materials a and b, which individually have refractive indices n_a and n_b, respectively. For the combination of these two materials, we assume that the resulting refractive index n is given by

$$n^2 = x_a n_a^2 + x_b n_b^2 , \qquad (5.154)$$

where x_a and x_b designate the mole fraction of material a and b, respectively, in the mixture, so that

$$x_a + x_b = 1 . \qquad (5.155)$$

With equations (5.154) and (5.155), the square of the refractive index of the combination

$$n^2 = x_a (n_a^2 - n_b^2) + n_b^2 \qquad (5.156)$$

changes linearly with the mole fraction of either one of its components. This appears as a reasonable assumption at least over limited ranges of material compositions. An index profile of circular symmetry, such as in equation (5.23), obtains from the combination of two materials when its mole fraction is varied with distance from the fibre axis. With x_{a1} for $r = 0$ on the fibre axis and $x_a(r)$, the relative permittivity may be written as

$$n^2 = [x_{a1}(n_a^2 - n_b^2) + n_b^2]\left\{1 + [x_a(r) - x_{a1}]\frac{n_a^2 - n_b^2}{x_{a1}(n_a^2 - n_b^2) + n_b^2}\right\} . \qquad (5.157)$$

When we compare it with equation (5.23) we note that

$$n_1^2 = x_{a1}(n_a^2 - n_b^2) + n_b^2 ,$$

$$-2\Delta = (x_a(a) - x_{a1})(n_a^2 - n_b^2)/n_1^2 , \qquad (5.158)$$

and the profile function is given by

$$g(r/a) = (x_a(r) - x_{a1})/(x(a) - x_{a1}) . \qquad (5.159)$$

In this form only n_1 and Δ depend on frequency but $g(r/a)$ remains independent of it.

Under this assumption for a two-component mixture or only one additive in a base glass with the linear dependence of n^2 on x, we can now evaluate $\partial D/\partial \omega$ in equation (5.152) and obtain for the frequency derivative of the phase integral

$$\frac{\partial}{\partial \omega} \int_{r_1}^{r_2} k_r \, dr = \int_{r_1}^{r_2} \frac{\partial k_r}{\partial \omega} \, dr - k_r(r_1,\omega) \frac{dr_1}{d\omega} + k_r(r_2,\omega) \frac{dr_2}{d\omega} . \quad (5.160)$$

Since r_1 and r_2 are the turning points, we have $k_r(r_1,\omega) = k_r(r_2,\omega) = 0$ and the last two terms on the right-hand side of equation (5.160) vanish. The differentiation of k_r with respect to frequency in the integrand of the first term leads to

$$\frac{\partial k_r}{\partial \omega} = \frac{n_1 k}{k_r} \left\{ \frac{n_1'}{c} [1 - 2\Delta g(r/a)] - n_1 k \frac{d\Delta}{d\omega} g(r/a) \right\} . \quad (5.161)$$

Partial differentiation of $D(\beta,\omega)$ with respect to β results likewise in

$$\frac{\partial}{\partial \beta} \int_{r_1}^{r_2} k_r \, dr = \int_{r_1}^{r_2} \frac{\partial k_r}{\partial \beta} \, dr \quad (5.162)$$

with

$$\frac{\partial k_r}{\partial \beta} = - \beta/k_r . \quad (5.163)$$

When both derivatives of the phase integral in equations (5.161) and (5.162) are substituted for $\partial D/\partial \omega$ and $\partial D/\partial \beta$, respectively, into equation (5.152), the group delay per unit fibre length is given by

$$\tau = \frac{n_1' n_1 k}{c\beta} \frac{\int_{r_1}^{r_2} [1 - 2\Delta g(r/a) - \frac{n_1 k}{n_1'} \frac{d\Delta}{dk} g(\frac{r}{a})] \frac{dr}{k_r}}{\int_{r_1}^{r_2} dr/k_r} . \quad (5.164)$$

According to this expression, material dispersion not only

affects the mode delay by the group index n_1' being different, in general, from the refractive index n_1. It contributes in addition through the term with $d\Delta/dk$ in the numerator integral. Whenever the material dispersion changes across the core profile of a graded-index fibre, the relative index difference will depend on frequency and $d\Delta/dk \neq 0$. Because it originates so directly from the index profile of the fibre, this contribution to the mode delay is called profile dispersion. Another designation for this part of the mode delay refers to the inhomogeneity of the graded-index core and calls it inhomogeneous material dispersion. While n_1' accounts for homogeneous material dispersion, that is dispersion in the homogeneous material, the term with $d\Delta/dk$ arises under present assumptions from the inhomogeneity of the material in the index profile, and is therefore designated as inhomogeneous material dispersion, or as profile dispersion.

To evaluate the mode delay from equation (5.163), we need to specify the profile function $g(r/a)$ of the fibre, but then also to solve the characteristic equation for the phase constant β of the respective mode. In order to obtain the delay characteristics of guided modes in an analytic form, we restrict attention at first to index profiles that are described by

$$n^2(r) = \begin{cases} n_1^2[1-2\Delta(r/a)^\alpha] & \text{for} \quad r \leq a \\ n_1^2[1-2\Delta] & \text{for} \quad r > a \end{cases} \qquad (5.165)$$

The square of the core index according to these equations follows a power-law in which the exponent α can have any value between 1 and ∞. Fig.5.18 shows several such power-law profiles with different exponents α. $\alpha = 2$ represents the truncated parabolic index profile of Section 5.2, while $\alpha \to \infty$ describes the step-index profile with a homogeneous core and unlimited cladding. In between $\alpha = 1$ and $\alpha \to \infty$, the power-law profiles can be the model for many types of graded-index fibres. This class of profiles is therefore well suited for the analysis of multimode operation of graded-index fibres.

A straightforward method of calculating the propagation

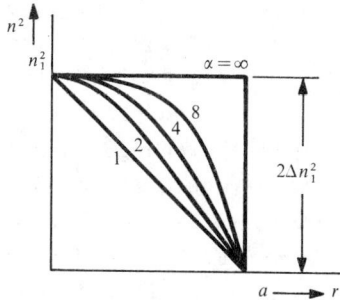

Fig.5.18. Power-law index profiles with different exponents α.

characteristics would first solve the characteristic equation (5.153) for the phase constant β of guided modes and then evaluate equation (5.164) or compute directly $\tau = d\beta/d\omega$ for the mode delay.

For the power-law profiles of equation (5.165), however, no general analytic solution of the phase integral is available. Gloge (1975) has found the expression

$$(n_1^2 k^2 - \beta^2)^{\frac{1}{\alpha}+\frac{1}{2}} = (\frac{\alpha}{\alpha+2})^{-\frac{1}{2}} (\frac{2}{\alpha+2})^{-\frac{1}{\alpha}} \left[l + 4 \left(\frac{2}{\alpha+2}\right)^{\frac{2}{\alpha}} p\right] v^{\frac{2}{\alpha}} \quad (5.166)$$

to approximate the exact solution quite satisfactorily as long as α is not too large. For the parabolic profile, it reduces to

$$n_1^2 k^2 - \beta^2 = V(l+2p)/2 , \quad (5.167)$$

which for large enough p or l agrees with equation (5.64) from the field solution for the unlimited parabolic profile. In general equation (5.166) solves the characteristic equation only approximately and its error increases with increasing exponent α of the power-law profile. The maximum error occurs for $\alpha \to \infty$ and can be determined by comparing β from equation (5.166) with the solution of the characteristic equation for the cladded-core fibre with homogeneous core.

We could, nevertheless, use equation (5.166) to evaluate the group delay. For the particular case of power-law profiles, however, this group delay can also be derived directly from the characteristic equation without using an approximation such as equation (5.166) for its solution (Gloge and Marcatili 1973). For large enough radial mode order p, the characteristic equation reduces to equation (5.20) which gives p for a particular mode in terms of its circumferential order l and its phase constant β. All modes of the same circumferential order l as in equation (5.20) but with larger phase constants than β in equation (5.20) have lower radial orders than p in equation (5.20). Altogether there are

$$4p = \frac{4}{\pi} \int_{r_2}^{r_1} [n^2(r)k^2 - \beta^2 - l^2/r^2]^{1/2} \, dr \qquad (5.168)$$

such modes which have the same order l but phase constants that are larger than β in equation (5.20). The factor 4 in equation (5.168) accounts for the two different field distributions, each with two orthogonal polarizations, in which most of the LP_{lp}-modes occur.

Let us now count the total number of guided modes with phase constants larger than a given value β. This mode count corresponds to the mode index ν with which we labelled the modes of a cladded-core fibre by equation (4.273) in the sequence of their cut-off values. It will lead here to an explicit presentation of β as a function of frequency for modes in the power-law profile, and allows us to evaluate the group delay. To count all the modes with phase constants larger than β, we need to sum the numbers $4p(l)$ for each circumferential order l over all l from $l = 0$ to the largest value l_{max}. The largest circumferential order l_{max} for a given β has the turning points converging on each other so that $r_1 = r_2$; it follows from

$$n^2(r)k^2 - \beta^2 - l^2/r^2 = 0 \qquad (5.169)$$

under the condition that both its solutions r_1 and r_2 coincide to form a double root. For multimode operation and except for the low-order modes which are farthest above cut-off, the value l_{max} is such a large number that we may consider l to be a continuous variable and replace the summation by an integration over l. The total number ν of modes with phase constants larger than β is then given by

$$\nu(\beta) = \frac{4}{\pi} \int_0^{l_{max}} \int_{r_1}^{r_2} [n^2(r)k^2 - \beta^2 - l^2/r^2]^{1/2} \, dr \, dl \, . \quad (5.170)$$

For $l = 0$ the integration over r extends from $r = 0$ to $r = r_2(0)$, where $r_2(0)$ designates the radius at which $n(r)k = \beta$. As l increases the turning points $r_1(l)$ and $r_2(l)$ as the limits of integration over r approach each other and converge for $l = l_{max}$. By changing the order of integration in equation (5.169), we obtain

$$\nu(\beta) = \frac{4}{\pi} \int_0^{r_2(0)} \int_0^{r[n^2(r)k^2-\beta^2]^{1/2}} [n^2(r)k^2 - \beta^2 - l^2/r^2]^{1/2} \, dl \, dr, \quad (5.171)$$

where the upper limit of the integration over l follows from equation (5.169). Integration of equation (5.171) with respect to l results in

$$\nu(\beta) = \int_0^{r_2(0)} [n^2(r)k^2 - \beta^2] \, r \, dr. \quad (5.172)$$

When β approaches the upper limit $n_1 k$ for guided modes, both the integrand and the upper limit of integration go to zero, indicating that only few modes have phase constants so close to $n_1 k$. On the other hand, when β assumes the cut-off value $\beta = n_2 k$, the upper limit of integration goes to the profile radius $r_2(0) = a$, and equation (5.172) counts the total number of guided modes just as in equation (5.33). Both the fraction of modes $\nu(\beta)$ with phase constants larger than β and the total number M of guided modes are given by equations (5.172) and (5.33), respectively, for any index profile as long as there are only two turning points r_1 and r_2.

For power-law profiles according to equation (5.165) the integrals of equations (5.172) and (5.33) may be evaluated to yield

$$\nu = n_1^2 k^2 a^2 \Delta \frac{\alpha}{\alpha+2} \left(\frac{n_1^2 k^2 - \beta^2}{2 n_1^2 k^2 \Delta} \right)^{(2/\alpha)+1} \qquad (5.173)$$

for the number of modes with phase constants smaller than β, and

$$M = \frac{\alpha}{\alpha+2} n_1^2 k^2 a^2 \Delta = \frac{\alpha}{\alpha+2} \frac{V^2}{2} \qquad (5.174)$$

for the total number of guided modes. Note that $M = V^2/2$ for $\alpha \to \infty$ in accordance with equation (4.21) for the homogeneous core, and $M = V^2/4$ for $\alpha = 2$ in accordance with equation (5.30) for the parabolic index profile.

To determine the group delay of individual modes, we solve equation (5.173) for β and obtain

$$\beta^2 = n_1^2 k^2 - 2 \left(\frac{\alpha+2}{\alpha} \frac{\nu}{a^2} \right)^{\frac{\alpha}{\alpha+2}} (n_1^2 k^2 \Delta)^{\frac{2}{\alpha+2}} . \qquad (5.175)$$

For a particular mode of orders p and l only $k = \omega/c$, n_1, and Δ depend on frequency. The exponent α of the power-law profile is independent of frequency under the conditions that led to equation (5.159). The mode number ν is a characteristic integer for any given mode.

With β given explicitly in terms of the frequency-dependent quantities n_1, k, and Δ, we can differentiate with respect to ω and obtain for the group delay per unit fibre length

$$\tau = \frac{n_1 n_1'}{cN} \left[1 - \frac{2(1 - N^2/n_1^2)}{\alpha+2} \left(1 + \frac{n_1 k}{2 n_1' \Delta} \frac{d\Delta}{dk} \right) \right] , \qquad (5.176)$$

where n_1' designates the group index of the core material on the fibre axis and $N = \beta/k$ the effective index of the mode. The term with $d\Delta/dk$ accounts for profile dispersion.

The effective index of guided modes ranges from $N = n_2$ at cut-off to $N = n_1$ very far above. For weakly guiding fibres with $n_2 \simeq n_1$, it differs only little from n_1, even at cut-off.

The second term within the brackets of equation (5.175) therefore always remains small compared to unity. With

$$\delta = (\tfrac{1}{2})(1 - N^2/n_1^2) \qquad (5.177)$$

as a small quantity and

$$N = n_1(1 - 2\delta)^{1/2}, \qquad (5.178)$$

we expand equation (5.176) into a power series of δ of which we need only the first few terms for sufficient accuracy:

$$\tau = \frac{n_1'}{c}\left[1 + \frac{\alpha-2-y}{\alpha+2}\delta + \frac{3\alpha-2-2y}{\alpha+2}\frac{\delta^2}{2}\right]. \qquad (5.179)$$

The quantity

$$y = 2\frac{n_1}{n_1'}\frac{k}{\Delta}\frac{d\Delta}{dk} \qquad (5.180)$$

accounts for the profile dispersion in this delay formula (Olshansky and Keck 1976). When the refractive indices n_1 and n_2 in the core centre and in the cladding are introduced for Δ in equation (5.180), the parameter y for profile dispersion appears as a function of n_1 and n_2 as well as of the group indices n_1' and n_2'

$$y = (2/\Delta)[1 - (n_2'/n_1')(n_2/n_1)].$$

Fig.5.19 shows y as a function of wavelength for fused silica that is doped either with GeO_2 or with TiO_2. Both dopants serve in graded-index fibres of fused silica to raise and grade the index. The curves in Fig.5.19 have been obtained by evaluating dispersion data as given by Fleming (1976) for the material combination SiO_2-GeO_2 and by Olshansky and Keck (1976) for the combination SiO_2-TiO_2. Fleming (1976) provides dispersion data for fused silica with three different concentrations of GeO_2. If the linear dependence of equation (5.154) on the material concentrations holds at all wavelengths, the profile dispersion parameter y is independent of material concentrations. Actually, however, the combina-

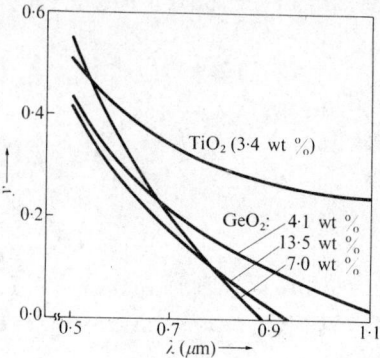

Fig.5.19. Profile dispersion parameter of fused silica doped with GeO_2 or with TiO_2 of the specified weight percentage.

tion of SiO_2 and GeO_2 follows the linear relationship of equation (5.154) only to a certain approximation. At some wavelengths in the range of Fig.5.19, it even deviates substantially from equation (5.154). We therefore obtain different characteristics of y as a function of wavelength for different GeO_2 concentrations in SiO_2.

Equation (5.179) gives the group delay of a particular mode in terms of the deviation δ of its effective index from the index n_1 in the centre of the profile. To evaluate this delay, we would need to know its phase constant or the deviation of its effective index from n_1. The total range of mode delay can, however, be determined without knowledge of the phase constants of individual modes. For the modes which are farthest above cut-off, the deviation is smallest and may be assumed as $\delta = 0$. These modes then are delayed by nearly

$$\tau_1 = n_1'/c \ . \tag{5.181}$$

The relative difference in delay with respect to τ_1 follows from equation (5.179) as

$$\frac{\Delta\tau}{\tau_1} = \frac{\tau - \tau_1}{\tau_1} = \frac{\alpha - 2 - y}{\alpha + 2}\delta + \frac{3\alpha - 2 - 2y}{\alpha + 2}\frac{\delta^2}{2} \ . \tag{5.182}$$

Any mode which is close to or at cut-off has the largest relative deviation $\delta = \Delta$. All other modes have deviations δ between 0 and Δ.

As long as α is not too close to $\alpha = 2 + y$, the term which is linear in δ dominates in equation (5.182), and the total delay spread of all guided modes is given by

$$\frac{\Delta\tau}{\tau} = \frac{\alpha-2-y}{\alpha+2} \qquad \text{for} \qquad \alpha \neq 2 + y . \qquad (5.183)$$

If, however,

$$\alpha = 2 + y , \qquad (5.184)$$

the total delay spread shrinks to

$$\frac{\Delta\tau}{\tau_1} = \frac{\Delta^2}{2} \qquad (5.185)$$

and becomes independent of the amount of profile dispersion y. For $\alpha \geq 2 + y$, including the case of equation (5.184), the low-order modes with δ near zero travel fastest while with increasing mode order (and hence increasing ν in equation (5.175)) their deviation δ also increases as well as their delay. For exponents α, however, which are somewhat smaller but still close to $\alpha = 2 + y$, the delay difference from equation (5.182) first becomes negative with increasing mode order and hence increasing δ. It then passes an extremum for modes with

$$\delta_{\text{extr.}} = \frac{2+y-\alpha}{3\alpha-2-2y} . \qquad (5.186)$$

For $\delta > \delta_{\text{extr.}}$, the negative delay difference diminishes again to end up at

$$\frac{\Delta\tau}{\tau_1} = \frac{\alpha-2-y}{\alpha+2} \Delta + \frac{3\alpha-2-2y}{\alpha+2} \frac{\Delta^2}{2} , \qquad (5.187)$$

which, depending on the exact value of α, may be positive or negative, when α is only a little smaller than $\alpha = 2 + y$.

We obtain the lowest possible delay spread when we let

$$\alpha = 2 + y - b\Delta ,\qquad (5.188)$$

and fix b so that the delay difference in equation (5.187) just becomes zero. The modes of highest order near cut-off travel just as fast under these conditions as the modes of lowest order very far above cut-off. All modes in between travel somewhat faster and have a negative delay difference. The condition $\Delta\tau = 0$ for $\delta = \Delta$ and α from equation (5.188) leads to $b = 2 + y/2$. The optimum exponent with respect to delay equalization is therefore given by

$$\alpha_{opt.} = 2 + y - (2+y/2)\Delta . \qquad (5.189)$$

For this exponent the modes with

$$\delta_{extr.} = \Delta/2 \qquad (5.190)$$

travel fastest and have the relative delay difference

$$\Delta\tau/\tau_1 = -\Delta^2/8 \qquad (5.191)$$

with respect to the slowest modes near and very far above cut-of

For α-values which are smaller than equation (5.189), the negative extremum of delay difference shifts to larger values of δ until, at $\alpha = 2 + y - (4+y)\Delta$, it reaches $\delta_{extr.} = \Delta$. For still smaller values of α, the mode delay decreases monotonically with δ and hence with increasing mode order. Eventually we return to the linear dependence of $\Delta\tau$ on δ, for which the slope is now negative.

For typical values of the relative index difference of the order of $\Delta = 0.01$ or even smaller, the relative delay spread shrinks from $\Delta\tau/\tau_1 = \Delta = 10^{-2}$ for the homogeneous core to $\Delta\tau/\tau_1 = \Delta^2/8 \simeq 10^{-5}$ for the optimum power-law profile. Such a drastic reduction in delay spread by nearly three orders of magnitude requires, of course, that the optimum profile be maintained within very tight tolerances. Small deviations in $n(r)$ from its optimum distribution will already lead to considerably more delay spread than in equation (5.191). To estimate these tolerances, we assume that the index profile still follows the power law but that the exponent deviates according to

$$\alpha = \alpha_{opt.} + \Delta\alpha \qquad (5.192)$$

from the optimum value $\alpha_{opt.}$ in equation (5.189). This change in exponent causes the refractive index to deviate by

$$\Delta n = n_1 \Delta (r/a)^{\alpha_{opt.}} \ln(r/a) \Delta\alpha \qquad (5.193)$$

from its optimum distribution. The maximum index deviation occurs at

$$r = a\, e^{-1/\alpha_{opt.}} \qquad (5.194)$$

and amounts to

$$\Delta n_{max} = - \frac{n_1 \Delta}{e\, \alpha_{opt.}} \Delta\alpha \qquad (5.195)$$

with e as the base of the natural logarithm. When we substitute $\alpha = \alpha_{opt.} + \Delta\alpha$ into equation (5.187) and use the optimum exponent from equation (5.189), the relative delay difference obtains as

$$\frac{\Delta\tau}{\tau_1} = \frac{\Delta\alpha - (2+y/2)\Delta}{4+y}\delta + \delta^2/2 \,. \qquad (5.196)$$

For a small deviation $\Delta\alpha$ this delay difference still has a negative extremum which, due to $\Delta\alpha$, has shifted from $\delta_{extr.} = \Delta/2$ to

$$\delta_{extr.} = \Delta/2 - \Delta\alpha/(4+y) \,. \qquad (5.197)$$

The modes with phase constants that correspond to $\delta_{extr.}$ differ in relative delay by as much as

$$\frac{\Delta\tau}{\tau} = -\frac{1}{8}[\Delta - \frac{2}{4+y}\Delta\alpha]^2 \qquad (5.198)$$

from the lowest order modes. Equation (5.198) represents the total delay spread only for negative deviations $\Delta\alpha$ from the optimum exponent. Only for $\Delta\alpha < 0$ do all modes - including those near cut-off - still travel faster than the lowest order modes, and all have negative delay differences with respect to them. For $\Delta\alpha > 0$, on the other hand, the modes near cut-off slow down with respect to the lowest order modes and assume positive delay differences. For the total delay spread, we would then have to take the difference in delay between the modes near cut-off and those with δ-values according to equation (5.197).

Positive errors Δn in the refractive index distribution correspond to negative deviations $\Delta\alpha$ in equations (5.192) and (5.193). Such a profile error causes the total relative delay spread to increase from $\Delta^2/8$ to

$$\frac{\Delta\tau}{\tau_1} = \frac{1}{8} [\Delta + \frac{2}{4+y} |\Delta\alpha|]^2 . \tag{5.199}$$

In terms of the maximum index deviation from equation (5.195), this total delay spread is given by

$$\frac{\Delta\tau}{\tau_1} = \frac{1}{8} [\Delta + \frac{4+2y}{4+y} \frac{e}{n_1 \Delta} \Delta n_{max}]^2 . \tag{5.200}$$

According to this formula, it needs only a maximum relative index deviation of

$$\frac{\Delta n_{max}}{n_1} = \frac{4+y}{4+2y} \frac{\Delta^2}{e} \tag{5.201}$$

to quadruple the total delay spread from its optimum value of $\Delta\tau/\tau = \Delta^2/8$. For a relative index difference of typically $\Delta = 0.01$ and zero profile dispersion ($y = 0$), the index deviation must stay below $\Delta n/n = (1/e) \times 10^{-4}$ to keep the total delay spread within these bounds. Even relative to the index difference $n_1\Delta = n_1 - n_2$, the maximum deviation should be only a fraction of one per cent.

The power-law profiles of Fig.5.18 serve well as a model for different index distributions in graded-index fibres. They include the parabolic profile and also the near-parabolic profile that, for small enough profile dispersion, minimizes the total delay spread for all guided modes. The homogeneous core appears as the limiting case of a power-law profile for $\alpha \to \infty$. To a certain extent, we can also analyse the effects of index deviations from a specified power-law distribution by accounting for the maximum index error with a corresponding change in the exponent of the power law. This error analysis assumes, however, that the actual index distribution again follows a power law with a different exponent. It is therefore limited to very special index deviations. As another limitation, the power-law profile with its truncation at the core boundary always has a break point at the core-cladding transition.

Actually the index distribution in graded-index fibres will always taper out smoothly from the core profile into the homogeneous cladding. It will always have at least one point of inflection within the core profile. To account for the effects of such index transitions but to also gain more versatility in the assessment of index deviations, we consider the following fourth-order index profile (Timmermann 1974)

$$n^2(r) = \begin{cases} n_1^2[1 - 2(1-d)\Delta(r/a)^2 - 2d\Delta(r/a)^4] & r \leq a \\ n_1^2[1 - 2\Delta] = n_2^2 . & r \geq a \end{cases} \quad (5.202)$$

For $d = 0$, this index profile reduces to the parabolic distribution, while for $d = 1$, it represents the power-law profile with $\alpha = 4$. For $d = -1$, its distribution has zero slope at $r = a$ and hence tapers into the homogeneous cladding index with no break at $r = a$. Fig.5.20 illustrates these three special cases of the fourth-order profile. When $d \ll 1$ the

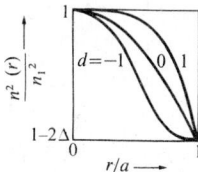

Fig.5.20. Special cases of the fourth-order profile of equation (5.202). $d = 1$: power-law profile with $\alpha = 4$. $d = 0$: parabolic profile. $d = -1$: smooth index transition at core-cladding boundary.

fourth-order term in the index distribution can account for small deviations of an actual index distribution from the parabolic profile.

An analytic solution of the characteristic equation (5.153) is not available for the general fourth-order profile in equation (5.202). We therefore evaluate equation (5.164) to obtain the group delay of individual modes. The profile function $g(r/a)$ in this delay formula is for the profile of equation (5.202) given by

$$g(r/a) = (1-d)(r/a)^2 + d(r/a)^4 . \tag{5.203}$$

We transform from the radius r to

$$\xi = (r/a)^2 \tag{5.204}$$

as the new variable of integration in the numerator and denominator integrals of equation (5.164), and designate the new limits of integration by

$$\xi_{t1,2} = (r_{1,2}/a)^2 . \tag{5.205}$$

The group delay relative to the delay very far above cut-off is then given by

$$\frac{\tau}{\tau_1} = \frac{n_1 k}{\beta} \frac{\int_{\xi_{t1}}^{\xi_{t2}} \{1 - (2\Delta + \frac{n_1 k}{n_1} \frac{d\Delta}{dk}) g(\xi)\} [P(\xi)]^{-1/2} d\xi}{\int_{\xi_{t1}}^{\xi_{t2}} [P(\xi)]^{-1/2} d\xi} \tag{5.206}$$

where

$$P(\xi) = \xi^3 + (1/d - 1) \xi^2 - \frac{\delta}{d\Delta} \xi + \frac{l^2}{dV^2} \tag{5.207}$$

constitutes a cubic polynomial in ξ. The coefficients of this polynomial depend on the profile parameters Δ and d as well as on the fibre parameter $V = n_1 k a \sqrt{2\Delta}$ and the phase parameter $\delta = \frac{1}{2}[1 - \beta^2/(n_1 k)^2]$. The delay formula (5.206) requires the polynomial $P(\xi)$ to have zero crossings ξ_{t1} and ξ_{t2} only in the interval $\xi = 0\ldots1$, that corresponds to the core region. The third-order polynomial $P(\xi)$ has three roots ξ_1, ξ_2, ξ_3 altogether. Only two of these roots are allowed in the core interval $\xi = 0\ldots1$. In order to meet this condition, the coefficients of the polynomial can only assume values within certain limits.

First we make sure that the index of the fourth-order profile decays monotonically from $r = 0$ to $r = a$ by requiring

$$1 \geq d \geq -1. \tag{5.208}$$

Under this condition, there are two, and only two zero crossings of $P(\xi)$ within $\xi = 0\ldots1$, if

$$0 < \delta/\Delta < 1 + (l/V)^2 \tag{5.209}$$

and

$$\left(\frac{l}{V}\right)^2 < \frac{2(1-d)^3}{27d^2}\left\{\left[1 + \frac{3d\delta}{(1-d)^2\Delta}\right]^{3/2} - \frac{9d\delta}{2(1-d)^2\Delta} - 1\right\}. \tag{5.210}$$

For the parabolic profile, with $d = 0$, the latter condition reduces to

$$l/V < \delta/(2\Delta), \tag{5.211}$$

while for the power-law profile with $\alpha = 4$, and hence $d = 1$, it has the form

$$l/V < \sqrt{2}[\delta/(3\Delta)]^{3/4}. \tag{5.212}$$

The fourth-order profile that tapers smoothly out into the cladding has $d = -1$ and the condition (5.210) results in

$$\left(\frac{l}{V}\right)^2 < \frac{16}{27}\left[(1 - \frac{3\delta}{4\Delta})^{3/2} + \frac{9\delta}{8\Delta} - 1\right]. \tag{5.213}$$

The areas in the plane of δ/Δ versus l/V to which these conditions for just two turning points limit δ/Δ and l/V values are illustrated in Fig.5.21 for the profiles with $d = 0$ and

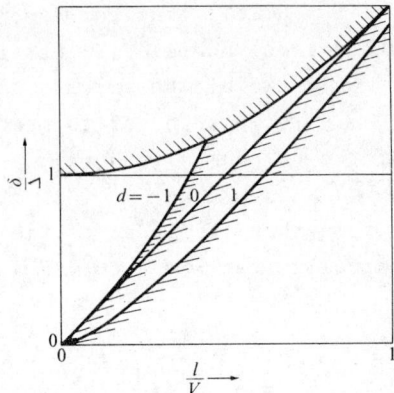

Fig.5.21. Limits of modes in the plane δ/Δ versus l/V for the three index profiles of Fig.5.20.

$d = \pm 1$. The guided modes have their cut-off at $\delta = \Delta$. Hence only those parts of the different regions in Fig.5.21 below $\delta/\Delta = 1$ actually represent guided modes. The lobes above $\delta/\Delta = 1$ contain the leaky modes which have only two turning points within the core, but a third turning point in the cladding. At this third turning point, the cladding fields turn from their evanescent character to a propagating character and radiate energy that tunnels through the evanescent field region of the cladding. When we determine the total delay spread, we will disregard these leaky modes and consider only delay differences between guided modes.

The cubic polynomial $P(\xi)$ in the integrals of the delay formula (5.206) renders them of the Jacobian or elliptic type which, in general, cannot be expressed in terms of elementary functions. For their evaluation we let the profile dispersion be zero. The relative difference in delay with respect to the delay per fibre length τ_1 of modes very far above cut-off can then be written as

$$\frac{\Delta\tau}{\tau_1} = \frac{n_1}{N} - 1 - \frac{n_1}{3N}\left\{2\Delta(1-d)\left[(\xi_3-\xi_1)\frac{E(\kappa)}{K(\kappa)} + \xi_1\right] + 2\delta\right\} \quad (5.214)$$

where $K(\kappa)$ and $E(\kappa)$ are complete elliptic integrals of the first and second kind, respectively, with their moduli according to

$$\kappa^2 = (\xi_3-\xi_2)/(\xi_3-\xi_1) \ . \qquad (5.215)$$

$\xi_{1,2,3}$ designate the three roots of the cubic polynomial $P(\xi)$ and are given by

$$\xi_1 = -2\rho \cos(\varphi/3) - (1-d)/(3d)$$

$$\xi_2 = 2\rho \cos[(\pi+\varphi)/3] - (1-d)/(3d) \qquad (5.216)$$

$$\xi_3 = 2\rho \cos[(\pi-\varphi)/3] - (1-d)/(3d)$$

with the following abbreviations

$$\rho = \frac{\text{sgn}(\nu)}{|d/(1-d)|} \left[\frac{1}{3} \left| \frac{1}{3} + \frac{d}{(1-d)^2} \frac{\delta}{\Delta} \right| \right]^{1/2}$$

$$\varphi = \text{arc} \cos(\nu/\rho^3) \qquad (5.217)$$

$$\nu = \frac{(1-d)^3}{d^3} \left[\frac{1}{27} + \frac{d}{6(1-d)^2} \frac{\delta}{\Delta} + \frac{d^2 \iota^2}{2(1-d)^3 v^2} \right] \ .$$

As long as the fourth-order coefficient d of the profile is small in absolute value or positive, the roots ξ_2 and ξ_3 of $P(\xi)$ occur in the core interval within $\xi = 0$ and 1. The third root ξ_1 is under these conditions either at negative ξ-values or above unity. If, however, d assumes large enough negative values, provided $d > -1$ of course, the root ξ_1 moves into the interval between $\xi = 0$ and 1, while at the same time ξ_3 shifts out of this interval to values $\xi > 1$. In this case ξ_3 and ξ_1 in equations (5.214) and (5.215) must formally be exchanged with each other.

For $d \ll 1$, the fourth-order profile deviates only slightly from the parabolic profile; in this case the relative delay difference is approximated by

$$\frac{\Delta\tau}{\tau_1} = \frac{\Delta^2}{2} \left[(\frac{\delta}{\Delta})^2 (1 + \frac{3}{4} \frac{d}{\Delta}) - \frac{\iota^2}{v^2} \frac{d}{\Delta} \right]. \qquad (5.218)$$

For the parabolic profile with $d = 0$, this expression reduces to

$$\Delta\tau/\tau_1 = \delta^2/2 , \qquad (5.219)$$

which agrees with the result from equation (5.182) for the power-law profile in the case of $\alpha = 2$ and $y = 0$.

We examine the delay differences according to equation (5.218), first for positive but still small values of the coefficient d. For $d > 0$, the modes near cut-off with $\delta = \Delta$, and of lowest circumferential order with $l = 0$, differ most in delay from the lowest-order modes farthest above cut-off with $\delta = 0$ and $l = 0$. For $d > 0$, therefore, the total delay spread from equation (5.218) is given by

$$\frac{\Delta\tau}{\tau_1} = \frac{\Delta^2}{2} (1 + \frac{3}{4} \frac{d}{\Delta}) . \qquad (5.220)$$

As soon as d turns negative, however, those modes near cut-off which have the highest circumferential order $l = V/2$, according to equation (5.211) and Fig.5.21, suffer most delay. The total delay spread then becomes

$$\frac{\Delta\tau}{\tau_1} = \frac{\Delta^2}{2} [1 + \frac{d}{2\Delta}] . \qquad (5.221)$$

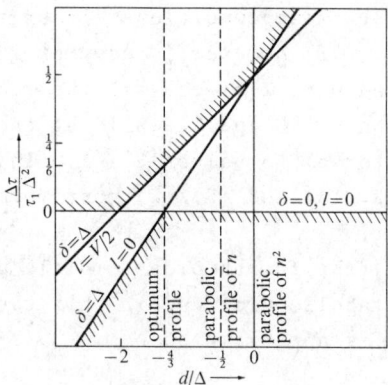

Fig.5.22. Relative delay difference and total relative delay spread in near-parabolic fourth-order profiles.

Fig.5.22 shows the two straight lines which represent equation (5.220) and equation (5.221); the shading on these lines and on the abscissa marks the boundaries for the delay differences of guided modes.

Equation (5.221) gives the total delay spread only for $0 \geq d \geq -4\Delta/3$. At $d = -4\Delta/3$, the straight line according to equation (5.220) crosses the abscissa and for $d < -4\Delta/3$, the modes at cut-off and with $l = 0$ travel faster than the lowest-order modes. In this range of d from $d = -4\Delta/3$ to $d = -2\Delta$ the largest delay difference occurs between modes near cut-off with $l = 0$ and with the highest circumferential order $l = V/2$. The total delay spread for $-4\Delta/3 > d > -2\Delta$ amounts to

$$\frac{\Delta\tau}{\tau_1} = -\Delta d/8 \; . \tag{5.222}$$

At $d = -2\Delta$, the straight line according to equation (5.221) crosses the abscissa, and also for $d < -2\Delta$ the modes with $\delta = \Delta$ and $l = V/2$ travel faster than the lowest-order modes. The largest delay difference now occurs again between the lowest-order modes and those modes near cut-off with $l = 0$. The total delay spread is again given by equation (5.220), or rather by the absolute value of that delay difference.

When we trace the total delay spread as a function of d over these four regions, we find its smallest value

$$\Delta\tau/\tau_1 = \Delta^2/6 \tag{5.223}$$

at $d_0 = -4\Delta/3$. This value of d gives the optimum fourth-order profile with respect to a minimization of the total delay spread. However, compared to the optimum power-law profile which has $\Delta\tau/\tau = \Delta^2/8$, it is slightly inferior.

To maintain the minimum total delay spread according to equation (5.223), or at least stay close to it, the profile must be manufactured to rather tight tolerances. A deviation of the fourth-order coefficient d from its optimum value $d_0 = -4\Delta/3$ of only

$$d - d_0 = (2/3)\Delta \tag{5.224}$$

already doubles the delay spread to $\Delta\tau = \tau_1\Delta^2/3$. For the index distribution

$$n = n_1 \left[1 - \Delta(1-d)\left(\frac{r}{a}\right)^2 - \Delta\left(\frac{\Delta}{2} + d\right)\left(\frac{r}{a}\right)^4\right] \qquad (5.225)$$

of the near-parabolic fourth-order profile, such a change in the fourth-order term causes the index to deviate from its specified distribution by

$$\Delta n = \Delta(d-d_0)n_1[(r/a)^2 - (r/a)^4] \: . \qquad (5.226)$$

The maximum index error occurs at

$$r/a = 1/\sqrt{2} \qquad (5.227)$$

and amounts to

$$\Delta n_{max} = n_1 \Delta(d-d_0)/4 \: . \qquad (5.228)$$

In the case of the deviation in equation (5.224), this maximum index error is only

$$\Delta n_{max} = n_1 \Delta^2/6 \qquad (5.229)$$

and is hence of the same order of magnitude as the maximum index error that doubles the total delay spread of the optimum power-law profile.

Equation (5.218) and the various formulae for total delay spread that follow from it in the different regions of d hold only for values of $d \ll 1$. When d is larger and the fourth-order profile deviates substantially from a parabolic distribution, the delay differences can only be obtained from a numerical evaluation, Fig.5.23 shows the total delay difference for $\Delta = 0 \cdot 016$ as a function of d. The broken line in this diagram represents equation (5.220) and approximates the actual delay spread over substantial ranges of d quite well. For $d = -1$, however, when the profile tapers into the cladding index without any break point, the total delay spread amounts to

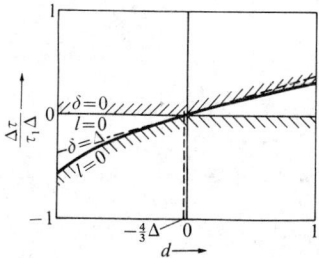

Fig.5.23. Relative delay difference and total relative delay spread in fourth-order profiles with $\Delta = 0 \cdot 016$ (Timmermann 1975).

$$\Delta\tau/\tau_1 = 0 \cdot 56 \, \Delta$$

and is markedly larger than the approximate value from equation (5.220). For such a profile with the smooth transition from core to cladding, the total delay spread is not even reduced to half of the delay spread in a homogeneous core with the same relative index difference Δ. On the other hand, we note from Fig.5.23 that in order to reduce the total relative delay spread from its value $\Delta\tau/\tau_1 = \Delta$ in the homogeneous core to $\Delta\tau/\tau_1 = \Delta/10$, the fourth-order coefficient d needs only be smaller than $d < 0 \cdot 26$. Thus a fourth-order profile which still deviates substantially from the parabolic profile already affords a remarkable reduction in delay spread.

So far we have examined delay differences only for two classes of index profiles. The power-law profiles of equation (5.165) with $d = 0$ have the minimum delay spread in equation (5.191) when their exponent is chosen according to equation (5.189). The fourth-order profiles of equation (5.202) have the minimum delay spread in equation (5.223) when, without profile dispersion, their fourth-order term has the coefficient $d = -4\Delta/3$. The question still remains open whether there is in absolute terms an optimum profile with respect to delay equalization, and to what residual amount it reduces the total delay spread. To answer this question, we refer to a computer-aided synthesis procedure that has been developed by Okamoto and Okoshi (1977) to find this optimum index profile. Starting with a profile function in the form of the even-order finite polynomial

$$g(r) = \sum_{\mu=1}^{m} d_\mu \left(\frac{r}{a}\right)^{2\mu}, \qquad (5.230)$$

Okoshi and Okamoto use a variational method to determine the phase constants of all M guided modes. They evaluate their delays τ_ν according to equation (5.152) and introduce their mean square deviation

$$\sigma^2 = \frac{1}{M} \sum_{\nu=1}^{M} \left[\tau_\nu^2 - \left(\frac{1}{M} \sum_{\nu=1}^{M} \tau_\nu\right)^2\right]. \qquad (5.231)$$

With this delay deviation they define

$$Q = \left(\int_{V_{cM}}^{V_{cM+1}} \sigma^2(V)\, dV\right) / \int_{V_{cM}}^{V_{cM+1}} dV \qquad (5.232)$$

as a measure of the total delay spread. The limits of integration V_{cM} and V_{cM+1} in this expression are cut-off values for the fibre parameter for the Mth mode and the $(M+1)$th mode, respectively. For large numbers M of guided modes, V_{cM} and V_{cM+1} are close together. They even coincide when the Mth mode is degenerate with the $(M+1)$th mode; under these conditions Q in equation (5.232) represents the delay deviation of equation (5.231) directly. Otherwise it constitutes an average of this delay deviation.

For a given number M of guided modes in a fibre of a given relative index difference Δ between core and cladding, there is a certain amount of profile dispersion y according to equation (5.180), and at a specific wavelength λ the average delay variance from equation (5.232) is a function only of the coefficients d_μ of the profile polynomial equation (5.230). By solving the system of equations

$$\frac{\partial Q}{\partial d_\mu} = 0 \qquad \mu = 1, 2, \ldots m \qquad (5.233)$$

for the d_μ, we obtain the polynomial profile that minimizes the delay spread. In order to cut down the amount of computer time, Okamoto and Okoshi (1977) limit the number of guided modes to $M = 34$ and the number of terms in the profile polynomial to $m = 5$. To compute the mode dispersion, they choose the following three values of fibre parameter:

$$V = 15 \cdot 69 , \quad 15 \cdot 725 , \quad 15 \cdot 76 ,$$

which are between the cut-offs of the two LP_{03}- and the next higher-order modes. They assume a profile dispersion according to $y = 0 \cdot 3$ and obtain for the optimum profile the following coefficients:

$$d_1 = 1 \cdot 1847 \qquad d_4 = 13 \cdot 418$$
$$d_2 = 3 \cdot 3960 \qquad d_5 = -7 \cdot 503 ,$$
$$d_3 = -9 \cdot 0498$$

Fig.5.24 shows the index distribution of this optimum profile. It has an annular core region next to the core-

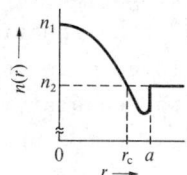

Fig.5.24. Optimum index distribution of polynomial profile (Okamoto and Okoshi 1977).

cladding boundary where the index decreases below the cladding index, very much as in a W-profile. Due to this index depression, the modes close to cut-off differ much less in delay from the lowest-order modes than they would without this depression. Any modes that have their second turning point close or even beyond the radius r_c in Fig.5.24 suffer frustration of their total reflection and have a third turning point in the cladding region. They are leaky modes and their delay difference does not contribute to the delay spread or delay deviation. In the core region between $0 \leq r \leq 0 \cdot 9 \, a$, the optimum index profile of Fig.5.24 can be approximated by a power-law profile according to

$$n^2(r) = n_1^2 [1 - 4\cdot 00 \ \Delta (r/a)^{2\cdot 27}] \ .$$

Without profile dispersion for $y = 0$, the optimum polynomial profile corresponding to Fig.5.24 would be approximated by

$$n^2(r) = n_1^2 [1 - 4\cdot 04\Delta \ (r/a)^{1\cdot 97}] \ .$$

Both these distributions are quite close to the parabolic profile with $\alpha = 2 + y$. The numerical example with $M = 34$, for which the optimum index distribution in Fig.5.24 was synthesized, reduces the total delay spread to 30 ps km^{-1} from 30 ns km^{-1} in the homogeneous core with the same values for M, Δ, and λ. This reduction in delay spread by a factor of 10^3 does not appear any more effective than in the optimum power-law and optimum fourth-order profile. However, those optimizations apply only to much larger mode numbers. For the relatively small number $M = 34$, the optimum power-law profile as well as the optimum fourth-order profile would have a larger delay spread than equations (5.191) and (5.223), respectively, due to the large mode delay near cut-off. These profiles would also need the annular region of depressed index as in Fig.5.24 to minimize the delay differences near cut-off.

5.7. IMPULSE RESPONSE

Signals which modulate optical carriers for transmission through fibres suffer distortion during transmission. This signal distortion is caused by the dispersion of the fibre material, as well as the dispersion of the fibre modes in which the signal propagates. If the signal travels in more than one mode of a multimode fibre, the delay differences between these modes cause additional distortion. To assess and compare the signal distortion by material and mode dispersion, and by delay differences, we need a specific time function as a standard signal at the fibre input.

For the uniform plane wave in section 1.3, we took the modulation of its amplitude by the Gaussian time function in equation (1.65) as a standard signal and with it examined the dispersion in homogeneous materials. If this Gaussian pulse has a 1/e-width t_0 of its intensity, the spectrum of the optical carrier is broadened, due to this modulation, to a line width which is given by equation (1.68) and amounts to $\Delta f = 2/(\pi\, t_0)$, or to $\Delta f/f = \Delta\lambda/\lambda = 2\lambda/(\pi c\, t_0)$ in relative terms. Even for pulses as short as $t_0 = 100$ ps and at light wavelengths near $\lambda = 1$ μm, the modulation broadens the carrier spectrum only to $\Delta\lambda = 0 \cdot 212$ Å. Light sources for optical communication, even those which are nominally coherent such as lasers, usually have a much wider emission spectrum. Their coherence is always more or less limited and their coherence time t_k in equation (1.82) will be so much smaller than the pulse duration that under practical conditions we can neglect all dispersion effects that are caused by the modulation broadening of the signal spectrum. We only need to consider the dispersion effects due to the finite line width of the partially coherent source. We can then even go to the limit of $t_0 \to 0$, but still neglect any broadening of the signal spectrum due to this impulse modulation. The standard time function that we now use as input signal for the analysis of signal distortion is the impulse function

$$P(t) = E\, \delta(t) \qquad (5.234)$$

with

$$\delta(t) = 0 \qquad \text{for} \quad t \neq 0$$

but

$$\int \delta(t)\, dt = 1.$$

With this impulse function, we violate the initial assumption that $t_0 \gg t_k$, but need not be concerned about it, because the actual signal pulses which we simulate with this impulse all still have $t_0 \gg t_k$. We can also argue that we decompose the actual time signal into impulses of the optical carrier and that the errors which we make when we neglect the infinite spectral width of the impulse modulated carrier will all cancel in superimposing the individual impulse responses to the overall signal response.

For the optical signal generator, we assumed an emission spectrum in section 1.3 that, according to equation (1.81), has the spectral distribution of a Gaussian function. Without specifying any such spectral distribution now, we let $g(f)$ with $f = \omega/2\pi$ be the line shape of the signal generator which we assume to be normalized according to

$$\int_0^\infty g(f) \, df = \frac{1}{2\pi} \int_0^\infty g(\omega) \, d\omega = 1 \ . \qquad (5.235)$$

With this normalized line shape of the source, its centre frequency follows as the weighted average from

$$f_c = \int_0^\infty f g(f) \, df \ . \qquad (5.236)$$

As a universal measure for the width of its spectral distribution, we introduce the root-mean-square width according to

$$B_{r.m.s.} = [\int_0^\infty (f-f_c)^2 \, g(f) \, df]^{1/2} \ . \qquad (5.237)$$

We assume now that the optical signal generator with its spectral emission $g(f)$ launches an impulse of energy $E_\nu g(f) df$ in the infinitesimal spectral range df into the mode ν of a fibre at time $t = 0$. This spectral component will suffer attenuation according to $\exp[-\alpha_\nu(f) z]$ and be delayed by $\tau_\nu(f) z$. It therefore contributes with

$$E_\nu \, g(f) \, df \, \delta[t - \tau_\nu(f) \, z] \, \exp[-\alpha_\nu(f) \, z] \qquad (5.238)$$

to the power flow along the fibre. Any interaction between modes is neglected here so that our considerations apply only to perfect fibres which do not depart from their straight cylindrical geometry.

When all components of the continuous emission spectrum are launched simultaneously with impulses at $t = 0$ into M modes, the power flow along the fibre shows the following impulse response

$$P(t,z) = \sum_{\nu=1}^{M} E_\nu \exp(-\alpha_\nu z) \int_0^\infty g(f) \, \delta[t-\tau_\nu(f)z] \, df. \quad (5.239)$$

We assume in this equation that each mode is excited with the same spectral distribution $g(f)$ of the source so that its excitation coefficient E_ν remains independent of frequency. We also neglect in equation (5.239) any dependence of the mode attenuation α_ν on frequency. In a perfect fibre without mode interaction, this frequency dependence of α_ν arises from the absorption spectra of the fibre material as well as from its bulk Rayleigh scattering. The latter loss contribution depends as f^4 on frequency according to equation (1.35). But these effects change α_ν so little in the narrow spectral ranges of optical signal sources that it may safely be assumed independent of frequency for all frequencies at which $g(f)$ contributes to the integral in equation (5.239).

Each spectral component $g(f)df$ contributes in each mode with the impulse (5.238) to the impulse response. When we integrate all spectral components in one particular mode, we obtain the impulse response from this mode alone. For a Gaussian line spectrum $g(f)$ according to equation (1.81), the impulse response from just one mode is a Gaussian time function which according to equation (1.97) has the 1/e-duration

$$t_{1\nu} = \frac{2}{\pi} \frac{d\tau_\nu}{df} \frac{z}{t_k} ; \quad (5.240)$$

and its peak passes at a distance z from the front end of the fibre at approximately the time

$$t_{0\nu} = \tau_\nu(f_c)z . \quad (5.241)$$

In the total power to which a photo-detector would respond, the Gaussian time functions from all the different modes with their respective peak amplitudes, their pulse widths $t_{1\nu}$, and their delays $t_{0\nu}$ superimpose to form the total impulse response.

Before we treat this very general case, let us first consider the signal distortion from delay differences alone and see for this simplified example what quantities we need for the characterization of impulse responses. We assume for this purpose that only a very narrow spectral line $g(f)$ with the small line width Δf excites all modes simultaneously with

$$E_\nu \, g(f) \, \Delta f \, \delta(t) \, ,$$

but let this line be wide enough, so that due to the Gaussian responses in each mode, the contributions from all modes, when superimposing in the photo-detector, will fuse in time, and that instead of the discrete sequence of impulses

$$\sum_\nu E_\nu \exp(-\alpha_\nu z) \, \delta(t-\tau_\nu z) \, g(f) \, \Delta f \, ,$$

we receive a continuous function in time. Its magnitude depends on the energy E_ν with which we excite each mode, on the attenuation α_ν which each mode suffers, and on how many Gaussian pulses arrive per unit time. To discern most clearly the effects of different delays, let all modes be excited equally strongly and let them all suffer the same attenuation. The impulse response will then depend in time only on the number of modes $\Delta\nu$ that arrive per time interval Δt. In the limit of a very large number M of modes ν, we obtain $(d\nu/dt)/M$ as the relative impulse response under these very specific conditions. The analysis of delay differences for the power-law profile in the previous section provides the relations which allow us to evaluate this impulse response. From equations (5.173), (5.174), and (5.177), we relate the number of modes ν up to a certain phase constant β_ν to the total wave number M and to the relative difference δ between $n_1 k$ and β. We obtain for the ratio of ν to M:

$$\nu/M = (\delta/\Delta)^{(2/\alpha)+1} \, . \qquad (5.242)$$

The group delay τ_ν per fibre length depends on the difference δ according to equation (5.179). By solving this equation for δ and substituting into equation (5.242), we obtain the fractional mode number ν as a function of the total delay time $t_\nu = \tau_\nu L$. Differentiating this function with respect to t_ν yields the relative impulse response.

As long as the exponent α of the power-law profile differs enough from $2+y$, only the first-order in δ term in equation (5.179) is of any significance. Therefore in the case of $\alpha \neq 2+y$ we obtain

$$\frac{\tau_1 L}{M}\frac{d\nu}{dt} = \frac{\alpha+2}{\alpha}\left[\frac{\alpha+2}{(\alpha-2-y)\Delta}\right]^{(2/\alpha)+1}[t/(\tau_1 L)-1]^{2/\alpha}. \quad (5.243)$$

This impulse response starts at $t = \tau_1 L$ and for $\alpha > 2+y$ lasts till $t = \tau_1 L + [(\alpha-2-y)/(\alpha+2)]\Delta\tau_1 L$. If $\alpha < 2+y$, the higher-order modes travel faster than the low-order modes. The impulse response begins under these conditions at $t_1 = \tau_1 L - [(2+y-\alpha)/(\alpha+2)]\tau_1 L$ and lasts till $t_1 = \tau_1 L$. For the near-parabolic profile with $\alpha = 2+y$, the term of second order in δ in equation (5.179) determines delay differences and yields the following impulse response:

$$\frac{\tau_1 L}{M}\frac{d\nu}{dt} = \frac{4+y}{(2+y)\Delta^2}\left[\frac{2}{\Delta}(t/(\tau_1 L)-1)^{\frac{1}{2}}\right]^{-y/(2+y)}. \quad (5.244)$$

This response begins again at $t = \tau_1 L$, but lasts only till $t = \tau_1 L + (\Delta^2/2)\tau_1 L$. The smallest total delay spread is obtained for the optimum power-law profile according to equation (5.189). In this profile, all modes travel faster than those at cut-off and those of very low order. The fastest modes arrive at $t = \tau_1 L - (\Delta^2/8)\tau_1 L$, and the impulse response lasts till $t = \tau_1 L$. During this time modes of different mode numbers ν always arrive at the same time. From the solution of equation (5.182) for $\alpha = 2+y-(2+y/2)\Delta$ these modes with equal delay have phase differences according to

$$\delta = \Delta/2 \pm [2\Delta\tau/\tau_1 + \Delta^2/4]^{1/2}. \quad (5.245)$$

Both contribute simultaneously to the impulse response. They add up to

$$\frac{\tau_1 L}{M}\frac{d\nu}{dt} = \frac{2+\alpha}{\alpha\Delta^2}\left[(\frac{t}{\tau_1 L}-1)\frac{2}{\Delta^2}+\frac{1}{4}\right]^{-1/2}\left\{\left[\frac{1}{2}+\left((\frac{t}{\tau_1 L}-1)\frac{2}{\Delta^2}+\frac{1}{4}\right)^{\frac{1}{2}}\right]^{(2/(2+y))} +\right.$$

$$\left. + \left[\frac{1}{2}-\left((\frac{t}{\tau_1 L}-1)\frac{2}{\Delta^2}+\frac{1}{4}\right)^{\frac{1}{2}}\right]^{(2/(2+y))}\right\}. \quad (5.246)$$

Fig.5.25 shows several impulse responses for power-law index profiles without any profile dispersion ($y = 0$). Both the homogeneous core and the parabolic index profile display rectangular impulse response, but the parabolic index profile only does so for $y = 0$. Note that in both Fig.5.25(a) and Fig.5.25(b), the vertical and the horizontal scales differ by

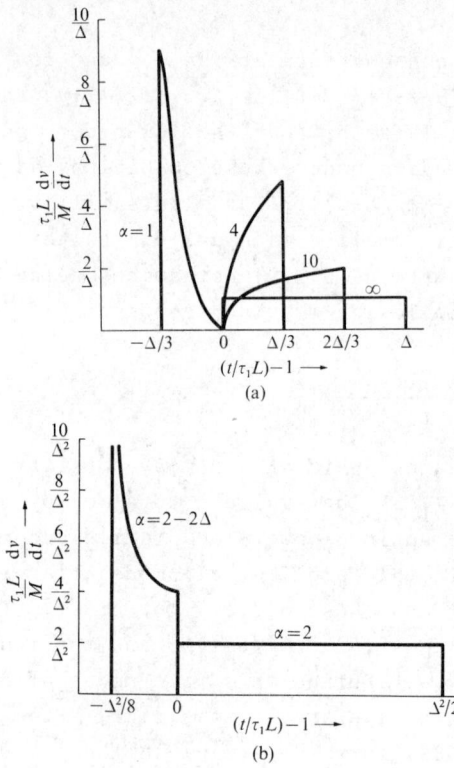

Fig.5.25. Impulse response of graded-index fibres with power-law profiles without profile dispersion for equal excitation and attenuation of all guided modes (Gloge and Marcatili 1973).

factors of $1/\Delta$ and Δ, respectively. All other power-law profiles change markedly in their impulse response from the rectangular pulses for $\alpha \to \infty$ and for $\alpha = 2$ with $y = 0$. The optimum profile, with its total pulse spread of only $\Delta\tau/\tau_1 = \Delta^2/8$, starts its impulse response with a peak that under present idealizations extends its amplitude beyond all limits. The modes which arrive at this time and form the peak turn with increasing mode number ν from decreasing to increasing delay. At the turning point in delay, infinitely many modes per unit time arrive in the limit of $d\nu/dt \to \infty$; they superimpose to the pulse peak of unlimited amplitude. We observe the same phenomenon whenever the $\tau(\nu)$-characteristic of the index profile has an extremum.

For signal responses of such widely varying shapes, the time between half-power points or between 1/e-power points is clearly not a good measure to characterize the pulse width. Any pulse with an infinite peak has zero pulse width when taken between half-power points. The total pulse duration as it follows from the total delay spread $\Delta\tau$ according to $t_1 = \Delta\tau L$ is likewise not a suitable quantity to describe the pulse width. If most modes have only small delay differences, but only a few modes depart much more from the average delay, the impulse response will last as long again as the total delay spread, but most of its energy concentrates in a much shorter time interval.

To obtain more suitable quantities for the characterization of signal response (Olshansky and Keck 1976), we consider the time moments $M_n(z)$ of the impulse response $P(t,z)$ as defined by

$$M_n(z) = \int_0^\infty t^n \, P(t,z) \, dt \, . \tag{5.247}$$

For $n = 0$, this definition of moments yields the total energy which the impulse response contains at z:

$$M_0(z) = E(z) = \int_0^\infty P(t,z) \, dt \, . \tag{5.248}$$

The first moment from equation (5.247), when normalized with respect to $E(z)$, yields the average delay of the impulse response

$$t_d(z) = M_1(z)/M_0(z) \, . \tag{5.249}$$

The second moment of the impulse response together with $M_0(z)$ and $t_d(z)$ defines a root-mean-square value of the pulse width according to

$$\sigma(z) = [M_2(z)/M_0(z) - t_d^2(z)]^{1/2} . \qquad (5.250)$$

Still higher moments of the impulse response may be combined to describe the power distribution in more detail. The first three moments suffice, however, to characterize pulse delay and broadening by equations (5.248), (5.249), and (5.250).

We understand best the meaning of these moments and the characteristic quantities which they define, when we evaluate them for the simple example of a multimode fibre in which an impulse from a coherent carrier excites all M guided modes simultaneously and with equal energy $E_\nu = E_0$. We assume all modes to suffer the same attenuation $\alpha_\nu = \alpha_0$ and, because of the coherent carrier, to have no material and mode dispersion. The impulse response is under these conditions given by

$$P(z,t) = E_0 \exp(-\alpha_0 z) \sum_{\nu=1}^{M} \delta(t-\tau_\nu z) \qquad (5.251)$$

and has the moments

$$M_n(z) = E_0 \exp(-\alpha_0 z) z^n \sum_{\nu=1}^{M} \tau_\nu^n . \qquad (5.252)$$

Its total energy follows from equation (5.248) as

$$M_0(z) = M E_0 \exp(-\alpha_0 z) . \qquad (5.253)$$

The impulse response suffers the average delay t_d from equation (5.249):

$$t_d(z) = \frac{z}{M} \sum_{\nu=1}^{M} \tau_\nu \qquad (5.254)$$

and has the root-mean-square width

$$\sigma(z) = \frac{z}{M} [M \Sigma \tau_\nu^2 - (\Sigma \tau_\nu)^2]^{1/2} \qquad (5.255)$$

from equation (5.250) which, under present conditions, equals the r.m.s. delay deviation from equation (5.231). If we express τ_ν according to

GRADED-INDEX FIBRES 543

$$\tau_\nu = \frac{1}{M} \sum_1^M \tau_\nu + \Delta \tau_\nu ,$$

using the average delay per fibre length and the deviation $\Delta \tau_\nu$ from this average, the r.m.s. pulse width of the impulse response in equation (5.255) reduces to

$$\sigma(z) = z \left[\frac{1}{M} \sum_1^M (\Delta \tau_\nu)^2\right]^{1/2} . \qquad (5.256)$$

It thus represents the standard deviation in delay from an average value according to equation (5.254). In optical transmission systems that utilize pulse-modulated light waves to carry the information, it is the r.m.s. width of these pulses at the receiver that determines the pulse spacing for a given error rate (Personick 1973). This r.m.s. pulse width therefore limits the pulse frequency and with it the rate at which the fibre can transmit information.

We return now to the impulse response in equation (5.239) of a multimode fibre in which all modes are excited simultaneously with the same spectral distribution $g(f)$ but with different energies E_ν. This general impulse response has moments $M_n(z)$ according to

$$M_n(z) = z^n \sum_{\nu=1}^M E_\nu \exp(-\alpha_\nu z) \int_0^\infty \tau_\nu^n(f) \, g(f) \, df . \qquad (5.257)$$

Material and mode dispersion manifest themselves in the dependence of mode delay τ_ν on frequency f. We account for these dispersion effects with sufficient accuracy, when we represent $\tau_\nu(f)$ by a Taylor series about the centre frequency f_c of the emission line, and consider only the first three terms of this expansion:

$$\tau_\nu(f) = \tau_\nu + \tau_\nu'(f-f_c) + \frac{1}{2} \tau_\nu''(f-f_c)^2 . \qquad (5.258)$$

τ_ν, τ_ν', τ_ν'' designate the delay per fibre length of mode ν and its first and second derivatives with respect to f, all evaluated at $f = f_c$. Upon substituting this Taylor series for $\tau_\nu(f)$ into equation (5.257), the nth moment of the impulse response is

$$M_n(z) = z^n \sum_{\nu=1}^{M} E_\nu \exp(-\alpha_\nu z) \int_0^\infty [\tau_\nu^n + n\,(f-f_c)\,\tau_\nu^{n-1}\,\tau_\nu' +$$

$$+ \frac{n}{2}(f-f_c)^2\,\tau_\nu^{n-1}\,\tau_\nu'' + \frac{n}{2}(n-1)(f-f_c)^2\,\tau_\nu^{n-2}(\tau_\nu')^2 + \dots]\,g(f)\,df.$$

(5.259)

The integrals $\int (f-f_c)^m g(f) df$ in this expression constitute the mth moment of the emission line with respect to f_c, and for $m = 1$ equal zero, while for $m = 2$ they represent the r.m.s. line width $B_{r.m.s.}$ according to equation (5.237). We can hence replace the integrals in equation (5.259) by these emission-line moments

$$M_n(z) = z^n \sum_{\nu=1}^{M} E_\nu \exp(-\alpha_\nu z) \left\{ \tau_\nu^n + \frac{1}{2} B_{r.m.s.}^2 [n\,\tau_\nu^{n-1}\tau_\nu'' + n(n-1)\tau_\nu^{n-2}(\tau_\nu')^2] \right\}$$

(5.260)

Any additional terms in this expansion are of third and higher orders in the relative line width $B_{r.m.s.}/f_c$ and need not be considered. To abbreviate the sums which appear in equation (5.260), we designate the average of a quantity Q_ν with respect to the energy distribution $E_\nu \exp(-\alpha_\nu z)$ by (Olshansky and Keck 1976)

$$\langle Q \rangle = \sum_\nu E_\nu \exp(-\alpha_\nu z)\,Q_\nu/M_0 .$$

(5.261)

We can then, with the first moment $M_1(z)$ from equation (5.260), write the average delay of the impulse response in equation (5.249) according to

$$t_d(z) = z[\langle \tau \rangle + \frac{1}{2} B_{r.m.s.}^2 \langle \tau'' \rangle] .$$

(5.262)

The r.m.s. pulse width from equation (5.250) appears in this notation as

$$\sigma(z) = z\{\langle \tau^2 \rangle - \langle \tau \rangle^2 + B_{r.m.s.}^2 [\langle \tau\tau'' \rangle - \langle \tau \rangle \langle \tau'' \rangle + \langle (\tau')^2 \rangle]\}^{1/2}.$$

(5.263)

If the signal travels in only one mode or if all modes which participate in signal transmission have the same delay characteristic, only the last term under the square root remains for the r.m.s. pulse width. The other terms arise from delay and dispersion differences between modes. We therefore call

$$\sigma_{IM} = z[\langle \tau^2 \rangle - \langle \tau \rangle^2 + B_{r.m.s.}^2 (\langle \tau\tau'' \rangle - \langle \tau \rangle\langle \tau'' \rangle)]^{1/2} \quad (5.264)$$

the root-mean-square pulse width from intermodal pulse broadening and

$$\sigma_M = z\, B_{r.m.s.} (\langle (\tau')^2 \rangle)^{\frac{1}{2}} \quad (5.265)$$

the root-mean-square pulse width from intramodal pulse broadening, i.e. pulse broadening within each mode. The total r.m.s. pulse width obtains from the sum of these two contributions squared as

$$\sigma = (\sigma_{IM}^2 + \sigma_M^2)^{\frac{1}{2}} . \quad (5.266)$$

The term in the intermodal pulse broadening σ_{IM} that depends on the mean square of the emission-line width remains small compared to the difference of the first two terms, even for emission lines as wide as those of luminescent diodes. The r.m.s. pulse width of intermodal pulse broadening is therefore adequately approximated by

$$\sigma_{IM} = z(\langle \tau^2 \rangle - \langle \tau \rangle^2)^{\frac{1}{2}} . \quad (5.267)$$

For equal excitation and attenuation of all modes, it reduces to the expression (5.256), which we derived to explain the definition of r.m.s. pulse width.

The r.m.s. pulse width of intramodal pulse broadening σ_M gives the average of material and mode dispersion of all modes. The contribution from each mode obtains from τ_ν' and, when τ_ν from equation (4.146) is written as

$$\tau_\nu = \frac{1}{c}\left[n_2' + (n_1' - n_2')\frac{\mathrm{d}(VB)}{\mathrm{d}V}\right] , \quad (5.268)$$

may, according to equation (4.270), be split into

$$\tau_\nu' = \tau_m' + \delta\tau_\nu' \quad (5.269)$$

with

$$\tau'_m = (2\pi/c^2)(dn'_2/dk) \qquad (5.270)$$

as the contribution from material dispersion and

$$\delta\tau_\nu = [(n'_1-n'_2)/cf]\, V\, d^2(VB)/dV^2 \qquad (5.271)$$

as the contribution from mode dispersion. In these equations n'_1 and n'_2 designate the group indices of the material in the core centre and the cladding of the fibre. They apply whenever core and cladding materials have similar dispersion characteristics.

When we take the square of equation (5.269) and average it over all modes with respect to their individual energies, we obtain the intramodal contribution to the r.m.s. pulse width in terms of material and mode dispersion

$$\sigma_M = zB_{r.m.s.}[(\tau'_m)^2 + 2\tau'_m\langle\delta\tau'\rangle + \langle(\delta\tau')^2\rangle]^{1/2}. \qquad (5.272)$$

It consists of one term from material dispersion alone and one term from mode dispersion alone, plus a mixed term from material as well as mode dispersion.

To study the combined effects of delay differences as well as material and mode dispersion on signal distortion in graded-index fibres, we take the power-law index profiles of Fig.5.13 as an example. The delay per fibre length of modes in these profiles follows from equation (5.179) and, when the relative phase difference δ is expressed by the mode numbers ν and M with equation (5.242), may be written as

$$\tau_\nu = \frac{n'_1}{c}[1 + C_1\Delta(\nu/M)^{\alpha/(\alpha+2)} + C_2\Delta^2(\nu/M)^{2\alpha/(\alpha+2)}], \qquad (5.273)$$

where we abbreviate

$$C_1 = \frac{\alpha-2-y}{\alpha+2}, \qquad C_2 = \frac{3\alpha-2-2y}{2(\alpha+2)}. \qquad (5.274)$$

The dispersion of individual modes follows from equation (5.273) when we differentiate it with respect to frequency and take account of the frequency dependence of M according to equation (5.174). We neglect all terms of second and higher order in Δ and obtain

$$\tau'_\nu = \frac{2\pi}{c^2}\frac{dn'_1}{dk} - \frac{n'_1}{c^2}\left(\frac{2\pi}{\alpha+2}\right)\left(\frac{\nu}{M}\right)^{\alpha/(\alpha+2)}\left[\frac{2C_1\alpha\Delta}{k} + \Delta\frac{dy}{dk} - (\alpha-2-y)\frac{d\Delta}{dk}\right].$$
(5.275)

The mode dispersion in this expression is usually negative; as long as the material dispersion is positive it compensates this material dispersion at least partially. Except for the vicinity of zero material dispersion in Fig.1.17, the first term in equation (5.275) is always so much larger than the second term, that mode dispersion remains of little significance in signal distortion. This holds in particular when for small delay differences the index profile is graded to a near-parabolic distribution and the factor C_1 becomes very small.

For the averages of the various delay and dispersion quantities, we assume all modes to be excited equally strongly, $E_\nu = E_0$, and that they all suffer the same attenuation $\alpha_\nu = \alpha_0$. The first of these two assumptions is met when we connect the multimode fibre to a multimode source. A laser that oscillates in a large number of modes of different transverse orders will very probably launch nearly the same power into most of the fibre modes. Incoherent sources, such as luminescent diodes, radiate so uniformly into the acceptance angle of multimode fibres that equal distribution of power among all fibre modes is nearly perfect. The second assumption holds for guided modes in fibres with the same bulk loss in core and cladding, and a wide enough cladding to prevent any power leakage into the surrounding jacket. Under these two assumptions all weighting factors in equation (5.261) are equal, and we obtain only direct averages.

For multimode fibres with a large number M of guided modes, the fractional mode number ν may be considered a continuous variable and the summation over ν replaced by integration. As a result of this integration, we obtain from the approximation (5.267) the following r.m.s. pulse width due to intermodal broadening of the impulse response:

$$\sigma_{IM} = \frac{z\, n_1' \Delta\alpha}{2c(\alpha+1)} \left(\frac{\alpha+2}{3\alpha+2}\right)^{1/2} \left[C_1^2 + \frac{4\Delta C_1 C_2(\alpha+1)}{2\alpha+1} + \frac{4\Delta^2 C_2^2 (2\alpha+2)^2}{(5\alpha+2)(3\alpha+2)}\right]^{1/2} .$$
(5.276)

The r.m.s. pulse width from intramodal pulse broadening also follows from an integration instead of the summation over ν and is given by

$$\sigma_M = z\, B_{r.m.s.} \frac{2\pi}{c^2} \left[\left(\frac{dn_1'}{dk}\right)^2 - 2\, \frac{n_1'}{k}\, \frac{dn_1'}{dk} \Delta\, C_1 \frac{\alpha}{\alpha+1} + (\Delta\, C_1\, n_1'/k)^2 \frac{4\alpha^2}{(\alpha+2)(3\alpha+2)}\right]^{1/2} . \quad (5.277)$$

Both σ_{IM} and σ_M depend critically on the exponent α of the power-law profile. An obvious choice to reduce the r.m.s. pulse width of the impulse response is

$$\alpha = 2 + y . \tag{5.278}$$

For this exponent the factor C_1 vanishes and, if $dy/dk = 0$, the intramodal contribution σ_M to the r.m.s. pulse width has only the material dispersion left. It amounts to

$$\sigma_M = z\, B_{r.m.s.} \frac{2\pi}{c^2} \frac{dn_1'}{dk} . \tag{5.279}$$

The intermodal contribution σ_{IM} reduces in this case to a small quantity of second order in Δ. For optical carriers with a high degree of coherence the r.m.s. bandwidth is so small that σ_{IM} might still dominate σ_M under these conditions. It will then be advisable to minimize σ_{IM} with a suitable choice of α. The minimum of σ_{IM} as a function of α occurs at

$$\alpha_{opt.} = 2 + y - \Delta\, \frac{(4+y)(3+y)}{5+2y} . \tag{5.280}$$

This optimum value of α with respect to r.m.s. pulse broadening from delay differences differs slightly from the $\alpha_{opt.}$ in equation (5.189), which minimizes the total delay spread in a power-law profile. The minimum value to which σ_{IM} reduces for $\alpha_{opt.}$ according to equation (5.280) is again of second order in Δ, just as σ_M has under these conditions a term of second order in Δ^2 from mode dispersion in addition to the contribution from material dispersion.

Fig.5.26 shows the total r.m.s. pulse width of the impulse response for power-law profiles without profile dispersion ($y = 0$) as a function of the exponent α. The three curves give the pulse width for typical r.m.s. bandwidths of the emission spectrum of three different classes of optical

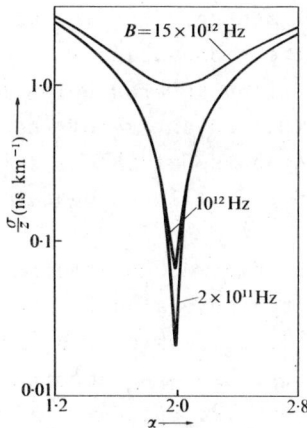

Fig.5.26. r.m.s. pulse width of the impulse response in a multimode quartz-glass fibre with power-law index profile; $\Delta = 0·01$, material dispersion coefficient $k(dn'/dk) = 1·8 \times 10^{-2}$ at $\lambda = 0·9$ µm, $y = 0$.

signal sources. Incoherent luminescent or light-emitting diodes typically have a 1/e-width of their nearly-Gaussian emission spectrum according to equation (1.82) of $B = 15 \times 10^{12}$ Hz, or $\Delta\lambda = 40$ nm at $\lambda = 900$ nm; this corresponds to an r.m.s. bandwidth of

$$B_{r.m.s.} = B/(2\sqrt{2}) \simeq 5 \times 10^{12} \text{ Hz} \qquad (5.281)$$

or $\Delta\lambda_{r.m.s.} = 13$ nm. Ordinary semiconductor injection lasers for continuous wave operation typically have $B = 10^{12}$ Hz or $\Delta\lambda \simeq 3$ nm at $\lambda = 900$ nm, corresponding in r.m.s. bandwidth to $B_{r.m.s.} = 3·5 \times 10^{11}$ Hz or $\Delta\lambda_{r.m.s.} \simeq 1$ nm. Semi-conductor lasers with high coherence can be reduced in their emission line width to $B = 2 \times 10^{11}$ Hz or $\Delta\lambda \simeq 0·5$ nm at $\lambda = 900$ nm, which corresponds to $B_{r.m.s.} = 7 \times 10^{10}$ Hz or $\Delta\lambda_{r.m.s.} = 0·26$ nm. The refractive index and material dispersion for the curves in Fig. 5.26 are taken for pure fused silica. Its dispersion coefficient follows from equation (1.102) or from Fig.1.17 and at $\lambda = 900$ nm amounts to $k(dn'/dk) = 1\cdot 8 \times 10^{-2}$; its refractive index is $n = 1\cdot 46$. The minima of the pulse broadening curves of Fig.5.26 represent essentially the r.m.s. pulse width due to material dispersion alone. When, for increasing coherence of the signal source, its line width decreases, this minimum becomes more pronounced; it goes to smaller values of residual r.m.s. pulse broadening from material dispersion. But to gain full advantage of the very deep and narrow minimum of r.m.s. pulse width for extremely coherent light, requires the index profile to be tailored to very tight tolerances.

If, in addition to homogeneous material dispersion, the glass mixture in the fibre core also has inhomogeneous material dispersion or profile dispersion, the curves in Fig.5.26 shift parallel to the abscissa by y according to equation (5.180). The minima in r.m.s. pulse width occur somewhat below $\alpha = 2 + y$. Except for this horizontal displacement, however, the curves maintain the same general shape as in Fig.5.26.

REFERENCES

Abramowitz, H. and Stegun, I.A. (1964). *Handbook of mathematical functions*. US Government Printing Office, Washington, DC.

Chan, K.B., Clarricoats, P.J.B., Dyott, R.B., Newns, G.R., and Savva, M.A. (1970). Propagation characteristics of an optical waveguide with a diffused core boundary. *Electron. Lett.* 6, 748-9.

Fleming, J.W. (1976). Material and mode dispersion in $GeO_2.B_2O_3.SiO_2$ glasses. *J.Amer.Ceram.Soc.* 59, 503-07.

Gloge, D. (1975). Propagation effects in optical fibres. *IEEE Trans. MTT* 23, 106-20.

────── and Marcatili, E.A.J. (1973). Multimode theory of graded-core fibres. *Bell Syst. tech. J.* 52, 1563-78.

Kirchhoff, H. (1973a). Wave propagation in gradient fibres with curvature. *Archiv Elektron. & Übertragungstech.* 27, 161-7.

Kirchoff, H. (1973b). Wave propagation along radially inhomogeneous glass fibres. *Archiv Elektron. & Übertragungstech.* 27, 13-18.

Okamoto, K. and Okoshi, T. (1977). Computer-aided synthesis of the optimum refractive-index profile for a multimode fibre. *IEEE Trans. MTT* 24, 213-21.

Okoshi, T. and Okamoto, K. (1974). Analysis of wave propagation in inhomogeneous optical fibres using a variational method. *IEEE Trans. MTT* 22, 938-45.

Olshansky, R. and Keck, D.B. (1976). Pulse broadening in graded-index optical fibres. *Appl. Opt.* 15, 483-91.

Payne, D.N. and Gambling, W.A. (1972). Low-loss liquid-core fibre waveguide. *Electr. Lett.* 8, 374-6.

Personick, S.D. (1973). Receiver design for digital fibre optic communication systems. *Bell Syst. tech. J.* 52, 843-905.

Petermann, K. (1975). The mode attenuation in general graded core multimode fibres. *Archiv Elektron. & Übertragungstech.* 29, 345-8.

Timmermann, C.C. (1974). The influence of deviations from the square law refractive index profile of gradient core fibres on mode dispersion. *Archiv Elektron. & Übertragungstech.* 28, 344-6.

Timmermann, C.C. (1975). Airy function approximation of graded index fibre modes. *Archiv Elektron. & Übertragungstech.* 29, 186-8.

Watson, G.N. (1922). *Theory of Bessel functions.* Cambridge University Press.

6
FIBRE IMPERFECTIONS

In previous chapters we assumed the cladded-core fibres to be perfectly straight and of circularly symmetric geometry with homogeneous material of constant refractive index in core, cladding, and jacket regions. Any deviations from the constant average values of refractive index were assumed to consist of random fluctuations correlated only over distances short compared to the wavelength of light. Such microscopic index fluctuations cause Rayleigh scattering of guided mode power mainly into the radiating field and could be accounted for as bulk loss by a contribution to the imaginary component α of the complex wave number of the material according to equation (1.35). The graded-index fibres were likewise assumed to be of perfectly straight and circularly symmetric geometry with their refractive index changing smoothly, and only with the distance from the fibre centre, but not in the circumferential direction or axial direction along the fibres. In such perfect fibres, all guided modes propagate independently from each other and suffer only the attenuation due to bulk loss from absorption and Rayleigh scattering. The leaky modes of such perfect fibres likewise propagate independently from each other and, in addition to bulk absorption and Rayleigh scattering, they are attentuated by the frustration of their total reflection and the associated radiation of some of their power.

Actual fibres are, however, never quite perfect in this sense. Aside from the microscopic fluctuations due to the random molecular structure of the material, the index also deviates macroscopically from its nominally constant or circularly symmetric distribution. It will then not only depend on the radial coordinate of the fibre but will also change circumferentially as well as in the longitudinal direction. The core cross-section, in particular of a cladded-core fibre, will also be not quite round and uniform in the axial direction. It will rather show some ellipticity or higher-order deviation from roundness, and its diameter as well as any deviation from roundness will change along the

fibre. In addition, a fibre will never be perfectly straight.
Pulling a glass fibre from a preform or out of a melt results
in a still fairly straight structure, but further processing
of this fibre such as coating and cabling introduces deviations of the fibre axis from a straight line. Fibres or fibre
cables can also not be installed along straight lines, but
need to be bent for accommodation in ducts and chutes.

All these deviations of the fibre from the perfect geometry and perfect index distribution lead to coupling among
modes. They no longer propagate independently from each other,
but exchange power. If the fibre guides only its fundamental
mode, fibre imperfections couple it to leaky modes and the
radiation field. The fundamental mode loses power due to
radiation and thus suffers additional attenuation. If the
fibre guides many modes, some or all modes exchange power and,
in addition, lose power to leaky modes and the radiation field.
The interaction between guided modes changes the overall dispersion characteristics of the multimode fibre, while the
radiation of power adds to its transmission loss. Therefore,
for single as well as multimode fibres, imperfections and the
mode interaction which they cause must be kept within certain
limits in order not to degrade their transmission characteristics too much. To determine tolerances for index deviations,
cross-sectional distortions, and fibre curvature, we need to
analyse mode coupling and its effects on transmission loss and
distortion.

6.1. COUPLED WAVE EQUATIONS

We restrict the analysis to weakly guiding fibres with
only small index differences between their cross-sectional
regions. Any index deviation from the nominal distribution
will then likewise amount to only a small index difference,
and imperfections of the cross-sectional geometry can also
be accounted for by small index deviations. We represent
the index distribution of the imperfect fibre by

$$n^2 = n_0^2(r) + n_p^2(r,\varphi,z) \;, \tag{6.1}$$

where $n_0(r)$ designates the circularly symmetric and longitudinally homogeneous index distribution of the perfect fibre and $n_p^2(r,\varphi,z)$ a perturbation of the nominal index distribution. This perturbation can describe any index deviation or deformation of the fibre cross-section. It can also account for curvature of the guide axis as we shall explain later.

According to our initial assumption, both n_0^2 and n_p^2 in equation (6.1), and hence also n^2, change only little over the fibre cross-section and along the fibre. The general vector wave equation (5.52) for the electric field may under these conditions be separated into its Cartesian components. Its transverse x-component appears, for example, as

$$\nabla^2 E_x + n^2 k^2 E_x = 0 \ . \tag{6.2}$$

The three component equations are independent of each other. Their solutions will be linearly and uniformly polarized just like the LP-modes of the weakly guiding perfect structure. No interaction between LP-modes of different polarization can be observed in this approximation.

For a general solution of equation (6.2) with n^2 according to equation (6.1), we represent the x-polarized electric field as a combination of all LP_{lp}-modes with this particular polarization

$$E_x = \sum_n V_n(z) \, E_n(r,\varphi) \ . \tag{6.3}$$

In this series representation, we have $E_n(r,\varphi)$ as the transverse electric field distribution of the nth mode and $V_n(z)$ as an amplitude factor for this mode, which in the perfect fibre would depend on z according to $\exp(\mp \gamma_n z)$.
$\gamma_n = \alpha_n + j\beta_n$ is the propagation constant of this mode where α_n and β_n are its attenuation and phase constants, respectively. The minus sign in the exponential dependence describes a forward-travelling mode, and the plus sign a backward-travelling mode. In the imperfect fibre, the index distribution depends on z, and $V_n(z)$ will then also differ from the simple exponential z-dependence of independent mode propagation.

The series representation (6.3) of the electric field in

terms of modal fields is only complete if it includes not only the guided and leaky modes, but also all radiation modes which the leaky modes do not properly account for. From our analysis of perfect fibres, we have only guided and leaky modes available for this field representation. We can, nevertheless, take equation (6.3) to include all radiation modes and thus be complete. Actual fibres have such a large cladding diameter and consequently so many cladding modes that a representation in terms of all core and cladding modes is complete enough for all practical purposes and no radiation modes are needed.

The transverse field distribution $E_n(r,\varphi)$ of a mode solves the transverse wave equation

$$\nabla_t^2 E_n + (n_0^2 k^2 + \gamma_n^2) E_n = 0 , \qquad (6.4)$$

which follows from equation (6.2) for $n^2 = n_0^2$ and a z-dependance according to $\exp(\pm \gamma_n z)$. As a solution of this wave equation, and since E_n decays exponentially for large r in the evanescent field region, it obeys an orthogonality condition which we now derive for use in our field expansion. By multiplying equation (6.4) with another of its solutions E_m and subtracting from this product equation (6.4) written for E_m and multiplied by E_n, we obtain

$$E_m \nabla_t^2 E_n - E_n \nabla_t^2 E_m + (\gamma_n^2 - \gamma_m^2) E_m E_n = 0 . \qquad (6.5)$$

When we integrate this equation over a fibre cross-section, we can apply Green's theorem to the first two terms according to

$$\int_0^{2\pi} \int_0^r (E_m \nabla_t^2 E_n - E_n \nabla_t^2 E_m) r dr\, d\varphi = \int_0^{2\pi} (E_m \frac{\partial E_n}{\partial r} - E_n \frac{\partial E_m}{\partial r}) r d\varphi. \qquad (6.6)$$

For the limit r of integration extending to infinity, this expression vanishes because of the exponentially decaying evanescent fields. The integral over the last term in equation (6.5) then leads to the orthogonality condition

$$\iint E_m E_n\, r dr\, d\varphi = 0 , \qquad (6.7)$$

for any two modes with $\gamma_n \neq \gamma_m$. The power which such a mode transmits axially with its electric field $E_x = V_n E_n$ and its magnetic field written as

$$H_y = I_n E_n \qquad (6.8)$$

follows from

$$P_n = V_n I_n^* \iint E_n^2 \, r \mathrm{d}r \, \mathrm{d}\varphi. \qquad (6.9)$$

For

$$\iint E_n^2 \, r \mathrm{d}r \, \mathrm{d}\varphi = 1, \qquad (6.10)$$

this power is simply given by the product

$$P_n = V_n I_n^*, \qquad (6.11)$$

just as if V_n and I_n were voltage and current on an ordinary transmission line. We therefore normalize the modal field according to equation (6.10) and obtain with equation (6.7)

$$\iint E_m E_n \, r \mathrm{d}r \, \mathrm{d}\varphi = \delta_{mn} \qquad (6.12)$$

as the orthonormality condition for fibre modes.

For a solution of the wave equation (6.2) of the imperfect fibre, we introduce the field expansion according to equation (6.3), multiply equation (6.2) by E_m, and integrate over the fibre cross-section with the limits of integration extending to infinity. With the wave equation (6.4) for modes of the perfect fibre and with the orthonormality condition (6.12) for these modes, the wave equation (6.2) converts to the following system of coupled ordinary differential equations:

$$\frac{\mathrm{d}^2 V_m}{\mathrm{d}z^2} - \gamma_m^2 V_m = - \sum_n V_n k^2 \iint n_p^2 E_n E_m \, r \mathrm{d}r \, \mathrm{d}\varphi. \qquad (6.13)$$

For an unperturbed index distribution of a perfect fibre, we have $n_p = 0$, and the coupled system reduces to independent second-order differential equations

$$\frac{d^2 V_m}{dz^2} - \gamma_m^2 V_m = 0 \qquad (6.14)$$

for the voltage coefficients V_m of each mode m. Their solutions

$$V_m = V_m^{(+)} \exp(-\gamma_m z) + V_m^{(-)} \exp(\gamma_m z) \qquad (6.15)$$

represent forward- and backward-travelling waves. If there are any imperfections, $n_p \neq 0$ and, according to the right-hand sides of equation (6.13), modes couple with each other. This coupling disturbs the exponential z-dependence of the uncoupled solution in equation (6.15).

The magnetic field component H_y of the nearly transverse electromagnetic field of weakly-guiding fibre modes relates to the electric field by

$$H_y = -(n_0/\gamma_m)(\varepsilon_0/\mu_0)^{\frac{1}{2}}(\partial E_x/\partial z) . \qquad (6.16)$$

This relation obtains as the appropriate approximation from Maxwell's equations, which for a forward- or backward-travelling wave reduces to $H_y = \pm n_0 (\varepsilon_0/\mu_0)^{\frac{1}{2}} E_x$. It leads to the following relation between voltage and current coefficients of the modal field expansion in equations (6.3) and (6.8):

$$\frac{dV_m}{dz} + \frac{\gamma_m}{n_0}(\mu_0/\varepsilon_0)^{\frac{1}{2}} I_m = 0. \qquad (6.17)$$

This set of equations corresponds to one of the differential equations for voltages and currents on an ordinary transmission line. A set of equations corresponding to the other transmission line equation is obtained from equation (6.13) when we differentiate equation (6.17) with respect to z and express $(d^2 V_m/dz^2)$ in equation (6.13) by dI_m/dz

$$\frac{dI_m}{dz} + \gamma_m n_0 (\varepsilon_0/\mu_0)^{\frac{1}{2}} V_m = 2n_0 (\varepsilon_0/\mu_0)^{\frac{1}{2}} \sum_n \kappa_{mn} V_n . \qquad (6.18)$$

Here we have used the abbreviation

$$\kappa_{mn} = \frac{k^2}{2\gamma_m} \iint n_p^2 E_m E_n \, r dr \, d\varphi \tag{6.19}$$

for the coupling coefficients from equation (6.13).

Better suited than the voltage and current coefficients of equivalent transmission lines are the amplitudes of forward- and backward-travelling waves for the analysis of wave propagation in optical fibres. With

$$V_m = [(\mu_0/\varepsilon_0)^{\frac{1}{2}}/n_0]^{1/2}(a_m+b_m)$$

and (6.20)

$$I_m = [n_0(\varepsilon_0/\mu_0)^{\frac{1}{2}}]^{1/2}(a_m-b_m) \, ,$$

we transform to a_m as the amplitude of the forward- and b_m of the backward-travelling wave. The power which a mode with these forward and backward amplitudes transmits is given by

$$P = \frac{1}{2}(V_m I_m^* + V_m^* I_m) = a_m^2 - b_m^2 \, . \tag{6.21}$$

By introducing a_m and b_m for V_m and I_m into the coupled line equations (6.17) and (6.18), they change to the following system of coupled wave equations

$$\begin{aligned}\frac{da_m}{dz} + \gamma_m a_m &= \sum_n \kappa_{mn}(a_n+b_n) \\ \frac{db_m}{dz} - \gamma_m b_m &= -\sum_n \kappa_{mn}(a_n+b_n) \, . \end{aligned} \tag{6.22}$$

Actually we need to consider only the first set of these two systems of equations because the second set for backward waves becomes identical to the first when we replace γ_m by $-\gamma_m$ on the left-hand side and in the coupling coefficient. Equation (6.22) may therefore be written as

$$\frac{da_m}{dz} + \gamma_m a_m = \sum_n \kappa_{mn} a_n \, , \tag{6.23}$$

where the summation in each of the equations extends now over all forward and backward coupled waves.

For a perfect fibre with $n_p = 0$ and all $\kappa_{mn} = 0$ as a consequence, the system reduces to independent equations for each of the forward- and backward- travelling waves. If the fibre has only small imperfections, the coupling remains weak so that

$$|\kappa_{mn}| \ll |\gamma_m| .$$

Under these conditions, a_m changes only gradually from its z-dependence without any coupling, and in

$$a_m = A_m(z)\exp(-\gamma_m z) \qquad (6.24)$$

$A_m(z)$ varies only slowly along the fibre. By transforming to $A_m(z)$ as the new wave amplitudes, the coupled wave equations (6.23) change to

$$\frac{dA_m}{dz} = \sum_n \kappa_{mn} A_n \exp\{(\gamma_m-\gamma_n)z\} . \qquad (6.25)$$

For interaction between two forward-travelling waves, the difference $\gamma_m - \gamma_n$ in their propagation constants appears in the exponential of the corresponding term in the right-hand side sum of equation (6.25). For interaction between a forward-travelling wave m and a backward-travelling wave n, on the other hand, we need to replace γ_n by $-\gamma_n$, so that the sum $\gamma_m + \gamma_n$ of their propagation constants appears in the exponential of the corresponding interaction term. To solve the equation system (6.25) by iteration, we let the A_n on the right-hand sides at first remain constant at their initial values $A_n(0)$. Integration along z then leads to

$$A_m(z) = A_m(0) + \sum_n A_n(0) \int_0^z \kappa_{mn} \exp\{(\gamma_m-\gamma_n)z'\}dz' . \qquad (6.26)$$

This first-order solution already indicates to what extent the amplitude $A_m(z)$ changes along the fibre when other modes n are launched with amplitude $A_n(0)$ and couple to it through imperfections. It shows the mode conversion, but does not yet show the effects of reconversion. If only A_m were launched and all other $A_n(0) = 0$, we obtain $A_m(z) = A_m(0)$ from this first-order solution. For a more accurate solution we perform another iteration by substituting the first-order solution for A_n into equation (6.25) and by integrating again along z to obtain the second-order solution:

$$A_m(z) = A_m(0) + \sum_n A_n(0) \int_0^z \kappa_{mn} \exp\{(\gamma_m - \gamma_n)z'\}dz' + \sum_n \sum_i A_i(0) \times$$

$$\times \int_0^z \kappa_{mn} \exp\{(\gamma_m - \gamma_n)z'\} \left[\int_0^{z'} \kappa_{ni} \exp\{(\gamma_n - \gamma_i)z''\}dz'' \right] dz' . \quad (6.27)$$

For most practical purposes, this second-order solution offers sufficient accuracy to study the effects of fibre imperfections. It shows, to first order at least, the effects of reconversion.

When with

$$A_n(0) = \delta_{mn} , \quad (6.28)$$

we launch only the mode m with unit amplitude and power, the amplitude of this mode changes in the second-order solution of the coupled wave equations according to

$$A_m(z) = 1 + \sum_n \int_0^z \kappa_{mn} \exp\{(\gamma_m - \gamma_n)z'\} \int_0^{z'} \kappa_{nm} \exp\{(\gamma_n - \gamma_m)z''\}dz'' \, dz' . \quad (6.29)$$

This expression shows directly the combined effects of mode conversion and reconversion due to wave coupling. All other modes are under these initial conditions excited according to

$$A_n(z) = \int_0^z \kappa_{nm} \exp\{(\gamma_n - \gamma_m)z'\} \, dz' . \quad (6.30)$$

To assess the effects of this wave coupling, we consider first guided modes in loss-less fibres. With $\gamma_n = j\beta_n$ and

$$\kappa_{mn} = -j \, c_{mn} , \quad (6.31)$$

where β_n and c_{mn} are now real quantities, the coupled wave equations

$$\frac{dA_m}{dz} = -j \sum_n c_{mn} A_n \exp\{j(\beta_m - \beta_n)z\} \qquad (6.32)$$

have for $A_n(0) = \delta_{1n}$ as initial conditions, the second-order solution

$$A_1(z) = 1 - \sum_n \int_0^z c_{1n}(x) \exp(-j\Delta\beta_n x) \int_0^x c_{n1}(y) \exp(j\Delta\beta_n y) \, dy \, dx$$

$$A_n(z) = -j \int_0^z c_{n1}(y) \exp(j\Delta\beta_n y) \, dy , \qquad (6.33)$$

where

$$\Delta\beta_n = \beta_n - \beta_1 . \qquad (6.34)$$

In order to evaluate the integrals in equation (6.33), we need to specify the character and distribution of fibre imperfections. Without being very specific about these imperfections, we can, however, draw a few general conclusions from equation (6.33). To this end, we develop the perturbation term n_p^2 in the index distribution of equation (6.1) into a Fourier series with respect to φ:

$$n_p^2(r,\varphi,z) = \sum_q [\varepsilon_q(r,z) \cos q\varphi + \varepsilon_q'(r,z) \sin q\varphi] . \qquad (6.35)$$

The Fourier coefficient ε_q and ε_q' in this expansion are each functions of r and z. With

$$\varepsilon_q(r,z) = R_q(r) \, f_q(z) ,$$

we describe index deviations of circumferential order q that for any cross-section of the fibre have the radial distribution $R_q(r)$ and change in magnitude along the fibre according to $f_q(z)$. To determine the coupling coefficients in equation (6.19), this perturbation multiplied by E_m and E_n must be integrated over the full fibre cross-section. The result for any two particular modes E_m and E_n can be written as

$$c_{mn} = C_{mn} f_q(z) , \qquad (6.36)$$

where C_{mn} differs from zero only for pairs of modes which differ by $|\Delta l| = q$ in their circumferential orders, and depends in its magnitude on the particular distribution R_q. But C_{mn} is independent of z.

The amplitude with which any coupled mode n is excited under these conditions by the incident mode 1 with $A_1(0) = 1$ follows from equation (6.33) for a fibre section of length L as:

$$A_n(L) = -j\, C_{n1} \int_0^L f(z)\, \exp(j\Delta\beta_n z)\, dz . \qquad (6.37)$$

Apart from the factor $-j\, C_{n1}$, it constitutes the limited Fourier transform

$$F(\Delta\beta_n) = \int_0^L f(z)\, \exp(j\Delta\beta_n z)\, dz \qquad (6.38)$$

of $f(z)$. To evaluate this transform, we decompose the axial distribution $f(z)$ of fibre imperfections into its spectral components according to

$$f(z) = \sum_p [F_p \cos(2\pi p\, z/L) + G_p \sin(2\pi p\, z/L)] . \qquad (6.39)$$

Any of these spectral components, for example F_p, contributes to $A_n(L)$ with

$$A_{np}(L) = \frac{C_{n1} F_p}{\Delta\beta_n} \frac{\{1-\exp(j\Delta\beta_n L)\}}{\{1-(2\pi p/(\Delta\beta_n L))^2\}} . \qquad (6.40)$$

Small imperfections with correspondingly small spectral components F_p of their axial distribution will contribute only little to mode conversion. If, however, the spatial frequency $\Omega_p = 2\pi p/L$ of a particular spectral component approaches the phase difference $\Delta\beta_n$ between the coupled modes, or even coincides with it, then the mode conversion from this component can become quite large.

For Ω_p close to $\Delta\beta_n$, the denominator in equation (6.40) is small and for

FIBRE IMPERFECTIONS 563

$$\Omega_p = 2\pi p/L = \Delta\beta_n \, , \qquad (6.41)$$

the mode conversion rises to

$$A_{np} = -j \, C_{n1} \, F_p \, L/2 \, . \qquad (6.42)$$

In this case, the beat wavelength $\Lambda_n = 2\pi/\Delta\beta_n$ of the two coupled modes equals the period length L/p of the particular spectral component of the imperfection. Contributions of mode conversion from each period of the periodic imperfection all add in phase to the coupled mode, and due to this constructive interference its amplitude grows linearly with length, while its power may even build up quadratically.

For long fibres with small imperfections, we need only to consider such spectral components which in their period length are very close to or even coincide with the beat wavelength of coupled modes. We then need only to sum these contributions either from equation (6.40), when $\Delta\beta_n$ differs from $2\pi p/L$, or directly from equation (6.42), when $\Delta\beta_n$ coincides with $2\pi p/L$.

6.2. RANDOM IMPERFECTIONS AND SINGLE-MODE LOSS

For imperfections in fibres, we will never know in advance exactly what their magnitude is and how they distribute along the fibre. Although for a given fibre in a cable installed stationary along a certain route these imperfections are fixed and do not change in time, it will hardly be possible to predict in advance the nature and extent of fibre imperfections. We will nevertheless have to design and install fibre cables that must fill certain specifications for their transmission characteristics, and therefore need to specify tolerances for their imperfections. Not knowing in detail what the distribution of imperfections along the fibre is, we assume it to be random and work with certain statistical quantities of this distribution. We will then, of course, at best be able to express only statistical quantities of transmission characteristics in terms of the statistics of imperfections.

For this purpose, we compare a large number K of fibres

with random imperfections which macroscopically are all identical. For the axial distribution $f_k(z)$ of one component of their imperfections in the fibre k, the ensemble average

$$\langle f(z)\, f(z+u) \rangle = \frac{1}{K} \sum_k f_k(z)\, f_k(z+u) \qquad (6.43)$$

indicates, to what extent the perturbation at z is correlated to the same perturbation at $z + u$. The auto-correlation function $\varphi(u)$ for any member of the ensemble is defined by

$$\langle f^2 \rangle\, \varphi(u) = \lim_{L \to \infty} \frac{1}{L} \int_0^L f(z)\, f(z+u)\, dz \; , \qquad (6.44)$$

where

$$\langle f^2 \rangle = \lim_{L \to \infty} \frac{1}{L} \int_0^L f^2(z)\, dz \qquad (6.45)$$

represents the mean square of $f(z)$. The Fourier transform of the auto-correlation function constitutes the power spectrum of the distribution

$$\phi(\Omega) = \langle f^2 \rangle \int_{u=-\infty}^{\infty} \varphi(u) \exp(-j\Omega u)\, du \; . \qquad (6.46)$$

The independent variable Ω in this spectral distribution represents the angular spatial frequency of the respective spectral component.

Let us now assume that $f(z)$ as a stochastic distribution is ergodic and stationary. Essentially, this means that its statistical properties are independent of the location along the fibre. For long fibres with imperfections of the same general nature over its full length, it seems quite safe to work with such an assumption. All ensemble averages will then be equal to the corresponding average over the full length of any particular member of the ensemble and, in particular,

$$\langle f(z)\, f(z+u) \rangle = \langle f^2 \rangle\, \varphi(u) \; . \qquad (6.47)$$

If only the one mode with $m = 1$ is launched into the fibre, its amplitude changes due to mode conversion and reconversion according to equations (6.29) or (6.33). For an evaluation of its propagation statistics, we let coupled modes be low enough in loss, so that the coupling coefficients c_{n1} are real quantities. With $\Delta\gamma_n = \gamma_n - \gamma_1$ and for wave coupling according to $c_{n1} = C_{n1} f(z)$ in which all coupling coefficients have the same distribution $f(z)$ along the fibre, the output amplitude for unit incident amplitude follows as

$$A_1 = 1 - \sum_n C_n^2 \int_{x=0}^{L} \int_{y=0}^{x} f(x)\, f(y)\, \exp\{\Delta\gamma_n (y-x)\}\, dy\, dx \; , \quad (6.48)$$

where $C_n^2 = C_{n1} C_{1n}$. Fig.6.1 illustrates the limits of integration in the xy-plane. Alternatively, we can also integrate over y and

$$u = x - y \qquad (6.49)$$

as shown in Fig.6.1. With u as one of the variables of inte-

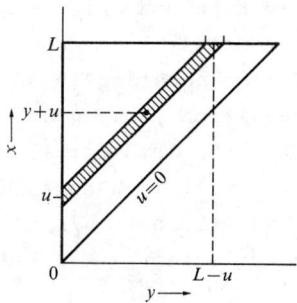

Fig.6.1. Limits of integration for equations (6.48) and (6.50) in the xy-plane.

gration, the output amplitude has the magnitude

$$|A_1| = \left| 1 - \sum_n C_n^2 \int_{u=0}^{L} \int_{y=0}^{L-u} f(y)\, f(y+u) \exp(-\Delta\gamma_n u)\, dy\, du \right|. \qquad (6.50)$$

We take the ensemble average of this expression and, because of the stationary character of the random distribution, introduce its auto-correlation function from equation (6.47).

We can then integrate readily over y and obtain

$$\langle |A_1| \rangle = |1 - \langle f^2 \rangle \sum_n c_n^2 \int_0^L (L-u) \, \varphi(u) \exp(-\Delta\gamma_n u) du| \, . \quad (6.51)$$

According to this amplitude average, we have to expect a decrease in amplitude and, as long as the amplitude does not change too much, we can account for this change in magnitude of the amplitude by an average attenuation constant

$$\langle \alpha \rangle = \langle f^2 \rangle \, \text{Re} \, [\sum_n c_n^2 \int_0^L (1-\frac{u}{L})\varphi(u)\exp(-\Delta\gamma_n u) du] \, . \quad (6.52)$$

If the absolute value of the second term in equation (6.51) is not small enough compared to unity, to allow this comparison with an exponential function of small argument, we can reduce the fibre length L to a shorter section and apply the argument that leads from equation (6.51) to the average attenuation in equation (6.52) to this section of the fibre. All we require from this fibre section is that the auto-correlation does not reach beyond this length. Under most practical conditions for small fibre imperfections, the fibre length L for which $\langle \alpha \rangle$ from equation (6.52) remains low enough to allow the transition from equation (6.51) to (6.52) is still much longer than the auto-correlation distance. The auto-correlation function will then contribute to the integral only for $u \ll L$, and the term u/L in the integrand of equation (6.52) may be neglected. This leads to the expression

$$\langle \alpha \rangle = \langle f^2 \rangle \, \text{Re}[\sum_n c_n^2 \int_0^L \varphi(u)\exp(-\Delta\gamma_n u) du] \quad (6.53)$$

for the average added loss.

The difference $\Delta\gamma_n$ in propagation constants between coupled modes consists of an attenuation difference $\Delta\alpha_n$ and a phase difference $\Delta\beta_n$

$$\Delta\gamma_n = \Delta\alpha_n + j\Delta\beta_n \, . \quad (6.54)$$

For interaction between guided core modes, the attenuation difference is quite often so small that

$$\Delta\alpha_n u \ll 1 \qquad (6.55)$$

for all distances u over which $\varphi(u)$ contributes to the integral in equation (6.53). Coupled modes in the perfect fibre suffer under these conditions nearly the same attenuation within the correlation distance of imperfections. For such low differential loss, the average added loss is

$$\langle \alpha \rangle = \langle f^2 \rangle \sum_n c_n^2 \int_0^L \varphi(u) \cos(\Delta\beta_n u) du \ . \qquad (6.56)$$

Because of the limited correlation distance, the integration in this expression may be extended to infinity and the integral replaced by the power spectrum of the perturbation from equation (6.46)

$$\langle \alpha \rangle = \frac{1}{2} \sum_n c_n^2 \ \phi(\Delta\beta_n) \ . \qquad (6.57)$$

According to this equation only those components of the power spectrum which have spatial frequencies $\Omega = \Delta\beta_n$ contribute to the added loss. This result for the average added loss agrees with the result for mode conversion due to the spectral component F_p of an arbitrary distribution of imperfections in equation (6.42). Only those spectral components F_p cause significant mode conversion that are close in their spatial frequency Ω_p to the phase difference of coupled modes.

For even shorter correlation when not only equation (6.55), but even

$$|\Delta\gamma_n u| \ll 1 \qquad (6.58)$$

holds for any distance u in which $\varphi(u)$ contributes to the integral, we obtain for the average added loss

$$\langle \alpha \rangle = \frac{1}{2} \phi(0) \ \mathrm{Re}\left[\sum_n c_n^2\right] \ . \qquad (6.59)$$

In this case, all beat wavelengths $\Lambda_n = 2\pi/\Delta\beta_n$ of coupled modes must be long compared to the auto-correlation distance of the random imperfections. It applies only for distributions of imperfections with very short correlation. The added loss depends only on the sum of the squares of all coupling factors C_n but not on the differences between propagation constants of coupled modes. Any power which under these conditions is converted to another mode remains lost for the incident mode. Because of the short correlation, it could never properly be reconverted to add in phase to the primary mode. Equation (6.53) or its approximate forms (6.56), (6.57), and (6.59) give the average added loss when one single mode is excited, and all other modes are of a parasitic nature only. It therefore serves to analyse the effects of imperfections in single-mode fibres or, when in a multimode fibre only one single mode (preferably the fundamental mode) is to be utilized for signal transmission. In a single-mode fibre, the fundamental mode couples to leaky core modes with their relatively high leaky-mode attenuation or to cladding modes which may suffer high attenuation from a lossy jacket. For single-mode transmission in a multimode fibre, the signal mode interacts with core and cladding modes. While many of the core modes of the perfect fibre are not much more attenuated than the signal mode, higher-order core modes and even more so the cladding modes may suffer so much attenuation that equation (6.53) must be evaluated to properly account for their higher loss. In both cases of single-mode operation, random imperfections cause transmission loss, and for an accurate analysis, all modes which contribute to equation (6.53), or its respective approximations, must be taken into account.

6.3. CURVATURE COUPLING IN FIBRES

If we take the straight and circular geometry of a perfect fibre as a reference structure, then only a slight bending of its axis already leads to a sizable deviation of the actual index distribution $n(r,\varphi,z)$ in equation (6.1) from the distribution $n_0(r)$ of the reference structure. Therefore, curvature of the fibre constitutes a very critical if not the most critical imperfection. As long as this curvature remains uniform, as in a fibre bend of constant curvature, or as long as it changes only gradually along the fibre, it causes only the leaky-mode loss that is associated with the frustration of total reflection at a curved interface. We analysed this leaky-mode loss in constant curvature bends for the slab waveguide in section 2.8, and found that it decreases exponentially with decreasing waveguide curvature. The leaky-mode curvature loss becomes excessive only for strong curvature. Corresponding results are obtained for fibres with constant curvature (Lewin 1975, Gloge 1975). The added loss in a uniform bend also increases exponentially with curvature, but fibres may be bent uniformly with substantial curvature before the curvature loss becomes excessive.

The situation is different when the curvature changes along the fibre, and in particular when the longitudinal curvature distribution contains spectral components with spatial frequencies Ω in the range of phase differences $\Delta\beta$ between modes that are coupled by curvature. Such changing curvature occurs not so much in the naked fibre, as it is pulled from a preform or out of a melt, but results more from coating and cabling of the fibre and from installation of fibre cables. It usually consists only of minute deviations of the fibre axis from a straight line and is therefore called microbending.

To analyse the effects of microbending, the actual index distribution in equation (6.1) must be taken into account by adding the deviation n_p^2 to the nominal distribution n_0^2 of the straight fibre. This representation leads, however, to extensive deviations n_p^2 and strong coupling between the modes of the straight reference fibre. We therefore establish another reference fibre by transforming the curved fibre into an equivalent straight fibre (Petermann 1976). The equivalent fibre must have such a perturbation n_p^2 of its index distribution in equation (6.1) that the transverse field distributions in the curved and the equivalent straight fibre equal each other. In the scalar wave equation

$$\nabla_t^2 E + k_t^2 E = 0 \qquad (6.60)$$

for the uniformly polarized field of a weakly guiding fibre, the transverse wave number k_t is in the uniformly curved fibre of Fig.6.2. given by

$$k_t^2 = n_0^2(x,y) k^2 - \beta^2 . \tag{6.61}$$

Just as in equations (2.200) and (2.201) for the curved slab, the field distribution of any mode in the curved fibre must depend as $\exp(-jl_z\phi_z)$ on the angle ϕ_z in Fig.6.2. For $E \sim \exp(-j\beta z)$ and $z = (\rho+x)\phi_z$, such a dependence is only possible if

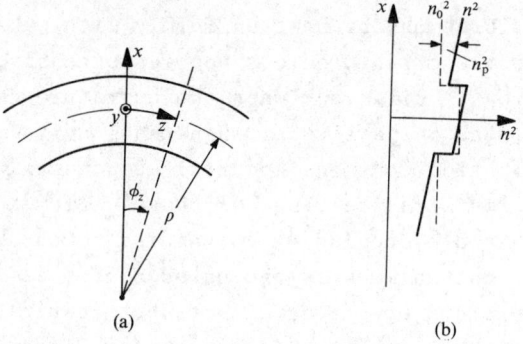

Fig.6.2. Curved fibre (a) and index distribution of equivalent straight fibre (b).

$$\beta = \beta_0 \, \rho/(\rho+x) \tag{6.62}$$

with the curvature radius ρ and $\beta(x=0) = \beta_0$ as the phase constant on the axis. When we substitute this expression for β into equation (6.61) and assume $x \ll \rho$, for any axial distance in the plane of curvature to which fibre fields extend we obtain

$$k_t^2 = n_0^2(x,y) k^2 + 2(x/\rho) \beta_0^2 - \beta_0^2 . \tag{6.63}$$

The small contribution $2(x/\rho)\beta_0^2$ to the transverse wave number from the fibre curvature changes only little when, for the weakly guiding fibre, we replace β_0 by $n_1 k$ with n_1 as the refractive index on the fibre axis

$$k_t^2 = [n_0^2(x,y) + 2(x/\rho) n_1^2] k^2 - \beta_0^2 . \tag{6.64}$$

For the equivalent straight fibre, we let the refractive index n deviate from n_0 according to equation (6.1) so that its transverse wave number follows from

$$k_t^2 = [n_0^2(x,y) + n_p^2(x,y,z)]k^2 - \beta_0^2 . \tag{6.65}$$

Equations (6.64) and (6.65) become identical when

$$n_p^2 = 2n_1^2 \, x/\rho . \tag{6.66}$$

Under this condition, the same transverse wave equation applies to the equivalent straight fibre with an index deviation n_p as to the bent fibre with radius of curvature ρ. The effects of fibre curvature on wave propagation are thus accounted for by an equivalent deviation in refractive index as shown in Fig.6.2(b) with n_p^2 according to equation (6.66). With this equivalent index deviation, the coupling coefficient for curvature in equation (6.19) can be determined and the solutions of coupled mode equations be evaluated

The transverse field distributions E_m and E_n in equation (6.19) must be normalized according to equation (6.12). If the field distributions which we find for modes of any particular index profile are not readily normalized, we can incorporate this normalization in equation (6.19) by substituting instead of E_m the normalized distribution $E_m (\iint E_m^2 \, dS)^{-1/2}$. The formula for coupling coefficients then generalizes to

$$c_{mn} = \frac{k^2}{2\beta_m} \frac{\iint n_p^2 \, E_m E_n \, dS}{(\iint E_m^2 \, dS \iint E_n^2 \, dS)^{1/2}} . \tag{6.67}$$

With the index deviation of the equivalent straight fibre in equation (6.66) and for $\beta_m \simeq n_1 k$, the coupling coefficient for a radius of curvature ρ is

$$c_{mn} = \frac{n_1 k}{\rho} \frac{\iint r^2 \cos\varphi \, E_m E_n \, dr \, d\varphi}{(\iint E_m^2 \, dS \iint E_n^2 \, dS)^{1/2}} . \tag{6.68}$$

We evaluate this expression first for the graded-index fibre with a parabolic index profile. If this profile extends to infinity, the transverse field distributions of its LP_{lp}-modes are given by equations (5.54) and (5.59) in terms of Laguerre polynomials and the Gaussian function. These field distributions of the unlimited profile approximate the actual distributions in a truncated profile quite well, unless the outer turning point r_2 is too close to the core radius and the fields extend too much into the cladding. This, however, happens only to modes of very low circumferential order close to cut-off. With the field distributions in equations (5.54) and (5.59) all integrals in equation (6.68) can be evaluated in closed form (Gradshteyn and Ryzhik 1965) and the following coupling coefficients are found for coupling of the LP_{lp}-mode to the $LP_{l+1,p}$-mode

$$c_{lp,(l+1)p} = \frac{n_1 k}{2} \frac{w}{\rho} [(p+l)/\varepsilon_l]^{\frac{1}{2}},$$

to the $LP_{l-1,p+1}$-mode

$$c_{lp,(l-1)(p+1)} = \frac{n_1 k}{2} \frac{w}{\rho} [p/\varepsilon_{l-1}]^{\frac{1}{2}},$$

to the $LP_{l-1,p}$-mode \hfill (6.69)

$$c_{lp,(l-1)p} = \frac{n_1 k}{2} \frac{w}{\rho} [(p+l-1)/\varepsilon_{l-1}]^{\frac{1}{2}},$$

and to the $L_{l+1,p-1}$-mode

$$c_{lp,(l+1)(p-1)} = \frac{n_1 k}{2} \frac{w}{\rho} [(p-1)/\varepsilon_l]^{\frac{1}{2}}$$

with the Neumann number

$$\varepsilon_n = \begin{cases} 1 & \text{for } n = 0 \\ 2 & \text{for } n \neq 0 \end{cases}. \qquad (6.70)$$

No other modes are coupled to the LP_{lp}-mode by curvature. All LP_{lp}-modes of the parabolic index profile with the same compound mode number

$$m = 2p + l \qquad (6.71)$$

have the same phase constant according to equation (5.64). The coupling coefficients in equation (6.69) couple only those modes which differ in their compound mode number m by $|\Delta m| = 1$. The first two coupling coefficients couple modes with m to modes with $m + 1$, and the last two coefficients modes with m to modes with $m - 1$. The phase difference between modes of the parabolic index profile that are coupled by curvature is always

$$\Delta \beta = 2/(n_1 k w^2), \qquad (6.72)$$

independent of the mode order.

Simple closed-form approximations can also be found for the curvature coupling coefficients of a weakly guiding fibre with a homogeneous core. An LP_{lp}-mode of this fibre which is not too close to cut-off confines its fields well inside the core and its transverse field distribution is to a good approximation given by

$$E_{lp} = J_l(u_{lp} r/a) \begin{cases} \cos l\varphi \\ \sin l\varphi \end{cases} \qquad (6.73)$$

where u_{lp} designates the pth zero of the Bessel function $J_l(u)$. With these field distributions we obtain coupling between all modes that differ by one in their circumferential order; the corresponding coupling coefficients

$$c_{lp,(l+1)q} = 2n_1 k \frac{a}{\rho} \left(\frac{2}{\varepsilon_l}\right)^{\frac{1}{2}} \frac{u_{lp} u_{l+1,q}}{(u_{l+1,q}^2 - u_{lp}^2)^2}$$

$$c_{lp,(l-1)q} = 2n_1 k \frac{a}{\rho} \left(\frac{2}{\varepsilon_{l-1}}\right)^{\frac{1}{2}} \frac{u_{lp} u_{l-1,q}}{(u_{lp}^2 - u_{l-1,q}^2)^2} . \qquad (6.74)$$

The Bessel function roots u_{lp} in these coefficients represent the limiting value which the phase parameter u of the respective LP_{lp}-mode approaches asymptotically far above cut-off.

For modes of high radial orders, in particular those with $p \gg l$, this asymptotic value is approximated by

$$u_{lp} \simeq (\pi/2)(2p+l-1/4) .$$

The coupling coefficients in equation (6.74) couple those modes most tightly that differ least in their asymptotic value u_{lp}. An LP_{lp}-mode thus couples most strongly to the $LP_{l\pm1,p}$-, $LP_{l+1,p-1}$-, and $LP_{l-1,p+1}$-modes. The difference between the asymptotic u-values of these modes and that of the LP_{lp}-mode amounts only to

$$u_{lp} - u_{l\pm1,p} \simeq \mp \pi/2$$

$$u_{lp} - u_{l+1,p-1} \simeq \pi/2 \qquad (6.75)$$

$$u_{lp} - u_{l-1,p+1} \simeq -\pi/2 .$$

All other modes that are coupled by curvature have at least twice this difference. Because of $(u_{lp}-u_{l\pm1,q})^2$ in the denominators of equation (6.74), their coupling coefficients are smaller at least by the factor (1/4). If we examine the compound mode number $m = 2p + l$ for the fibre with a homogeneous core, we find that the modes which curvature couples most tightly to the LP_{lp}-modes are all those which differ by one in their compound mode number from $m = 2p + l$. Compared to their curvature coupling coefficients, all other curvature-coupled modes which differ by $|\Delta m| > 1$ from $m = 2p + l$ have coupling coefficients which are by $1/(\Delta m)^2$ smaller. Next-neighbour coupling with respect to the compound mode number m therefore dominates curvature coupling to all other modes quite strongly in the homogeneous core. Except for coupling to modes with $l = 0$ and $l = 1$, and under conditions for which equation (6.75) applies, all coefficients for curvature coupling between these next neighbours in m are given approximately by the following simple formula:

$$c_{lp,(l\pm1)p} = c_{lp,(l+1)(p-1)} = c_{lp,(l-1)(p+1)} = (2/\pi^2)(n_1 ka/\rho). \qquad (6.76)$$

FIBRE IMPERFECTIONS 575

In this approximation the curvature coupling between next neighbours is independent of mode orders. But also for curvature coupling between modes that differ by $\Delta m = 2(p-q)+ l - k$ in their compound mode numbers $2p + l$ and $2q + n$, the coupling coefficients follow uniquely from

$$c_{lp,nq} = -\delta_{l,n\pm 1} \, [2/(\pi^2 (\Delta m)^2)] \, (n_1 k a/\rho) \qquad (6.77)$$

and depend only on the difference in compound mode number. The Kronecker symbol in equation (6.77) renders the coupling different from zero only for modes that differ by one in their circumferential order.

6.4. MICROBENDING LOSS IN SINGLE-MODE FIBRES

For a single-mode fibre, the curvature coupling to cladding modes, or to leaky modes and to the rest of the radiation field causes additional loss. To analyse this added curvature loss accurately, we need to sum the contributions from a large number of coupled cladding modes (Kuhn 1975). For an approximate but much simpler analysis of curvature coupling in single-mode fibres (Petermann 1976), we utilize the result in equation (6.69) that for an unlimited parabolic profile the lowest-order fundamental mode with its Gaussian field distribution couples in the curved fibre directly only to an LP_{11}-mode. For other profiles, the fundamental mode also couples directly to other modes of circumferential order $l = 1$. In single-mode fibres, these coupled modes are guided by the cladding. But as long as the fundamental and first-order modes resemble the Gaussian field distributions of the unlimited parabolic profile, curvature coupling between these modes dominates the curvature coupling of the fundamental mode to all other modes. We may then lump the fields of all these other modes with the field of the first-order coupled mode and assume this field distribution E_p to propagate as a quasi mode. E_p superimposes on to the field E_0 of the fundamental mode according to

$$E = V_0(z) \, E_0(x,y) + V_p(z) \, E_p(x,y) \, . \qquad (6.78)$$

To make the best possible choice for the field distribution $E_p(x,y)$ of this quasi mode, we let the fibre be uniformly curved with constant radius of curvature ρ and require equation (6.78) to represent the field $E_b(x,y)$ of the fundamental mode of the equivalent straight fibre with the index deviation according to equation (6.66). We determine this field and the phase constant β_b of the fundamental mode of the equivalent straight fibre by expanding both in terms of ascending powers of $(1/\rho)$:

$$E_b(x,y) = E_0 + \frac{1}{\rho} E^{(1)} + \frac{1}{\rho^2} E^{(2)} + \ldots \quad (6.79)$$

$$\beta_b = \beta_0 + \frac{1}{\rho} \beta^{(1)} + \frac{1}{\rho^2} \beta^{(2)} + \ldots \quad (6.80)$$

and by solving the wave equation

$$\nabla_t^2 E_b + [(n_0^2 + 2n_1^2 \frac{x}{\rho}) k^2 - \beta_b^2] E_b = 0 \quad (6.81)$$

for the $E^{(n)}$ and $\beta^{(n)}$ of this expansion with perturbation methods. In the case of a circularly symmetric index profile $n_0(x,y) = n_0(r)$ and the circularly symmetric field distribution $E_0(x,y) = E_0(r)$ of its fundamental mode, the first order perturbation of β_b vanishes,

$$\beta^{(1)} = 0 , \quad (6.82)$$

and the following differential equation obtains for the first-order perturbation of E_b

$$\nabla_t^2 E^{(1)} + [n_0^2(r)k^2 - \beta_0^2] E^{(1)} = -2x\, n_1^2 k^2 E_0(r) . \quad (6.83)$$

We solve this equation for $E_p(x,y) = (2n_1^2 k^2)^{-1} E^{(1)}(x,y)$ by setting

$$E_p(x,y) = x\, h(r) E_0(r) . \quad (6.84)$$

The factors x and $(h)^{1/2}$ in this formulation have the dimension of length. But by taking the light wavelength λ as the unit of length and by considering all other distances in terms of λ, we ensure that E_p has the same dimension as E_0. We thus avoid any additional factors for these dimensions. Substituting E_p from equation (6.84) into equation (6.83) leads to the following differential equation for $h(r)$:

$$r \frac{d^2h}{dr^2} + \frac{dh}{dr}[3 + 2rg(r)] + 2h\,g(r) + r = 0 \qquad (6.85)$$

with

$$g(r) = (dE_0/dr)/E_0 . \qquad (6.86)$$

For the Gaussian field distribution

$$E_0 = \exp[-r^2/w^2] \qquad (6.87)$$

of a parabolic index profile, we obtain $g(r) = -2r/w^2$, and equation (6.85) is solved by the constant quantity

$$h(r) = w^2/4 . \qquad (6.88)$$

The field distribution of the fundamental mode in other single-mode fibres was initially assumed to resemble the Gaussian field of equation (6.87). The function $h(r)$ which follows for them from equation (6.85) will also be nearly constant across the fibre. We therefore let $dh/dr = d^2h/dr^2 = 0$ in equation (6.85) and determine $h(r)$ from

$$2h\,dE_0/dr = -r\,E_0 . \qquad (6.89)$$

A constant average value for $h(r)$ obtains when we multiply this equation by rE_0 and integrate it over the fibre cross-section

$$4h = 2 \frac{\int_0^\infty r^3 E_0^2\,dr}{\int_0^\infty r E_0^2\,dr} = w_0^2 . \qquad (6.90)$$

This average value defines the effective spot size or mode

radius w_0 of the circularly symmetric field distribution E_0. The actual solution $h(r)$ of equation (6.85) differs only little from this average value. Fig.6.3 shows $h(r)$ for the fundamental mode of cladded-core fibres with different frequency parameters V. This function, or its average value from equation (6.90), determines the distribution $E_p(x,y)$ according to equation (6.84).

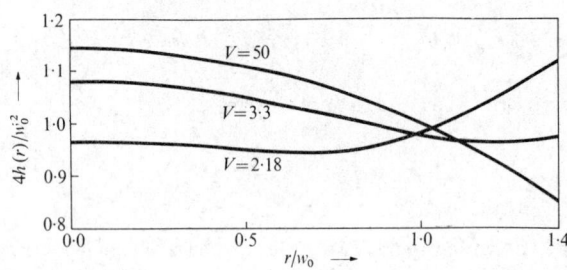

Fig.6.3. Factor $h(r)$ by which the curvature term of the fundamental mode field in the stepped-index fibre with $V = ka(n_1^2-n_2^2)^{\frac{1}{2}}$ differs according to equation (6.84) from $x\,E_0(r)$ with $E_0(r)$ as the fundamental mode field of the straight fibre (Petermann 1976).

$E_p(x,y)$ is the transverse field distribution of a quasi mode which, according to equation (6.78), lumps all the fields which are excited by curvature coupling in addition to the field $E_0(x,y)$ of the fundamental mode in the straight fibre. If this field distribution $E_p(x,y)$ were launched into a straight and perfect fibre, it would excite cladding modes or radiation fields that would all propagate with their individual phase and attenuation constants. In equation (6.78) we postulate, however, that $E_p(x,y)$ remains unchanged along the fibre, except for the variation in amplitude according to $V_p(z)$. To find a phase constant β_p with which this quasi mode propagates along the fibre, we assume $V_p(z)$ in the straight fibre to vary with z according to $\exp(\mp j\beta_p z)$, and substitute $V_p(z)\,E_p(x,y)$ into the scalar wave equation of the straight fibre:

$$\nabla_t^2 E_p + (n_0^2 k^2 - \beta_p^2) E_p = 0 \,. \qquad (6.91)$$

E_p, of course, does not solve this equation, but we obtain an average or effective value for β_p by multiplying it by E_p, and integrating it over the fibre cross-section. If we then solve for β_p^2 and substitute for $\nabla_t^2 E_p$ from equation (6.83) and for E_p from equation (6.84), we obtain the following expression for the phase constant of this quasi-mode

$$\beta_p^2 = \beta_0^2 - \frac{4}{w_{p0}^2} \qquad (6.92)$$

with

$$\frac{4}{w_{p0}^2} = \frac{\int_0^\infty r^3 \, h(r) \, E_0^2 \, dr}{\int_0^\infty r^3 \, h^2(r) \, E_0^2 \, dr} \, . \qquad (6.93)$$

The phase constant of this quasi-mode is always smaller than the phase constant β_0 of the fundamental mode. For the Gaussian distribution of the fundamental mode field in a parabolic index profile, we have $h = w^2/4$, so that w_{p0} reduces to

$$w_{p0} = w \, . \qquad (6.94)$$

In this case, the phase difference between the fundamental mode and the quasi-mode of a bent fibre follows from

$$\beta_0^2 - \beta_p^2 = 4/w^2 \, . \qquad (6.95)$$

For $\beta_p \simeq \beta_0 \simeq n_1 k$, it is simply given by

$$\beta_0 - \beta_p = 2/(n_1 k \, w^2)$$

in accordance with equation (6.72). For other index profiles with nearly Gaussian distribution for E_0, equation (6.93) defines an effective spot size w_{p0} for the interaction between the fundamental mode of field distribution E_0 and the quasi curvature mode with a field distribution E_p.

The two-mode approximation in equation (6.78) for fundamental mode propagation and coupling in a curved fibre can now be introduced into the wave equation (6.60) for the equivalent straight fibre with n_p^2 according to equation (6.66), and coupled wave equations can be derived for voltage and current coefficients $V(z)$ and $I(z)$ of the transverse electric and magnetic fields. The voltage and current coefficients may then be replaced by the amplitudes $a(z)$ and $b(z)$ of forward- and backward-travelling waves. The orthogonality condition

$$\int_S E_p E_0 \, dS = 0 ,$$

which we require in order to derive the coupled mode equations, is satisfied by the circularly symmetric field of the fundamental fibre mode and the quasi-mode field from equation (6.84). The coupling coefficients in equation (6.19) assume, in addition, that the modal field distributions are also normalized according to equation (6.10). Without this normalization, the coupling coefficients still follow from equation (6.19) but to obtain $c_{mn} = j \kappa_{mn}$, we must now divide the right-hand side of equation (6.19) by $\int E_m^2 \, dS$ as a normalization factor. Evaluation of equation (6.19) with the respective normalization factor thus leads to the following coupling coefficients for the travelling wave amplitudes

$$c_{0p} = \frac{k^2}{2\beta_0} \frac{\int_S n_p^2 E_p E_0 \, dS}{\int_S E_0^2 \, dS} = \frac{n_1^2 k^2}{\beta_0 \rho} \frac{\int_0^\infty r^3 h(r) E_0^2 \, dr}{2 \int_0^\infty r E_0^2 \, dr} \qquad (6.96)$$

$$c_{p0} = \frac{k^2}{2\beta_p} \frac{\int_S n_p^2 E_p E_0 \, dS}{\int_S E_p^2 \, dS} = \frac{n_1^2 k^2}{\beta_p \rho} \frac{\int_0^\infty r^3 h(r) E_0^2 \, dr}{\int_0^\infty r^3 h^2(r) E_0^2 \, dr} . \qquad (6.97)$$

The coupling coefficient c_{p0} contains the effective spot size w_{p0} from equation (6.93) for the interaction between E_0 and E_p. Another such spot size w_{0p} is defined by w_{p0} together with the ratio of integrals in equation (6.96)

$$w_{0p}^2 = \frac{8}{w_{p0}^2} \frac{\int r^3 h(r) E_0^2 \, dr}{\int r E_0^2 \, dr} \, . \qquad (6.98)$$

When E_0 equals the Gaussian field distribution, and $h = w^2/4$, as well as $w_{p0} = w$, this effective spot size also reduces to $w_{0p} = w$. In this case, all spot sizes, w_0, w_{p0}, and w_{0p}, become identical to the spot size w of the Gaussian field distribution.

For field distributions of the fundamental mode in other index profiles, the three effective spot sizes differ only very little from each other. Fig.6.4 illustrates these small

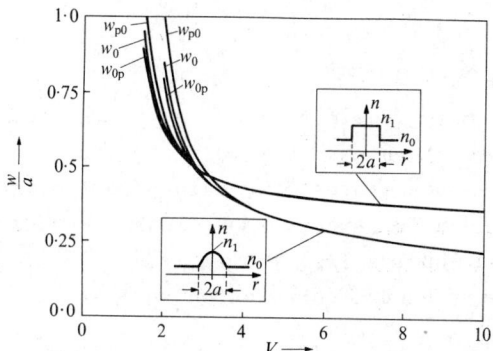

Fig.6.4. Effective spot sizes for the fundamental mode and for its curvature coupling in fibres with stepped-index profiles and with truncated parabolic profiles (Petermann 1976).

differences for the stepped-index profile and for the truncated parabolic profile. For large enough values of the fibre parameter V, all three spot sizes are nearly equal to each other. But even at the limit of the single-mode region, where for $V_c = 2 \cdot 405$ in the stepped index profile and for $V_c = 3 \cdot 53$ in the truncated parabolic profile, the LP_{11}-modes go beyond cut-off, there is only little difference between w_0, w_{p0}, and w_{0p}. In terms of these effective spot sizes, the coupling coefficients are given by

$$c_{0p} = \frac{n_1^2 k^2}{\beta_0 \rho} \frac{w_{p0}^2 \, w_{0p}^2}{16}$$

$$c_{p0} = \frac{n_1^2 k^2}{\beta_p \rho} \frac{4}{w_{p0}^2}$$

(6.99)

and for $\beta_p \simeq \beta_0 \simeq n_1 k$ reduce to

$$c_{0p} = \frac{n_1 k}{\rho} \frac{w_{p0}^2 \, w_{0p}^2}{16}$$

$$c_{p0} = \frac{n_1 k}{\rho} \frac{4}{w_{p0}^2}$$

(6.100)

With these coupling coefficients, any of the solutions of the coupled wave equations may now be evaluated.

In single-mode fibres, the main effect of curvature coupling is the added fundamental mode loss. We assume the random curvature distribution $1/\rho(z)$ of microbending to be stationary and have a short enough correlation so that

$$\Delta \alpha z < 1 \qquad (6.101)$$

holds for any distance z within which its auto-correlation function $\varphi(z)$ differs from zero. Alternatively, equation (6.101) implies a low enough differential loss between the fundamental mode and the quasi mode of curvature interaction. Under this condition, the average added loss follows from equation (6.56) and with equations (6.92) and (6.100) may be written in terms of the effective spot sizes and the power spectrum $\phi(\Delta\beta)$ of the random distribution at $\Delta\beta = 2/(n_1 k \, w_{p0}^2)$

$$\langle \alpha \rangle = \frac{1}{8} \langle (1/\rho)^2 \rangle (n_1 k \, w_{0p})^2 \, \phi(2/(n_1 k \, w_{p0}^2)) \, . \qquad (6.102)$$

To examine how accurate this loss formula is, we compare results from its evaluation with results of a more accurate theory of curvature loss in single-mode fibres. Petermann and Storm (1976) have analysed curvature losses in W-fibres under single-mode and multimode conditions. Instead of the one quasi-mode of our present approximation, they have considered interaction with the continuous and complete spectrum of radiation modes. Instead of the summation in equation (6.57), their loss formula contains an integration over the coupling function to the continuous spectrum of radiation modes. For the power spectrum of curvature distribution, we consider the inverse power law

$$\phi(\Omega) = K/\Omega^{2p} \qquad (6.103)$$

with integer numbers $p = 0,1,2$ and the constant K to be adjusted so that $\phi(\Omega)$ corresponds to the actual curvature power spectrum of fibres over limited ranges of the spatial frequency Ω. $p = 0$ represents the flat power spectrum of a curvature distribution with zero correlation distance. With increasing p, the power spectrum narrows in width and the correlation extends over longer distances. As a typical result, Fig.6.5 compares the more accurate curvature loss from the

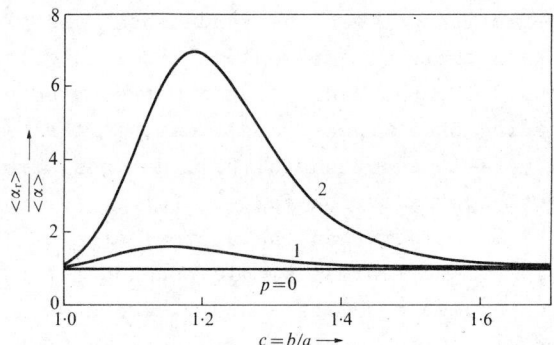

Fig.6.5. Ratio of accurate curvature loss $\langle \alpha_r \rangle$ from interaction with radiation modes to the approximate curvature loss $\langle \alpha \rangle$ according to equation (6.102) in single-mode W-fibres with $V = ka(n_1^2-n_2^2)^{\frac{1}{2}} = 7 \cdot 1$ and $V_3 = ka(n_1^2-n_3^2)^{\frac{1}{2}} = 2 \cdot 4$. The parameter p designates different power spectra of the random curvature distribution according to equation (6.103). $c = b/a = 1$ represents a homogeneous core of index n_1 with a single unlimited cladding of index n_3 (Petermann and Storm 1976).

interaction with the complete spectrum of radiation modes to our present approximation from the interaction with only one quasi-mode by plotting the ratio of both values versus the diameter ratio $c = b/a$ of the W-profile in Fig.4.33(b) for three different power spectra of the curvature distribution. The ratio of permittivity differences between core and inner and outer cladding as well as the fibre parameter V have been chosen in Fig.6.5 to maintain the single-mode condition even in the limit $c = 1$. We conclude from Fig.6.5 that our loss approximation agrees with the more accurate curvature loss for $c = 1$ and for large c-values as well as for small p-values. For the present values of V and $(n_1^2-n_2^2)/(n_1^2-n_3^2)$, the accurate curvature loss is, however, much larger at intermediate values of c when p simultaneously has substantial values. This difference between accurate and approximate curvature loss is obviously caused by the peculiar transverse field distributions which the fundamental mode and the decisive radiation modes exhibit in the W-fibre for intermediate c-values in Fig.6.5. The effects of these peculiar field distributions and in particular the weakly evanescent fields of the fundamental mode in the outer cladding are not properly accounted for when interaction between the fundamental and a quasi-mode is employed with their effective spot sizes to calculate the curvature loss. The phase difference between these modes loses its meaning under these conditions, so that as soon as $\phi(\Omega)$, with $\Omega = \Delta\beta$ in equation (6.102), depends strongly on $\Delta\beta$, the present approximation fails. For $p > 0$, the actual curvature loss is larger than the approximate loss values from equation (6.102). For normal single-mode operation of W-fibres, fibre parameter values V and V_3 are chosen which for a given diameter ratio c correspond to or are near to the LP_{11} cut-off. Such a W-fibre provides effective fundamental mode guidance and low or even negative fundamental mode dispersion. It also has transverse field distributions of its HE_{11}- and LP_{11}-modes that allow meaningful spot sizes to be calculated from equations (6.90), (6.93), and (6.98), which give accurate results for the added curvature loss when substituted into equation (6.102). Except for some peculiar and not very practical situations in W-fibres and fibres with similar index profiles, the approximate loss formula (6.102) holds quite well under all conditions in single-mode fibres with uniform as well as with graded-index cores. It may even be applied to calculate the added curvature loss of the fundamental mode in multimode fibres, because curvature couples this mode so much more strongly to the LP_{11}-mode than to higher-order modes that the two-mode approach, with the HE_{11}- and one quasi-mode, appears entirely adequate. Once we have such normal conditions that equation (6.102) gives accurate results for the added curvature loss, it is not even necessary to compute all three effective spot sizes from equations (6.90), (6.93), and (6.98). These spot sizes differ so little from each other under normal conditions that we may let $w_{p0} = w_{0p} = w_0$ and just substitute w_0 from equation (6.90) for all spot sizes into equation (6.102).

Taking a particular example to analyse the added curvature loss and its dependence on fibre characteristics and curvature statistics, we assume the curvature to have an exponential auto-correlation according to

$$\varphi(u) = \exp(-|u|/L_k) \qquad (6.104)$$

with the auto-correlation distance L_k. The power spectrum of this curvature distribution is given by

$$\phi(\Omega) = \langle (1/\rho)^2 \rangle \frac{2L_k}{1+\Omega^2 L_k^2} . \qquad (6.105)$$

It has a Butterworth characteristic with a half-power bandwidth of spatial frequencies $\Omega = 1/L_k$.

As a function of correlation distance, the Butterworth power spectrum in equation (6.105) assumes its maximum value $\phi_{max} = 1/\Omega$ at $L_k = 1/\Omega$. As a consequence, the added loss will also have its maximum for this particular correlation distance. Since for the added loss in equation (6.102), the power spectrum must be taken at $\Omega = \Delta\beta$, the loss maximum occurs when the correlation distance is

$$L_k = \Lambda/2\pi ,$$

with $\Lambda = 2\pi/\Delta\beta$ as the beat wavelength between the fundamental mode and the quasi-mode in the curved fibre. In terms of the effective spot size, this critical value of correlation distance is given by

$$L_k = n_1 k\, w_0^2/2 \,. \tag{6.106}$$

With this correlation, the added curvature loss amounts to

$$\langle \alpha \rangle = \frac{1}{16} \langle (1/\rho)^2 \rangle (n_1 k)^3 w_0^4 \tag{6.107}$$

and is proportional to w_0^4. For $L_k \ll n_1 k\, w_0^2/2$, the average added loss depends as

$$\langle \alpha \rangle = \frac{1}{4} \langle (1/\rho)^2 \rangle (n_1 k\, w_0)^2 L_k \tag{6.108}$$

on the effective spot size squared, while for $L_k \gg (n_1 k\, w_0^2)/2$, it is given by

$$\langle \alpha \rangle = \frac{1}{16} \langle (1/\rho)^2 \rangle (n_1 k)^4 w_0^6/L_k \,. \tag{6.109}$$

Under any of these conditions therefore, the added loss increases more or less critically with spot size. The larger the spot radius of the fundamental mode, the weaker will be its guidance by the single-mode fibre and the more sensitive will it be to microbending. These results hold quite generally for any index profile of the fibre.

6.5. SIGNAL DISTORTION IN SINGLE-MODE TRANSMISSION

Mode interaction in imperfect fibres not only causes additional transmission loss, it also distorts signals which modulate the optical carrier. This signal distortion from mode interaction adds to the distortion from material and mode dispersion and, in the case of multimode transmission, modifies the signal distortion due to delay differences between modes. For single-mode transmission in a perfect fibre only material and mode dispersion distort the signal. If in an imperfect fibre the signal mode interacts with other modes, it converts power to other modes, which, when reconverted, adds to the signal mode with phase and delay differences and thus changes the initial time dependence of the signal-mode amplitude and power.

To analyse this distortion, we apply a statistical method that was developed for microwave propagation in waveguides with random imperfections (Unger 1961). We assume the optical carrier of angular frequency ω_0 to excite only the signal mode 1 at the fibre input with a rectangular pulse of width T. Its input amplitude may be written as the following Fourier integral

$$A_1(0,t') = \frac{1}{\pi} \int_{-\infty}^{\infty} \sin\omega \frac{T}{2} \exp\{j(\omega_0+\omega)t'\} \frac{d\omega}{\omega} . \qquad (6.110)$$

At the time $t' = 0$, according to this expression, the first half of the pulse has entered the fibre.

For the sake of simplicity we disregard any material and mode dispersion and assume that the signal mode as well as all coupled modes n have frequency- independent attenuation and group delays. The output amplitude which results from equation (6.48) can then also be written in terms of Fourier integrals according to

$$A_1(L,t) = \exp(-\alpha_1 L)[A_1(0,t) - \frac{1}{\pi} \sum_n \int_{-\infty}^{\infty} \sin\omega \frac{T}{2} \exp\{j(\omega_0+\omega)t\}$$
$$Q_n(\omega_0+\omega) \frac{d\omega}{\omega}] . \qquad (6.111)$$

The time scale in this equation has been changed to

$$t = t' - \tau_1 L \qquad (6.112)$$

and accounts for the delay $\tau_1 L$ of the undistorted signal in the signal-mode. The factor $Q_n(\omega_0+\omega)$ in the Fourier integrals of reconverted wave components follows from equation (6.48) as

$$Q_n(\omega_0+\omega) = c_n^2 \int_{x=0}^{L} \int_{y=0}^{x} f(x) f(y) \exp[(\Delta\alpha_n + j\Delta\beta_n)(y-x)] \, dy \, dx . \qquad (6.113)$$

These reconverted wave components trail the primary wave

$A_1(0,t)$ when the delay of coupled modes $\tau_n L$ is larger than the delay $\tau_1 L$ of the primary wave; otherwise they arrive ahead of the primary wave. They form a pulse tail that follows or precedes the primary pulse. The nth component of this tail is given by

$$q_n(t) = \frac{1}{\pi} \int_{-\infty}^{\infty} \sin(\omega T/2) Q_n(\omega_0 + \omega) \exp(j\omega t) d\omega/\omega . \qquad (6.114)$$

We need to analyse only one of these components and can sum all significant components later on. For this one component, we drop the index n and let

$$\Delta\alpha + j\Delta\beta = \Delta\gamma(\omega_0 + \omega) = \Delta\gamma + j\Delta\tau\omega ,$$

where

$$\Delta\gamma = \Delta\gamma(\omega_0) ,$$

and

$$\Delta\tau = \tau_n - \tau_1 \qquad (6.115)$$

is the delay difference per fibre length between the parasitic mode n and the signal mode 1. By substituting from equation (6.113) into equation (6.114), our pulse tail component is

$$q(t) = \frac{c^2}{2\pi j} \int_{x=0}^{L} f_1(x) \, dx \left\{ \int_{\omega=-\infty}^{\infty} [\exp\{j\omega(t+T/2-\Delta\tau x)\} - \exp\{j\omega(t-T/2-\Delta\tau x)\}] \frac{d\omega}{\omega} \int_{y=-\infty}^{\infty} f_2(y) \times \exp(j\omega\Delta\tau y) dy \right\} , \qquad (6.116)$$

where

$$f_1(x) = f(x) \exp(-\Delta\gamma x) \qquad (6.117)$$

and

FIBRE IMPERFECTIONS   589

$$f_2(y) = \begin{cases} 0 & \text{for} \quad -\infty < y < 0 \\ f(y)\exp(\Delta\gamma y) & \text{for} \quad 0 < y < x \\ 0 & \text{for} \quad x < y < \infty \end{cases} \quad (6.118)$$

To further evaluate the double integral in the parentheses of equation (6.116), we either let

$$t_1 = t + T/2 \quad \text{or} \quad t_1 = t - T/2 . \quad (6.119)$$

The inverse Fourier transform of $f_2(y)$ is written as

$$F_2(\xi) = \frac{1}{2\pi} \int_{-\infty}^{\infty} f_2(y)\exp(j\xi y)\,dy , \quad (6.120)$$

so that

$$f_2(y) = \int_{-\infty}^{\infty} F_2(\xi)\exp(-j\xi y)\,d\xi . \quad (6.121)$$

We integrate $f_2(y)$ from $y = x$ to $y = x - (t_1/\Delta\tau)$

$$\int_{y=x}^{x-(t_1/\Delta\tau)} f_2(y)\,dy = \int_{+\infty}^{x-(t_1/\Delta\tau)} f_2(y)\,dy = j\int_{-\infty}^{\infty} \exp\{j\xi(t_1/\Delta\tau - x)\}$$

$$F_2(\xi)\,d\xi/\xi, \quad (6.122)$$

substitute for $F_2(\xi)$ from equation (6.120) and transform to $\omega = \xi/\Delta\tau$ as the new variable of integration

$$-\int_{y=x}^{x-(t_1/\Delta\tau)} f_2(y)\,dy = \frac{1}{2\pi j} \int_{\omega=-\infty}^{\infty} \exp j\omega(t_1 - \Delta\tau x)\,\frac{d\omega}{\omega} \int_{y=-\infty}^{\infty}$$

$$f_2(y)\exp(j\omega\Delta\tau y)\,dy . \quad (6.123)$$

The right-hand side of this equation appears in parentheses in equation (6.116). We substitute its left-hand side and obtain for the pulse tail

$$q(t) = c^2 \int_{x=0}^{L} f_1(x) \, dx \int_{y=x-t/\Delta\tau-T/(2\Delta\tau)}^{x-t/\Delta\tau+T/(2\Delta\tau)} f_2(y) \, dy. \qquad (6.124)$$

According to equation (6.118) $f_2(y)$ differs from zero only in the range $0 < y < x$. The double integral for the pulse tail,

$$q(t) = c^2 \int_{x(t)} \int_{y(t)} f(x) \, f(y) \, \exp\{(\Delta\alpha+j\Delta\beta)(y-x)\} dy \, dx, \qquad (6.125)$$

therefore extends only over the shaded area of the xy-plane in Fig.6.6.

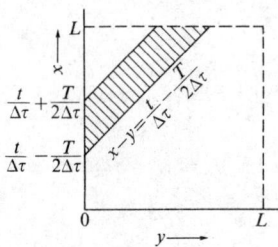

Fig.6.6. Area of integration to calculate the pulse tail according to equation (6.125).

When imperfections are randomly distributed along the fibre, we can only determine certain of the statistical quantities of the pulse tail. We assume again that the distribution of imperfections along the fibre is stationary and, for the ensemble average of $q(t)$ introduce the autocorrelation function φ of $f(x)$ according to equation (6.44)

$$\langle q(t) \rangle = c^2 \langle f^2 \rangle \int_{x(t)} \int_{y(t)} \varphi(x-y) \exp\{(\Delta\alpha+j\Delta\beta)(y-x)\} dy \, dx \qquad (6.126)$$

According to Fig.6.6, the integration extends only over regions in which

$$x - y > (t/\Delta\tau) - T/(2\Delta\tau) . \qquad (6.127)$$

If, for a given correlation distance L_k of the stationary distribution, we have

$$(t/\Delta\tau) - T/(2\Delta\tau) > L_k \qquad (6.128)$$

then the autocorrelation function $\varphi(u)$ is zero within the range of integration of equation (6.126), and the pulse tail vanishes from these times t on. Any contributions to the pulse tail which arrive later than $t = T/2 + \Delta\tau\, L_k$ at the fibre end are on average zero. Imperfections which cause these contributions are so far apart that they are no longer correlated with each other. Actually the delay difference $\Delta\tau\, L_k$ which directly coupled modes accumulated within the correlation distance is, under practical conditions, so short compared to the pulse width T, that we are only interested in what happens to the pulse tail at times $t > T/2 + \Delta\tau\, L_k$, when its direct average value is zero.

For the r.m.s. value of the pulse tail, we consider first its magnitude squared

$$|q^2(t)| = |c^4| \int_x \int_{x_1} f(x)f(x_1)\exp\{-(\Delta\alpha+j\Delta\beta)x - (\Delta\alpha-j\Delta\beta)x_1\}dx_1\, dx \times$$

$$\times \int_y \int_{y_1} f(y)f(y_1)\exp\{(\Delta\alpha+j\Delta\beta)y + (\Delta\alpha-j\Delta\beta)y_1\}dy\, dy_1. \quad (6.129)$$

We assume here that for a specific imperfection $f(z)$ one parasitic mode dominates the contributions to the pulse tail. This assumption is justified in most index profiles for those imperfections which couple the fundamental mode mainly to one other neighbouring mode. Curvature in the square-law profile is a typical example where the fundamental mode couples directly only to an LP_{11}-mode.

At times later than $t = T/2 + \Delta\tau\, L_k$ and for the range of integration in equation (6.130), $f(x)$ and $f(y)$ are not correlated. The average of equation (6.129) may therefore be taken independently for $f(x)$, and then for $f(y)$.

$$\langle |q(t)|^2 \rangle = |C|^4 (\langle f^2 \rangle)^2 \int_x \int_{x_1} \varphi(x-x_1) \exp\{-\Delta\alpha(x+x_1) - j\Delta\beta(x-x_1)\} \, dx_1 \, dx \times$$

$$\times \int_y \int_{y_1} \varphi(y-y_1) \exp\{\Delta\alpha(y+y_1) - j\Delta\beta(y_1-y)\} \, dy_1 \, dy \, . \quad (6.130)$$

Fig.6.7 shows the area of integration in the xx_1-plane and in the yy_1-plane. In the yy_1-plane the lower limit of integration over y is the larger of the two values 0 and

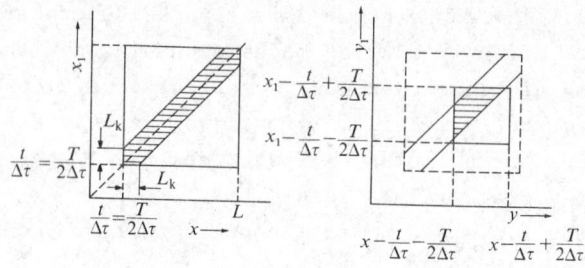

Fig.6.7. Integration ranges in the xx_1- and yy_1-planes for the mean square of the pulse tail.

$x - t/\Delta\tau - T/(2\Delta\tau)$, and for the integration over y_1 the lower limit is the larger of the two values 0 and $x_1 - t/\Delta\tau - T/(2\Delta\tau)$. To evaluate the integrals in equation (6.130), we transform variables according to

$$\begin{aligned} x + x_1 &= v \\ y_1 + y &= v' \\ x - x_1 &= u \\ y_1 - y &= u' \, . \end{aligned} \quad (6.131)$$

This transformation has the functional determinant 1/4 and

$$dx \, dx_1 \, dy \, dy_1 = (1/4) \, dv \, du \, dv' \, du' \, .$$

It changes equation (6.130) to

$$\langle |q(t)|^2 \rangle = \frac{1}{4} |C|^4 (\langle f^2 \rangle)^2 \int_u \varphi(u) \exp(-j\Delta\beta u) \, du \int_v \exp(-\Delta\alpha v) dv \times$$

$$\times \int_{u'} \varphi(u') \exp(-j\Delta\beta u') \, du' \int_{v'} \exp(\Delta\alpha v') \, dv'. \quad (6.132)$$

The new limits of integration are shown in Fig.6.8; in these limits we have abbreviated by

$$\xi = t/\Delta\tau \quad \text{and} \quad \chi = T/(2\Delta\tau) . \quad (6.133)$$

Note that always $v' > |u'|$.

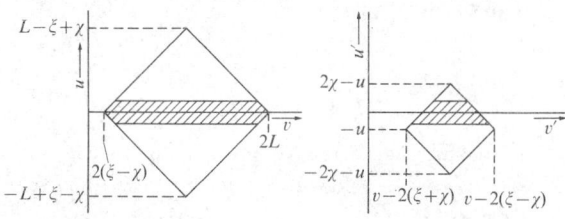

Fig.6.8. Integration ranges in the uv- and $u'v'$-planes for the mean square of the pulse tail.

The shaded regions in Fig.6.7 and 6.8 mark the autocorrelation range of the imperfections. The autocorrelation function differs from zero only in these regions. Since the correlation distance L_k may safely be assumed to always be short compared to the fibre length, the integration in the vu-plane may be extended from $u = -\infty$ to $u = \infty$. The differential loss $\Delta\alpha L_k$ within the correlation distance L_k is usually so small that $\exp(-\Delta\alpha L_k) \simeq 1$ and the integration over v may be extended from $2(\xi-\chi)$ to $2L$. Thus these integration limits become independent of u.

In the $v'u'$-plane we need to integrate only between those limits for which $u < L_k$, because only within these limits will the subsequent integration over u contribute anything. Under the above assumption that the delay difference $\Delta\tau L_k$ within the correlation distance is short compared to the pulse width, we have $\chi \ll L_k$, and the limits of integration over u'

may likewise be extended from $-\infty$ to ∞, as well as the integration limits of v' from $v - 2(\xi+\chi)$ to $v - 2(\xi-\chi)$. Finally, with these approximations, we need to restrict the integration over v' to positive values of v'. We obey this restriction by splitting the integration over v from $2(\xi-\chi)$ to $2(\xi+\chi)$, for which v' ranges between 0 and $v - 2(\xi-\chi)$, and from $v = 2(\xi+\chi)$ to $v = 2L$, for which v' ranges between $v - 2(\xi+\chi)$ and $v - 2(\xi-\chi)$. With these limits of integration the mean square pulse tail appears as

$$\langle |q(\xi)|^2 \rangle = \frac{1}{4} |C|^4 (\langle f^2 \rangle)^2 \int_{-\infty}^{\infty} \varphi(u) \exp(-j\Delta\beta u) \, du \times$$

$$\times \left[\int_{2(\xi-\chi)}^{2(\xi+\chi)} \exp(-\Delta\alpha v) dv \int_{-\infty}^{\infty} \varphi(u') \exp(-j\Delta\beta u') du' \int_{0}^{v-2(\xi-\chi)} \exp(\Delta\alpha v') dv' \right. +$$

$$\left. + \int_{2(\xi+\chi)}^{2L} \exp(-\Delta\alpha v) dv \int_{-\infty}^{\infty} \varphi(u') \exp(-j\Delta\beta u') du' \int_{v-2(\xi+\chi)}^{v-2(\xi-\chi)} \exp(\Delta\alpha v') dv' \right]. \quad (6.134)$$

This expression holds only for $\xi < L - \chi$; for $\xi > L - \chi$ the upper limit of integration over v exceeds $2L$ and thus extends beyond the total fibre length. Therefore, if $L - \chi < \xi < L + \chi$, the mean square of the pulse tail follows simply from

$$\langle |q(\xi)|^2 \rangle = \frac{1}{4} |C|^4 (\langle f^2 \rangle)^2 \int_{-\infty}^{\infty} \varphi(u) \exp(-j\Delta\beta u) du \int_{-2(\xi-\chi)}^{2L} \exp(-\Delta\alpha v) dv \times$$

$$\times \int_{-\infty}^{\infty} \varphi(u') \exp(-j\Delta\beta u') du' \int_{0}^{v-2(\xi-\chi)} \exp(\Delta\alpha v') \, dv'. \quad (6.135)$$

The integration in equations (6.134) and (6.135) over v' and v can now readily be performed. For the integration over u' and u, we introduce the power spectrum of imperfections according to equation (6.46) and obtain the mean square of the pulse tail as

FIBRE IMPERFECTIONS

$$\langle |q(\xi)|^2 \rangle = |C|^4 \frac{\phi^2(\Delta\beta)}{4\Delta\alpha^2} \exp\{-2\Delta\alpha(\xi-\chi)\}\{4\Delta\alpha\chi + (1-\exp(-4\Delta\alpha\chi)) \times$$
$$\times [2\Delta\alpha(L-2\chi) - 2\Delta\alpha(\xi-\chi) - 1]\} \quad (6.136)$$

during the time interval $\chi + \tau L_k < \xi < L - \chi$. For the remaining time, during $L - \chi < \xi < L + \chi$, the mean square pulse tail obtains as

$$\langle |q(\xi)|^2 \rangle = |C|^4 \frac{\phi^2(\Delta\beta)}{4\Delta\alpha^2} \exp\{-2\Delta\alpha(\xi-\chi)\}[2\Delta\alpha(L-\xi+\chi)-1 +$$
$$+ \exp\{-2\Delta\alpha(L-\xi+\chi)\}] . \quad (6.137)$$

We now resubstitute for ξ and χ and write the r.m.s. pulse tail directly as a function of time t and pulse width T:

$$\sqrt{\langle |q(t)|^2 \rangle} \equiv q_{r.m.s.}(t) = |C|^2 \frac{\phi(\Delta\beta)}{2\Delta\alpha} \exp\left(-\frac{\Delta\alpha}{\Delta\tau}\left(t - \frac{T}{2}\right)\right) \times$$
$$\times \left\{2 \frac{\Delta\alpha}{\Delta\tau} T + (1-\exp(-2\frac{\Delta\alpha}{\Delta\tau} T))\left[2\Delta\alpha\left(L - \frac{T}{\Delta\tau}\right) - 2\frac{\Delta\alpha}{\Delta\tau}\left(t-\frac{T}{2}\right) - 1\right]\right\}^{1/2}$$
$$(6.138)$$

for

$$T/2 + \Delta\tau L_k < t < \Delta\tau L - T/2 , \quad (6.139)$$

and

$$q_{r.m.s.}(t) = |C|^2 \frac{\phi(\Delta\beta)}{2\Delta\alpha} \exp\left(-\frac{\Delta\alpha}{\Delta\tau}\left(t - \frac{T}{2}\right)\right)\left[2\frac{\Delta\alpha}{\Delta\tau}\left(\Delta\tau L - t + \frac{T}{2}\right) +\right.$$
$$\left. + \exp\left\{-2\frac{\Delta\alpha}{\Delta\tau}\left(\Delta\tau L - t + \frac{T}{2}\right)\right\} - 1\right]^{1/2} \quad (6.140)$$

for

$$\Delta\tau L - T/2 < t < \Delta\tau L + T/2 . \quad (6.141)$$

Before we discuss this result, let us first summarize all the conditions under which only equations (6.138) and (6.140) apply:

1. Of all the parasitic modes which the fibre imperfection couples to the signal mode, one must dominate all the others, so that it contributes much more to the pulse tail than all the others.
2. The differential loss of this parasitic mode must be small enough, so that within the correlation distance, it decays not much more than the signal mode ($\Delta\alpha\, L_k \ll 1$).
3. The pulse should last longer than the delay difference between this parasitic mode and the signal mode in the correlation distance ($T > \Delta\tau\, L_k$).
4. Equations (6.138) and (6.140) apply to the time intervals indicated by equations (6.139) and (6.141), respectively. In particular, following the signal pulse $A_1(0,t)$, the time $\Delta\tau\, L_k$ must pass before equation (6.138) applies.
5. For $L_k \ll L$, the delay difference $\Delta\tau L$ over the full fibre length must be larger than the pulse width T for the time interval of equation (6.139) to exist. If $\Delta\tau L$ is only larger than half this pulse width equation (6.140) describes the r.m.s. pulse tail in the indicated range.

Fig.6.9 shows a typical response for such an imperfect fibre to the rectangular input pulse. The pulse tail follows

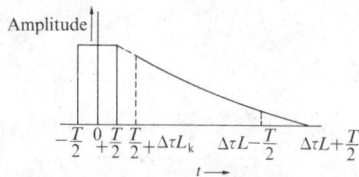

Fig.6.9. Rectangular output pulse for an imperfect fibre. From $t = T/2 + \Delta\tau L_k$ to $t = \Delta\tau L - T/2$, equation (6.138) determines the r.m.s. pulse tail and from $t = \Delta\tau L - T/2$ to $t = \Delta\tau L + T/2$ equation (6.140).

the primary signal pulse and distorts it. Altogether, this tail lasts for the delay difference between parasitic and signal mode over the full length of the fibre. In two special cases, the formulae for the r.m.s. pulse tail simplify to some extent:

1. The dominant parasitic mode has the same attenuation as the signal mode ($\Delta\alpha = 0$)

$$q_{r.m.s.}(t) = |C|^2 \phi(\Delta\beta)[(T/\Delta\tau)(L - t/\Delta\tau)]^{\frac{1}{2}} \qquad (6.142)$$

for

$$T/2 + \Delta\tau L_k < t < \Delta\tau L - T/2 , \qquad (6.143)$$

$$q_{r.m.s.}(t) = |C|^2 \frac{\phi(\Delta\beta)}{\sqrt{2}} (L - t/\Delta\tau + T/2\Delta\tau) \qquad (6.144)$$

for

$$\Delta\tau L - T/2 < t < \Delta\tau L + T/2 . \qquad (6.145)$$

2. The parasitic mode suffers so much more attenuation than the signal mode that $2\Delta\alpha L$ is the dominant term under the square root of equation (6.138)

$$q_{r.m.s.}(t) = |C|^2 \phi(\Delta\beta)\exp\{-\frac{\Delta\alpha}{\Delta\tau}(t-\frac{T}{2})\}[\frac{L}{2\Delta\alpha}(1-\exp(-2\frac{\Delta\alpha}{\Delta\tau}T))]^{\frac{1}{2}} \qquad (6.146)$$

for

$$T/2 + \Delta\tau L_k < t < \Delta\tau L - T/2 . \qquad (6.147)$$

During the time interval of equation (6.141), the pulse tail has decayed to insignificant values for such high parasitic mode loss.

According to equation (6.146), the r.m.s. pulse tail of the output signal decays exponentially; its initial amplitude increases directly proportional to the square root of fibre length but decreases with nearly the square root of differential loss. For equal attenuation of signal and parasitic mode, the r.m.s. pulse tail decays at first according to the square root dependence on time and later on linearly with time. For $\Delta\tau L$ much larger than $T/2$, its initial amplitude also increases with the square root of fibre length, and is in addition proportional to the square root of the ratio of pulse width T to delay difference $\Delta\tau$ per unit fibre length. Longer pulses accumulate more reconverted power in their tails. At any instant of time during which the pulse tail lasts, its amplitude is directly proportional to the magnitude squared of the coupling factor C and to the value of the power spectrum ϕ at the phase difference. All these characteristics allow to evaluate the pulse tail and assess its detrimental effects on single-mode signal transmission in multimode fibres with imperfections.

6.6. STEADY-STATE POWER FLOW IN MULTIMODE FIBRES

In a multimode fibre, where many modes participate in signal transmission, imperfections couple these modes and cause them to interact. Optical communication systems that employ multimode fibres as the transmission medium have detectors in their receivers, that respond not to the amplitude of individual modes but to the total power out of the fibre. Rather than analysing the amplitudes of modes along a multimode fibre, it is therefore more appropriate to determine the power in different modes and how fibre imperfections convert power between modes in propagation. For random imperfections moreover, no computations of the actual power are feasible but only predictions of what power we can expect in certain modes at certain times along the fibre. We will attempt to determine the average power under certain initial conditions for a fibre with certain statistics of its imperfections. Differential equations for the average power in randomly perturbed waveguides were first derived by Young (1963) and solved for the propagation of optical waves in hollow dielectric and metallic tubes by Bergeest and Unger (1969). In this presentation, we will follow a derivation by Marcuse (1974).

Differential equations for the change of average power in the different modes of a randomly perturbed fibre are derived when we take the ensemble average

$$P_m = \langle |A_m^2| \rangle \tag{6.148}$$

of the power in mode m, differentiate it with respect to z

$$\frac{dP_m}{dz} = \langle \frac{dA_m}{dz} A_m^* \rangle + \langle A_m \frac{dA_m^*}{dz} \rangle , \qquad (6.149)$$

and, for (dA_m/dz), substitute from the coupled wave equations (6.32):

$$\frac{dP_m}{dz} = -j \sum_n C_{mn} \langle A_n(z) A_m^*(z) f(z) \rangle \exp\{j(\beta_m - \beta_n)z\} + c.c. \quad (6.150)$$

The second term c.c. on the right-hand side denotes the complex conjugate of the first term as in equation (6.149). We assume in this differential equation an imperfection that allows a representation according to

$$c_{mn} = C_{mn} f(z) \qquad (6.151)$$

with one unique z-dependence for all coupling coefficients in equation (6.32). We furthermore let the differential attenuation between all coupled modes be low enough in the perfect reference fibre that only the phase differences need to be accounted for in the difference of their propagation constants. Any mode attenuation in the perfect fibre can be accounted for later on if it is not too high. To express the ensemble average of the triple product in equation (6.150) by average powers we start with the amplitude $A_n(z')$ of mode n at a location $z' < z$ which is separated from z by more than the autocorrelation distance L_k of the random distribution $f(z)$, i.e.

$$z - z' > L_k . \qquad (6.152)$$

From the first-order solution (6.26) of the coupled wave equation, we obtain $A_m(z)$ in terms of $A_m(z')$ and all other $A_n(z')$

$$A_m(z) = A_m(z') - j \sum_n C_{mn} A_n(z') \int_{z'}^{z} f(z'') \exp\{j(\beta_m - \beta_n)z''\} dz''. \quad (6.153)$$

When we substitute this solution for $A_n(z)$ and $A_m(z)$ into the triple product of equation (6.150), and take the ensemble average, we can disregard all terms which contain $\langle f(z) \rangle$ as a factor. This direct average of $f(z)$ can always be made zero if we only choose an appropriate reference fibre. We can furthermore disregard any terms of the triple product that contain $\langle A_m(z') A_n^*(z') \rangle$ with $m \neq n$. Different modes in the randomly perturbed fibre will have random phase shifts with respect to each other with hardly any correlation; its ensemble average will then vanish:

$$\langle A_m(z') A_n^*(z') \rangle = 0 \qquad \text{for} \quad m \neq n. \qquad (6.154)$$

Finally in the triple product of equation (6.150), we can neglect all products of the second terms of equation (6.153). These products contain $C_{mn} C_{ij} f^2(z)$ and for the weak coupling of slight imperfections are of higher order small. With these three simplifications, we arrive at the following expression for the rate of change of average power

$$\frac{dP_m}{dz} = \sum_n C_{mn} [\sum_j C_{mj} \langle A_n(z') A_j^*(z') \int_{z'}^{z} f(x) f(z) \exp\{-j(\beta_m - \beta_j)x\} dx \rangle -$$

$$- \sum_i C_{ni} \langle A_m^*(z') A_i(z') \int_{z'}^{z} f(x) f(z) \exp\{j(\beta_n - \beta_i)x\} dx \rangle] \exp\{j(\beta_m - \beta_n)z\} +$$

$$+ \text{c.c.} \quad (6.155)$$

For the ensemble averages in each term of the double summation on the right-hand side, we argue similarly as for equation (6.154). For a random distribution not only $A_n(z')$ and $A_j(z')$ are at random phase and hardly correlated to each other for $n \neq j$, but there is also hardly any correlation between $A_n(z')$ and $A_j(z')$ on one hand, and the integral as the third factor in the triple product on the other hand. What little correlation there is will diminish even further when we extend the integration further beyond the autocorrelation distance of $f(z)$. Because of this lack of correlation, it is safe to write

$$\langle A_n(z') A_j^*(z') \int_{z'}^{z} f(x) f(z) \exp\{-j(\beta_m - \beta_j)x\} dx \rangle$$

$$= \delta_{nj} \langle |A_n^2(z')| \rangle \int_{z'}^{z} \langle f(x) f(z) \rangle \exp\{-j(\beta_m - \beta_n)x\} dx. \quad (6.156)$$

For any stationary distribution of imperfections, the ensemble average $\langle f(x)\, f(z) \rangle$ equals the mean square of $f(z)$ times its autocorrelation function $\varphi(z-x)$. Since the integrals were assumed to extend beyond the autocorrelation distance, we can let their lower limit go to $-\infty$, and obtain

$$\int_{z'}^{z} \varphi(z-x)\exp\{-j(\beta_m-\beta_n)x\}\, dx = \exp\{-j(\beta_m-\beta_n)z\} \int_{0}^{\infty} \varphi(u)\exp\{-j(\beta_m-\beta_n)u\}\, du. \tag{6.157}$$

Upon introducing these simplifications into each of the double summations of equation (6.155), they reduce to single summations, and we can write for the change of average power

$$\frac{dP_m}{dz} = \sum_n c_{mn}^2\, \phi(\beta_m-\beta_n)\, [\langle |A_n^2(z')| \rangle - \langle |A_m^2(z')| \rangle]. \tag{6.158}$$

Here we have also introduced the power spectrum $\phi(\beta_m-\beta_n)$ of the random distribution as it is related by equation (6.46) to the Fourier transform of the autocorrelation function $\varphi(u)$ with $\varphi(u) = \varphi(-u)$. The differential equations (6.158) relate the rate of change of P_m at z to the average modal powers at z'.

For imperfections with short correlation distance L_k, however, z' need only be a little smaller than z without violating equation (6.152). Along the short section of fibre from z' to z, corresponding to the correlation distance L_k, the average power in any of the modes presumably changes very little, in particular for small imperfections with their weak mode coupling. We can then let

$$P_n(z) \simeq \langle |A_n^2(z')| \rangle \tag{6.159}$$

and obtain

$$\frac{dP_m}{dz} = \sum_n c_{mn}^2\, \phi(\beta_m-\beta_n)\, [P_n(z) - P_m(z)]. \tag{6.160}$$

The attenuation which each of the modes suffers in the perfect fibre due to bulk loss and power leakage in the case of leaky modes can also be accounted for by supplementing equation (6.160) with the appropriate loss term $-2\alpha_m P_m$. This term on the right-hand side will cause the average power to decay exponentially with $2\alpha_m$, which corresponds to an amplitude decay with the attenuation constant α_m. The average power of modes with only moderate attenuation in fibres with random imperfections therefore follows from the following system of coupled power equations:

$$\frac{dP_m}{dz} = -2\alpha_m P_m + \sum_n d_{mn} [P_n - P_m] , \qquad (6.161)$$

where

$$d_{mn} = C_{mn}^2 \phi(\beta_m - \beta_n) . \qquad (6.162)$$

These equations agree with what we expect physically. For a particular mode m, the factor $-2\alpha_m$ describes the relative power loss per unit distance due to mode attenuation, while $-\sum_n d_{mn}$ gives the average relative power loss per unit distance due to mode conversion into all other modes. It agrees under present assumptions with equation (6.57) for the average added loss to a single mode due to mode conversion into other modes. The additional terms which the sum $\sum_n d_{mn} P_n$ in equation (6.161) contains account for the power which on the average is added to P_m per unit distance from all other coupled modes due to mode conversion.

The coupled power equations form a system of first order differential equations with constant coefficients. Writing this system in matrix notation

$$\underline{P}' = \underline{\underline{d}} \, \underline{P} , \qquad (6.163)$$

we note that $\underline{\underline{d}}$ represents a square matrix with $-2\alpha_m - \sum_n d_{mn}$ as diagonal elements while d_{mn} are its off-diagonal elements. Because of $C_{nm} = C_{mn}$, this matrix is symmetrical with respect to its main diagonal. $\underline{\underline{d}}$ describes the average power coupling of the random imperfections; it is therefore referred to as the statistical matrix. We get the general solution of equation (6.163) when we diagonalize the statistical matrix $\underline{\underline{d}}$ by trans-

FIBRE IMPERFECTIONS

forming from the vector $\underline{P}$ of average modal power to another vector $\underline{W}$ with the eigenvector matrix $\underline{\underline{U}}$ of $\underline{\underline{d}}$ according to

$$\underline{P} = \underline{\underline{U}}\ \underline{W}\ . \tag{6.164}$$

The coupled power equations (6.163) change by this transformation to the following system of uncoupled differential equations

$$\underline{W}' = -\underline{\underline{\Gamma}}\ \underline{W}\ , \tag{6.165}$$

where $\underline{\underline{\Gamma}}$ represents a diagonal matrix. The diagonal and only elements of $\underline{\underline{\Gamma}}$ are eigenvalues of the statistical matrix $\underline{\underline{d}}$; the elements of the new power vector $\underline{W}$ correspond to the normal modes of the randomly coupled system. As normal modes of the statistical matrix, we call them statistical modes. The elements of $\underline{\underline{\Gamma}}$ as well as of $\underline{W}$ have physical meaning only as expectation values. W_n is the power which we can expect to travel in the statistical mode n and which on the average is distributed according to equation (6.164) among the actual fibre modes. The eigenvalue Γ_n is the attenuation with which we can expect the power in the statistical mode n to decay exponentially along the fibre.

In order that equation (6.164) transforms the coupled power equations (6.163) into the uncoupled equations (6.165) for the statistical modes, the following characteristic equation must be solved by any of the eigenvalues Γ_n of $\underline{\underline{d}}$

$$\det(\underline{\underline{\Gamma}} + \underline{\underline{d}}) = 0\ . \tag{6.166}$$

The determinant which this characteristic equation contains has the form

$$\begin{vmatrix} \Gamma - 2\alpha_1 - \sum_n d_{1n} & d_{12} & d_{13} & \cdots \\ d_{21} & \Gamma - 2\alpha_2 - \sum_n d_{2n} & d_{23} & \\ d_{31} & d_{32} & \Gamma - 2\alpha_3 - \sum_n d_{3n} & \\ \vdots & \vdots & \vdots & \end{vmatrix} = 0\ . \tag{6.167}$$

For interaction between a total of M modes, the determinant is of Mth degree and the characteristic equation an Mth order polynomial in Γ which has M solutions Γ_n. Not all these eigenvalues need to be different from each other. In the case of equal eigenvalues, the corresponding statistical modes are degenerate in that they have the same expectation value for loss. The elements of the eigenvector matrix $\underline{U}$ which, by equation (6.164), relate the statistical mode power to the average power in the fibre modes, follow as solutions of the linear equations

$$\underline{\underline{d}}\,\underline{U} + \underline{U}\,\underline{\Gamma} = 0 \qquad (6.168)$$

from

$$U_{kj} = q_{ik}^{(j)} \left[\sum_{k} (q_{ik}^{(j)})^2 \right]^{-1/2}, \qquad (6.169)$$

where $q_{ik}^{(j)}$ is the adjoint to the element (i,k) in the determinant of equation (6.167) with $\Gamma = \Gamma_j$. The index i is arbitrary in equation (6.169); it must only remain the same for the determination of all elements U_{kj} of one column j of the eigenvector matrix. The solutions of the homogeneous equation system (6.168) are determined only up to an arbitrary common factor. In equation (6.169), this common factor has been chosen so that all column vectors $\underline{U}_j$ of $\underline{U}$ are unit vectors. As eigenvectors of the symmetrical matrix $\underline{\underline{d}}$, the column vectors $\underline{U}_j$ are orthogonal to each other and by equation (6.169) are so normalized that they satisfy the orthonormality condition

$$\underline{U}_j\,\underline{U}_k \equiv \sum_{i} U_{ij}\,U_{ik} = \delta_{jk}. \qquad (6.170)$$

Eigenvectors $\underline{U}_j$ from equation (6.169) which belong to degenerate eigenvalues are not automatically orthogonal to each other. But we can always find linear combinations of them which will satisfy the orthonormality condition (6.170).

With

$$W_n(z) = W_n(0)\exp(-\Gamma_n z) \qquad (6.171)$$

FIBRE IMPERFECTIONS 605

as the statistical-mode solution of equation (6.165) in terms of its initial conditions $W_n(0)$, the general solution for the average modal power along the fibre can be written as

$$P_m(z) = \sum_n U_{mn} W_n(0) \exp(-\Gamma_n z) . \qquad (6.172)$$

The initial values for the statistical mode powers are obtained in terms of the actual power distribution at the fibre input when we multiply equation (6.172) by U_{mj}, sum over all m, and use the orthonormality condition (6.170)

$$W_j(0) = \sum_m U_{mj} P_m(0) . \qquad (6.173)$$

Equation (6.172) together with equation (6.173) gives the average power in each mode along the imperfect fibre in terms of the initial power distribution $P_m(0)$.

By numbering the statistical modes in the order of increasing attenuation according to

$$\Gamma_1 < \Gamma_2 < \Gamma_3 \ldots , \qquad (6.174)$$

the first statistical mode W_1 with its eigenvector $\underline{U}_1$ will have the lowest attenuation Γ_1. Whatever the initial power distribution $P_m(0)$, in a fibre of length

$$L \gg L_s = (\Gamma_2 - \Gamma_1)^{-1} , \qquad (6.175)$$

this mode will dominate and determine the distribution of average output power according to

$$P_m(L) \simeq U_{m1} \exp(-\Gamma_1 L) \sum_n U_{n1} P_n(0) . \qquad (6.176)$$

Once an imperfect fibre exceeds L_s from equation (6.175), the distribution of average power reaches a stationary state, which contains only the statistical mode of lowest attenuation. In this stationary state, the effective multimode attenuation and transmission loss is determined by the lowest eigenvalue Γ_1 of the characteristic equation (6.167).

Simple solutions of this equation exist for two limiting

cases. The first case requires a statistical matrix $\underline{d}$ which is almost diagonal with only small elements d_{mn} off the diagonal. In this case, the eigenvalues are approximately given by

$$\Gamma_m = 2\alpha_m + \sum_n d_{mn} + \frac{1}{2} \sum_{n \neq m} \frac{d_{mn} d_{nm}}{\alpha_m - \alpha_n + \sum_i (d_{mi} - d_{ni})} \quad . \quad (6.177)$$

In addition to the requirement of an almost diagonal matrix $\underline{d}$, this approximation also assumes that none of the coupled modes have equal attenuation; it even requires that the off-diagonal elements d_{mn} of the statistical matrix are small compared to the difference between the diagonal elements which belong to the modes that are coupled by d_{mn}:

$$d_{mn} \ll |\alpha_m - \alpha_n + \sum_i (d_{mi} - d_{ni})| \quad . \quad (6.178)$$

Under these conditions of weak interaction or large differential attenuation, we can expect an attenuation for the statistical modes that differs only a little from the average attenuation $2\alpha_m$ of modal power in a perfect fibre. This difference in loss consists of the sum $\sum d_{mn}$ of all average loss factors due to power conversion into other modes plus additional contributions from all coupled modes which depend on the differential loss $\alpha_m - \alpha_n$. Interaction with a mode n which has $\alpha_n > \alpha_m$ lowers the statistical mode attenuation by this contribution, while a coupled mode n with $\alpha_n < \alpha_m$ raises the attenuation of the statistical mode m. If the mode with $m = 1$ has the lowest attenuation constant α_1 of all guided modes, then the first effect of weak interaction on the corresponding statistical mode attenuation Γ_1 will be an increase of $2\alpha_1$ by $\sum d_{1n}$. As a second-order effect, however, this increased statistical mode attenuation will be lowered by all the negative terms of the second sum in equation (6.177). Altogether, therefore, the lowest loss statistical mode has a loss constant that is lower than the corresponding diagonal element in the statistical matrix.

The distribution of power among the different modes changes only slightly along the fibre for this case of weak

coupling or large differential loss. Whatever power was
launched into the modes of lowest attenuation maintains its
distribution along the fibre in these modes. Only a small
amount of energy converts to other modes; but the respective
mode attenuation absorbs this energy at a higher rate than it
converts. If only one mode, preferably the dominant HE_{11}-
mode, is initially launched into the fibre, most of the energy
will remain in this mode, and the weak interaction with other
modes will only raise its loss according to equation (6.177).

Another limiting case for which the characteristic equation (6.167) can be solved approximately to determine attenuation and power distribution of statistical modes is that of
strong interaction. For this case, we assume that the off-
diagonal interaction terms of the statistical matrix are all
much larger than the largest difference in attenuation
$2(\alpha_{max}-\alpha_{min})$ between any of the modes that participate in
interaction due to the fibre imperfections. Usually the
lowest-order mode which is involved in the interaction process will also have the lowest attenuation. In all practical
cases, this will be the fundamental HE_{11}-mode of the fibre.
On the other hand, the highest-order mode in the interaction
process will likewise have the highest attenuation. We will
now assume

$$d_{nm} \geqslant 2(\alpha_{max}-\alpha_{min}) \qquad (6.179)$$

and call it the condition for strong interaction. Under practical conditions, this assumption will actually be never
satisfied for all modes. Random imperfections in fibres will
always convert energy from lower- to higher-order modes and
eventually to leaky modes with their high attenuation. We
will nevertheless assume equation (6.179) to be satisfied
and disregard the interaction with leaky modes or the radiation field. To solve the characteristic equation (6.167)
under the condition (6.179), we sum all columns except the
first in the determinant of equation (6.167), and add this
sum to the first column:

$$\begin{vmatrix} \Gamma-2\alpha_1 & d_{12} & d_{13} & \cdots \\ \Gamma-2\alpha_2 & \Gamma-2\alpha_2-\sum_n d_{2n} & d_{23} & \cdots \\ \Gamma-2\alpha_3 & d_{31} & \Gamma-2\alpha_3-\sum_n d_{3n} & \cdots \\ \vdots & \vdots & \vdots & \end{vmatrix} = 0 \ . \quad (6.180)$$

In this process, we did not change the value of the determinant, but in its first column have now only differences $\Gamma-2\alpha_n$ between the statistical mode attenuation Γ and mode attenuation $2\alpha_n$. We continue now with the assertion that a Γ-value of the same order as the mode attenuation $2\alpha_n$ will solve equation (6.180). The solution which we presently obtain will prove this assertion to be true. For $\Gamma \simeq 2\alpha_n$ and under the condition (6.179) for strong interaction, the attenuation differences $\Gamma-2\alpha_n$ will all be small compared to $\sum_n d_{mn}$ and may be neglected in the diagonal terms of the second and all following columns. The characteristic equation now reduces to

$$\begin{vmatrix} \Gamma-2\alpha_1 & d_{12} & d_{13} & \cdots \\ \Gamma-2\alpha_2 & -\sum d_{2n} & d_{23} & \cdots \\ \Gamma-2\alpha_3 & d_{3n} & -\sum d_{3n} & \cdots \\ \vdots & \vdots & \vdots & \end{vmatrix} = 0 \ . \quad (6.181)$$

The determinant in this equation has identical adjoints to all elements of the first column. This statement can easily be verified by adding suitable rows to other rows in any of the subdeterminants to the first column, without changing the value of these determinants in the process. With all these adjoints equal to each other, we truncate the determinant at M rows and M columns and extract the factor $M\Gamma - 2\sum_1^M \alpha_n$ from it. From this factor, we obtain

$$\Gamma_1 = \frac{2}{M} \sum_{n=1}^{M} \alpha_n \qquad (6.182)$$

as one solution of the characteristic equation. It constitutes the arithmetic mean of the attenuation constants of all modes that we take into consideration, and satisfies our assertion $\Gamma \simeq 2\alpha_n$. Under present conditions, this is the smallest eigenvalue of the characteristic equation and hence the attenuation to be expected for the statistical mode of lowest loss. The eigenvector $\underline{U}_1$ of this lowest-order statistical mode W_1 has also a particularly simple form. Since all the adjoints to the first column of the determinant in equation (6.181) are equal, we obtain from equation (6.177) identical values

$$U_{ki} = 1/\sqrt{M} \qquad (6.183)$$

for all elements of the eigenvector.

Whatever may be the distribution of power among the different modes of the fibre from its initial excitation, eventually, for a long enough fibre, this statistical mode with its lowest attenuation will dominate all the other modes. The average power P_n for this statistical mode of lowest loss is equal for all fibre modes. No matter in what distribution the power is initially launched into the fibre, and how the different statistical modes are excited, the strong interaction between modes eventually leads to an even distribution of power among all coupled modes. Any statistical modes of higher loss have died out at such distance from the excitation.

Under practical conditions mode attenuation and mode coupling have values in between these two extreme cases of weak and strong interaction. These more general conditions require a numerical solution of the characteristic equation for the attenuation of statistical modes and an evaluation of the eigenvector matrix $\underline{U}$ in equation (6.164) to determine their average modal powers. Most significant in these calculations will always be the statistical mode of lowest loss and the fibre length which this mode needs to become dominant for a given initial distribution of power.

6.7. POWER TRANSIENTS IN MULTIMODE FIBRES

The steady-state solution of coupled power equations for an imperfect multimode fibre only gives information about the transmission loss that we can expect in such a fibre. To also determine the effects which fibre impefections have on signal distortion, we need to consider the transient state and, for this purpose, must generalize the coupled power equations. In a perfect fibre without any coupling between modes, the signal power in a mode propagates with the group velocity v_g and suffers the group delay τ_m of this mode m per unit distance along the fibre. If we neglect mode and material dispersion, or if the signal power concentrates in a very narrow spectral range around the frequency of the optical carrier, the signal will not be distorted. Only the frequency dependence of group delay would cause such distortion and this we will neglect here, and let the group delay be independent of frequency. A signal in mode m with an initial form of its power as a function of time $P_m(0,t) = P_m(t)$ will maintain this form along the fibre and will only be delayed by $\tau_m z$. At z the signal form is therefore given by

$$P_m(z,t) = P_m(t - \tau_m z) . \qquad (6.184)$$

The differential equation which has this particular dependence on z and t as its solution and therefore describes this undistorted signal propagation properly, is

$$\frac{\partial P_m}{\partial z} + \tau_m \frac{\partial P_m}{\partial t} = 0 . \qquad (6.185)$$

If, in addition to the delay τ_m, the signal suffers power loss according to $2\alpha_m$, we must add the corresponding loss term to equation (6.185) and obtain

$$\frac{\partial P_m}{\partial z} + \tau_m \frac{\partial P_m}{\partial t} = -2\alpha_m P_z \qquad (6.186)$$

as the differential equation which, with its solution

$$P_m(z,t) = P_m(t-\tau_m z)\exp(-2\alpha_m z) , \qquad (6.187)$$

describes undistorted signal propagation with delay τ_m and power loss $2\alpha_m$ per unit fibre length.

To include this transient behaviour in the coupled power equations for the average modal powers of an imperfect fibre, we let this average power be not only a function of position but also of time, and add the term $\tau_m \, \partial P_m/\partial t$ to each of its left-hand sides. This leads from equation (6.161) to the following system of partial differential equations:

$$\frac{\partial P_m}{\partial z} + \tau_m \frac{\partial P_m}{\partial t} = -2\alpha_m P_m + \sum_n d_{mn}(P_n - P_m) . \qquad (6.188)$$

This equation system can, of course, describe transient phenomena only approximately. On its right-hand sides it still uses the average steady-state solutions of the coupled wave equations to account for the rate of change of average power along the fibre. It will therefore only hold for transients that differ not too much from the steady state, and only for interaction between modes which all travel with nearly the same delay. If coupled modes differ too much in delay, and if, for example, short pulses are launched into these modes simultaneously, they would spread apart along the fibre, and the steady-state interaction process, which the right-hand sides of equation (6.161) imply, no longer applies. Very fortunately, however, and as the solution of equation (6.188) will reveal to us presently, the interaction between modes has the tendency to keep pulses in coupled modes from spreading apart. Equation (6.188) is therefore useful for a wider range of delay differences and faster transients, and for stronger interaction between modes, than we might first expect.

For the solution of equation (6.188), we generalize the steady-state solution (6.172) of equation (6.161) in terms of statistical modes and try the following series representation in terms of Fourier integrals (Marcuse 1974)

$$P_m(z,t) = \frac{1}{2\pi} \sum_i \int_{-\infty}^{\infty} U_{mi}(\omega) W_i(\omega) \exp[j\omega t - \Gamma_i(\omega)z] \, d\omega . \qquad (6.189)$$

Any one term of this series representation has the Fourier transform

$$p_{mi}(z,j\omega) = U_{mi}(\omega) W_i(\omega) \exp[-\Gamma_i(\omega)z] \qquad (6.190)$$

and constitutes the contribution to the average power in mode m from the statistical mode i. The contributions from this statistical mode i to all of the modes m must together solve the Fourier transform of the equation system (6.188) with respect to t. This Fourier transform of equation (6.188) reads

$$\frac{dp_m}{dz} + j\omega \tau_m p_m = -2\alpha_m p_m + \sum_n d_{mn}(p_n - p_m) . \qquad (6.191)$$

When we substitute for p_m the Fourier transforms p_{mi} from equation (6.190), we obtain the following system of linear equations:

$$\sum_n \left\{ d_{mn} + [\Gamma(\omega) - j\omega \tau_m - 2\alpha_m - \sum_n d_{mn}] \delta_{mn} \right\} U_{mn}(\omega) = 0 \qquad (6.192)$$

for the components $U_{mn}(\omega)$ of the eigenvector $\underline{U}_n(\omega)$ of the statistical mode n. It is a homogeneous system, so that for non-trivial solutions the determinant of its coefficients must vanish:

$$\det \left\{ d_{mn} + [\Gamma(\omega) - j\omega \tau_m - 2\alpha_m - \sum d_{mn}] \delta_{mn} \right\} = 0. \qquad (6.193)$$

The $\Gamma(\omega)$-values which solve this characteristic equation are the propagation constants of statistical modes. If all modes m propagate with equal delay $\tau_m = \tau_0$, then eigenvalues which follow for $\Gamma(\omega) - j\omega \tau_0$ from equation (6.192) are identical to the steady-state eigenvalues Γ_i of statistical modes from equation (6.166). In this case, the statistical modes propagate with

$$\Gamma_i^{(0)} = \Gamma_i + j\omega \tau_0 , \qquad (6.194)$$

where Γ_i is the steady-state attenuation of the statistical mode i. Under this condition, the statistical modes all propagate with the same delay as the actual modes. In general, however, the eigenvalues from equation (6.193) will be more complicated functions of jω which differ from eigenvalue to eigenvalue. Once these eigenvalues have been determined as solutions of equation (6.193), the components $U_{mi}(\omega)$ of the eigenvectors can also be found by solving equation (6.192) according to the same formula (6.169) as for eigenvector components of steady-state statistical modes. The matrix of coefficients in equation (6.192) is symmetrical, as is the statistical matrix in equation (6.168) for the steady-state solutions. The eigenvectors $\underline{U}_m(\omega)$ are therefore orthogonal to each other and, with the normalization of equation (6.169), obey the orthonormality condition

$$\underline{U}_i(\omega) \, \underline{U}_k(\omega) = \delta_{ik} \,. \tag{6.195}$$

This condition allows a convenient expansion of the average mode power as a function of time and distance along the fibre in terms of the initial time function $P_m(0,t)$ of power in each fibre mode.

For this expansion, we take the series representation of mode power in equation (6.189) at $z = 0$ and invert the Fourier integral to obtain the Fourier transform

$$\sum_i U_{mi}(\omega) \, W_i(\omega) = \int_{-\infty}^{\infty} P_m(0,t) \exp(-j\omega t) \, dt \,. \tag{6.196}$$

We now multiply this equation by $U_{mk}(\omega)$, sum over all m, and utilize the orthonormality condition (6.195):

$$W_k(\omega) = \sum_m U_{mk}(\omega) \int_{-\infty}^{\infty} P_m(0,t) \exp(-j\omega t) \, dt \,. \tag{6.197}$$

With these expressions for the power in each statistical mode, we can, in principal, evaluate the series representation (6.189) for average power in each of the modes. Any attempt to determine the propagation of signals from these general expressions would, however, meet with considerable difficulties. Not only must we solve the polynomial in $\Gamma(\omega)$ of the characteristic equation (6.193), we must also determine the eigenvector components from the linear equation system (6.193) and execute the double summation and double integration which equations (6.197) and (6.189) involve.

We therefore resort to a perturbation method to obtain an approximate solution (Marcuse 1974). We base the perturbation solution on the steady-state solution of the coupled power equations and expand eigenvalues and eigenvector components into series of terms with ascending powers of ω

$$\Gamma_i(\omega) = \Gamma_i^{(0)} + j\omega\, \Gamma_i^{(1)} + \omega^2\, \Gamma_i^{(2)} + \ldots \qquad (6.198)$$

$$U_{mi}(\omega) = U_{mi} + j\omega\, U_{mi}^{(1)} + \omega^2\, U_{mi}^{(2)} + \ldots \qquad (6.199)$$

The zero-order term $\Gamma_i^{(0)}$ in the series expansion of $\Gamma_i(\omega)$ follows from equation (6.194) with the eigenvalue Γ_i of the steady-state statistical mode and τ_0 as a suitable average value for the delay per fibre length of all coupled modes. The zero-order term U_{mi} in the expansion of $U_{mi}(\omega)$ represents the eigenvector component i for the steady-state statistical mode m. All important characteristics of the transient solution are still obtained when we approximate $U_{mi}(\omega)$ by just this zero-order term:

$$U_{mi}(\omega) = U_{mi}. \qquad (6.200)$$

The eigenvalue $\Gamma_i(\omega)$, however, requires a more accurate representation, including in addition to $\Gamma_i^{(0)}$ the first- and second-order terms. $\Gamma_i(\omega)$ enters the exponent in equation (6.189) where for large z small errors in $\Gamma_i(\omega)$ make a substantial difference.

As a specific signal form to study the effects of mode interaction, we take the Gaussian time function of equation (1.65) and excite all modes simultaneously, but with different peak powers P_{m0}, with the signal

$$P_m(0,t) = P_{m0}\, \exp[-(2t/t_0)^2]. \qquad (6.201)$$

For this initial excitation, the statistical modes have from equation (6.197) the following Fourier transforms:

$$W_i(\omega) = (W_{i0} \, t_0 \, \sqrt{\pi}/2) \exp[-(\omega t_0/4)^2] \qquad (6.202)$$

where

$$W_{i0} = \sum_m U_{mi} \, P_{m0} \qquad (6.203)$$

from the initial power distribution. The average power in each mode along the fibre is given by equation (6.189) when these Fourier transforms $W_i(\omega)$ are substituted:

$$P_m(z,t) = \frac{t_0}{4\sqrt{\pi}} \sum_i W_{i0} U_{mi} \exp(-\Gamma_i z) \int_{-\infty}^{\infty} \exp\{j\omega[t-(\tau_0+\Gamma_i^{(1)})]\} \exp\{-\omega^2[t_0^2/16+\Gamma_i^{(2)} z]\} d\omega. \qquad (6.204)$$

The inverse Fourier transform of this Fourier integral representation extends its limits to infinity. The exponent in its integrand, however, contains the series expansion of $\Gamma_i(\omega)$ only to second order in powers of ω. Hence, in this perturbation approximation results will only be accurate for sufficiently large distances along the fibre where, due to the factor $\exp[-\omega^2 \Gamma_i^{(2)} z]$, the integrand decays to small enough values, when ω becomes larger in the course of integration. The perturbation solution cannot be applied to short fibres or to fibres close to the point of excitation. z must be large or the fibre long enough for this condition to prevail. This is no real handicap, because for signal transmission, we are mainly interested in the response at the end of long fibres. Evaluation of the definite integrals in equation (6.204) leads to the following average powers at a large enough distance from the point of excitation

$$P_m(z,t) = \sum_i W_{i0} U_{mi} \exp(-\Gamma_i z) \frac{t_0}{(t_0^2+16\Gamma_i^{(2)} z)^{\frac{1}{2}}} \exp\left\{-4 \frac{[t-(\tau_0+\Gamma_i^{(1)})z]^2}{(t_0^2+16\Gamma_i^{(2)} z)}\right\}. \qquad (6.205)$$

Each statistical mode i contributes to the average power according to its initial strength of excitation W_{i0} and its eigenvector component U_{mi}. Each of these contributions has the Gaussian time function $\exp\{-[2(t-\tau_i z)/t_i]^2\}$ according to which it is delayed by

$$\tau_i z = (\tau_0 + \Gamma_i^{(1)}) z \qquad (6.206)$$

and broadened to a pulse width

$$t_i = (t_0^2 + 16\Gamma_i^{(2)} z)^{1/2} \qquad (6.207)$$

from the pulse width t_0 of the initial excitation. It has also decayed to $(t_0/t_i)\exp(-\Gamma_i z)$ of its initial peak power, due mainly to the steady-state attenuation Γ_i of the statistical mode.

At the end of a long fibre, the first statistical mode with the lowest steady-state attenuation Γ_1 will dominate all other statistical modes, no matter what the initial excitation. The average output power in each of the fibre modes m is then simply given by

$$P_m(L,t) = W_{10} U_{m1} (t_0/t_1) \exp(-\Gamma_1 L) \exp\left\{-[2(t-\tau_1 L)/t_1]^2\right\}. \qquad (6.208)$$

With respect to the input pulse, it is delayed according to the delay per fibre length

$$\tau_1 = \tau_0 + \Gamma_1^{(1)} \qquad (6.209)$$

of the lowest-loss statistical mode, and the pulse has broadened to

$$t_1 = (t_0^2 + 16\Gamma_1^{(2)} L)^{1/2} . \qquad (6.210)$$

For short input pulses, this response becomes independent of the input pulse width and, obviously, also of the input pulse shape. In the limit

$$P_m(0,t) = \lim_{t_0 \to 0} 2/(t_0 \sqrt{\pi}) \, E_{m0} \, \exp[-(2t/t_0)^2] , \qquad (6.211)$$

the input signal into each mode consists of the impulse

$$P_m(0,t) = E_{m0} \, \delta(t) \qquad (6.212)$$

of energy E_{m0} with $\delta(t)$ as Dirac's delta function. The impulse response due to the statistical mode of lowest loss is

$$P_m(L,t) = (\sum_k U_{k1}E_{k0})U_{m1}(1/2)(\pi\Gamma_1^{(2)}L)^{-1/2}\exp(-\Gamma_1 L)\exp\left\{-[(t-\tau_1 L)/(2(\Gamma_1^{(2)}L)^{\frac{1}{2}})]^2\right\}.$$

(6.213)

It has a Gaussian time function of pulse width

$$t_1 = 4(\Gamma_1^{(2)}L)^{\frac{1}{2}}.$$

(6.214)

In the perfect multimode fibre, in which all modes propagate independently from each other, the impulse response follows under corresponding conditions from equation (5.250). Due to the delay differences, the pulse spreads apart, and the width of the impulse response increases linearly with length. The total delay spread $\Delta\tau$ of all modes in the perfect fibre directly determines the rate of pulse spreading. In an imperfect fibre for strong interaction between modes, the power distributes more or less evenly among all modes. While propagating along the fibre, it changes from mode to mode and on the average has travelled for an equal time fraction in all modes. On the average then, it has also suffered the average delay $\tau_0 + \Gamma^{(1)}$ per fibre length of the lowest-loss statistical mode. Pulse broadening occurs only due to the variance of modal delays, and as such is proportional not to fibre length, but only to the square root of length.

The factor $\Gamma_1^{(2)}$ of the second-order term in the eigenvalue expansion of the lowest-loss statistical mode is responsible for the pulse broadening in equations (6.210) and (6.214). This factor has been determined by Marcuse (1974). His general result is, however, difficult to discuss and allows hardly any general conclusions. We therefore take as a simple example the interaction between only two modes, to recognize how $\Gamma_1^{(2)}$ depends on the magnitude and distribution of imperfections. For only two modes, the characteristic equation (6.192) has the form

$$\begin{vmatrix} \Gamma(\omega)-j\omega(\tau_0-\Delta\tau)-2\alpha_1-d_{12} & d_{12} \\ d_{12} & \Gamma(\omega)-j\omega(\tau_0+\Delta\tau)-2\alpha_2-d_{12} \end{vmatrix} = 0 \quad (6.215)$$

with

$$\tau_0 = (\tau_1+\tau_2)/2 \quad (6.216)$$

as the average group delay per fibre length and

$$\Delta\tau = (\tau_2-\tau_1)/2 \quad (6.217)$$

as the deviation of mode delay from this average. The quadratic equation (6.215) is solved by

$$\Gamma_{1,2}(\omega) = \alpha_1+\alpha_2+d_{12}+j\omega\tau_0 \mp [d_{12}^2+(\alpha_2-\alpha_1)^2+j\omega 2\Delta\tau(\alpha_2-\alpha_1)-\omega^2\Delta\tau^2]^{\frac{1}{2}}. \quad (6.218)$$

The expansion of this expression in terms of ascending powers of ω leads to

$$\Gamma_{1,2} = \alpha_1+\alpha_2+d_{12} \mp [d_{12}^2+(\alpha_2-\alpha_1)^2]^{\frac{1}{2}} \quad (6.219)$$

as the steady-state eigenvalues of statistical modes, and

$$\Gamma_{1,2}^{(1)} = \mp \Delta\tau(\alpha_2-\alpha_1)[d_{12}^2 + (\alpha_2-\alpha_1)^2]^{-1/2} \quad (6.220)$$

for the deviation per fibre length of the statistical mode delay from τ_0. The second-order term in the expansion of $\Gamma_{1,2}(\omega)$ has the factor

$$\Gamma_{1,2}^{(2)} = \pm(d_{12}\,\Delta\tau/2)^2[d_{12}^2 + (\alpha_2-\alpha_1)^2]^{-3/2}. \quad (6.221)$$

Under normal conditions, the lower-order mode 1 of the fibre suffers less delay and loss than the higher-order mode 2 ($\tau_1 < \tau_2$ and $\alpha_1 < \alpha_2$). Under these conditions, the low-loss statistical mode with the loss constant Γ_1 has $\Gamma_1^{(1)} < 0$ and travels with less than the average delay τ_0. Its dispersion coefficient $\Gamma_1^{(2)}$ results from equation (6.221) as a positive real quantity. The high-loss statistical mode with $\Gamma_2 > \Gamma_1$ travels with more than the average delay. But its coefficient $\Gamma_2^{(2)}$ is negative, and hence no real quantity obtains from equation (6.221) for the broadening of its impulse response. This result indicates that this high-loss statistical mode has not the physical reality of the low-loss statistical mode; it only serves to describe the transition from an initial excitation into the stationary distribution of average mode power according to the low-loss statistical mode.

For strong enough interaction between the two modes so that

$$d_{12} \gg |\alpha_2 - \alpha_1|,$$

the statistical mode losses reduce to

$$\Gamma_1 = \alpha_1 + \alpha_2 - \tfrac{1}{2}(\alpha_2 - \alpha_1)^2/d_{12} \qquad \Gamma_2 = 2d_{12} + \alpha_1 + \alpha_2 + \tfrac{1}{2}(\alpha_2 - \alpha_1)^2/d_{12}.$$
(6.222)

In this case, we have $\Gamma_2 \gg \Gamma_1$, and it takes only a short distance along the fibre to establish the stationary distribution of the low-loss statistical mode. The first-order coefficients of the eigenvalue expansion reduce for strong interaction to

$$\Gamma_{1,2}^{(1)} = \mp \Delta\tau(\alpha_2 - \alpha_1)/d_{12}.$$
(6.223)

This quantity is extremely small so that both statistical modes propagate with nearly the average mode delay τ_0. The second-order coefficients of the eigenvalue expansion reduce for strong interaction to

$$\Gamma_{1,2}^{(2)} = \pm (\Delta\tau/2)^2/d_{12}$$
(6.224)

and the impulse response broadens now to

$$t_1 = 2\Delta\tau (L/d_{12})^{\frac{1}{2}}.$$
(6.225)

This pulse width is proportional to the deviation $\Delta\tau$ of mode delays per fibre length from their average delay τ_0 and, according to equation (6.162), also increases linearly with the reciprocal of the coupling factor C_{12} but with the square root of fibre length L and the reciprocal of the power spectrum ϕ at the spatial frequency $\Omega = \Delta\beta$. Compared to the spacing of the two impulses in the impulse response for two uncoupled modes

$$t_{10} = 2\Delta\tau L, \qquad (6.226)$$

it has changed by the factor

$$t_1/t_{10} = 1/\sqrt{d_{12}L} . \qquad (6.227)$$

If we define a coupling length by

$$L_c = 1/d_{12} , \qquad (6.228)$$

then we have

$$t_1 = 2\Delta\tau (L_c L)^{\frac{1}{2}} , \qquad (6.229)$$

and at $L = L_c$, the impulse response with mode interaction lasts just as long as the spacing between two impulses in the impulse response for uncoupled modes. For $L > L_c$, we obtain $t_1 < t_{10}$ from $t_1 = t_{10}(L_c/L)^{\frac{1}{2}}$. The mode mixing due to fibre imperfections then narrows the output pulses as compared to the perfect fibre. These results have been obtained for interaction between only two modes but in principle they also hold for interaction between the many modes of a multimode fibre. The delay deviation $\Delta\tau$ and power coupling d_{12} as well as coupling length L_c must then only be replaced by suitable average values.

6.8. POWER DIFFUSION APPROXIMATION

The solution of coupled power equations for interaction between only two modes already reveals important transmission characteristics of multimode fibres with random imperfections. More accurate information still can be expected when instead

of just two modes, we consider all guided modes but let only
those modes couple directly with each other which have neighbouring mode orders. A typical case for next-neighbour coupling is the graded-index fibre with a parabolic index profile
and with curvature as its imperfection. According to equation (6.69), only those coupling coefficients differ from
zero which couple the LP_{lp}-mode to the $LP_{l\pm1,p}$-modes and to
the $LP_{l+1,p-1}$- and $LP_{l-1,p+1}$-modes. Curvature in fibres of
other index profiles also couples LP_{lp}-modes to $LP_{l\pm1,q}$-modes
with other radial orders q. But the coupling factor $C_{\mu\nu}$ in
equation (6.162) decreases quite rapidly with increasing difference Δm in the compound mode number $m = 2p + l$ of coupled
modes. The same decay in coupling strength between modes with
increasing difference in their radial and circumferential mode
orders can also be expected for other fibre imperfections, as
long as they involve only smooth deviations in their refractive index from the nominal distribution of the perfect fibre.
The other factor in equation (6.162) that determines mode
interaction for random distributions of imperfections is the
value of their power spectrum at the spatial frequency $\Omega = \Delta\beta$.
If imperfections are correlated over longer fibre distances,
the power spectrum drops off rapidly with increasing Ω and
coupling between modes of some phase difference causes only
weak interaction in the coupled power equations.

For these reasons, we arrive at a good approximation for
the coupled power equations when we consider interaction only
between modes that are next to each other in mode order and
phase constant. Guided modes of graded-index fibres with a
parabolic index profile propagate with phase constants according to equation (5.64). All LP_{lp}-modes in these fibres with
the same compound mode number

$$m = 2p + l$$

have the same phase constant and form a group of degenerate
modes. Each such group has $m - 1$ modes of one polarization
with circumferential orders

$$l = 0, 2, 4 \ldots m-2 \qquad \text{for } m \text{ even}$$

and

$$l = 1,3,5 \ldots m-2 \qquad \text{for } m \text{ odd}.$$

In graded-index fibres with index profiles not too different from the parabolic distribution, the LP_{lp}-modes with the same compound mode number $m = 2p + l$ are not exactly degenerate but they differ only very little in phase constant. This may be inferred from the approximate characteristic equation (5.166) for power-law profiles. The factor $4[2/(2+\alpha)]^{2/\alpha}$, by which p on the right-hand side of this equation is multiplied, equals 2 for the parabolic index profile. If α deviates from 2 according to $\alpha = 2 + \nu$, this factor deviates from 2 only by terms of order ν^2 and higher. Even in the fibre with a homogeneous core, the characteristic equation (4.15) reduces to

$$u[1 + l^2/(2u^2)] \simeq (\pi/2)(2p+l) \qquad (6.230)$$

for all circumferential orders l that are small compared to the transverse phase parameter $u = (n_1^2 k^2 - \beta^2)^{1/2} a$. Hence under both these conditions, for graded-index fibres of nearly parabolic profile and for low enough circumferential orders in the homogeneous core, the compound mode number $m = 2p + l$ characterizes groups of nearly degenerate modes. Since the near degeneracy holds for these limiting cases of practical fibres, we can also expect it to apply for all intermediate index distributions. We therefore let each compound mode number m designate a group of $m - 1$ nearly degenerate modes.

We assume all modes of each nearly degenerate group to have the same loss α_m in the perfect fibre, and let this loss differ, if at all, only from mode group to mode group. Furthermore, we assume the interaction between these nearly degenerate modes to be strong enough, so that they all have the same average power $\bar{P}_m$ along the fibre. One of these assumptions supports the other, because for equal loss α_m the power coupling according to equation (6.162) needs only be small to establish the conditions for strong interaction. Alternatively, we arrive at an equal power distribution among the modes

of a nearly degenerate group m when we assume the initial
excitation to launch the same amount of power into each of
them. Power coupling between modes of degenerate groups m
which are not next to each other in their compound mode number
m is neglected in this approximate analysis. Under these conditions, we can sum all coupled power equations (6.188) for
the $(m-1)$-modes of the degenerate group m and obtain for the
average power $\overline{P}_m$ in any of the modes of a degenerate mode
group m

$$(m-1)\left(\frac{\partial \overline{P}_m}{\partial z} + \tau_m \frac{\partial \overline{P}_m}{\partial t} + 2\alpha_m \overline{P}_m\right) = m\, d_m(\overline{P}_{m+1} - \overline{P}_m) - (m-1)d_{m-1}(\overline{P}_m - \overline{P}_{m-1}). \tag{6.231}$$

The average coupling factors d_m in this sum of equations designate the sum of all coupling factors $d_{\mu\nu}$ with which a mode
of compound mode number m couples to modes of compound mode
number $m + 1$. According to this sum of coupled power equations, the $(m-1)$-modes of one polarization in the degenerate
group m interact with the m modes of the degenerate group
$m + 1$, while the modes of group $m - 1$ interact with the $(m-1)$-
modes of group m. Although this approximation relies on the
specified degeneracy of modes in fibres with a parabolic index profile or on the corresponding near degeneracy of modes
in other power-law profiles and the step-index fibre, it
may also be applied to fibres with similar index profiles.
The degeneracy will likewise not be complete then, but we can
still count on the mode groups m to be nearly degenerate, at
least to the degree to which the modes of compound number m
are nearly degenerate in the homogeneous core fibre.

For large mode numbers in multimode fibres, the right-
hand side of equation (6.231) may well be represented by
taking m as a continuous variable and by expressing the differences in average power and power coupling by derivatives
(Gloge 1972 and 1973). In the first step, we have

$$\overline{P}_{m+1} - \overline{P}_m = \frac{\partial \overline{P}_m}{\partial m} \Delta m \tag{6.232}$$

with $\Delta m = 1$. The second step leads to

$$m\, d_m\, \frac{\partial \bar{P}_m}{\partial m} - (m-1)\, d_{m-1}\, \frac{\partial \bar{P}_{m-1}}{\partial m} = \frac{\partial}{\partial m}\left(m\, d_m\, \frac{\partial \bar{P}_m}{\partial m}\right)\Delta m, \qquad (6.233)$$

again with $\Delta m = 1$. We substitute these expressions on the right-hand side of equation (6.188) and divide by $m - 1 \simeq m$:

$$\frac{\partial \bar{P}(m)}{\partial z} + \tau(m)\, \frac{\partial \bar{P}(m)}{\partial t} + 2\alpha(m)\, \bar{P}(m) = \frac{1}{m}\frac{\partial}{\partial m}[m\, d(m)\, \frac{\partial \bar{P}(m)}{\partial m}]. \qquad (6.234)$$

Since the compound mode number m now appears as a continuous variable, all quantities which depend on m are written as functions of m in equation (6.234). A more suitable variable than m is its ratio

$$x = m/m_c,$$

where m_c designates the largest compound mode number for guided modes near cut-off. We relate this largest compound mode number to the total number M of guided modes by adding all numbers $m - 1$ for each degenerate group from $m = 2$ to $m = m_c$. Since we let m be a continuous variable, this summation may be replaced by integration and results in

$$M = 2 \int_2^{m_c} (m-1)\, dm = m_c^2 - 2m_c \simeq m_c^2.$$

The factor 2 at the integral accounts for the two polarizations in which all modes of a nearly degenerate group occur, and of which we took only one for the interaction in the coupled power equations. The variable x of relative compound mode numbers ranges between $0 < x < 1$. With it equation (6.234) transforms to

$$\frac{\partial \bar{P}(x,z,t)}{\partial z} + \tau(x)\, \frac{\partial \bar{P}(x,z,t)}{\partial t} + 2\alpha(x)\, \bar{P}(x,z,t) =$$

$$\frac{1}{M}\frac{1}{x}\frac{\partial}{\partial x}[x\, d(x)\, \frac{\partial \bar{P}(x,z,t)}{\partial x}]. \qquad (6.235)$$

This partial differential equation relates essentially the second-order derivative of average mode power with respect to the relative mode number x to the first-order derivatives of $\overline{P}(x,z,t)$ with respect to time t and distance z along the fibre. It constitutes a diffusion equation that describes how the average power diffuses in time along the fibre from mode group to neighbouring mode group.

For solutions of this diffusion equation, we consider first the steady state in which the excitation of modes at the front end of the fibre is constant in time. The average power distribution along the fibre is then likewise independent of time, and the time derivative in equation (6.235) vanishes

$$\frac{\partial \overline{P}(x,z)}{\partial z} + 2\alpha(x)\ \overline{P}(x,z) = \frac{1}{M}\frac{1}{x}\frac{\partial}{\partial x}\ [x\ d(x)\ \frac{\partial \overline{P}(x,z)}{\partial x}]. \quad (6.236)$$

This diffusion equation for the steady state describes the diffusion of average power from an initial distribution $\overline{P}(x,0)$ among the degenerate mode groups to a z-dependent distribution along the fibre. For its solution, we need to know how the attenuation $\alpha(x)$ in the perfect fibre and the power coupling $d(x)$ due to fibre imperfections depend on the compound mode number. For the sake of simplicity, we assume core and cladding materials of the fibre to have the same bulk loss α_0. All guided modes will then have the same attenuation in the perfect fibre equal to this bulk loss, i.e.

$$\alpha(x) = \alpha_0\ .$$

Most of the leaky modes suffer much more loss due to radiation. We assume therefore that the attenuation of all leaky modes is infinitely large. The power in these leaky modes is then radiated at an infinitely higher rate than it diffuses at cut-off from guided into leaky modes. No power can ever accumulate in these leaky modes. Infinitely high leaky mode attenuation hence implies the following boundary condition for the solution of equation (6.236) in the range $0 < x < 1$:

$$\overline{P}(1,z) = 0\ . \quad (6.237)$$

A boundary condition for $\overline{P}(x,z)$ at $x = 0$ follows when we consider the total power flow in the steady state (Olshansky 1975). For this total power, we integrate the power $m\,\overline{P}(m/m_c,z)$ in one polarization of the mode group m with its m-fold degeneracy, over all mode groups from $m = 0$ to $m = m_c = \sqrt{M}$ and obtain as total power flow

$$P(z) = M \int_0^1 x\,\overline{P}(x,z)\,dx\;. \qquad (6.238)$$

We then find the diffusion equation for $P(z)$ by also performing this integration for equation (6.236)

$$\frac{dP(z)}{dz} + 2\alpha_0\,P(z) = \left\{x\,d(x)\,\frac{\partial \overline{P}(x,z)}{\partial x}\right\}_0^1\;. \qquad (6.239)$$

The one term on the right-hand side for $x = 1$ represents the total power diffusion per fibre length from the guided modes near cut-off to the leaky modes. The term for $x = 0$ must vanish, since there are no modes below $x = 0$ into which power could diffuse:

$$[x\,d(x)\,\partial \overline{P}(0,z)/\partial x]_{x=0} = 0\;. \qquad (6.240)$$

For the power coupling $d(x)$, we choose the following class of power-law functions (Olshansky 1975):

$$d(x) = d_0\,x^{-2q}\;. \qquad (6.241)$$

By adjusting the parameter q of this power law, it can account for mode interaction in fibres with different index profiles and for typical distributions of fibre imperfections. Microbending in graded-index fibres will be discussed in the next section, and we will find $d(x)$ according to equation (6.241) to adequately describe the mode interaction for this very important practical case.

We introduce the constant attenuation α_0 and the coupling function $d(x)$ from equation (6.241) into the diffusion equation and try to solve it with the following product

$$\bar{P}(x,z) = U(x)\exp-(\Gamma+2\alpha_0)z \ . \qquad (6.242)$$

For the function $U(x)$ in this product, we obtain the differential equation

$$\frac{d_0}{M} \frac{1}{x} \frac{d}{dx} [x^{1-2q} \frac{dU}{dx}] + \Gamma U = 0 \ . \qquad (6.243)$$

This equation, together with the boundary conditions (6.237) and (6.240), constitutes a boundary value problem, to which there are then an infinite set of eigenfunctions $U_j(x)$ with eigenvalues Γ_j. For these eigenfunctions, we find as solutions of equation (6.243)

$$U_j(x) = N_j \, x^q \, J_v(u_{vj} \, x^{1+q}) \ , \qquad (6.244)$$

where the order v of the Bessel function is given by

$$v = \pm \, q/(1+q) \ . \qquad (6.245)$$

To satisfy the boundary condition (6.237), the u_{vj} must be zeros of $J_v(u)$. The other boundary condition (6.240) requires the negative sign for the order of the Bessel function v in equation (6.245). The infinitely many zeros u_{vj} of $J_v(u)$ provide us with the infinite set of eigenfunctions. Their respective eigenvalues are

$$\Gamma_j = (d_0/M)(1+q)^2 \, u_{vj}^2 \ . \qquad (6.246)$$

Our boundary value problem is not self-adjoint, and the eigenfunctions are therefore not directly orthogonal to each other. But when multiplied by x and integrated over the full range from $x = 0$ to $x = 1$, the product $U_j(x) \, U_k(x)$ of any two different eigenfunctions vanishes. We use this characteristic to formulate the following orthonormality condition:

$$\int_0^1 x \, U_j(x) \, U_k(x) \, dx = \delta_{jk} \ . \qquad (6.247)$$

It requires the normalization factor in equation (6.244) to

have the value

$$N_j = (2(1+q))^{1/2}/|J_{\nu+1}(u_{\nu j})| \quad . \tag{6.248}$$

The eigenfunctions U_j constitute the diffusion approximation for the eigenvectors $\underline{U}_j$ of the statistical matrix $\underline{\underline{d}}$ in the coupled power equations (6.163). They specify how the power of a statistical mode j distributes among the continuum of nearly degenerate mode groups with compound mode number xm_c. The statistical mode j propagates with its eigenvalue Γ_j as attenuation constant.

An initial steady-state excitation of the fibre with a power distribution $\overline{P}(x,0)$ at its front end launches each of the statistical modes with

$$W_j = \int_0^1 x \, U_j(x) \, \overline{P}(x,0) \, dx \quad . \tag{6.249}$$

This initial power distribution diffuses along the fibre into

$$\overline{P}(x,z) = \sum_j U_j(x_j) \, W_j \, \exp[-(2\alpha_0 + \Gamma_j)z] \quad . \tag{6.250}$$

The statistical modes in this distribution suffer different losses along the fibre. Mode 1 with its eigenvalue Γ_1 determined by the first zero $u_{\nu 1}$ of $J_\nu(u)$ suffers least loss. For a long enough fibre, it eventually dominates all the other modes and leads to the stationary distribution

$$\overline{P}(x,z) = U_1(x) \, W_1 \, \exp[-(2\alpha_0 + \Gamma_1)z] \quad . \tag{6.251}$$

Under typical conditions for fibres with imperfections, the parameter q in equation (6.241) has such values that the order of the Bessel functions lies in the range $-1 < \nu < 1$. Their first zero then occurs between $0 < u_{\nu 1} < 3\cdot 83$, while the jth zero is approximately given by $u_{\nu j} \simeq u_{\nu 1} + (j-1)\pi$. The eigenvalues of the first and second statistical modes differ under these conditions by

$$\Gamma_2 - \Gamma_1 = (d_0/M)(1+q)^2 \, \pi(2u_{\nu 1}+\pi) \quad , \tag{6.252}$$

and it takes only a distance

$$L_s = (M/d_0)/[(1+q)^2 \pi(2u_{v1}+\pi)] \qquad (6.253)$$

along the fibre to establish the stationary distribution of the first statistical mode for the average power.

We turn now to an initial excitation of the fibre with signals that vary in time. To solve the diffusion equation (6.235) for time transients, we let again all nearly degenerate groups m of guided modes have the same average attenuation α_0 and require with equation (6.237) that all leaky modes have infinite loss. For the power coupling $d(x)$, we introduce again the same dependence on relative compound mode number according to the power law in equation (6.241). The initial excitation at the front end of the fibre is assumed to have the same time function $\overline{P}(x,0,t)$ for all modes. Its Laplace transform with respect to time is

$$p(x,0,s) = \int_0^\infty \overline{P}(x,0,t)\exp(-st)\,dt \quad . \qquad (6.254)$$

The Laplace transform

$$p(x,z,s) = \int_0^\infty \overline{P}(x,z,t)\exp(-st)\,dt \qquad (6.255)$$

of the transient power distribution along the fibre must then satisfy the differential equation

$$\frac{\partial p}{\partial z} + s\tau(x)p + 2\alpha_0 p = \frac{d_0}{M}\frac{1}{x}\frac{\partial}{\partial x}[x^{1-2q}\frac{\partial p}{\partial x}] \quad . \qquad (6.256)$$

For its solution, we let

$$p = U(x,s)\exp\{-(\Gamma(s)+2\alpha_0)z\}, \qquad (6.257)$$

which leads to the differential equation

$$\frac{1}{M}\frac{1}{x}\frac{\partial}{\partial x}[x\,d(x)\frac{\partial U}{\partial x}] + \Gamma U = s\tau U \qquad (6.258)$$

for the eigenfunctions $U_j(x,s)$ of statistical modes. Compared to equation (6.243), this differential equation is perturbed by $s\tau U$ on its right-hand side. To determine the perturbed eigenfunctions $U_j(x,s)$ and eigenvalues $\Gamma_j(s)$, we expand them into series of ascending powers in s:

$$U_j(x,s) = U_j(x) + s\, U_j^{(1)}(x) + s^2\, U_j^{(2)}(x) + \ldots$$
$$\Gamma_j(s) = \Gamma_j + s\, \Gamma_j^{(1)} + s^2\, \Gamma_j^{(2)} + \ldots \qquad (6.259)$$

Substituting these series expansions into equation (6.258), we obtain for $U_j(x)$ and Γ_j the differential equation (6.243), but for the terms of first and second order in s the two differential equations

$$\frac{d_0}{Mx}\frac{d}{dx}\left[x^{1-2q}\frac{dU_j^{(1)}}{dx}\right] + \Gamma_j\, U_j^{(1)} = (\tau - \Gamma_j^{(1)})\, U_j \qquad (6.260)$$

$$\frac{d_0}{Mx}\frac{d}{dx}\left[x^{1-2q}\frac{dU_j^{(2)}}{dx}\right] + \Gamma_j\, U_j^{(2)} = (\tau - \Gamma_j^{(1)})\, U_j^{(1)} - \Gamma_j^{(2)}\, U_j\,. \qquad (6.261)$$

To solve these equations, we express $U_j^{(1)}(x)$ in terms of $U_i(x)$

$$U_j^{(1)} = \sum_i a_{ji}\, U_i\,. \qquad (6.262)$$

The coefficients a_{ji} of this expansion follow with equation (6.247) from

$$a_{ji} = \int_0^1 x\, U_j^{(1)}\, U_i\, dx\,. \qquad (6.263)$$

We now expand $U_j^{(1)}$ in equation (6.260) according to equation (6.262), multiply equation (6.260) by xU_k, and integrate it from $x = 0$ to $x = 1$. We then apply the orthonormality condition (6.247) and obtain

$$a_{jk}[\Gamma_j - \Gamma_k] = M_{jk} - \Gamma_j^{(1)}\, \delta_{jk} \qquad (6.264)$$

with

$$M_{jk} = \int_0^1 x\, \tau\, U_j\, U_k\, dx\,. \qquad (6.265)$$

The set of equations (6.264) yields for $j = k$ the first-order perturbation

$$\Gamma_j^{(1)} = M_{jj} \qquad (6.266)$$

for the eigenvalue $\Gamma_j(s)$. For $j \neq k$, equation (6.264) yields

$$a_{jk} = \frac{M_{jk}}{\Gamma_j - \Gamma_k} . \qquad (6.267)$$

The coefficient a_{jj} of the expansion follows when we normalize according to

$$\int_0^1 x \, U_j^2(x,s) \, dx = 1 , \qquad (6.268)$$

which together with the normalization in equation (6.247) requires $\int_0^1 x \, U_j \, U_j^{(1)} dx = 0$ and hence $a_{jj} = 0$. The expansion (6.262) thus leads to the following first-order perturbation of the eigenfunctions $U_j(x,s)$ from their steady-state values

$$U_j^{(1)}(x) = \sum_{k \neq j} \frac{M_{jk}}{\Gamma_j - \Gamma_k} U_k . \qquad (6.269)$$

The second-order perturbation of eigenfunctions is likewise expressed in terms of U_i:

$$U_j^{(2)} = \sum_{i=1}^{\infty} b_{ji} U_i \qquad (6.270)$$

with its expansion coefficients according to

$$b_{ji} = \int_0^1 x \, U_j^{(2)} \, U_i \, dx . \qquad (6.271)$$

The same procedure that leads from equation (6.260) to equation (6.264) now leads from equation (6.261) to

$$b_{jk}[\Gamma_j - \Gamma_k] = \sum_i a_{ji} M_{ik} - \Gamma_j^{(1)} a_{jk} - \Gamma_j^{(2)} \delta_{jk} . \qquad (6.272)$$

For $j = k$, we obtain the second-order perturbation of the eigenvalue

$$\Gamma_j^{(2)} = \sum_i a_{ji} M_{ij} \qquad (6.273)$$

and for $j \neq k$, the coefficients b_{jk} of the expansion in equation (6.270)

$$b_{jk} = \frac{1}{\Gamma_j - \Gamma_k} [\sum_i a_{ji} M_{ik} - \Gamma_j^{(1)} a_{jk}] . \qquad (6.274)$$

The coefficient b_{jj} of this expansion follows from the normalization conditions (6.247) and (6.268) which require

$$b_{jj} = -\frac{1}{2} \sum_i a_{ji}^2 . \qquad (6.275)$$

With these first- and second-order perturbations of the steady-state solution, we are now ready to construct the general though approximate solution for power diffusion in the transient state. We assume for this solution an impulse excitation of all modes simultaneously, and let the initial power be distributed over all modes according to

$$\bar{P}(x,0,t) = \bar{E}(x) \, \delta(t) . \qquad (6.276)$$

This impulse function has the Laplace transform

$$p(x,0,s) = \bar{E}(x) . \qquad (6.277)$$

We represent the Laplace transform of the transient power distribution along the fibre in terms of the statistical modes for the transient state

$$p(x,z,s) = \sum_i W_i \, U_i(x,s) \exp\{-[2\alpha_0 + \Gamma_i(s)]z\}. \qquad (6.278)$$

For the transient eigenvalues $\Gamma_i(s)$ and eigenfunctions $U_i(x,s)$. we use the expansions of equation (6.259). For the eigenvalue expansion, we need at least all terms up to second order in s because these eigenvalues appear in the exponent of equation (6.278), and small perturbations modify $p(x,z,s)$ significantly. For the eigenfunctions, however, we content ourselves with the first term in equation (6.259) which is just the steady-state eigenfunction from equation (6.244). The orthonormality (6.247) of the steady-state eigenfunctions enables us to determine the initial excitation W_i of each statistical mode. Equation (6.278) reduces with $U_i(x,s) = U_i(x)$ and for $z = 0$ to

$$p(x,0,s) = \sum_i W_i\, U_i(x) \ . \tag{6.279}$$

We multiply this equation by $xU_k(x)$ and integrate over x from $x = 0$ to $x = 1$. By virtue of equation (6.247), this integration leads to

$$W_i = \int_0^1 x\, \overline{E}(x)\, U_i(x)\, dx \ . \tag{6.280}$$

For the transient power diffusion along the fibre, we now take the Laplace integral of equation (6.278)

$$\overline{P}(x,z,t) = \sum_i W_i U_i(x)\exp\{-(2\alpha_0+\Gamma_i)z\}\frac{1}{2\pi j}\int_{-\infty}^{\infty}\exp\{s(t-z\Gamma_i^{(1)})-s^2 z\Gamma_i^{(2)}\}ds. \tag{6.281}$$

With $s = j\omega$ and

$$\tau_i = \Gamma_i^{(1)} \tag{6.282}$$

as well as

$$t_i^2 = -16\, z\Gamma_i^{(2)} \ , \tag{6.283}$$

the Laplace integrals in each term of this sum have the same form as the Fourier integrals in equation (6.204). They may hence be evaluated accordingly and the same argument applies as to the z-range for which this evaluation is accurate: $-\Gamma_i^{(2)} z$ must be real and positive and large enough to render all higher-order terms of the perturbation expansion for $\Gamma_i(s)$ insignificant. We thus arrive at the following expression for transient power diffusion along the imperfect fibre:

$$\overline{P}(x,z,t) = \frac{2}{\sqrt{\pi}}\sum_i \frac{W_i}{t_i}\, U_i(x)\exp\{-(2\alpha_0+\Gamma_i)z\}\exp\{-4(t-\tau_i z)^2/t_i^2\} \ . \tag{6.284}$$

Just as in the steady state, the power distribution starts in many of the statistical modes. After a length L_s as given by equation (6.253), however, all these modes except the first, with the lowest attenuation, have practically died out. The power assumes the stationary distribution $U_1(x)$ of this first statistical mode, and the impulse response is dominated by the first term of the sum in equation (6.284):

$$\overline{P}(x,z,t) = \frac{2}{\sqrt{\pi}} \frac{W_1}{t_1} U_1(x) \exp\{-(2\alpha_0+\Gamma_1)z\} \exp\{-4(t-\tau_1 z)^2/t_1^2\}. \quad (6.285)$$

It has the time dependence of a Gaussian pulse with an 1/e-width t_1, delayed by τ_1 per unit fibre length, and attenuated according to $2\alpha_0+\Gamma_1$, the steady-state attenuation constant of the statistical mode of lowest loss. This result agrees in principal with the impulse response that we obtained in equation (6.208) from the solution of coupled power equations. It shows in particular a pulse broadening that increases only with the square root of fibre length, and is not proportional to the fibre length; it also shows the loss penalty Γ_1 that we pay for this reduction in pulse broadening. In terms of eigenvalues and eigenfunctions of statistical modes, the delay τ_1 and width t_1 of the impulse response are given by

$$\tau_1 = \int_0^1 x\, \tau(x)\, U_1^2(x)\, dx \quad (6.286)$$

$$t_1 = 4\left[\sum_{i=2}^\infty \frac{\left[\int_0^1 x\, \tau(x)\, U_1(x)\, U_i(x)\, dx\right]^2}{\Gamma_i - \Gamma_1} z\right]^{1/2}. \quad (6.287)$$

All eigenvalues Γ_i, including the attenuation constant Γ_1 with which the impulse response decays, are according to equation (6.246) proportional to the factor d_0 of power coupling. t_1 is therefore proportional to $1/\sqrt{d_0}$. d_0 on the other hand is mainly determined by the mean square magnitude of the fibre imperfection. If a particular kind of imperfection, for example microbending, changes its average magnitude without changing its other statistical quantities, the additional loss

Γ_1 increases in proportion to d_0 while t_1^2 decreases in proportion to $1/d_0$. The product $\Gamma_1 z (t_1/z)^2$, however, stays constant and is independent of z. When such a fibre imperfection increases so much in magnitude that it doubles the additional loss $\Gamma_1 z$, it will at the same time reduce the width of the impulse response by the factor $1/\sqrt{2}$. According to this general law, we must pay with additional transmission loss for the reduction in signal distortion and the possibility to transmit more information through an imperfect fibre with higher pulse rates.

To compare the impulse response of the imperfect fibre with that of its perfect replica, we need an equivalent measure for the pulse broadening in both cases. For this purpose, we refer to the r.m.s. pulse width which equation (5.250) defines in terms of moments of the impulse response. For a perfect fibre, this r.m.s. pulse width increases linearly with fibre length according to

$$\sigma_u = \Delta \tau_u z , \qquad (6.288)$$

where $\Delta \tau_u$ defines the r.m.s. width of the impulse response per unit length of the perfect fibre. The imperfect fibre, on the other hand, shows a Gaussian impulse response, only when it is long enough for the mode interaction to establish the steady-state power distribution of the lowest-loss statistical mode. The r.m.s. width of this Gaussian response is

$$\sigma_1 = t_1/(2\sqrt{2}) ,$$

and with equation (6.287) may be written as

$$\sigma_1 = \Delta \tau_1 \sqrt{z/\Gamma_1} , \qquad (6.289)$$

where

$$\Delta \tau_1 = \sqrt{2} \left[u_{v1}^2 \sum_{i=2}^{\infty} \frac{\left\{ \int_0^1 x \, \tau(x) \, U_1(x) \, U_i(x) \, dx \right\}^2}{u_{vi}^2 - u_{v1}^2} \right]^{1/2} . \qquad (6.290)$$

For a given distribution of imperfections $\Delta\tau_1$ is independent of its mean square magnitude. $\Delta\tau_1$ is mainly determined by the variation in delay $\tau(x)$ with the relative compound mode number $x = m/m_c$. For small delay differences between modes, $\Delta\tau_1$ will likewise be small and for equal delay of all modes, it would even vanish.

When an impulse is simultaneously launched into a number of or into all guided modes of an imperfect fibre, the impulse response widens initially according to equation (6.288) and only after a sufficient length will it gradually change into the stationary response and widen according to equation (6.289). The gradual transition from the linear dependence of pulse width on fibre length to the square root dependence takes place near a length L_c that follows from $\sigma_u(L_c) = \sigma_1(L_c)$ according to

$$L_c = (\Delta\tau_1/\Delta\tau_u)^2/\Gamma_1 \ . \qquad (6.291)$$

Since both $\Delta\tau_u$ and $\Delta\tau_1$ depend similarly on the delay differences between modes, this effect cancels to a certain extent and the so-called coupling length L_c in equation (6.291) is mainly determined by the statistical mode loss Γ_1 due to fibre imperfections. This conclusion agrees with equation (6.228) for the coupling length when only two modes interact in an imperfect fibre.

When we take the ratio $R = \sigma_1/\sigma_u$ of r.m.s. width for the impulse response in the imperfect and perfect fibre and multiply its square with the additional steady-state power loss $\Gamma_1 z$, we obtain the z-independent quantity

$$R^2 \Gamma_1 z = (\Delta\tau_1/\Delta\tau_u)^2 \ . \qquad (6.292)$$

The effects of delay differences between modes also cancel for this quantity so that it remains nearly constant for a particular distribution of imperfections independent of their magnitude. Imperfections which cause an additional transmission loss $\Gamma_1 z$ will reduce the r.m.s. pulse width by the factor $1/\sqrt{\Gamma_1 z}$ as compared to the perfect fibre.

Equations (6.284) and (6.285) give the impulse response

in each of the degenerate mode groups $m = m_c$ of the fibre for simultaneous excitation of these mode groups with different impulse energies $\overline{E}(x)$ according to equation (6.276). A photo-detector in a receiver at the fibre output $(z=L)$ responds to the total power in all modes. In order to obtain this total power, we have to perform the integration in equation (6.238) which takes account of the m-fold degeneracy but considers only all modes of one polarization. We assume here that both polarizations are excited with equal intensity; the total output power is then given by

$$P(L,t) = 2M \int_0^1 x\, \overline{P}(x,L,t)\, dx\ , \qquad (6.293)$$

where for $\overline{P}(x,L,t)$ we must substitute from equation (6.284) or for $L > L_s$ from equation (6.285). If an impulse of unit energy excites all modes in both polarizations equally strong, each polarization of each mode starts with the impulse energy

$$\overline{E}(x) = 1/M\ . \qquad (6.294)$$

The statistical modes have under these conditions amplitudes W_i from equation (6.280) according to

$$W_i = \frac{1}{M} \int_0^1 x\, U_i(x)\, dx\ . \qquad (6.295)$$

When we substitute equation (6.244) for $U_i(x)$, we can determine the value of the definite integral in equation (6.295) and find

$$W_i = (-1)^{i+1} \sqrt{2} / \{(M\, u_{vi}\, (1+q)^{\frac{1}{2}}\}\ . \qquad (6.296)$$

It shows us that, for equal excitation of all fibre modes, the first and lowest-loss statistical mode has the largest amplitude, and that higher-order statistical modes decrease in initial amplitude according to their respective Bessel function root in the denominator of equation (6.296). Under normal excitation conditions, we are therefore well justified in **neglecting** all higher-order statistical modes for $z > L_s$ and take only the lowest-loss mode according to equation (6.285).

With present excitation conditions, this lowest-loss statistical mode carries in both polarizations the total power

$$P(z,t) = \frac{8}{\pi^2(1+q)\,u_{v1}^2\,t_1} \exp\{-(2\alpha_0+\Gamma_1)z - 4(t-\tau_1 z)^2/t_1^2\}\,. \tag{6.297}$$

The factor in front of the exponential accounts with its dependence on z according to $1/t_1 \sim (z)^{-1/2}$ for the diminishing peak while the pulse spreads. It also accounts for the initial excitation of higher-order statistical modes so that only a fraction of the initial power is launched into the first statistical mode. The total energy of this impulse response for unit input energy follows from equation (6.297) as

$$E(z) = \{4/[(1+q)u_{v1}^2]\}\exp[-(2\alpha_0+\Gamma_1)z]\,. \tag{6.298}$$

It decays exponentially along the fibre due to the bulk loss α_0 of the material and due to the power diffusion and radiation loss Γ_1 as caused by the fibre imperfections.

6.9. MICROBENDING OF MULTIMODE FIBRES

As an example to evaluate the impulse response from the power diffusion equation, we again treat the case of microbending, since this constitutes the most critical imperfection, also in multimode fibres. To include different graded-index fibres and the cladded-core fibre with a stepped index profile, we take the class of power-law profiles in Fig.5.18. The phase constants of their guided modes are given by equation (6.174) and their delay per fibre length by equation (5.273), both in terms of the fractional mode number ν. To relate this fractional mode number ν to the compound mode number $m = 2p + l$, we recall that for the parabolic index profile all $2(m-1)$ modes with the same compound mode number m are degenerate, and that in other power-law profiles, including even the stepped-index distribution, these $2(m-1)$ modes maintain a near degeneracy. All modes with compound mode numbers smaller than a given m have then also a smaller phase constant.

Altogether there are

$$\nu = 2 \int_{2}^{m} (m-1)\,dm = m^2 - 2m \simeq m^2 \qquad (6.299)$$

such modes. The group m of nearly degenerate modes has in this approximate solution of the characteristic equation the phase constant

$$\beta = n_1 k\ [1 - 2\Delta(m/m_c)^{2\alpha/(\alpha+2)}]^{1/2} . \qquad (6.300)$$

The delay per fibre length for this group of modes amounts to

$$\tau_m = \frac{n_1'}{c}\left[1 + \Delta C_1 (m/m_c)^{2\alpha/(\alpha+2)} + \Delta^2\ C_2 (m/m_c)^{4\alpha/(\alpha+2)}\right], \qquad (6.301)$$

where C_1 and C_2 are given by equation (5.274). The phase difference between modes in group m and those in $m+1$ follows from equation (6.300) for $m \gg 1$

$$\Delta\beta = \frac{2}{a}\left[\frac{\alpha\Delta}{\alpha+2}\right]^{\frac{1}{2}} x^{(\alpha-2)/(\alpha+2)} . \qquad (6.302)$$

For the parabolic index profile, it is $\Delta\beta = (2\Delta)^{\frac{1}{2}}/a$, just as in equation (6.72). For the stepped index profile, we obtain

$$\Delta\beta = (2\Delta^{\frac{1}{2}}/a)x , \qquad (6.303)$$

which increases in proportion to $x = m/m_c$.

The average power coupling d_m, which appears first in equation (6.231), represents the sum of all coupling factors $d_{\mu\nu}$ between one particular mode μ of the group m and all modes ν of the group $m+1$, which the respective imperfection (curvature in the present example) couples to the mode μ. To determine this average coupling factor, we add the square of all coupling coefficients $c_{lp,kq}$ for fixed orders l and p and varying orders k and q whose difference in compound mode number is equal to unity:

$$\Delta m = 2q + k - 2p - l = 1 \, .$$

For the parabolic index profile, the coefficients for curvature coupling appear in equations (6.69). The first two represent the coupling between modes with $m = 2p + l$ and modes with $m + 1$. If we just add the square of these two coefficients, we obtain the desired average. For all $l > 1$, it is simply given by

$$d_m = (1/8) \, n_1^2 \, k^2 \, w^2 m \, \phi(\Delta\beta) \, . \tag{6.304}$$

In this expression, $\phi(\Delta\beta)$ represents the power spectrum of the curvature distribution $\kappa(z) = 1/\rho(z)$ and must be evaluated at the spatial frequency $\Omega = \Delta\beta$. When we replace m by the relative mode number $x = m/m_c$ and express m_c as well as the spot radius w by the fibre parameter $V = n_1 k a (2\Delta)^{\frac{1}{2}}$, the average coupling factor for curvature in the parabolic profile appears as

$$d(x) = (1/8) \, n_1^2 \, k^2 \, a^2 \, x \, \phi(\Delta\beta) \, . \tag{6.305}$$

For the stepped profile of the cladded core fibre with a homogeneous core, the coefficients for curvature coupling are given by equation (6.74) and, as an adequate approximation, by equations (6.76) and (6.77). Equation (6.76), in particular, contains all coupling coefficients for modes that differ only by one in their compound mode number m. They represent the strongest curvature coupling, and the diffusion equation takes only this next-neighbour coupling into account. These coupling coefficients are independent of mode order. We therefore obtain for the average power coupling between modes of any group m

$$d(m) = 2(2/\pi^2)^2 (n_1 k a)^2 \, \phi(\Delta\beta) \, . \tag{6.306}$$

Here only the power spectrum $\phi(\Delta\beta)$ depends on the compound mode number because $\Delta\beta$ changes with m according to equation (6.303).

The parabolic index profile and the homogeneous core

represent two special cases of the general power-law profile for which we have field solutions of guided modes available to evaluate the coefficients of curvature coupling. An approximation for the power coupling in power-law index profiles with intermediate exponents α that interpolates between equation (6.305) and equation (6.306) may be constructed according to Olshansky (1975) as

$$d(x) = (n_1^2 k^2 a^2/12)(\tfrac{3}{2} x)^{4/(\alpha+2)} \phi(\Delta\beta) . \qquad (6.307)$$

The power law in which x appears in this interpolation formula complies with equation (6.241) and therefore allows the general solution (6.244) of the diffusion equation to be applied. The phase difference at which the power spectrum $\phi(\Delta\beta)$ in this formula must be evaluated follows from (6.302).

For the solution of the diffusion equation, we next need to specify the power spectrum $\phi(\Omega)$ of the curvature distribution, also in accordance with equation (6.241), so that we obtain the actual solution in the form of equation (6.244).

When fibres are pressed against a rough surface such as by packaging in a cable they will to a certain extent conform with this surface by bending. A soft surface will yield to the pressure so that not all its roughness transfers to the fibre. We will deal with this problem in section (7.8) when we discuss the packaging of fibres in cables. The laws of elasticity lead to a curvature of the fibre with a spectral distribution according to equation (7.22). If a uniform force presses the fibre against the rough but soft surface, then the power spectrum of curvature depends on frequency as

$$\phi(\Omega) \sim \Omega^4/[1 + \Omega^4 (H/D)]^2 ,$$

where H according to equation (7.15) represents the flexural rigidity of the fibre, and D is nearly equal to the elastic modulus E_s of the compressed surface. With a characteristic length L_k according to

$$L_k^4 = (H/D) \simeq I(E_g/E_s) , \qquad (6.308)$$

where E_g designates the elastic modulus of the fibre and

$$I = (\pi/4)\, b^4 \tag{6.309}$$

its moment of inertia, the curvature power spectrum appears as

$$\phi(\Omega) \sim \Omega^4 / [1 + (\Omega\, L_k)^4]^2 \tag{6.310}$$

and the characteristic length L_k may be interpreted as a correlation distance. The same dependence of power spectrum on spatial frequency also obtains when the fibre is embedded in a soft medium of elastic modulus E_s and bumps in a supporting structure displace the fibre laterally at irregular intervals (Olshansky 1975).

The spatial frequency characteristic of equation (6.310) has its maximum at

$$\Omega = 1/L_k \,. \tag{6.311}$$

To avoid excessive curvature loss, we should aim for

$$\Delta\beta \gg 1/L_k \,, \tag{6.312}$$

so that only small spectral components of $\phi(\Delta\beta)$ contribute to the average mode coupling in equation (6.307). In order to satisfy equation (6.312), the fibre should have a large relative index difference Δ, large outer radius b, and be embedded in a material with a low modulus of elasticity E_s. Once equation (6.312) holds the power spectrum decreases with $\Delta\beta = \Omega$ according to

$$\phi(\Omega) \sim \Omega^{-4} \,. \tag{6.313}$$

At low spatial frequencies additional components contribute to the power spectrum. Equation (6.310) does not account for these spectral components. Gradual changes in the direction of fibre cables along the route cause the power spectrum to remain essentially constant at a substantial mag-

nitude down to zero spatial frequency. We therefore choose
the general function

$$\phi(\Omega) = \langle \kappa^2 \rangle \frac{2p\, L_k\, \sin(\pi/2p)}{1 + (\Omega\, L_k)^{2p}} \qquad (6.314)$$

to describe the power spectrum of any actual curvature distribution. Equation (6.314) has been so normalized, that

$$\frac{1}{\pi} \int_0^\infty \phi(\Omega)\, d\Omega = \langle \kappa^2 \rangle \qquad (6.315)$$

represents the mean square of curvature κ and $\phi(\Omega)$ the power spectral density of $\kappa(z) = 1/\rho(z)$ per unit spatial frequency $\Omega/2\pi$. Under the condition (6.312), the power spectrum in equation (6.314) reduces to

$$\phi(\Omega) = \langle \kappa^2 \rangle\, 2p\, L_k\, \sin(\pi/2p)\, (\Omega\, L_k)^{-2p} \qquad (6.316)$$

and models the characteristic (6.313) for $p = 2$. For $p = 1$, equation (6.314) reduces to the Butterworth characteristic of the power spectrum in equation (6.105) for an exponential autocorrelation function. For $2p \gg 1$, it reduces to

$$\phi(\Omega) = \langle \kappa^2 \rangle\, \pi\, L_k / [1 + (\Omega\, L_k)^{2p}] \,. \qquad (6.317)$$

To comply with the power law in equation (6.241), let us henceforth use the high-frequency approximation (6.316). We then obtain for the average power coupling from equation (6.307) the expression

$$d(x) = d_0\, x^{-2q}, \qquad (6.318)$$

where

$$d_0 = \langle \kappa^2 \rangle\, p\, \sin(\pi/2p)\, \frac{n_1^2\, k^2\, a^2}{6\, L_k^{2p-1}} \left[\frac{(\alpha+2)a^2}{4\alpha\Delta}\right]^p \left(\frac{3}{2}\right)^{4/(\alpha+2)} \qquad (6.319)$$

and

$$q = [p(\alpha-2) - 2]/(\alpha+2) . \qquad (6.320)$$

Let us first examine the eigenvalues Γ_i of the steady-state solution of the diffusion equation. With equations (6.246), (5.174), and (6.319), these eigenvalues may be expressed as

$$\Gamma_i = \langle \kappa^2 \rangle L_k (a/L_k)^{2p} \Delta^{-p-1} p \sin(\pi/2p) G_i(\alpha,p) , \quad (6.321)$$

where for a particular eigenvalue the factor

$$G_i(\alpha,p) = \left(\frac{2}{3}\right) \left(\frac{\alpha+2}{4\alpha}\right)^{p+1} \left(\frac{3}{2}\right)^{\frac{4}{\alpha+2}} (1+q^2) u_{vi}^2 \qquad (6.322)$$

depends only on the characteristic exponents α and p of the index profile and curvature power spectrum, respectively. The smallest eigenvalue Γ_1 represents the additional power loss in the statistical mode of lowest attenuation. For distances $z > L_s = 1/(\Gamma_2 - \Gamma_1)$, the steady-state power distribution is given by

$$U_1(x) = N_1 x^q J_\nu(u_{\nu 1} x^{1+q}) ;$$

it remains stationary along the fibre and decays with $\alpha_0 + \Gamma_1/2$ as attenuation constant. According to equation (6.321), the added curvature loss $\Gamma_1/2$ increases with the core radius as a^{2p} but decreases with the relative index Δ^{-p-1}. A fibre with a larger core and a smaller index difference becomes in this proportion more sensitive to microbending. Fig. 6.10 shows the factor $G_1(\alpha,p)$ of the added curvature loss as a function of the profile parameter α for three different exponents p of the curvature power spectrum. For $p > 0$, the added curvature loss under the stationary conditions of the statistical mode of lowest loss changes only little with the characteristic exponent α of the power-law index profile. It increases slightly towards $\alpha = 2$, when the profile becomes near parabolic. In case of a flat power spectrum for the curvature distribution, with $p = 0$, this increase in added curvature loss towards $\alpha = 2$ is somewhat more pronounced.

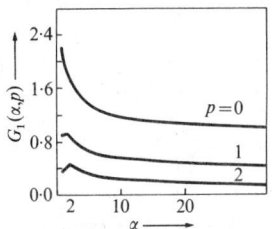

Fig.6.10. Normalized microbending loss in multimode fibres $G_1(\alpha,p)$ with power-law index profile of characteristic exponent α for power spectra of curvature distribution with different exponents p (Olshansky 1975).

This added curvature loss under stationary conditions is governed by the diffusion of power from the modes of highest compound mode numbers m into the radiation field. The phase differences between modes are independent of m for the parabolic profile but increase with m according to equation (6.302) when $\alpha > 2$. Accordingly, the added loss decreases when α increases for $\alpha > 2$, in particular for a flat curvature spectrum with $p = 0$.

The impulse response of the randomly curved fibre widens initially according to equation (6.288) in proportion to the distance z from the fibre input. For distances larger than the coupling length L_c from equation (6.291), the impulse response approaches a Gaussian time function which according to equation (6.289) widens only with the square root of distance z. To determine the coupling length L_c and the width of the impulse response in the stationary power distribution, we need to evaluate the delay divergence in equation (6.290). The integrals which appear in its series representation contain the delay $\tau(x)$ as a function of relative compound mode number x. For the power-law profile, this function is given by equation (6.301). Except for index profiles with $\alpha \simeq 2 + y$, the term linear in Δ provides an adequate approximation for $\tau(x)$. The eigenfunctions $U_i(x)$ of the steady-state diffusion equation appear in equation (6.244) normalized according to equation (6.248). Substituting from these equations, we obtain for the integrals in the sum of equation (6.290)

$$\int_0^1 x\tau U_1 U_i \, dx = N_1 N_i (n_1'/c) C_1 \Delta \int_0^1 x^{2q+1+2\alpha/(\alpha+2)} J_\nu(u_{\nu 1} x^{1+q}) J_\nu(u_{\nu i} x^{1+q}) \, dx. \tag{6.323}$$

Transforming to $s = x^{1+q}$ as the new variable of integration leads to

$$\int_0^1 x\tau U_1 U_i \, dx = N_1 N_i (n_1'/c) C_1 [\Delta/(1+q)] \int_0^1 s^r J_\nu(u_{\nu 1} s) J_\nu(u_{\nu i} s) \, ds \tag{6.324}$$

with the exponent of s given by

$$r = [3\alpha + p(\alpha-2)]/[\alpha + p(\alpha-2)] . \tag{6.325}$$

These integrals must in general be evaluated numerically. The delay divergence $\Delta\tau_1$ in equation (6.290) is then obtained by including all significant terms of its series representation. Further, for the coupling length L_c according to equation (6.291), we need the delay difference $\Delta\tau_u$ per fibre length of the impulse response of the perfect fibre without any mode interaction. For power-law profiles, the r.m.s. pulse width of the impulse response is given by equation (5.276). In our present evaluation, we disregard all terms of second and higher orders in Δ. In this approximation, we obtain from equation (5.276)

$$\Delta\tau_u = \frac{n_1' \Delta}{2c} \frac{C_1 \alpha}{\alpha+1} \left(\frac{\alpha+2}{3\alpha+2}\right)^{\frac{1}{2}} . \tag{6.326}$$

The ratio $\Delta\tau_1/\Delta\tau_u$ determines the additional loss $\Gamma_1 L_c$ within the coupling length according to equation (6.291) and the reduction R in pulse broadening according to equation (6.292). $\Delta\tau_1/\Delta\tau_u$ is in the present approximation independent of n_1', Δ, and C_1, as well as of the core diameter a. It depends only on the exponent α of the index profile and the exponent p of the power spectrum of imperfections. Fig.6.11 shows the square of this ratio as a function of α. for three different values of the curvature spectrum parameter p.

Equation (6.288) defines $\Delta\tau_u$ as the r.m.s. widening per

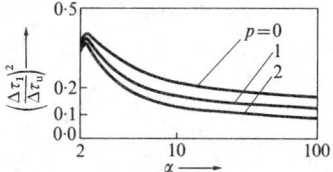

Fig.6.11. $(\Delta\tau_1/\Delta\tau_u)^2$ represents the average microbending loss per coupling length as well as the reduction in pulse width per microbending loss for power-law index profiles with exponent α and curvature power spectra with exponent p (Olshansky 1975).

fibre length of the impulse response of the perfect fibre. $\Delta\tau_1$, on the other hand, represents from equation (6.289) for the imperfect fibre the r.m.s. width of the impulse response relative to the fibre length, times $(\Gamma_1 z)^{\frac{1}{2}}$, with $\Gamma_1 z$ as the added curvature loss. The square of their ratio in Fig.6.11 represents according to equation (6.291) the added curvature loss $\Gamma_1 L_c$ in a length of fibre that corresponds to the coupling length L_c. At the same time, this ratio $\Delta\tau_1/\Delta\tau_u$ divided by $(\Gamma_1 z)^{\frac{1}{2}}$ also gives the reduction in r.m.s. width of the impulse response compared to the perfect fibre. Under the present assumptions for the index profile, for the guided-mode attenuation, and for the curvature power spectrum, the ratio $\Delta\tau_1/\Delta\tau_u$ depends only on the characteristic exponents α and p. This dependence is, however, relatively weak and manifests itself mainly in a decrease with increasing α for $\alpha \gtrsim 2\cdot 3$. At $\alpha \simeq 2$, the same value for $\Delta\tau_1/\Delta\tau_u$ results for all values of p. The dependence on α and p reflects the interplay of phase differences between modes and spectral distribution of imperfections as they influence the mode interaction. For $\alpha = 2$, the phase differences do not depend on the compound mode number m and neither does the average coupling. $\Delta\tau_1/\Delta\tau_u$ is then also independent of the spectral distribution of curvature. For $\alpha > 2$, on the other hand, $\Delta\beta$ increases with m and, with increasing p or increasing α, higher-order modes will interact less in the coupling process.

Altogether, however, the ratio $\Delta\tau_1/\Delta\tau_u$ varies only little and slowly with p and α and for the range of practical p- and α-values lies between 0·3 and 0·6. It represents a characteristic quantity with which we conveniently estimate the coupling length L_c and the reduction in pulse spreading for a given added loss Γ_1.

REFERENCES

Bergeest, R. and Unger, H.-G. (1969). Optische Wellen in Hohlleitern. *Arch.elekt. Übertr.* 23, 529-538.

Gloge, D. (1972). Optical power flow in multimode fibres. *Bell Syst. tech. J.* 51, 1767-83.

─────── (1973). Impulse response of clad optical multimode fibres. *Bell Syst. tech. J.* 52, 801-16.

─────── (1975). Propagation effects in optical fibres. *IEEE Trans.* MTT 23, 106-20.

Gradshteyn, I.S. and Ryzhik, I.M. (1965). *Table of integrals, series, and products.* Academic Press, New York and London.

Kuhn, M.H. (1975). Curvature loss in single mode fibres with lossy jacket. *Archiv Elektron. & Übertragungstech.* 29, 400-02. (Erratum (1976) in *Archiv Elektron. & Übertragungstech.* 30, 124).

Lewin, L. (1975). *Theory of waveguides.* Newnes-Butterworths, London.

Marcuse, D. (1974). *Theory of dielectric optical waveguides.* Academic Press, New York and London.

Olshansky, R. (1975). Mode coupling effects in graded index optical fibres. *Appl. Opt.* 14, 935-45.

Petermann, K. (1976). Theory of microbending loss in monomode fibres with arbitrary refractive index profile. *Archiv Elektronik & Übertragungstech.* 30, 337-42.

─────── and Storm, H. (1976). Microbending loss in single-mode W-fibres. *Electr. Lett.* 12, 537-8.

Unger, H.-G. (1961) Regellose Störungen in Wellenleitern. *Archiv Elektronik & Übertragungstech.* 15, 393-401.

Young, D.T. (1963). Model for relating coupled power equations to coupled amplitude equations. *Bell Syst. tech.J.* 42, 2761-4.

7
FIBRE FABRICATION AND CABLING

Optical fibre guides are mainly used for signal transmission over medium and long distances. These distances range from tens of metres for the internal wiring with optical fibres of buildings or aircraft and ships, to tens of kilometres in wide-band long-distance optical communication systems for multiplex telephone and television and data transmission. In all these applications fibres must have much less loss than we ordinarily encounter in the bulk of normal transparent material. In particular, for long-distance signal transmission we require the loss to measure only in a few decibels per kilometre. For such low loss values the materials for the fibre core must have extremely little absorption and scattering. Only high quality glasses but hardly any other materials meet these requirements of utmost transparency. We therefore discuss first the properties of these glasses and their preparation for fibre fabrication.

7.1. GLASSES FOR FIBRES

Glass is an amorphous solid which consists of elements or chemical compounds that also occur in the crystalline state. In a crystal, these components arrange themselves in a regular lattice; as a glass, on the other hand, they form an irregular network. Fused silica, for example, consists of SiO_4 tetrahedra as basic units. These tetrahedra join at their O-corners to form the three dimensional irregular network of the glassy state. Any O^{2-} ion in this network that forms a bridge between two SiO_4 tetrahedra is tied to two Si^{4-} ions. The network has relatively wide meshes of irregular size.

Systems with several components can also go into the glassy state, but the mixture of components will lead to a irregular network which is more loosely knit than with only one basic unit. In the binary system of Na_2O-SiO_2, for example, each Na^+ ion ties to an O^{2-} ion and thus breaks the bridge between SiO_4 tetrahedra. Other alkali oxides act similarly. These breaks in the network weaken the glass structure and are responsible for many changes in its characteristics. Most important, they lower the softening point and melting temperature of glasses. Fused silica with its relatively closely-knit network of SiO_4 tetrahedra melts at 2000°C, while soda lime silicate glasses consisting of Na_2O, CaO, and SiO_2 melt near 1400°C. Alkali lead silicate glasses with the composition $Na_2O-PbO-SiO_2$ also melt near 1400°C, while soda aluminosilicate with Na_2O, Al_2O_3, and SiO_2 as constituents melts near the somewhat higher temperature of 1450°C. Another important compound glass system is soda borosilicate. Its constituents are NaO_2, B_2O_3, and SiO_2, and it melts near the even lower temperature of 1250°C. These relatively low melting temperatures allow these glasses to be prepared by melting, with little contamination from the crucibles if proper care is exercised.

Fused silica in its pure and homogeneous state shows the lowest absorption of most glasses in the visible and near-infrared range of light wavelengths. It also compares favourably to low melting glasses with respect to its intrinsic scattering losses. Its high melting point, however, requires special methods for its preparation and the fabrication of fibres. But the high temperatures which are necessary to produce pure fused silica by chemical reactions and to soften it for fibre pulling will also let many contaminations evaporate and keep it relatively pure without too many additional precautions. With all these favourable characteristics, fused silica has to its disadvantage for optical fibres one of the lowest refractive indices of any glass. At $\lambda = 0.589$ μm, it amounts to only $n = 1.4585$. If we want to use fused silica in its pure form as the core material of a low-loss fibre, we need to lower its index for the cladding. There are only very few possibilities of lowering the index of SiO_2 glass by doping it with other materials and at the same time maintaining its other characteristics similar to those of the pure material. One doping agent that effects a lower refractive index is boric oxide (Van Uitert *et al.* 1973). Pure B_2O_3 has an index of refraction of 1.4582 at $\lambda = 0.589$ μm, which is only slightly less than the index of pure SiO_2 at this wavelength. If the index of the composite glass would vary monotonically between the limiting values of the pure constituents, it could not be low enough to be useful for the cladding. Actually, however, the index of the B_2O_3-SiO_2 system varies with composition according to Fig.7.1 and shows a definite minimum near the composition $6SiO_2:1B_2O_3$. This minimum index is nearly 0·3 per cent lower than that of pure fused silica. It is, however, partly due to quenching of a metastable state that occurs, in particular during fibre drawing (Wemple et al. 1973). Long term annealing of the borosilicate may raise the index again and thus reduce the index difference compared to that of pure fused silica.

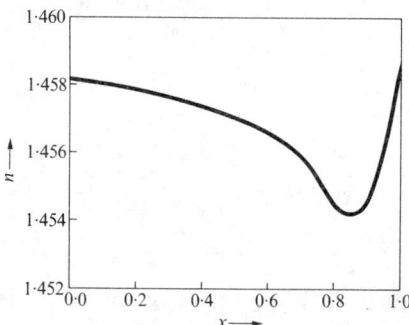

Fig.7.1. Index of refraction of the binary glass system B_2O_3-SiO_2 as a function of the mole fraction x of SiO_2 in the composition (Van Uitert et al. 1973).

Incorporating fluorine into fused silica is another possibility for lowering its refractive index (Mühlich et al. 1975). In contrast to the metastable character of the index variation in borosilicate, the index reduction in fluorine-doped silica is an intrinsic property of the fluorine atom in the SiO_2 matrix. The index difference between pure SiO_2 and the fluorine-doped material increases linearly with the fluorine concentration up to and beyond several per cent of fluorine mole fraction in fused silica. The SiO_2 index is lowered by 0·2 per cent when the mole fraction of fluorine amounts to 1 per cent. Thus by incorporating a mole fraction of 5 per cent of fluorine, the index is lowered by 1 per cent. The high optical quality of pure fused silica is not degraded by such small amounts of fluorine doping.

All other additives to fused silica that are known to modify its refractive index without degrading its optical quality too much raise the index as compared to pure SiO_2. The oxides GeO_2, P_2O_5, TiO_2, and Al_2O_3 have proved quite suitable for this purpose. The mole fractions with which these oxides are added to SiO_2 can be between 1 and 25 per cent. Fig.7.2 shows the refractive index of GeO_2-doped fused silica

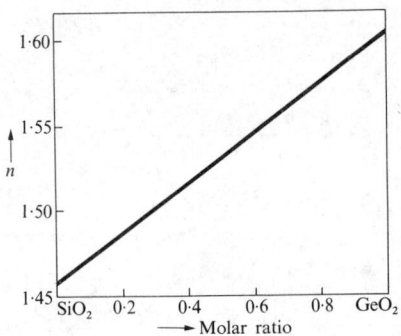

Fig.7.2. Refractive index of GeO_2-doped fused silica as a function of the GeO_2 mole fraction.

as a function of the dopant mole fraction. The index increases linearly with GeO_2 concentration at a rate of nearly $\Delta n = 0.001$ for one per cent of increase in GeO_2 mole fraction. For a GeO_2 concentration with 20 per cent mole fraction therefore, the index increases by 1.4 per cent.

Germanium is commonly used as a semiconductor and is available in pure form. Nevertheless, it is not abundant and therefore quite expensive. P_2O_5 might hence be preferred as a doping material, since phosphorus is a common element and relatively inexpensive. When phosphorus pentoxide is added to fused silica to form a binary glass system, neither the intrinsic material absorption nor the Rayleigh scattering increases significantly (Payne and Gambling 1974, Gambling et al. 1976). Although P_2O_5 sublimes at 300°C, is hygroscopic, and has an expansion coefficient some 25 times larger than fused silica, it nevertheless forms a stable binary glass system with silica, which has thermal expansion compatible with that of pure fused silica at least for concentrations up to nearly 25 mole per cent. The resultant glass exhibits hardly any tendency either for phase separation or for devitrification. It is also quite resistant to attack by water. The refractive index of the binary phosphosilicate glass increases linearly at least initially with the concentration of P_2O_5. The rate of this initial increase is 0·043 per cent of index change for each per cent of increase in P_2O_5 molar fraction; it seems to fit the purely additive and linear relationship of equation (5.156) between the refractive indices of pure silica with $n = 1·458$ and pure phosphorus pentoxide with $n = 1·52$. The differences in thermal expansion and viscosity between P_2O_5 and fused silica ultimately limit the amount of P_2O_5 which can be incorporated into silica for cladded-core and graded-index fibres.

Titanium dioxide is also incorporated into fused silica to raise its refractive index (Maurer 1973). By adding one weight per cent of TiO_2, the refractive index increases by about 0·026 per cent. A binary glass system with TiO_2 doping in fused silica has the disadvantage, however, that titanium can occur in the glass matrix in some states of ionization which absorb in the spectral range of interest. Plus-three titanium (Ti^{3+}) is a particularly strong absorber and it is difficult to completely oxidize it. A special temperature treatment is necessary to minimize the concentration of reduced titanium by oxidation with water into titanium dioxide and hydrogen.

In contrast to fused silica and the binary glass systems with a high content of silica, the ternary or compound glass systems, of which soda lime silicate and soda borosilicate are notable examples, have much lower melting temperatures and therefore lend themselves more readily to continuous production techniques such as pulling fibres out of the molten glass. They also have a higher index of refraction and can be modified to yield substantially lower index values for the cladding region. Soda borosilicate allows a relative index difference of 3 per cent with mechanical and thermal characteristics of the modified glass that are still suitable for a fibre cladding.

The corresponding index difference for soda lime silicate can be as large as 4 per cent and for alkali lead silicate glasses, index differences as high as 10 per cent are possible between modifications that still match as core and cladding materials in fibres. All these low melting compound glasses are, however, more easily contaminated than fused silica, and it is more difficult to purify them in order to reduce absorption and scattering from contaminations. Because of their low softening and melting temperatures, recontamination occurs at all stages of glass preparation and fibre fabrication.

7.2. CHEMICAL VAPOUR DEPOSITION FOR QUARTZ-GLASS FIBRES

Fibres with extremely low transmission loss and distortion are produced with pure fused silica and binary glass systems with a high silica content. Natural quartz in the form of rock crystal is only of limited use for quartz-glass fibres because its exact composition and its impurities depend very much on chance. Usually, therefore, only the fused silica for outer cladding regions of fibres is produced from natural quartz. The material in these regions serves more to support and give strength to the structure rather than to guide the light. The low-loss core and inner cladding regions are made of synthetic quartz, in which any contaminations are kept below critical concentrations. A well proven method to fabricate a preform for the fibre uses a controlled chemical vapour-deposition technique to produce pure fused silica as well as doped silica layers on a quartz-glass substrate. According to Fig.7.3 this quartz substrate has the form of a tube and layers of pure or doped quartz-glass are deposited

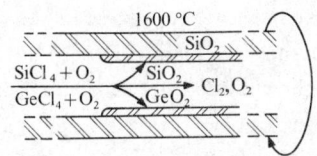

Fig.7.3. Chemical vapour deposition of GeO_2-doped fused silica.

on its inner wall. The starting materials for the chemical
vapour deposition are in many cases purified silicon tetrachloride and chlorides of the particular dopant. Both are
vapourized, mixed with oxygen and passed through the tubular
substrate. The tube is heated over a fraction of its length
to a high enough temperature for the chlorides in the gas
mixture to oxidize, flocculate out of the gas stream, and
fuse on the inner walls as a vitreous layer. In the case of
silicon tetrachloride and germanium tetrachloride for the
dopant, the gases in the mixture react as follows:

$$SiCl_4 + O_2 \rightarrow SiO_2 + 2Cl_2$$

$$GeCl_4 + O_2 \rightarrow GeO_2 + 2Cl_2 \; .$$

Chlorine as well as residual oxygen leave the tubular substrate at the opposite end as waste gases. In the case
of P_2O_5 as a dopant, phosphorus oxychloride is added to the
gas mixture which then oxidizes according to

$$4POCl_3 + 3O_2 \rightarrow 2P_2O_5 + 6Cl_2 \; .$$

For B_2O_3 as a dopant, to lower the refractive index as compared
to pure fused silica, B_2Cl_3 is added to the gas mixture and re-
acts with oxygen as follows:

$$4BCl_3 + 3O_2 \rightarrow 2B_2O_3 + 6Cl_2 \; .$$

For fluorine to lower the refractive index of fused silica
either SF_6, CF_4, or SiF_4 may be used as dopant sources (Abe
1976).

The quartz tube may either be mounted vertically or
horizontally for the deposition process. When mounted horizontally, it should be rotated about its axis in order to ensure axially symmetric deposition layers. To achieve a homogeneous chemical reaction, yielding oxides not only near the
inner walls of the tube but in the whole cross-section, and
to fuse the reaction products to a glassy layer on the wall,
a temperature of 1400°C or higher is needed. An oxygen-

hydrogen burner provides this temperature in a sufficiently
wide zone along the tube. The oxides flocculate in the hot
zone and deposit downstream from this zone on the wall at
first as soot. By moving the burner or furnace slowly in the
direction of the gas stream, the soot subsequently fuses to a
bubble-free glass layer. The speed at which the burner moves
determines the temperature in the hot zone, and with it this
temperature may be controlled. The thickness of the glass
layer which forms during one run of the burner along the tube
depends on the rate at which the gas mixture flows through
the tube and on the temperature of the hot zone. For high
flow rate and high temperature it amounts to several micrometres and even goes up beyond 10 µm. Thicker depositions
require repeated runs of the burner.

Instead of an oxygen-hydrogen burner, a resistance-heated
furnace or induction heating may be employed to raise the temperature in the hot zone and control it accurately. For more
controlled deposition conditions, such a resistance-heated
furnace is to be preferred.

Another possibility is to activate the chemical vapour deposition in a microwave-generated plasma at reduced gas pressure of 1 to 20 Torr (Küppers and Koenings 1976). Fig.7.4

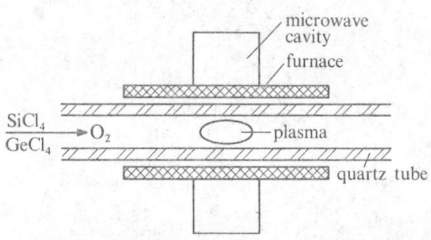

Fig.7.4. Chemical vapour deposition activated by a mircowave plasma
(Küppers and Koenings 1976).

illustrates this process schematically. An electric furnace
heats the tube to nearly 1000°C. A microwave cavity absorbs
about 200 W to generate a plasma inside the tube. At low
enough pressure (1 - 20 Torr), the reaction products diffuse

to the wall and deposit as a glass layer without any soot formation in the gas phase. The rapid movement of the deposition zone which this process allows and its high efficiency yield many deposition layers in short times with only small amounts of the useful reaction products wasted.

The quartz tube with layers of doped and pure silica deposited on its inner walls will be collapsed to a solid preform and drawn into a fibre as shown schematically in Fig.7.5.

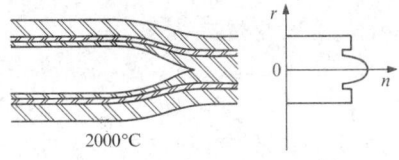

Fig.7.5. Collapsing of tube to a solid preform and index distribution.

In this process, the deposited layers become the core and inner cladding region of the fibre. Depending now on what index profile is specified for the fibre, the type of doping and its concentration in the preform layers will be chosen. To obtain a W-profile in the fibre, one starts to deposit layers first which are doped either with boric oxide or with fluorine. Boric oxide for the outermost layers has, in addition, the desirable characteristic that it forms a buffer zone between the outer support tube of low-grade quartz and the inner cladding and core layers. It inhibits diffusion of any contaminations from the outer quartz glass into the inner regions of cladding and core. To raise the index again from the inner cladding layers to the core layers, the amount of B_2O_3 or F-doping is decreased and, if an even larger index difference than that between the inner cladding layers and pure silica is specified, one of those dopants is added that increases the index above that of pure silica. By raising the dopant concentration from layer to layer, any index distribution in the core region can be approximated. In particular, it is thus possible to form the near-parabolic index profile of minimum delay spread and least widening of the impulse response in a multimode fibre.

When all layers are deposited inside the tube, it is collapsed into a solid rod. For this process, the temperature is raised beyond the softening point of the outer quartz glass. Its surface tension then contracts the softened material and the tube collapses into a solid rod. The oxygen-hydrogen burner, an electric furnace, or an induction heater as used for the chemical vapour deposition reach the required softening temperature of nearly 2000°C when they move at lower speed along the tube. The tube can therefore be collapsed in the same arrangement that was used for layer deposition. Depending on the traverse speed at which the proper temperature for collapse is reached on the hot zone, only a few passes are normally sufficient to effect the total collapse of the tube.
The plasma-activated deposition with the microwave cavity requires extra equipment, for example, an oxygen-hydrogen burner to collapse the tube.

During the collapsing process, the preform is liable to suffer two different defects. As one of these defects, the tube tends to deform into a solid preform of elliptical cross-section when collapsed. With some positive gas pressure inside the tube, it is possible, however, to counteract this tendency and maintain circular symmetry for collapse to a round rod.

The other defect appears as a dip in the refractive index profile near the axis of the collapsed preform. The high temperature to soften the outer quartz glass lets some of the dopant diffuse out of the innermost layers of the tube and evaporate into the inner tube space before total collapse. With the lower dopant concentration on the axis of the collapsed preform, its index dips to lower values there. This dip in the centre of the profile can be reduced by increasing the concentration of the more volatile dopant in the innermost layers or by substituting, or adding, a less volatile dopant to these layers. Both defects, the ellipticity and the index dip on the axis, are transferred to the fibre when it is pulled from the preform. Depending on the dimensions of the silica support tube and on the number and thickness of layers that are deposited, the solid preforms have an outer diameter between 5 and 15 mm and are from 0·3 m to 1·5 m long.

As the final step in the fibre fabrication, the preform is drawn into a fibre on a fibre-drawing machine. Fig.7.6 illustrates schematically the main parts of a fibre drawing arrangement. The preform is mounted into a support that lowers it slowly at a constant rate and feeds it into the neck-down region where it is heated over only a short distance beyond its softening point. Different heating arrangements can serve this purpose. One possibility is again to use an oxygen-hydrogen burner; but it must be carefully constructed in the form of a ring burner with several nozzles to heat the preform uniformly from all sides. Diffusion burners with separate nozzles

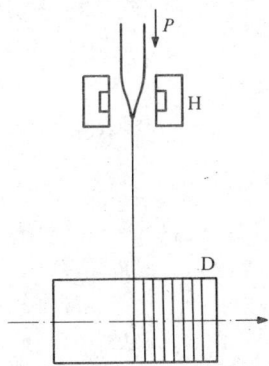

Fig.7.6. Fibre-drawing machine. P, preform; H, heating furnace or burner; D, drum.

for oxygen and hydrogen close together or even surrounding each other are to be preferred for this purpose. The two gases diffuse into each other right outside the nozzles and a soft flame is produced, thus minimizing fibre diameter variations caused by gas turbulence. Still less diameter variations in the fibre are enocuntered when instead of the ring burner a resistance-heated furnace or induction heating is used. These heating arrangements have a hot zone of perfect circular symmetry, which is confined to a narrow region along the preform. Even better control of a heating zone with perfect symmetry can be obtained when the high-energy light beam from a CO_2 laser is focused on to the fibre preform by the rotating lens arrangement in Fig.7.7. Such a laser beam heats the fibre preform with little danger of contamination, and fibres of extremely uniform diameter are drawn with it (Krawarik 1975).

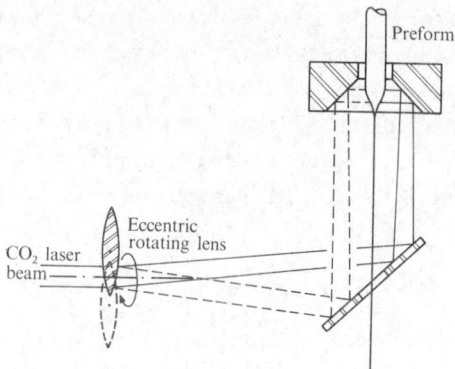

Fig.7.7. Fibre drawing by laser heating with rotating lens.

The fibre is pulled and wound on to a drum that rotates at the appropriate speed and moves slowly in its axial direction so that the fibre is wound on to it, turn next to turn, in a single layer. With the temperature at the transition from preform to fibre adjusted for a proper formation of the fibre, the final fibre diameter is determined by the diameter of the preform, the rate at which the preform is fed into the heater, and the pulling speed of the drum. By continuously monitoring the fibre diameter during the pulling operation and automatically controlling the temperature of the heater as well as the feeding rate of the preform and the pulling speed of the drum, fibres of a specified and constant diameter are drawn and wound on to the drum. Fibre drawing from quartz preforms usually proceeds at a temperature near 1900°C and at pulling speeds in the range of 0.3 ms^{-1} to 2 ms^{-1}. In this drawing process the ratios of core to cladding to outer diameter of the collapsed preform reproduce very nearly in the fibre. Also any index profiles for core and inner cladding regions are scaled down in pulling by the same ratio but maintain their shape and index differences.

The length up to which fibres can be drawn by this process depends in an obvious fashion on the diameter and length of the collapsed preform and the diameter down to which the fibre is drawn. For long fibres one should start with a preform of large diameter. Chemical vapour deposition, however, works well only in a limited range of diameters. For proper deposition of glass layers inside the tube, its inner diameter should be around 10 mm and for adequate diameter ratios of core and cladding, its wall should be only a few millimetres thick. The collapsed preform has under these conditions an outer diameter around 10 mm and, when 1 m long, can be drawn into a 10 km fibre of 100 µm diameter.

Chemical vapour deposition produces the layers which, in the final fibre, form core and inner cladding of the guiding structure. The formation of these layers takes place in a region inside the substrate tube which is well shielded against any contaminations and, in addition, heated to a temperature at which many such contaminations simply evaporate. It therefore allows preforms for fibres to be fabricated which have nearly as little absorption and scattering loss as the constituent materials have intrinsically in their pure and homogeneous state. Fig.7.8 shows the loss spectrum that is typically

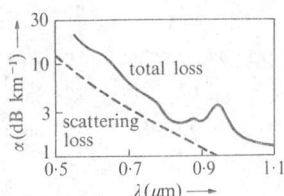

Fig.7.8. Typical loss spectrum of a quartz-glass fibre with a GeO_2-doped core produced by chemical vapour deposition. The solid line represents the total transmission loss, the broken line is the total loss contribution from scattering (Grabmaier et al. 1976).

measured for a quartz-glass fibre whose preform was fabricated by chemical vapour deposition. The solid line represents the total transmission loss and the broken line all loss contributions from scattering. In the spectral range $\lambda = 0\cdot 6$ to 1 µm and beyond, the total scattering loss very nearly follows the $1/\lambda^4$-law of Rayleigh scattering, and its level also corresponds to the intrinsic loss from Rayleigh scattering that we expect in perfect quartz glass. The absorption loss, as the difference between the solid and broken line, increases monotonically with frequency for wavelengths below $\lambda = 0\cdot 8$ μm. It corresponds to the intrinsic absorption loss of pure quartz glass with only slight additions between $\lambda = 0\cdot 6$ and $0\cdot 8$ μm due to small contaminations and material imperfections. In pure fused silica one observes quite often a loss peak near $\lambda = 0\cdot 63$ μm that is due to imperfections in the fused silica network. These imperfections are induced by the fibre drawing process during which some of the Si-O-Si bonds in the network are broken up (Kaiser 1974). The drawing-induced absorption can be eliminated by annealing the broken Si-O-Si bonds subsequent to fibre drawing. More critical in the example of Fig.7.8 are the absorption peaks at $\lambda = 942$ nm and $\lambda = 880$ nm. These absorption peaks are caused by the hydroxyl ion OH^- which has a vibrational resonance at $\lambda = 2790$ nm with a second overtone at $\lambda = 942$ nm, and a combination tone at $\lambda = 880$ nm from the third overtone of the OH^--vibration and the first overtone of the SiO_2-matrix resonance at $\lambda = 12\,500$ nm. A concentration of only $0\cdot 83$ p.p.m. of water in fused silica will raise the loss by 1 dB km^{-1} in the absorption peak of the hydroxyl ions at $\lambda = 942$ nm. For low-loss fibres, we therefore require fused silica which has an extremely low concentration of these hydroxyl ions. High quality quartz-glass fibres have minima of their total transmission loss below 2 dB km^{-1} between $\lambda = 0\cdot 8$ and $0\cdot 9$ μm and below 1 dB km^{-1} near the wavelength $\lambda = 1\cdot 06$ μm of the neodymium YAG laser.

Multimode quartz-glass fibres with a large fibre parameter V can be constructed by chemical vapour deposition to an index profile of their core that is very close to the optimum for minimum delay spread and minimum pulse broadening. Deviations of the index profile from the optimum distribution arise in chemical vapour deposition in the form of index steps from layer to layer and, from the collapsing process, in form of the index dip near the fibre axis.

When we attempt to produce the near-optimum parabolic profile according to

$$n^2(r) = n_1^2 [1 - 2\Delta(r/a)^2] \tag{7.1}$$

by chemical vapour deposition, we obtain at first the staircase profile on the left-hand side of Fig.7.9. The N steps of this staircase correspond to N deposition layers of different refractive index. Actually the index will not change abruptly; diffusion during depositon, collapsing, and fibre

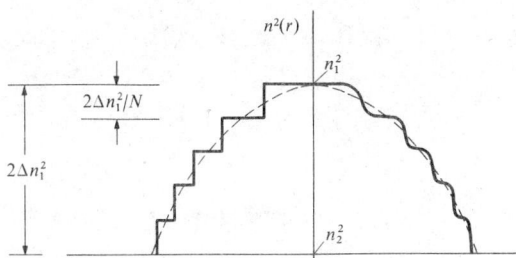

Fig.7.9. Staircase index profile of graded-index fibre with layer structure before diffusion (left) and after diffusion (right) (Behm 1976).

drawing will rather smooth the index profile to the distribution on the right-hand side of Fig.7.9. Such a diffused staircase profile may be described by adding a sine perturbation to the perfect distribution of equation (7.1) (Behm 1976)

$$n^2(r) = n_1^2 [1 - 2\Delta(r/a)^2 + d\frac{\Delta}{N}\sin(2N\pi r^2/a^2)] . \tag{7.2}$$

The coefficient d accounts for the amount of diffusion. If n^2 always decreases with increasing r, d is never larger than $1/\pi$. The sine perturbation of the index profile changes phase constants and the delay of guided modes. For a first-order approximation of this phase and delay change, we start from the field distribution of guided modes in the unlimited parabolic profile in equations (5.54) and (5.59) and the phase constants in equation (5.64). The phase parameter $B = (\beta^2/k^2 - n_2^2)/(n_1^2 - n_2^2)$ for these modes is consequently given by

$$B_0 = 1 - 2(2q+l+1)/V \qquad (7.3)$$

in terms of the fibre parameter $V = n_1 ka\sqrt{2\Delta}$.

A perturbation solution of the wave equation (5.53) for the perturbed index distribution in equation (7.2) based on these unperturbed field distributions leads to a phase parameter

$$B = B_0 + B_1 \qquad (7.4)$$

with a first-order perturbation B_1 according to

$$B_1 = \frac{d}{2N}\frac{(l+2q)!}{(l+q)!l!}(-1)^q \frac{x^{l+1}}{(1+x^2)^{(2l+q+1)/2}}\sin[(2q+l+1)\tan^{-1}(1/x)] \times$$

$$\times F[-q,-q,-2q-l,1+x^2]. \qquad (7.5)$$

In this expression F designates the Gaussian hypergeometric series and x is

$$x = V/(2\pi N). \qquad (7.6)$$

Evaluation of this perturbed phase parameter with respect to the group delay of guided modes leads to explicit expressions for the delay of modes in the parabolic index profile with sine perturbations (Behm 1976). From these expressions the total delay spread of all guided modes has been calculated for the maximum sine perturbation corresponding to $d = 1/\pi$, and plotted in Fig.7.10 relative to the total delay spread $\Delta\tau L = \tau_1 L\Delta$ in a step profile as a function of the number N of layers. The profile dispersion has been assumed to be zero in this evaluation. The curves in Fig.7.10 are for different values of the relative index difference Δ and the core radius a relative to the wavelength λ. The curves tend to unity when the profile has only few layers. For a large number of layers they approach $\Delta/2$ which corresponds to the total delay spread $\Delta\tau L = \tau_1 L\Delta^2/2$ in the parabolic index profile. For large core sizes ($a > 10\lambda$), the graded-index core must have more than 40 layers in order to keep the total delay spread close

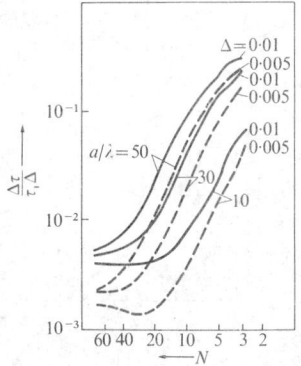

Fig.7.10. Delay spread in a parabolic index profile with sine perturbation relative to delay spread in stepped-index fibres versus number of layers N for different core sizes and index differences (Behm 1976).

to that of the unperturbed parabolic profile. For this reason a method of chemical vapour deposition is to be preferred that puts on many layers in a short time. The plasma-activated deposition method which uses a microwave cavity fills this requirement quite well. Alternatively, one might also prefer binary silica systems in which the index-changing dopant diffuses more easily to smooth the index staircase during the different stages of production. P_2O_5 and B_2O_3 seem to fill this requirement, because only little index deviation due to the layer structure is observed in graded-index fibres made out of phosphosilicate or borosilicate glass.

The index dip in the core centre reduces the number of guided modes according to equation (5.33) by the amount by which the solid of revolution of the index profile changes its volume. However, for shallow or narrow dips, this relative change in volume remains quite small and the number of guided modes nearly the same as in the profile without dip. The guided modes experience some change in phase constant due to the index dip. This change in phase constant is significant for those modes which in the ray picture have rays that come close or even cross the dip. For a narrow dip on the axis, these are modes with nearly meridional rays and consequently low circumferential order l. For index dips according to Fig.7.11 with a round centre region of homogeneous but depressed index and an abrupt step to a parabolic profile, equation (5.164) has been evaluated and explicit expressions

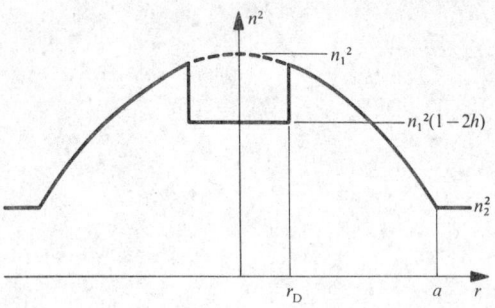

Fig.7.11. Parabolic index profile with abrupt dip on axis.

have been found for the delay of all guided modes (Behm 1977). The power-law profile with a Gaussian dip according to

$$n^2 = \begin{cases} n_1^2 [1 - 2\Delta(r/a)^\alpha - \hbar \exp(-r^2/r_D^2)] & r < a \\ n_1^2 [1 - 2\Delta] & r > a \end{cases} \quad (7.7)$$

has also been analysed (Olshansky 1976), and the delay of its modes evaluated numerically. As a result of these calculations, Fig.7.12 shows the total delay spread and the r.m.s. width of the impulse response according to equation (5.254) for equal excitation of all guided modes both relative to their respective values in the cladded-core fibre with a homogeneous core. The solid curves in Fig.7.12 (a) and (b) give relative delay spread and r.m.s. pulse width, respectively, for the abrupt dip with a homogeneous bottom. The circles indicate corresponding values for the Gaussian dip in a parabolic profile and the squares hold for the Gaussian dip in the optimum power-law profile with a characteristic exponent $\alpha = 2 - 2\Delta$. All profiles are assumed to have no profile dispersion.

By comparing the circles for the Gaussian dip with the

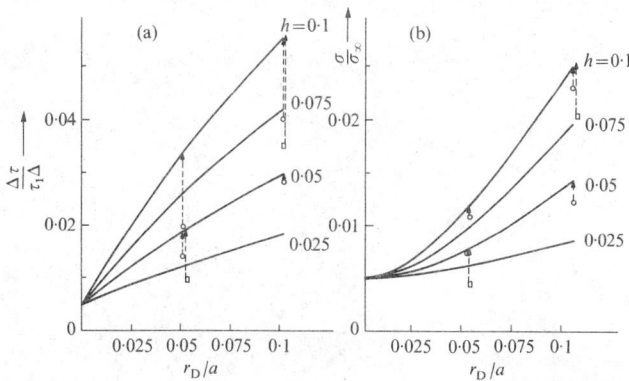

Fig.7.12. Total delay spread (a) and r.m.s. width (b) of impulse response for parabolic index profile with abrupt dip relative to their values for a homogeneous core (Behm 1977). Circles: parabolic-profile with Gaussian dip. Squares: optimum power-law profile with Gaussian dip.

curves in Fig.7.12, we note that a Gaussian dip of the same depth and equivalent width as the abrupt dip has lower values for total delay spread and r.m.s. pulse width. However, these differences are not so large that we could not infer all important degradations from the results for the equivalent abrupt dip. Before we draw any conclusions from Fig.7.12, we should examine the change that the dip effects on the overall impulse response. Fig.7.13 compares for equal excitation of all guided modes the impulse response of a perfect parabolic profile to that of a parabolic profile which is perturbed by either an abrupt or a Gaussian index dip on axis. The impulse responses in Fig.7.13 take account only of the effect of delay differences between modes but disregard both material dispersion and mode dispersion. The main effects of the index dips are to lower the maximum amplitude of the rectangular impulse response of the parabolic profile and to cause a pulse tail. The pulse width, however, when measured at the half power points, hardly changes at all. It therefore depends very much on the kind of signal that the fibre is to transmit whether such a centre dip in the index profile is really critical. For any applications in which it would be objectionable, appropriate

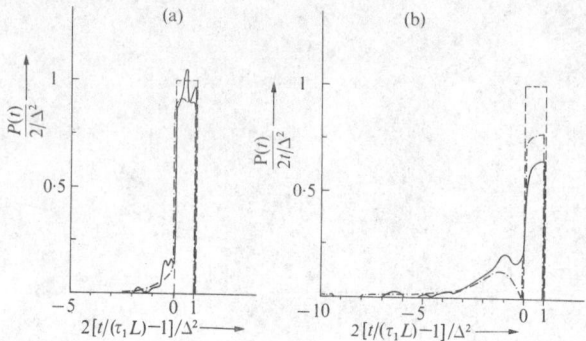

Fig.7.13. Impulse response for equal excitation of all guided modes of a parabolic profile without (-----) and with an abrupt (-·-··-) dip and with a Gaussian (———) index dip on its axis (Behm 1977). a) $\Delta = 0\cdot 01$, $h = 0\cdot 05$, $r_D/a = 0\cdot 05$. b) $\Delta = 0\cdot 01$, $h = 0\cdot 1$, $r_D/a = 0\cdot 1$.

measures have to be taken to reduce the dip by increasing the dopant concentration for the last few layers during chemical vapour deposition, or by using a less volatile dopant at least for the last few centre layers, to reduce any out-diffusion and evaporation during the collapsing process.

7.3. DOUBLE CRUCIBLE PROCESS FOR COMPOUND GLASS FIBRES

Of the compound glass systems with lower melting temperatures than fused silica, the soda lime silicate, soda borosilicate and alkali lead silicate glasses are best suited for optical fibres. Soda aluminosilicate glass deserves special attention also, because it has shown particularly low loss from intrinsic scattering (Pinnow et al. 1975). Near $\lambda = 1$ μm, this scattering loss was measured below $0\cdot 2$ dB/km^{-1}. Aside from their intrinsic absorption, additional absorption loss in these glasses in the visible and near-infrared range of wavelength is mainly caused by trace quantities of a number of transition metal ions and by water, which is dissolved in the glass as hydroxyl ions. The loss due to Rayleigh scattering in these glasses is composition dependent and reaches its lower limit only when these glasses are near to their perfectly homogeneous state with no phase separation or devitrification. Preparation of these glasses and the fabrication of fibres from them has therefore to aim at the highest degree of purity and homogeneity and avoid any recontamination or phase separation at all process stages.

To obtain any such glass in its near perfect form in the bulk, the high-purity base materials are pulverized and melted in a silica crucible in a resistance-heated furnace or by high-frequency induction heating. In case of induction heating, the molten glass itself acts as the susceptor. At temperatures above 1000°C, most alkali glasses develop enough ion conductivity to couple the melt to the high-frequency field of nearly 5 MHz. To prevent or at least reduce any dissolution of the silica crucible, it is cooled either by water or a gas stream. On the inner wall of the cooled crucible, the steep temperature gradient lets a thin layer of pure glass remain solid, which then inhibits the diffusion of any impurities from the crucible into the molten glass. This process thus eliminates nearly all contamination during glass melting. The induction heating is started with a graphite cylinder underneath the bottom of the crucible, to preheat a small quantity of the mixture of pulverized base materials. Once this priming quantity starts to melt, it is moved into the high-frequency field instead of the graphite heater, and the remaining mixture is added. When all the glass is molten, it is homogenized by either stirring the melt or by bubbling pure gases through the melt. Bubbling of a reducing gas mixture through the melt also serves to further refine the glass. Eventually the homogenized and refined glass melt is cooled down to reduced temperatures at which rods may directly be pulled from its surface. These rods have a diameter of 5 to 10 mm. They should be subjected to only a minimum of handling and exposure to the atmosphere in order that they suffer least recontamination before they are further processed into fibres.

To pull cladded-core fibres from these glass rods, two cylindrical crucibles are arranged concentrically as in Fig. 7.14 with one crucible surrounding the other. These crucibles consist of pure platinum or a platinum alloy. Pure silica has also been used for the inner crucible. Both crucibles taper

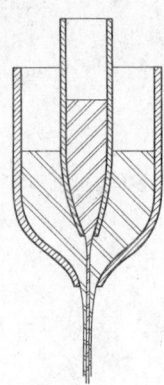

Fig.7.14. Double crucible for drawing compound glass fibres.

down to concentric nozzles out of which the glasses are pulled into a fibre. High-index glass rods homogenized and refined by the previous process are fed at low speed into the inner crucible to form the core of the fibre. Low-index glass rods of the same basic glass system as for the fibre core are fed into the outer crucible for the fibre cladding. The rods should enter the crucibles at a fixed position and a a concave meniscus should form between the rod and the melt; this prevents any gas bubbles from being drawn into the melt and trapped by it. The double crucible is heated to the melting temperature of both glasses by a resistance-heated furnace. This furnace together with the double crucible take the place of the heating arrangement and preform in the fibre pulling machine of Fig.7.6. For the actual fibre pulling, the furnace temperature is adjusted to render the glasses in both crucibles of a suitable viscosity. The cladding glass which flows from the nozzle of the outer crucible will then pull along the core glass, which flows from the nozzle of the inner crucible. Both together form the cladded-core fibre. The diameters of both nozzles must be so chosen and matched to each other so that the specified core and cladding size can be obtained for the fibre. With core and cladding nozzle diameters of 0·5 mm and 3 mm, fibre core and cladding diameters are typically 10 to 20 μm and 70 to

100 μm, respectively (Day et al. 1975). Within certain limits, the core diameter can be varied and the outer fibre diameter kept constant by just changing the diameter of the inner nozzle, but keeping the diameter of the outer nozzle constant. Thus, it is even possible to draw single-mode and multimode fibres from the same double crucible, in which only the nozzle diameter of the inner crucible has been changed.

The double crucible method produces cladded-core fibres with homogeneous cores and a relatively abrupt index transition between core and cladding. Some smoothing of the index transition occurs during fibre drawing due to diffusion of core and cladding glass agents into each other in the molten states and at the high temperature near the nozzles of the double crucible. For multimode dimensions these fibres have a rather large delay spread and show much pulse widening.

For the production of graded-index fibres out of compound glasses with low melting temperatures, the double crucible method has been modified (Koizumi et al. 1974). It uses ion exchange between core and cladding glass during the fibre pulling to form the desired index profile across the core. The best results with this modified double crucible method for graded-index fibres, in particular with respect to transmission loss, have been obtained with soda borosilicate glasses. Raw materials are boric acid, sodium nitrate, synthetic silica, a small quantity of arsenic trioxide and, for the core glass, an additional quantity of thallium nitrate (Koizumi and Ikeda 1975). The arsenic trioxide serves to fine the glass by controlling harmful states of oxidation of trace impurities. The thallium nitrate raises the index of the core glass and grades the index profile by diffusion of thallium ions from the core to the cladding glass.

The inner crucible for the core material is loaded with those soda borosilicate glasses to which thallium ions at a molar concentration of 1·5 to 4 per cent have been added. Thallium ions have the largest electronic polarizability among most monovalent ions in glass but only a small ionic radius. They therefore increase the refractive index very effectively but also diffuse easily at the high termperature of the softened or molten glass. The outer crucible is loaded with a pure soda borosilicate glass for the cladding. Fig.7.15 indicates the form and relative sizes for the nozzles of the inner and outer crucible.

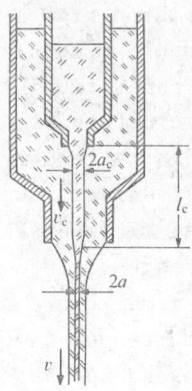

Fig.7.15. Modified double crucible for graded-index fibres by ion exchange (Koizumi et al. 1974).

If the fibre of core diameter $2a$ is drawn at a speed v, the core glass flows out of the inner crucible through the lower part and nozzle of the outer crucible at approximately the speed

$$v_c = a^2 v / a_c^2 , \qquad (7.8)$$

where $2a_c$ designates the enlarged core glass diameter just below the inner nozzle in the double crucible. It takes the core glass the time

$$t_c = l_c / v_c \qquad (7.9)$$

to move the length l_c of this region. Mainly during this time, it exchanges thallium ions for sodium ions with the cladding glass flowing out of the outer crucible. If D is the diffusion constant for the diffusing ions, then the parameter

$$K = D \, t_c / a_c^2$$

measures the amount of ions which are exchanged by diffusion. With t_c and a_c from equations (7.9) and (7.8), this parameter may also be expressed as

$$K = D \, l_c / (a^2 v) \qquad (7.10)$$

in terms of the ion exchange length l_c, the pulling speed v, and the core radius a. The solution of this diffusion problem shows that the thallium concentration near the axis changes to an almost parabolic distribution at the end of the ion exchange length l_c, when the ion exchange parameter K ranges from 0·01 to 0·1. Since the thallium ions have a far greater electronic polarizability than the other ions that diffuse in the process, their concentration determines the refractive index profile in the core region. Adjusting l_c, a, and, with the temperature, the diffusion coefficient D, for the proper value of K, will therefore lead to the near-parabolic index distribution that reduces delay spread and pulse widening considerably.

Fig. 7.16 shows the fibre size and index distribution that

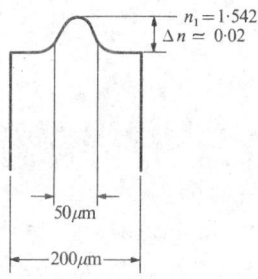

Fig.7.16. Typical size and index distribution of graded-index fibre fabricated by the modified double crucible method (Koizumi et al. 1974).

typically result from this modified double crucible method. It has the almost ideal parabolic distribution near the fibre axis, but then tapers gradually into the uniform cladding index. The formula

$$n(r) \simeq n_2 + \frac{n_1 - n_2}{\cosh^2(r/a)} \tag{7.11}$$

provides an analytic approximation for this index profile. The maximum index difference $\Delta n = n_1 - n_2$ that can be obtained in this fibre depends on the concentration of Tl_2O in the soda borosilicate that remains on the axis of the fibre subsequent to ion exchange and fibre drawing. Fig.7.17 shows

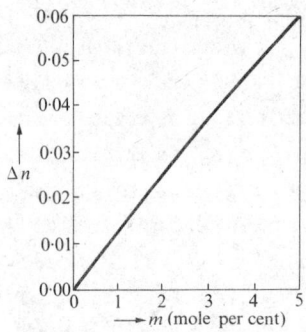

Fig.7.17. Increase of refractive index in soda borosilicate of composition $m\, Tl_2 \cdot (20-m)Na_2O \cdot 20B_2O_3 \cdot 60SiO_2$ as a function of Tl_2O molar fraction (Uchida 1976).

this index difference as a function of the Tl_2O mole fraction. In actual fibres, the amount of Tl_2O is typically near 1·5 mole per cent (Uchida 1975).

Alternatively to equation (7.11), the fourth-order index profile in Fig.5.19 with the coefficient d chosen according to $d = -1$ models the ion exchange profile of the double crucible method quite well. The total delay spread of guided modes in this profile is, however, quite large due to the relatively small delay of modes with large circumferential but small radial orders. These modes concentrate their fields in the transition region between core and cladding where in the present case the index is considerably lower than it should be for the minimum delay spread in the near parabolic profile. Except for these modes of very high circumferential and low radial orders, the delay differences remain small. If excitation of and coupling to these fast modes can be avoided, the graded-index fibre from ion exchange out of the double crucible shows very little delay spread.

Compound glass fibres which are produced by the double crucible method either with a homogeneous core or by ion exchange with a graded-index core also show very low transmission loss. If the materials and fibres are carefully handled under clean room conditions, if all crucibles are likewise kept free from traces of any harmful impurities, in particular free from water, and if the surrounding atmosphere is free of harmful gas components, the resulting fibre will hardly be contaminated and its absorption loss will be close to the intrinsic absorption of the pure glass. In addition, homogenizing and fining of the molten glass and the rapid cooling of the glass from the molten phase in the double crucible to the fibre result in good homogeneity of the fibre. Therefore any scattering loss will also stay close to the intrinsic Rayleigh scattering of the perfect glasses and very little mode conversion and radiation due to index fluctuations occur. Fig.7.18 shows the loss spectrum of a soda

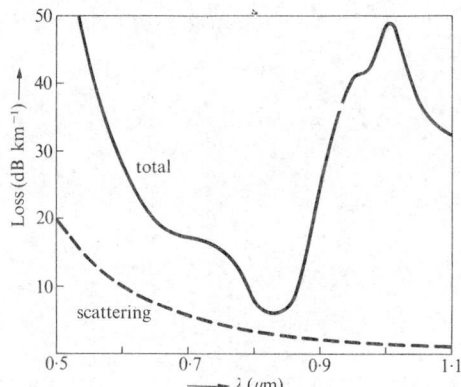

Fig.7.18. Total loss (———) and scattering loss (-----) of a soda borosilicate fibre with graded-index core (Ikeda and Yoshiyagawa 1976).

borosilicate fibre with a graded-index core from the modified double crucible method. At the loss minimum near λ = 830 nm, the total loss of 6 dB km^{-1} (Ikeda and Yoshiyagaw 1976) consists of 3 dB km^{-1} of intrinsic scattering and 1 dB km^{-1} of intrinsic absorption both of the homogeneous and pure base glass. The absorption due to trace impurities of iron and copper adds nearly 2 dB km^{-1} to the total loss at this wavelength. The absorption peak near λ = 942 nm again represents the second overtone of vibrational resonance of the hydroxyl ion OH$^-$. Concluding the discussion of the double crucible method, it should be noted that its continuous operation allows fibres of arbitrary length to be fabricated.

7.4. ROD-IN-TUBE METHOD

The rod-in-tube method lends itself to fabricate cladded-core glass fibres with homogeneous cores. Both quartz-glass fibres as well as fibres from low-melting compound glasses may be produced by this method. The homogenized and refined core material of higher refractive index is prepared as a round rod with a smooth and clean surface which, for better results, may also be polished. The cladding material of lower index of refraction is given a tubular form for the core rod to fit into it. The inner wall of this cladding tube is likewise cleaned and polished. The rod is then inserted into the tube as in Fig.7.19, and both are mounted as the preform on to the

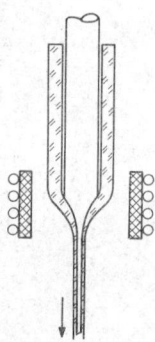

Fig.7.19. Fibre drawing from rod in tube preform.

fibre pulling machine of Fig.7.6. Just as in case of the quartz-glass preform from chemical vapour deposition, the rod-in-tube preform is now lowered into the heater where it begins to soften. It is then drawn into the fibre and wound on to the drum of the fibre pulling machine.

Special care must be exercised with the rod-in-tube method that the interface between core and cladding remains free of any particles or voids, which would act as scattering centres and degrade the transmission characteristics of the fibre. As an example of what the rod-in-tube method is able to accomplish, Fig.7.20 shows the loss spectrum of a quartz-

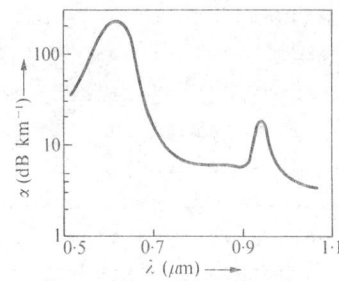

Fig.7.20. Spectral loss of quartz-glass fibre with fused silica core and fluorine-doped cladding fabricated by the rod-in-tube method (Mühlich *et al.* 1975).

glass fibre that has been produced by this method (Mühlich *et al.* 1975). The core rod for this fibre was made from water-free fused silica and had a bulk absorption of 2 dB km^{-1} at λ = 800 nm. For the cladding synthetic fused silica was used, which during its production was doped with fluorine of a concentration of 5 per cent to obtain an index which is by 2 per cent lower than the core index of pure fused silica. The bulk absorption loss of this fluorine-doped cladding tube was measured at λ = 800 nm to be 8 dB km^{-1}. The rod-in-tube preform was heated by an oxygen-hydrogen burner and drawn to a fibre with 50 µm core and 120 µm outer diameter. The loss spectrum of the fibre in Fig.7.20 shows minima of 6 dB km^{-1} at λ = 850 nm and 4 dB km^{-1} at λ = 1060 nm. The absorption peak near λ = 945 nm represents the second overtone of the hydroxyl vibrational resonance, while the broad loss maximum near λ = 630 nm can be attributed to drawing induced imperfections of the glass matrix and to imperfections at the core-cladding boundary.

7.5. POLYMER CLAD FIBRES

Optical fibre guides are also fabricated with a low-loss glass as core material but plastics for the cladding material. This combination of low-loss glass core and plastic cladding offers several advantages. If fused silica is used as core material, there are only few glass systems with an index of refraction lower than that of fused silica, and even these have only minute index differences compared to fused silica. On the other hand, a number of polymers have substantially lower refractive indices and, in the visible and near infrared, show sufficiently small absorption to be suitable as a cladding material. They do not, of course, reach the low bulk loss that is obtained in glasses of good optical quality. Nevertheless plastic-clad fibres with a fused silica core and a polymer cladding can have transmission loss values suitable for applications in signal transmission over short or intermediate distances. With the substantial index differences between the fused silica core and the polymer cladding, these fibres can have very large numerical apertures. Another advantage is the relative ease with which these polymer-clad glass fibres are manufactured.

Of the various polymers that appear suitable as cladding materials for a fused silica core, perfluoronated-ethylene-propylene with the trade name Teflon and the designation FEP 100 has a particularly low refractive index; it amounts to only $n = 1 \cdot 338$ at $\lambda \simeq 0 \cdot 9$ μm. FEP is also easy to process in the molten state. Its bulk loss is, however, quite high due, in part, to the large scattering of its semi-crystalline structure. The loss was measured at near 5 dB cm^{-1} (Kaiser et al. 1975). Not quite as low in refractive index is Teflon PFA, a copolymer of tetrafluoroethylene and perfluoro-vinyl-methyl ether, with $n = 1 \cdot 42$. As a cladding material, it has better extrusion characteristics than Teflon FEP (Blyler et al. 1975). Also suitable for cladding fibres is hexafluoropropylene mixed with vinylfluoride (Suzuki and Kashiwagi 1974) with an index of refraction of $n = 1 \cdot 415$. Both Teflon PFA and the latter mixture have the same high scattering loss as Teflon FEP. A polymer with much lower bulk loss and also a low enough refractive index can be found among the silicone resins (Tanaka et al. 1975). Fig.7.21 shows the bulk loss spectrum of a particularly low-loss silicone. Its loss minimum of 0·9 dB m^{-1} occurs near λ = 0·77 μm. Loss peaks are observed at λ = 0·74, 0·91, 1·02, and 1·09 μm, but the highest of them at λ = 0·91 μm goes to only 70 dB m^{-1}.

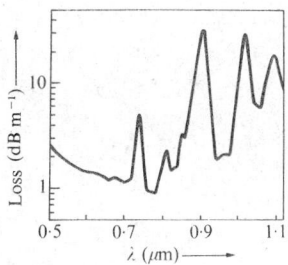

Fig.7.21. Bulk loss spectrum of a low-loss silicone resin (Tanaka et al. 1975).

Polymer-clad fibres are fabricated by first drawing the fused silica core from a solid preform on a precision fibre drawing machine. The polymer cladding is usually applied on-line with the fibre drawing. The method of cladding application depends on the cladding material and how it can be processed. Teflon as a cladding must be extruded in the form in which it is mounted on the fibre. For this purpose, a tube extruder is attached on the fibre drawing machine and, immediately following the drawing process, the fibre core is passed through the centre orifice of this tube extruder. The polymer is first extruded as a large diameter tube around the fibre core and, while still molten, it is drawn down to its final size outside the fibre. The dimensions, extrusion rate, and drawing speed are so adjusted that the polymer tube fits only loosely around the fibre. This loose fit of the cladding tube reduces the loss which the core modes would suffer by the high bulk loss of the cladding material.

The silicone resin as well as other cladding materials that are available in liquid form are applied by dipping and then must be cured for solidification. The unclad fibre core

coming out of the drawing process is dipped continuously into
a bath of the low-loss silicone resin which is already mixed
with its hardener. It passes subsequently through a curing
furnace. For protection of the low-loss polymer cladding, the
fibre is dipped once more into a high-loss resin with a higher
modulus of elasticity than that of the low-loss resin.

The dipping of the core into the cladding bath, and the
subsequent curing of the cladding material, provide a much
more intimate contact between core and cladding than the extrusion of the cladding as a loosely fitting tube around the
core. It is therefore best suited for low-loss cladding
materials. Fibres which are drawn and clad with the low-loss
silicone resin whose bulk losses are shown in Fig.7.21 yield
the transmission loss spectrum of Fig.7.22. This transmission

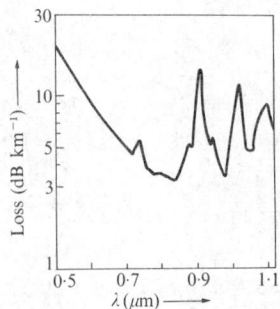

Fig.7.22. Transmission loss spectrum of a silicone-clad fibre excited by
a light source with a numerical aperture of 0·2 (Tanaka *et al.*
1975).

loss was measured with a light source of the relatively large
numerical aperture of 0·2. Such a light source excites so
many of the low-loss propagating modes that, together with
mode interaction, a stationary power distribution is reached,
and Fig.7.22 represents quite closely the stationary transmission loss of the statistical mode of lowest loss. The main
peaks of the bulk cladding loss of Fig.7.21 also appear in the
fibre transmission loss in Fig.7.22, but at a much lower level.
With respect to their minimum loss values of nearly 3 dB km^{-1},

such polymer-clad fibres represent a very economical alternative to low-loss glass fibres with glass cladding. Their homogeneous core, on the other hand, with its large numerical aperture, causes large delay differences between modes and a large overall delay spread. In the stationary impulse response, however, the large cladding loss of higher-order core modes limits the pulse broadening to values much below what we would expect without the added cladding loss. Instead of the 200 ns km^{-1} of impulse widening calculated according to equation (4.280) from the index difference between core and cladding, only 20 ns of pulse widening are actually measured for such a fibre with a length of 1 km (Tanaka et al. 1975).

In production, the polymer-clad fibre offers important advantages. Drawing, cladding, and coating of the fibre can follow each other in line and produce fibres at a rate of up to 100 m min^{-1}. The starting materials are commercially available and of relatively low cost.

7.6. SINGLE FIBRES OR FIBRE BUNDLES IN CABLES

Optical fibre cables contain one or several fibres. Also large numbers of fibres are bundled together and put into a cable. Fibre cables with a large number of fibres are either constructed for each fibre to be optically isolated from all the others so that it serves as a separate and individual transmission line, or all fibres or only groups of them, form bundles in which all fibres belong to the same transmission channel and transmit the same signal. The latter kind of optical fibre cable serves mainly in control and process applications were it bridges only short distances. It usually can have relatively high transmission loss and is subject to only small mechanical forces and other environmental stress. Therefore, the cable construction needs to provide only a minimum of mechanical protection. If one or several of the fibres of the large number in a bundle should rupture, the transmission of the bundle as a whole will not degrade noticeably. Such fibre cables therefore provide a high degree of reliability even with modest expenditure for the cable construction and fabrication.

The situation is quite different for optical fibre cables

in which each individual fibre serves as a separate transmission line. It is particularly critical when the fibre cable uses low-loss fibres to transmit wide-band signals over long distances. The high reliability which is usually required for signal transmission must then be met by each individual fibre. Each fibre must be so prepared and the whole fibre cable so constructed that all fibres are optically isolated from each other. The cable construction must also provide absolute protection against mechanical forces during installation and all environmental influences during operation on site. Only a minimum of degradation in transmission characteristics can be allowed. In particular, the cabling itself and installation along the route should keep any microbending low enough and exclude any critical spectral components of its curvature distribution, so that transmission loss remains as low as these applications usually require. Since the design and production of these cables with separate fibres need so much more consideration and effort than is necessary for cables with one or several fibre bundles, the following sections on the preparation and cabling of fibres deal only with cables which contain separate fibres.

7.7. FIBRE COATING AND JACKETING

Glass fibres are quite straight and inherently of high strength in their pristine condition immediately after drawing. However, the high tensile strength degrades very soon after fabrication if the bare fibre is left exposed to the normal atmosphere and handled in this state. Under the action of chemical attack by surface moisture and due to mechanical abrasion, microcracks form at its surface and propagate. They reduce the tensile strength substantially. As the first step in fibre cabling, the fibres are therefore coated with a protective layer or surrounded by a plastic jacket. In order to prevent any abrasion of the pristine fibre and to provide a barrier against chemical attack and hence maintain its initial strength, the fibre is coated or jacketed on-line with the drawing process. Quite often this first coat or jacket protects and conserves not only the glass surface of the fibre but is designed to provide additional strength in order to increase the tensile stress and also to insulate the fibre optically from its neighbours in a bundle in order to suppress any crosstalk in signal transmission. To meet the first objective of additional strength, the jacket itself should be made of a material of high tensile strength and high modulus of elasticity. For the second objective of optical isolation, the jacket material should absorb all light at the wavelength of signal transmission. The coating or jacket should furthermore not cause any microbending of the fibre and even strengthen it agains microbending by external forces that arise in the subsequent steps of fibre cabling and installation.

If only fibre abrasion is to be prevented and surface protection by a barrier against chemical attack to be provided, it suffices to coat the fibre on-line with the drawing operation with a plastic film of only a few micrometres thickness. A co-polymer in an acetone solution may serve such a purpose (France, Dunn, and Newns 1976). Fused silica as well as compound glass fibres may be coated with this plastic film. Within seconds after the fused silica fibre is pulled from the preform or the compound glass fibre out of the double crucible, it runs in the vertical direction through a coating bath and out of a nozzle at its bottom. The forces from surface tension in the nozzle nearly balance the hydrostatic pressure from the coating solution in the bath. The hydrodynamic Navier-Stokes equations reduce under these conditions to the static Laplace equation for the fluid velocity. When this equation is solved subject to the appropriate boundary conditions, the mass flow through the nozzle can be determined, and the coating thickness follows as (France, Dunn, and Newns 1976)

$$t = \{b^2 + \frac{l}{\ln(c/b)} [(c^2-b^2)/2 - b^2 \ln(c/b)]\}^{1/2} - b \quad (7.12)$$

where $2b$ is the glass fibre diameter, $2c$ the nozzle diameter, and l a volume loading factor of the solution. This loading factor is typically $l = 0.14$. The coating thickness is independent of the viscosity of the solution and of fibre drawing speed; it varies only slowly with the fibre diameter.

A two-stage drying process after coating improves its

quality. The first drying oven drives the solvent from the solution at a temperature below the solvent boiling point. In the second stage a high temperature oven polishes the coating surface by heating the plastic briefly above its melting point. Such plastic coating can be applied to low-loss fibres without degrading their loss spectrum.

If, in addition to surface protection the jacket is to lend additional strength to the fibre a material of the appropriate tensile strength must be selected that can also be processed into a fibre jacket. Nylon 12 has been found to be particularly well suited as a jacket material (Rokunohe *et al*. 1976). The nylon jacket is mounted on the fibre by extrusion on-line with fibre drawing. Such a nylon jacket causes hardly any degradation of fibre transmission loss when its thickness remains smaller than 300 μm and when it fits tightly on to the fibre. Furthermore, to prevent any increase in loss, the fibre with the nylon jacket extruded on to it should be slowly cooled down to room temperature by air cooling.

Even better results for the tensile strength of a fibre with jacket were obtained when the fibre was first coated with silicone resin and then surrounded by a jacket of Nylon 12 (Nakahara *et al*. 1976). Fibres with a fused silica core and a polymer cladding of low-loss silicone resin have also been jacketed with nylon (Tanaka *et al*. 1976).

As noted above, the nylon jacket should fit tightly around the fibre and not surpass an upper limit in thickness so as not to increase the transmission loss. These conditions limit the cross-section of the nylon jacket and hence the additional tensile strength it can give to the fibre. To obtain more strength than can be provided by the simple nylon jacket, a reinforcement might be supplied for the jacket with a plastic material of higher strength and modulus of elasticity. A prime candidate for such a high strength and high modulus reinforcement is Kevlar$^{(R)}$, a product of E.I. Du Pont De Nemours & Co. (USA). Fig.7.23 shows as an example the cross-section of a low-loss silicone-clad fibre with a Kevlar reinforced nylon jacket. While the low-loss silicone-clad fibre with a nylon jacket of 1 mm outer diameter has 8 N tensile strength, the Kevlar reinforcement in the nylon

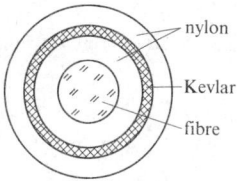

Fig.7.23. Cladded-core fibre with Kevlar reinforced nylon jacket (Tanaka et al. 1976).

jacketed fibre of Fig.7.23 raises the tensile strength to 24 N (Tanaka et al. 1976).

7.8. FIBRE PACKAGING

A fibre cable packs one fibre together with others and encloses them in a sheath. Usually a cable structure also includes other members to increase its strength and flexibility. In fibre packaging for cabling, the individual fibres are bundled and pressed together or pressed against other members of the cable structure. This packaging subjects the fibres to lateral forces which deform and displace the fibre laterally. Most critical for the maintenance of good transmission characteristics are any lateral displacements and the microbending that is associated with them. Cable deformation and bending will cause additional forces to act upon the fibres.

The fibre coating and jackets which have been described in the previous section were all designed to protect the surface of the bare fibre and stabilize or increase its tensile strength. They meet these objectives without degrading the transmission of the individual fibre. No consideration has yet been given to any external lateral forces to which the coated or jacketed fibre might be subjected, and to the microbending that these lateral forces might cause. Although the cable structure should be designed to minimize any such lateral forces, it can nevertheless not eliminate them. To reduce their degrading influence on transmission characteristics through the microbending which they cause, either the fibre jacket might be designed appropriately or additional members with which the fibres are packed into a cable might keep lateral deformations and microbending at a minimum.

As a model to study the lateral displacement of a packaged fibre when lateral forces act upon it, we consider the arrangement in Fig.7.24 (Gloge 1975). In this model, the bare glass fibre is pressed against an elastic surface which is

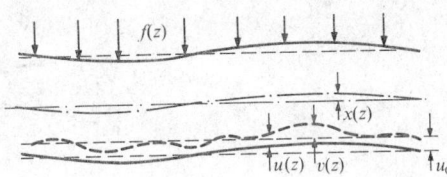

Fig.7.24. Fibre pressed against rough elastic surface as model for fibre packaging (Gloge 1975).

rough and, without the fibre, deviates parallel to the fibre axis by $v(z)$ from a perfect plane. When the force $f(z)$ per fibre length presses the fibre down, the surface yields and the fibre bends, resulting in a fibre displacement $x(z)$ and a displacement of the rough surface underneath the fibre axis by $u(z)$. We assume here that the surface is elastic or smooth enough and the fibre so flexible or the pressure so strong to produce a continuous contact between fibre and surface. Fibre and surface displacements are then related by

$$x = v - (u-u_0) , \qquad (7.13)$$

where $u_0 = \langle u \rangle$ is the average of $u(z)$ along z. According to the theory of elasticity for the glass fibre as a thin elastic beam, its displacement x under the force f follows from

$$\frac{d^4 x}{dz^4} = \frac{f}{H} , \qquad (7.14)$$

where

$$H = E\,I \qquad (7.15)$$

represents the flexural rigidity or stiffness of the fibre with E as its modulus of elasticity and

$$I = (\pi/4)\, b^4 \qquad (7.16)$$

as the moment of inertia of its circular cross-section of radius b. The surface displacement underneath the fibre increases linearly with f according to

$$u(z) = \frac{f(z)}{D} \qquad (7.17)$$

as long as f changes only gradually along z. The constant of proportionality D differs from the modulus of elasticity E_s of the surface material only by a factor close to unity. When we express the force f in equation (7.14) by surface and fibre displacements according to equations (7.17) and (7.13), we obtain

$$\frac{H}{D}\frac{d^4 x}{dz^4} + x = v \,. \qquad (7.18)$$

Important characteristics of solutions to this fourth-order differential equation become apparent when we take the Fourier transforms $X(\Omega)$ and $V(\Omega)$ of $x(z)$ and $v(z)$ and with them transform equation (7.18) into the domain of angular spatial frequencies

$$\Omega = 2\pi/l_v \qquad (7.19)$$

with l_v as the period length of the spectral components of $V(\Omega)$. In the Ω-domain, equation (7.18) has the form

$$X = \frac{V}{1 + \Omega^4\, H/D} \,. \qquad (7.20)$$

The curvature of the fibre for a slight displacement $x(z)$ from a straight axis is given by

$$\kappa(z) = \frac{d^2 x}{dz^2} \,. \qquad (7.21)$$

Its Fourier transform $K(\Omega)$ is hence $\Omega^2 X(\Omega)$ and from equation

(7.20) follows as

$$K = \frac{\Omega^2 V}{1+\Omega^4 \; H/D} \; . \qquad (7.22)$$

According to equations (7.20) and (7.22) a spectral component $V(\Omega)$ of surface roughness causes spectral components of displacement and curvature components that in their amplitudes depend strongly on the period length l_v. As a function of l_v, the curvature transform K has its maximum at

$$l_v = R \equiv 2\pi (H/D)^{1/4} \; , \qquad (7.23)$$

the so-called retention length (Gloge 1975) of the fibre on the elastic surface. Periodic disturbances with a period shorter than this retention length hardly affect the fibre. The fibre on the elastic support has enough stiffness for the support to yield under the pressure. For period length l_v longer than R the disturbance transfers fully to the fibre. A good design for fibre packaging should therefore aim for large values of the retention length.

There are different approaches to meet the design objective of a large retention length for fibre packaging. As one approach the jacket of the individual fibre might be designed to combine high flexural rigidity with good lateral compressibility. In the terms of equations (7.15) and (7.23), the fibre jacket should lend to the structure a large product of moment of inertia and effective modulus of elasticity in the longitudinal direction combined with a low modulus of elasticity in the transverse direction. One possible solution for a fibre jacket with a large retention length is the double jacket in Fig.7.25. A hard inner jacket with a high modulus of elasticity lends stiffness to the fibre, a soft outer jacket of a low modulus of elasticity provides lateral compressibility. The soft shell absorbs any displacements by elastic deformation when the fibre is pressed against a rough surface. Another solution for a fibre jacket with a large retention length utilizes anisotropic materials to obtain in a single jacket a high elastic modulus in the longitudinal direction

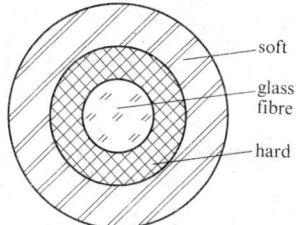

Fig.7.25. Fibre with hard inner jacket and soft outer shell for high retention length.

and a low elastic modulus in the transverse direction. As a suitable material, orientated polypropylene has been used as a loose-fitting tube around the fibre (Jackson et al. 1976). The method of packaging is to extrude a large bore tube at first with the fibre passing freely inside it and then, in an orientation oven, draw this tube down to its final size by a reduction factor of 10:1 and more. The isotropic material has a modulus of elasticity of 10^8 kg m^{-2}. The orientation of the polymer raises this elastic modulus in the longitudinal direction by a factor of ten. In the transverse direction, it decreases at the same time also by a factor of ten. The structure thus becomes quite stiff in bending and yet absorbs small-magnitude axial distortions of short correlation length quite well by surface compression.

Another approach to meet the design objective of a large retention length in fibre packaging is already suggested by Fig.7.24. As indicated by this figure, the support with an uneven surface to which the fibre might be pressed by external forces should yield as much as possible to these forces so that the fibre in its jacket with its inherent stiffness stays straight. In equation (7.17), the factor $1/D$ of lateral compressibility should be as large as possible or the modulus of elasticity E_s of the supporting surface very small. The fibres should be packed between cushions. If not all around, such cushions should at least be provided in the radial direction on both sides of the fibre. Fig.7.26 shows a cable with coated or jacketed fibres arranged around a cir-

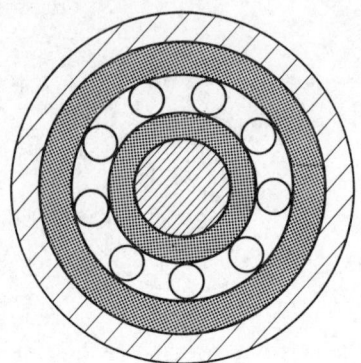

Fig.7.26. Fibre cable with fibres packed on a circumference between cushions.

cumference and tubular cushions on both sides of the ring of fibre. As a material for the cushion a soft plastic or better still a soft plastic foam is suitable. Polyurethane in compact form or as a foamed material is used for these buffer layers. In particular, the foam with its extremely high compressibility provides nearly perfect cushioning.

7.9. CABLE STRENGTH AND STRAIN RELIEF

Glass has a higher modulus of elasticity then most of the other materials with which it is combined in a fibre cable. If a longitudinal force strains the cable and all its members equally, the highest tensile strain builds up in the glass fibre. If the glass fibre is not in some way relieved from this strain, it will be by far the most endangered element in the cable and the first one to rupture. The forces which the cable with all its elements must safely resist arise when the cable is bent to be wound on drums, handled during installation, pulled into ducts, and bent again along the route. During all these steps, it might in addition be compressed sideways, if not actually exposed to crushing. The cable flexibility which most of these operations require is achieved by stranding the fibres either individually or in groups together with spacers or buffer elements. The fibres in the circumferential arrangement of Fig.7.26 are thus stranded along a helical path of very steep pitch with one or a few turns of the helix per metre. If the fibres are not packed too tightly so that they can slip freely against each other and against other supporting members, the bending forces distribute evenly and put only little strain on the individual fibres.

To avoid too much tensile strain on the individual fibre when the whole cable is pulled by an external force, it must be furnished with additional strength members. The first choice for placing such a strength member in a cable with several fibres is usually the cable centre. Other strength members may be placed on a circumference outside of the stranded fibres. The cable sheath may also be reinforced with strength members and may serve to strengthen the cable against tension forces. The material for the strength members should combine a high modulus of elasticity and high tensile strength with low specific gravity. It will, under these conditions, strengthen the cable, reduce the strain and tension of the fibres, without adding too much to its weight.

The weight factor is of particular significance because the tension which is required to pull a cable into a duct increases with the weight of the cable. Experience shows that to pull a small diameter cable into a fairly straight duct made of polyvinylchloride, a pulling force is needed that nearly equals the weight of the cable. Under less favourable conditions as much as five times this force may be required. This force increases linearly with the length of the cable, and for long cable sections, may grow to large values. On the other hand, to reduce the number of splices, one wants to pull in at one time sections of cable which are as long as possible. Cable sections which are one or even several kilometres long must be considered. For cable sections of such lengths, five times their own weight means a substantial pulling force that requires the cable to withstand quite high tensions. The cable must be furnished with sufficient strength members. But in addition the elongation which it suffers under these extreme pulling forces cannot be allowed to strain the fibres by the same amount. Under such a large strain, the tension in the individual fibre would come close to or even surpass its tensile strength and the fibre would be susceptible to rupture. We rather need a substantial strain relief for the fibres so that when external pulling forces strain the cable as a whole up to a certain limit, the individual fibres are either not strained at all, or strained only substantially less than the cable.

An obvious method to relieve the fibres of any strain up to a certain limit of cable elongation and bending is to support the stranded fibres only loosely in their helical position around the circumference. Such supports may be provided by spacers on rims as illustrated schematically in Fig.7.27. This loose support leaves enough leeway for the

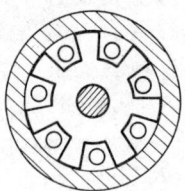

Fig.7.27. Fibres supported loosely by spacers on rims, for strain relief.

individual fibre to remain completely unstrained when the cable elongates elastically under pulling forces.

Another method that partially relieves the fibres of the cable strain is illustrated by the cable cross-section in Fig.7.28 (Jackson 1976). The cable consists of a small diameter

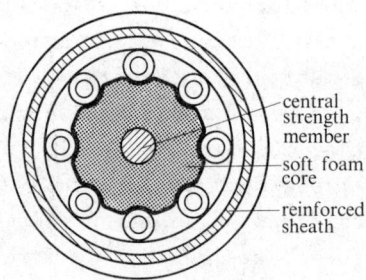

Fig.7.28. Fibre cable with soft foam core for strain relief (Jackson 1976).

central strength member which is surrounded by a layer of soft
flexible foam. The individual glass fibres are each jacketed
by a suitable material of high elastic modulus in the longitudinal direction for additional flexural rigidity. They are
then wound helically on to the foam and lightly affixed to
its surface so that they retain their relative positions under
stresses which bending and installation of the cable might
induce. The stranded fibres on the deformable central core
are surrounded and protected by a polymer sheath in which
additional strength members are embedded.

When the cable is strained by pulling forces the jacketed fibres in their helical position exert an inward pressure
on the foam core. If this core material is soft enough to
have sufficient compressibility, the helix diameter may be
substantially reduced while only minor tensions build up in
each of the packaged fibres. The strain of the individual
fibre remains thus much lower than in the cable as a whole.

7.10. CABLE DESIGN

The preceding three sections on fibre coating and cabling
have considered the more important aspects and requirements
that are specific for cables with glass fibres. Typical solutions to the specific problems of glass fibres in cables have
also been discussed in these sections. Practical designs of
optical fibre cables must pay attention to all these aspects
and requirements and should incorporate suitable solutions to
the specific problems of fibre cabling. In addition, these
designs should also consider all requirements that the fabrication, handling, installation, and operation of cables for
signal transmission in general impose. The cable structure
should be of a design that can be fabricated by conventional
cable-making processes. It should be handled and installed by
similar methods and with similar equipment as for other communication cables. The cable construction should be rugged enough
to withstand all the external forces and other environmental
stress that other cables are similarly exposed to.

A typical structure for a fibre cable that fills many
of these requirements is shown in Fig.7.29. In the cable
core a central strength member is positioned, which not only

694   FIBRE FABRICATION AND CABLING

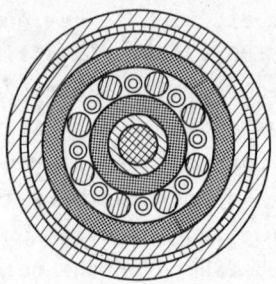

Fig.7.29. Fibre cable with eight jacketed fibres stranded with plastic
buffer string spacers between cushion layers and with strength
members in core and sheath.

strengthens the fibre against tension by pulling forces but
also helps to prevent any buckling of the fibre. A soft
plastic surrounds the core to provide a central support for
the fibres in the form of a cushion. Around this plastic
layer, the jacketed fibres are loosely stranded with plastic
strings between them as spacers and buffers. Another layer
of soft plastic around the fibres is held by the plastic
strings at an inner diameter that leaves the fibres much
freedom to move and provides an outer cushion for them.
The surfaces of inner and outer cushion layers are of such a
type and smoothness that the fibres easily slip on them. The
cable core with the inner cushion layer, the fibres, and the
outer cushion layer, is encapsulated by an outer sheath. To
increase the tensile strength of the fibre, this outer sheath
is reinforced by embedding additional tension members of
extra strength polymer fibres, or as a whole is made of a
material of high tensile strength. The outer sheath is also
designed to prevent any buckling of the cable and protect it
against lateral forces and crushing. Additional requirements
for the outer cable sheath, which depend on the location of
installation, are resistance against flames or against chemical attack. Such special requirements are met by the choice
of suitable sheath materials.

    In discussing the various requirements for optical cable
design, no consideration has yet been given to the problems
of cable connectors and fibre splices. In the next chapter,
one section will be devoted to the connection and splicing
of single fibres. Coupling of optical cables which contain

a large number of fibres is eased considerably if each pair
of fibres need not be connected individually, but if instead,
a cable connector is constructed that joins all pairs of
fibres collectively. For such multi-fibre connectors, the
cable design should package fibres inside the cable in a
geometrically regular fashion. Fig.7.30 shows, as an example,

Fig.7.30. Rectangular array of 4 fibre ribbons with 4 fibres each.

a rectangular array of fibre ribbons in which each ribbon
contains a certain number of fibres in parallel. The ribbon
structure sandwiches the fibres between laminates that provide sufficient protection and mechanical strength for
handling and cabling. These laminates should also not degrade the transmission characteristics of the fibre, for example by microbending. In addition, the laminates should peel
off easily from the fibres in order to insert and align the
naked fibre ends coming out of the stacked ribbons in precision grooves of a multi-fibre connector.

In the fibre cable, the stack of ribbons forms the core
and is twisted to improve the cable bending characteristics.
A plastic jacket is then extruded loosely over the twisted
stack with a soft material to fill some of the space in between ribbons and jacket. This arrangement isolates the cable
core mechanically from its outer sheath. More such isolation
may be provided by plastic twines around the jacket before
it is surrounded by an outer sheath with additional strength
members. Fibre cables with such stacked ribbons have been
fabricated with twelve fibres in each ribbon and a total of
twelve ribbons (Schwartz *et al.* 1976). The outer diameter of
such a cable with 144 fibres is only 12 mm. The average
transmission loss at $\lambda = 0.82$ μm increased from 5 dB km^{-1} for
the unpackaged fibre to 6 dB km^{-1} for the installed cable.

REFERENCES

Abe, K. (1976). Fluorine doped silica for optical waveguides. 2nd European Conference on Optical Fibre Communication. *Société des Electriciens, des Electroniciens et des Radio-électriciens (SEE)* Paris, 59-61.

Behm, K. (1976). Group delay in CVD-fabricated fibres with diffused stair-like index profile. *Archiv Elektron. & Übertragungstech.* 30, 329-331.

—————— (1977). Dispersion in CVD-fabricated fibres with a refractive index dip on the fibre axis. *Archiv Elektron. & Übertragungstech.* 31, 45-8.

Blyler, Jr., L.L., Hart, A.C., Jaeger, R.E., Kaiser, P., and Miller, T.J. (1975). *Low-loss polymer clad silica fibres produced by laser drawing.* Paper presented at Topical Meeting on Optical Fibre Transmission, Williamsburg. Opt. Soc. Amer.

Day, C.R., Beales, K.J., Midwinter, J.E., and Newns, G.R. (1975). *The development of optical fibre using sodium borosilicate glasses and the double crucible technique.* 1st European Conference on Optical Fibre Communication. IEE, London, 33-5.

France, P.W. and Dunn, P.L. (1976). *Optical fibre protection by solution plastic coating.* 2nd European Conference on Optical Fibre Communication. SEE, Paris, 177-181.

Gambling, W.A., Payne, D.N., Hammond, C.R., and Norman, S.R. (1976). Optical fibres based on phosphosilicate glasses. *Proc. IEE* 123, 570-6.

Gloge, D. (1975). Optical fibre packaging and its influence on fibre straightness and loss. *Bell Syst. Tech. J.* 54, 245-62.

Grabmaier, J., Kinshofer, G., Plättner, R., Eisenrith, K.-H. and Deserno, U. (1975). Growth of quartz glass rods for fibre optics in a plasma torch using powdered SiO_2 starting material. *Siemens Forschungs- und Entwicklungsberichte* 4, 310-17.

Grabmeier, J., Schneider, H., Lebetzki, E., and Douklias, N. (1976). Preparation of low loss optical fused-silica fibres by modified chemical vapour deposition technique. *Siemens Forschungs- und Entwicklungsberichte* 5, 171-5.

Ikeda, Y. and Yoshiyagawa, M. (1976). *Development of low loss glasses for Selfoc fibres.* 2nd European Conference on Optical Fibre Communication. SEE, Paris, 27-31.

Inada, K., Akimoto, T., Kojima, M. and Sanada, K. (1975). *Transmission characteristics of a low loss silicone-clad fused silica-core fibre.* 1st European Conference on Optical Fibre Communication. IEE, London, 57-59.

Jackson, L.A. (1976). *Strain relief in optical fibre cable design.* (Private Communication).

─── Reeve, M.H., and Dunn, A.G. (1976). *Orientated polymer for optical fibre packaging.* 2nd European Conference on Optical Fibre Communication. SEE, Paris, 175-6.

Kaiser, P. (1974). Drawing induced coloration in vitreous silica fibres. *J. opt. Soc. Amer.* 64, 475-81.

Kaiser, P., Hart, A.C.Jr., and Blyler, L.L. Jr. (1975). Low loss FEP-clad silica fibres. *Appl. Opt.* 14, 156-62.

Koizumi, K., Ikeda, Y., Kitano, I., Furukawa, M., and Sumimoto, T. (1974). New light-focussing fibres made by a continuous process. *Appl. Opt.* 13, 255-60.

─── ─── (1975). *Low loss light-focussing fibres made by a continuous process.* 1st European Conference on Optical Fibre Communication. IEE London, 24-26.

Krawarik, P.H. (1975). *Power spectral measurements for optical fibre outer diameter variation.* Topical Meeting on Optical Fibre Transmission, Williamsburg. Optical Soc. Amer.

Küppers, D. and Koenings, J. (1976). *Preform fabrication by deposition of thousands of layers with the aid of plasma activated CVD.* 2nd European Conference on Optical Fibre Communication, SEE Paris, 49-54.

Maurer, R.D. (1973). Glass fibres for optical communications. *Proc. IEEE* 61, 452-62.

Mühlich, A., Rau, K., Simmat, F., and Treber, N. (1975). *A new doped synthetic fused silica as bulk material for low loss optical fibres.* 1st European Conference on Optical Fibre Communication. IEE London, post-deadline paper.

Nakahara, T., Hoshikawa, M., Yoshida, M., and Suzuki, S. (1976). *Transmission properties of ring-type optical fibres.* 2nd European Conference on Optical Fibre Communication. SEE Paris, 149-55.

Olshansky, R. (1976). Pulse broadening caused by deviations from the optimal index profile. *Appl. Opt.* 15, 782-8.

Payne, D.N. and Gambling, W.A. (1974). New silica-based low loss optical fibre. *Electr. Lett.* 10, 289-90.

Pinnow, D.A., Van Uitert, L.G.F., Rich, T.C., Ostermayer, F.W., and Grodkiewicz, W.H. (1975). *Investigation of the soda aluminosilicate glass system for application to fibre optical waveguides.* Topical Meeting on Optical Fibre Transmission, Williamsburg, Opt. Soc. Amer.

Rokunohe, M., Shintani, T., Yamija, M., and Utsumi, A. (1976). *Stability of transmission properties of optical fibre cables.* 2nd European Conference on Optical Fibre Communication, SEE Paris. 183-188.

Schwartz, M.I., Kempf, R.A., and Gardner, W.B. (1976). *Design and characterization of an exploratory fibre optic cable.* 2nd European Conference on Optical Fibre Communication. SEE Paris, 311-14.

Suzuki, Y. and Kashiwagi, M. (1974). Polymer-clad fused-silica optical fibre. *Appl. Opt.* 13, 1-2.

Tanaka, S., Inada, K., Akimoto, T., and Kozima, M. (1975). Silicone-clad fused-silica-core fibre. *Electr. Lett.* 11, 153-4.

——— Naruse, T., Osanai, H., Inada, K. and Akimoto, T. (1976). *Properties of cabled low-loss silicone clad optical fibres.* 2nd European Conference on Optical Fibre Communication. SEE, Paris, 189-92.

Uchida, T. (1975). *Preparation and properties of compound glass fibres.* Paper presented at URSI General Assembly, Commision VI, Lima.

Van Uitert, L.G., Pinnow, D.A., Williams, J.C., Rich, T.C., Jaeger, R.C., and Grodkiewicz, W.H. (1973). Borosilicate glasses for fibre optical waveguides. *Mat. Res. Bull.* 8, 469-76.

Wemple, S.H., Pinnow, D.A., Rich, T.C., Jaeger, R.C., and Van Uitert, L.G. (1973). Binary SiO_2-B_2O_3 glass system: Refractive index behaviour and energy gap considerations. *J. appl. Phys.* 44, 5432-7.

8
FIBRE JUNCTIONS AND TRANSITIONS

Fibres in optical systems for signal transmission must be connected to light sources for optical signals to be launched into the fibres, and to optical detectors that receive these signals. If the distance between optical transmitters and receivers cannot be bridged by one single length of fibre, additional connectors are needed to join two or more fibre sections along the transmission path. For a permanent installation, fibres are spliced with each other once and for all; in other applications, detachable connectors are needed to join fibres only temporarily. The connectors between fibres and transmitters and receivers are in general also detachable in order that the latter may be exchanged in the case of any faults. For such a detachable connection between a fibre cable and transmitter and receiver ports, short sections of fibre of the same kind and size as in the cable are at one end attached permanently to the optical signal source, and to the photo-detector, respectively, and at the other end have a detachable connector which can be joined to the fibre cable. In an optical communication system, we therefore have permanent joints and detachable connectors between fibres as well as transitions from the output of light sources to fibres and from fibres into photo-detectors. Permanent fibre splices and detachable fibre connectors join fibres of the same cross-section. They form simple butt joints. Transitions between the output ports of optical transmitters and fibres or between fibres and the input ports of optical receivers depend in their form on the nature of these input and output ports. If, in the output port of the transmitter, the wave to be launched into the fibre has a similar transverse field distribution as the transverse field distribution of fibre modes, then a butt joint of the fibre to the port suffices for good launching efficiency. If, however, these field distributions differ considerably in transverse extension or in shape or even in both, then a transition must transform one field distribution into the other to

achieve a better launching efficiency. For the design of
fibre joints and transitions, we analyse first the excitation
of fibre waves at the junction between fibres or between the
fibre and another optical waveguide. We also include in this
analysis fibre wave excitation by optical beam waves. In
subsequent sections, we consider transitions to optical fibre
guides.

8.1. ANALYSIS OF FIBRE MODE EXCITATION

In optical waveguide junctions, both waveguides are
either directly jointed together or have only a narrow spacing
between their end faces. As an example, Fig.8.1 shows such a

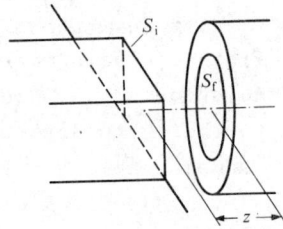

Fig.8.1. Junction of a rectangular strip guide with a cladded core fibre.

junction between a rectangular strip guide and a round
cladded-core fibre. The front face of the fibre reflects
some of the light that is incident upon it. To reduce
this reflection, the junction is immersed into a fluid
which matches the indices of refraction of both waveguides at the junction. Instead of this matching fluid,
a permanent junction may also be embedded into a material
of the appropriate matching index that solidifies, for example, by curing subsequent to immersion. If both waveguides
differ in their refractive index, the matching fluid should
have a refractive index in between both. As an alternative to a

complete immersion of the junction into a matching fluid, the end faces of one or both waveguides could be coated with quarter-wave layers of a refractive index that matches the waveguide indices to that of the space in between. If it is possible to join both waveguides without any gap, only one quarter-wave layer of index n_m is needed to match the indices of refraction n_i and n_f of both waveguides according to

$$n_m^2 = n_i\, n_f \,. \tag{8.1}$$

If under these circumstances both waveguides have the same or nearly the same refractive index $n_i \simeq n_f$, then, of course, no matching layer is needed between them.

We assume that the junction in Fig.8.1 is immersed in a matching fluid and consequently neglect all residual reflections in the analysis. The wave which is incident from the primary waveguide or as a beam has an electric field distribution in the front face of the fibre which we designate by $\vec{E}_0$. We assume this wave to carry unit power. If both waveguides join directly together without any gap, then $\vec{E}_0$ equals the electric field $\vec{E}_i$ of the incident wave of unit power in the primary waveguide,

$$\vec{E}_0 = \vec{E}_i \,. \tag{8.2}$$

If both waveguides are spaced by a distance z as in Fig.8.1, then $\vec{E}_0$ constitutes the near field of radiation from the end face of the primary waveguide. Kirchhoff's diffraction theory leads to the following approximation for this radiation field (Silver 1949):

$$\vec{E}_0 = \frac{1}{4\pi} \iint_{S_i} \vec{E}_i \exp(-jk_m r)[-jk_m + (-jk_m + 1/r)(z/r)]\,dS/r \,, \tag{8.3}$$

where $k_m = n_m k$ designates the wave number of the matching medium between both waveguides and r is the distance between a source point on the end face of the primary waveguide and the field point on the front face S_f of the fibre. Equation (8.3) assumes the spacing z not to be too narrow. For very narrow spacings between the two waveguides, the field $\vec{E}_0$ obtains by interpolating between $\vec{E}_0$ from equations (8.2) and (8.3).

The incident wave excites guided and radiation modes of the fibre, the electric fields of which superimpose in the front face of the fibre to give the total field

$$\vec{E} = \sum_\mu \vec{E}_\mu .$$

The summation extends over all guided core and cladding modes of the fibre, but is meant to also include an integration over the continuous spectrum of radiation modes. The orthogonality between all guided modes may for weakly guiding fibres according to equation (6.7) be expressed as

$$\left| \iint_{S_f} E_\mu E_\nu \, dS_0 \right| = [(\mu_0/\varepsilon_0)^{\frac{1}{2}}/n_1] |P_\mu| \, \delta_{\mu\nu} , \qquad (8.4)$$

where P_μ designates the power which the mode μ with the electric field E_μ transmits. A condition similar to equation (8.4) exists also for the orthogonality between guided modes and radiation modes as well as between radiation modes (Marcuse 1974). With the many core and cladding modes of fibres, however, we have no need to include radiation modes. Furthermore, practical interest concentrates on the power with which the incident field excites core modes. We therefore need not evaluate the excitation of radiation modes.

Since the total field $\vec{E} = \sum_\mu \vec{E}_\mu$ in the front face of the fibre is excited by the field $\vec{E}_0$ of the incident field, it must equal this field:

$$\vec{E}_0 = \sum_\mu \vec{E}_\mu . \qquad (8.5)$$

If we form the inner product of this equation with $\vec{E}_\nu$, integrate over the full fibre cross-section and apply the orthogonality condition (8.4), the only terms non-zero on the right-hand side are those for which $\mu = \nu$ and whose $\vec{E}_\nu$ has the same uniform polarization as $\vec{E}_\mu$. With these terms, the power with which a mode ν in its particular polarization is excited follows from

$$P_\nu = |n_1 (\varepsilon_0/\mu_0)^{\frac{1}{2}} \iint_{S_f} \vec{E}_0 \vec{E}_\nu \, dS| \,. \tag{8.6}$$

$\vec{E}_0$ was assumed to represent the incident field for unit incident power. Because of this normalization, we obtain with P_ν the efficiency with which $\vec{E}_0$ launches its power into the mode ν of the fibre.

As the first example, we consider a Gaussian beam wave in Fig.8.2 that is incident on a graded-index fibre, with the beam inclined by $\Delta\alpha$ with respect to the fibre axis and offset by Δx at its front face. The beam waist with a spot radius w_i is spaced at Δz from the fibre front face so that according to equation (3.13), its phase front at the fibre front face is curved with a radius

$$R = \Delta z + w_i^4 \, k_m^2/(4\Delta z) \,. \tag{8.7}$$

Such excitation of a graded-index fibre by a Gaussian beam wave is provided by lasers which have optical cavities with curved mirrors and oscillate in their lowest transverse order. It also occurs when a graded-index fibre with parabolic index profile is used as a primary waveguide spaced at a distance Δz from the front face and when only its lowest-order fundamental mode is incident. The arrangement of Fig. 8.2 also serves as a model for other waveguide junctions in which the incident wave has nearly the field distribution of the lowest-order Gaussian beam mode.

To evaluate the launching efficiency according to equation (8.6), we assume the lowest-order fundamental mode of the graded-index fibre to have the field distribution as in an unlimited parabolic profile according to equations (5.54) and (5.59) with spot radius w_0. For small offset Δx and small tilts of the incident beam mode, the evaluation of equation (8.6) leads to the following launching efficiency for the excitation of this fundamental mode (Kogelnik 1964):

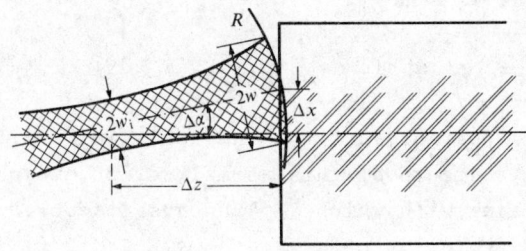

Fig.8.2. Gaussian beam wave incident upon a graded-index fibre.

$$p_{00} = \frac{4}{\left(\frac{w_0}{w}+\frac{w}{w_0}\right)^2 + \left(\frac{k_m w w_0}{2R}\right)^2} - \frac{\Delta x^2}{2}\left(\frac{1}{w_0^2}+\frac{1}{w_i^2}\right) - \frac{k_m^2 \Delta \alpha^2}{2}(w_0^2+w_i^2) \; .$$

(8.8)

Without any offset or tilt, the first term alone remains on the right-hand side of this equation. Fig.8.3 shows this term as a function of w/w_0 for different values of the parameter $\sigma = k_m w_0^2/R$. For 100 per cent launching efficiency,

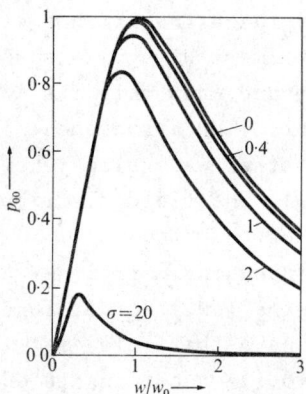

Fig.8.3. Launching efficiency of the fundamental mode of the parabolic index profile excited by the fundamental Gaussian beam mode of spot radius w and curvature radius R of the phase fronts at the fibre front face. w_0, spot radius of fundamental fibre mode; $\sigma = k_m w_0^2/R$.

the incident beam mode must be matched to the fundamental fibre mode by placing its beam waist at the front face of the fibre and by adjusting its spot radius w_i to the spot radius w_0 of the fundamental fibre mode. Any deviations from these perfect launching conditions reduce p_{00} below unity. In the case of such deviations, the incident beam mode also excites higher-order fibre modes and most of the power difference is launched into these higher-order modes.

If two different graded-index fibres with parabolic index profiles are directly joined to each other without any tilt and offset and with no spacing between their end faces, the fundamental mode with spot radius w_i incident in the primary fibre excites the fundamental mode with spot radius w_0 in the secondary fibre with a launching efficiency

$$p_{00} = 4/(w_0/w_i + w_i/w_0)^2 . \tag{8.9}$$

This formula follows from equation (8.8) with $w = w_i$ and $R \to \infty$; it is represented by the curve in Fig.8.3 with $\sigma = 0$.

The effects of tilts and offsets of the incident beam mode on launching efficiency are accounted for by the second and third terms on the right-hand side of equation (8.8). Any tilts are particularly critical; the reduction of launching due to such tilts increases with the square of the ratio of spot radius to wavelength. Single-mode fibres with large spot sizes of their fundamental mode are very sensitive to such tilts. It is the same kind of critical sensitivity that they also exhibit with respect to microbending.

We consider next a cladded-core fibre with a stepped index profile instead of the graded-index fibre in Fig.8.2. The field distribution of LP_{lp}-modes in the stepped index profile with core radius a, core index n_1, and cladding index n_2 is given by equations (4.125) and (4.128). By substituting the transverse electric field from these equations into equation (8.6) and by using for $\vec{E}_0$ the field distribution of the incident fundamental beam mode, the launching efficiency for different fibre modes may be evaluated. Fig.8.4 shows this launching efficiency for offset Δx, tilt $\Delta \alpha$, and displacement Δz of the incident beam mode all equal to zero. A typical index difference between core and cladding has been chosen for Fig.8.4, and the beam mode is incident from free space with

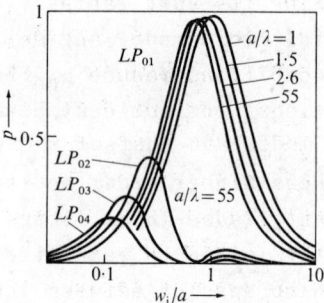

Fig.8.4. Launching efficiency of the fundamental mode and LP_{0p}-modes of next higher orders excited by a Gaussian beam mode (Heyke 1970). $n_1 = 1\cdot55$; $n_2 = 1\cdot50$; $n_m = n_0 = 1$.

$n_m = n_0 = 1$. The launching efficiency has been calculated for different values of a/λ in the single-mode range, and the typical multimode case with a fibre parameter value $V = 134$. For this multimode case not only the launching efficiency for the fundamental mode appears in Fig.8.4, but also for three of the next higher-order LP_{0p}-modes. As a function of w_i/a, the fundamental mode launching efficiency has a maximum value. This maximum value is nearly independent of the ratio a/λ, but with increasing a/λ shifts to lower values of w_i/a. Fig. 8.5 shows this optimum ratio w_i/a for maximum LP_{01}-launching

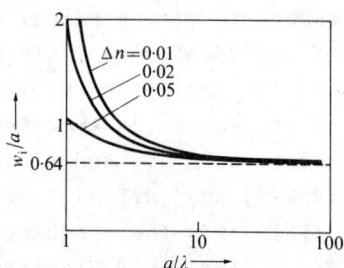

Fig.8.5. Optimum spot radius of Gaussian beam mode for maximum launching efficiency of fundamental fibre mode (Heyke 1970); $n_0 = 1\cdot50$.

efficiency for various index differences between core and cladding as a function of a/λ. In the multimode limit of very large ratios a/λ, the optimum value approaches

$$(w_i/a)_{opt} = 0.64 \tag{8.10}$$

asymptotically, no matter what the index difference between core and cladding. The efficiency with which higher order LP_{0p}-modes are launched also becomes independent of Δn for large values of a/λ or the fibre parameter V. The LP_{0p}-launching efficiencies which are plotted for $V = 134$ in Fig. 8.4 resemble this asymptotic limit quite closely. For $w_i/a = 0.64$, when the LP_{01}-launching efficiency is at its maximum, only very little power goes into higher-order modes.

If the beam mode is offset by Δx from the fibre axis, the LP_{01}-excitation decreases. Fig.8.6 shows the LP_{01}-launching

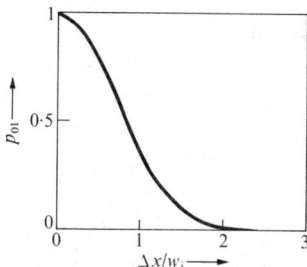

Fig.8.6. Launching efficiency of fundamental fibre mode for excitation by an offset Gaussian beam mode (Heyke 1970). $a/\lambda \gg 1$, $w_i/a = 0.64$.

efficiency as a function of relative beam mode offset $\Delta x/w_i$ under multimode condition for the optimum ratio w_i/a. At $\Delta x = w_i$, when the beam is offset by its spot radius w_i, the LP_{01}-launching efficiency drops to $p_{01} = 0.36$. Nearly the same reduction in LP_{01}-launching efficiency to $p_{01} = 0.36$ obtains for $\Delta x = w_i$ in the single-mode range of fibre parameter values, if only w_i/a has its respective optimum value according to Fig.8.5 (Marcuse 1970a).

An axial displacement Δz of the incident beam mode from $\Delta z = 0$ is far less critical for the LP_{01}-launching efficiency than the offset Δx of the beam. Fig.8.7 demonstrates the effect of this displacement by plotting the LP_{01}-launching efficiency versus $\Delta z/a$ under multimode conditions and for the optimum ratio w_i/a. On the other hand, the LP_{01}-launching

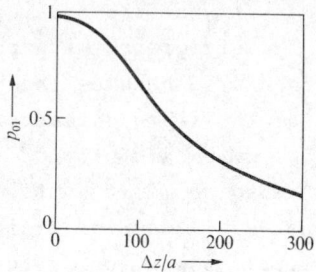

Fig.8.7. Launching efficiency of fundamental fibre mode for excitation by a Gaussian beam mode with axial displacement Δz (Heyke 1970). $a/\lambda \gg 1$, $w_i/a = 0\cdot 64$.

efficiency is quite sensitive with respect to any tilts of the incident beam mode. Fig.8.8 shows p_{01} as a function of tilt angle $\Delta\alpha$ for different values of a/λ in the single-mode region of the fibre and for weakly overmoded fibres, where the core

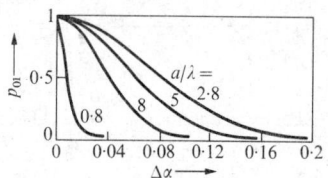

Fig.8.8. Launching efficiency of fundamental fibre mode for tilt angle $\Delta\alpha$ of the incident beam mode (Marcuse 1970a). $w_i/a = (w_i/a)_{\text{optim}}$ for LP_{01}-excitation; $n_1 = 1\cdot 01$, $n_2 = 1$.

guides only a few higher-order modes. The spot radius w_i of the incident mode has been adjusted for each value of the parameter a/λ to its respective optimum value for maximum LP_{01}-launching efficiency without any tilt. In the case of thin cores ($a/\lambda = 0\cdot 8$), the LP_{01}-excitation decreases quite rapidly with tilt angle, because such cores guide the LP_{01}-mode only weakly and its fields extend relatively far out into the cladding. Near the single-mode limit at $a/\lambda = 0\cdot 28 \ (n\Delta n)^{-1/2}$, the LP_{01}-excitation shows its least degradation by the tilt angle. For thicker cores, p_{01} decreases again more rapidly with increasing tilt angle, because the LP_{01}-field distribution increases in size together with the core size, and beam mode power is now also launched into higher-order fibre modes.

8.2. FIBRE CONNECTORS AND SPLICES

If fibres of the same nominal core size and index distribution are to be connected or spliced together, a butt joint provides the best coupling efficiency. To minimize any coupling losses, the two fibres should join together with little spacing between their end faces and without any axial offsets and tilts. Butt joints of single-mode fibres are particularly sensitive to any offsets and tilts. They must be adjusted within tight tolerances in order that their thin cores align well with each other. The coupling efficiency of such single-mode fibre joints is defined as the ratio of power that is launched through the joint into the fundamental fibre mode to the power incident in this mode from the primary fibre. By substituting for $\vec{E}_i$ in equations (8.2) or (8.3) the field distribution of this fundamental fibre mode for unit power, the coupling efficiency is found from equation (8.6). For $\vec{E}_\nu$ in this equation the fundamental mode field for unit power must also be substituted. From an evaluation of these equations, Fig.8.9 shows the coupling efficiency of a single-mode

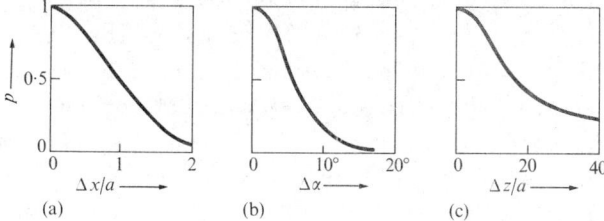

Fig.8.9. Coupling efficiency of single-mode fibre joints with $\Delta = 0\cdot012$, $n_1 = 1\cdot535$, and $a/\lambda = 1\cdot6$ as a function of offset Δx, tilt angle $\Delta \alpha$, and spacing Δz between fibre ends (Guttmann and Krumpholtz 1973).

fibre joint for typical values of relative core size of the fibre and the relative index difference between its core and cladding (Guttmann and Krumpholtz 1973). In Fig.8.9(a), the coupling efficiency appears as a function of the offset Δx between both fibres for zero tilt between fibres and zero spacing between their end faces. To avoid excessive coupling losses, the fibre cores must be aligned within small fractions of their core radius a. The coupling efficiency as a function of tilt angle appears in Fig.8.9(b) for zero offset and zero spacing between the fibres at the joint. For low coupling losses, the fibres should not be tilted against each other by much more than one degree. The coupling efficiency for zero tilt and zero offset is plotted in Fig.8.9(c) as a function of the spacing between fibre ends. From this diagram it appears that the coupling efficiency suffers only little from moderate spacing of fibres at the junction. The end faces may be more than several core diameters apart before the coupling loss becomes excessive.

If a multimode fibre is required to transmit signals only in its fundamental mode, any fibre connectors or splices must be designed to couple power mainly between the fundamental modes of both fibres and convert only a minimum of power to higher-order modes. In this case the absolute tolerances for offset, tilt, and spacing between fibres at the joint are similarly tight as for single-mode fibres. In normal operation of multimode fibres, the signal travels in many if not all of the guided modes of the fibre core. This multimode operation does not restrict the power conversion between core modes at fibre joints; it only requires that for little coupling loss most of the power that is incident in core modes be launched into core modes again and not converted into cladding modes or radiation. The coupling efficiency for multimode operation is defined as the ratio of total power launched into all core modes to the total incident power in core modes of the primary fibre. For an imperfect joint with offset, tilt or spacing between fibres, this coupling efficiency depends not only on the nature and degree of imperfection, but also on the distribution of incident power among the core modes of the primary fibre. For an analysis of this coupling efficiency, we assume here that the incident power distributes uniformly among all core modes. Such even power distribution prevails when an incoherent source excites the fibre. Incoherent sources illuminate the small numerical apertures of weakly guiding fibre cores uniformly across the whole fibre core if only their aperture of uniform radiance extends at least over the core cross-section. It thus launches the same fraction of power into all modes. A perfect fibre with equal attenuation for all modes and no interaction between modes maintains this uniform power distribution over any distance. In an imperfect fibre with mode interaction between guided modes only, and with equal attenuation for all these modes, a uniform power distribution between modes will also prevail, either from the initial excitation or, independent of the initial excitation, from mode-mixing, in the stationary power distribution of the statistical mode of lowest loss.

To analyse the coupling efficiency for the uniform distribution of incident power among all modes, we calculate first the intensity across the fibre core which is associated with this uniform excitation of modes. From the uniform illumination by an incoherent source, the fibre core accepts as much radiation as is incident within its numerical aperture. Equation (5.47) gives this numerical aperture for a graded-index fibre as a function of radial distance r from the core centre. The intensity accepted at r is proportional to the square of the numerical aperture at r and is hence given by (Gloge and Marcatili 1973)

$$S(r) = S(0) \frac{n^2(r) - n_2^2}{n_1^2 - n_2^2} . \qquad (8.11)$$

It equals directly the distribution of the square of the refractive index.

We consider now the butt joint between two fibres in Fig. 8.10 with an offset Δx (Neumann and Weidhaas 1976). Of the four regions in Fig. 8.10, the fibres overlap only in regions I and II. If the power is incident in fibre i, only its fractions that regions I and II transmit at the joint contribute to the excitation of modes in fibre o. Throughout

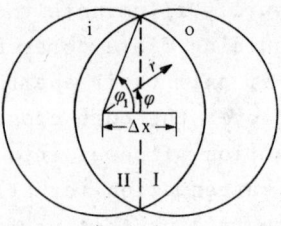

Fig.8.10. Butt joint of fibres with offset Δx.

region I, in particular, the numerical aperture of fibre i is smaller than the numerical aperture of fibre o. Therefore all the power which is incident through this region will be accepted by fibre o. We obtain this power from the following integral across region I:

$$P_{II} = \frac{2S(0)}{n_1^2 - n_2^2} \int_0^{\varphi_1} \int_{r_1}^{a} [n^2(r) - n_2^2] \, r\, dr \, d\varphi \,, \qquad (8.12)$$

where the limits of integration are

$$r_1 = \Delta x / (2 \cos \varphi) \qquad (8.13)$$

and

$$\varphi_1 = \cos^{-1}(\Delta x / 2a) \,. \qquad (8.14)$$

Throughout region II, fibre i has a larger numerical aperture than fibre o. Fibre o therefore accepts only part of the power that is incident from fibre i in this region. At any point of region II, fibre o accepts radiation within the same numerical aperture as fibre i radiates it at the symmetrical point in region I. The power which fibre o accepts in region II is therefore equal to the power that fibre i radiates in region I. Altogether fibre o accepts $2P_{II}$, and the coupling efficiency is

$$p = 2P_{II}/P \,, \qquad (8.15)$$

where P designates the total incident power according to

$$P = \frac{2\pi S(0)}{n_1^2 - n_2^2} \int_0^a (n^2(r) - n_2^2)\, r\,dr \ . \qquad (8.16)$$

To determine the coupling efficiency, the integrals in equations (8.12) and (8.16) must in general be evaluated numerically. For the parabolic index profile according to equation (5.57), the following analytic expression results from the evaluation of these integrals:

$$p = \frac{2}{\pi} \arccos\left(\frac{\Delta x}{2a}\right) - \frac{\Delta x}{\pi a}\left[4 - \left(\frac{\Delta x}{a}\right)^2\right]^{\frac{1}{2}} \left\{1 - \frac{1}{12}\left[2 + \left(\frac{\Delta x}{a}\right)^2\right]\right\} \qquad (8.17)$$

In the case of small offsets, this expression is approximated by

$$p \simeq 1 - \frac{8\Delta x}{3\pi a} , \qquad (8.18)$$

which is accurate within 1 per cent for $\Delta x/a < 0\cdot 4$. A step index fibre has a constant numerical aperture across its core, and for uniform power distribution among modes also has a constant intensity across the core. The coupling efficiency for an offset fibre joint follows, under these conditions, simply as the ratio of the overlap area to the core cross-section:

$$p = \frac{2}{\pi} \arccos\left(\frac{\Delta x}{2a}\right) - \frac{\Delta x}{\pi a}\left[1 - \left(\frac{x}{2a}\right)^2\right]^{\frac{1}{2}} . \qquad (8.19)$$

In the case of small offsets, this expression is approximated by

$$p \simeq 1 - \frac{2\Delta x}{\pi a} , \qquad (8.20)$$

which is accurate within 1 per cent for $\Delta x/a < 0\cdot 6$. A graphical representation of equations (8.17) and (8.19) and their respective approximations (8.18) and (8.20) is shown in Fig. 8.11. When we compare Fig.8.11 with Fig.8.9(a), we note that nearly the same relative offset with respect to core size may be allowed in the multimode fibre as in the single-mode fibre, if the same coupling efficiency is required. In

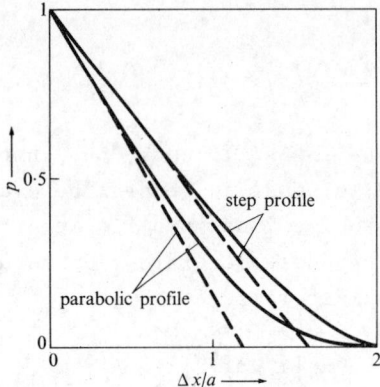

Fig. 8.11. Coupling efficiency of multimode fibre joint with offset λx for parabolic index profile and step index profile (Neumann and Weidhass 1976). Solid lines from equations (8.17), (8.19); broken lines from equations (8.18), (8.20).

absolute terms, of course, the offset of the large-core multimode fibres may be much larger.

Before fibres are connected or spliced to each other, their end faces must be prepared to be optically flat and perpendicular to the fibre axis. Conventional grinding and polishing techniques consume much time and are quite costly; they can hardly be applied during field installation. It is known on the other hand, that glass fibres below a certain outer diameter break with flat and perpendicular end faces when they are stressed close to their tensile strength and then scored by a diamond or carbide scorer to initiate the break. The fracture originates under these conditions at the point of the fibre surface where it is scored and a plane crack propagates with increasing velocity perpendicularly through the fibre. During this process strain energy converts to kinetic energy, until the crack reaches a critical velocity, where the excess energy creates subsurface cracks first, and then four new surfaces by forking the crack. When the fracture forks into several cracks, the fibre separates at the break into three or more pieces with jagged surfaces. The single plane crack accelerates to this critical velocity when it has travelled the distance

$$d_c = K/\sigma_B^2 \tag{8.21}$$

from the point of origin (Gloge et al. 1973). In this relation, σ_B designates the tensile stress at which the fibre breaks when it is scored. K depends only on material constants but is best determined experimentally. For fused silica it was found to be $K = 56$ kg^2 mm^{-3}, and for soda borosilicate glass $K = 37$ kg^2 mm^{-3} has been obtained. The distance d_c in equation (8.21) represents a critical diameter up to which a straight fibre breaks with flat and perpendicular end faces under tensile stress and by scoring. Fused silica fibres show a tensile strength of $\sigma_B = 25$ kg mm^{-2} when a diamond scorer is used to initiate the break. For them, therefore, the critical distance d_c and limiting diameter is 90 µm.

When fibres of larger diameters than this limiting value d_c are fractured under the tensile stress σ_B and by scoring their surface, only part of the end faces remain mirrorlike and the rest beyond the distance d_c from the origin of the fracture in Fig.8.12 appears as the so-called 'mist and hackle zones' (Gloge et al. 1973). In order to break fibres

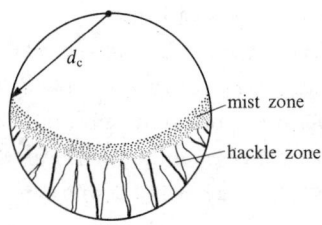

Fig.8.12. End face of fractured fibre with mist and hackle zones.

of larger diameter than d_c with flat and perpendicular end faces, its cross-section beyond d_c in Fig.8.12 must be relieved at least partially of the tension which is required at the surface where the scoring initiates the break. The crack as it propagates through the fibre will then not convert so much strain energy into kinetic energy and will not reach the critical velocity of propagation that sets the limit of equation (8.21). A way to accomplish this strain relief is to bend the fibre while the average tension σ_0 is on it. The stress then distributes according to

$$\sigma = \sigma_0 + E\, x/\rho \qquad (8.22)$$

across the fibre, where E designates the elastic modulus of the glass. The maximum tension occurs at the outer point $x = a$ of the bent fibre circumference and must be equal to the tensile strength σ_B to initiate the break. The minimum tension occurs at the inner point and should have a finite positive value. If compression develops at this point, the crack ceases to propagate through the fibre and a lip forms at the end of the break. The radius of curvature for which $\sigma = \sigma_B$ at $x = a$ but $\sigma = 0$ at $x = -a$ follows from equation (8.22) as

$$\rho_0 = E\, 2a/\sigma_B . \qquad (8.23)$$

Fused silica has $E = 7 \cdot 2 \cdot 10^{-3}$ kg mm^{-2} so that with $\sigma_B = 25$ kg mm^{-2}, such fibres may be bent to $\rho_0 = 288\, d$ before any compression develops under the stress which is required to initiate the break by scoring.

By bending fibres to a curvature radius as small as ρ_0 and scoring them under the appropriate stress, they may be fractured even if their diameter exceeds d_c. They will break with flat and perpendicular end faces as long as the product $d\sigma^2$ does not exceed the material constant K in equation (8.21) for any distance d from the origin of the break and the corresponding local tensile stress σ. For σ according to equation (8.22) and with $\rho = \rho_0$, the maximum value of $d\sigma^2$ occurs on the surface of the fibre at the distance $d = \sqrt{4/5}\, a$ from the origin of the break. When we require this maximum value to remain smaller than K, we find that the fibre diameter cannot be larger than

$$d_m = 3 \cdot 5\, d_c . \qquad (8.24)$$

For fibres of larger diameter, even the maximum strain relief by bending cannot prevent the formation of mist and hackle zones on the end faces. In the case of fused silica fibres with $d_c \simeq 90$ µm, we obtain $d_m \simeq 320$ µm.

Fig.8.13 shows the principal arrangement with which to break bent fibres under stress by scoring. The fibre under controlled tension is bent to a radius of curvature $\rho \geq \rho_0$ by a form which has a suitable coating on its surface for the fibre to slide on it freely. A scoring blade is then lowered on to the fibre, its pressure adjusted, and it is then pulled across the fibre.

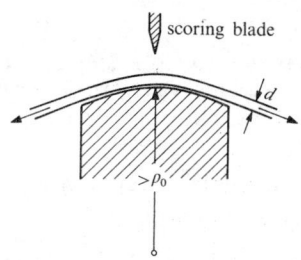

Fig.8.13. Principal arrangement for breaking fibres with flat and perpendicular end faces by bending under stress and scoring (Gloge et al. 1973).

To connect fibres, their end faces must be properly set against each other so that their cores align. Core alignment is particularly difficult for single-mode fibres. Their core diameter is small compared to the outer diameter and, depending on the method of fabrication, they have more or less eccentricity with respect to each other. The outer diameter of single-mode fibres can usually not be used as a reference to align the fibre cores. For connecting such fibres, one must instead provide adjustments to move one fibre end face with respect to the other and then align the cores by setting these adjustments for maximum coupling efficiency of the connector. During adjustment, a signal must be transmitted in the fundamental fibre mode through the connector, and its coupling loss observed. Fibre connectors which are easily adjusted manually utilize the eccentric principle of Fig.8.14 (Guttmann, Krumpholtz, and Pfeiffer 1975). The fibre ends are positioned and glued into wedge-shaped grooves, which are machined longi-

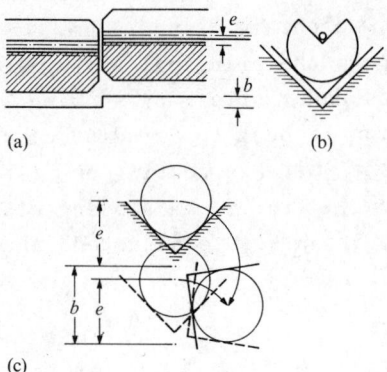

Fig.8.14. Alignment of fibre cores for single mode fibres (Guttmann *et al.* 1975). a) longitudinal section; b) cross-section; c) adjustment by rotation of pins.

tudinally into pins to such a depth that each fibre axis has the eccentricity *e* with respect to the pin axis. The pins in turn are inserted into two parallel grooves of different depth in the common mount. Their axes are thus displaced with respect to each other by an eccentricity *b*. If both pins are rotated, they can be moved into a position where the fibre cores align.

Alignment of multimode fibres is accomplished much more easily and without an optical monitoring signal. In contrast to single-mode fibres, they have large-diameter cores and claddings of intermediate thickness. Usually, the core is concentric enough with the outer fibre diameter to use the latter as reference for core alignment. The alignment is accomplished by placing both fibre ends against each other in a common groove.

Alternatively, the fibre ends may also be mounted in a common sleeve. Such sleeves to align and connect fibres are much smaller in size and more readily accommodated in multi-fibre cable splices than the normal arrangement with an alignment groove. The sleeve must, however, fit quite tightly around the fibre ends to exclude any displacement due to fibre clearance inside the sleeve. Fibres are, of course, difficult to insert into a sleeve with little clearance. To avoid such mounting problems and at the same time gain

the accuracy of a groove alignment, a loose sleeve with large
clearance but a square cross-section might be used (Miller
1975). The fibre ends are biased to one corner of the cross-
section by bending the fibres outside the tube. This corner
then serves as a groove to align the fibre ends. The alignment is self-adjusting and when the fibres are bent, their
own stiffness rotates the sleeve into the proper position for
fibre alignment.

Another form of sleeve that eases the insertion of fibres
for splicing has a large bore on both sides and tapers down
to the actual fibre diameter only near its centre. Such a
tapered sleeve may even be wide enough near its ends for a
short length of the coated or jacketed fibre to fit into it.
The coating or jacket will then be stripped off the fibre only
over part of the end sections inside the sleeve. The mechanical connection between fibre jacket and sleeve will lend much
additional strength to the fibre splice.

For a permanent connection of fibres in the sleeve, it
is first filled with a thermo-setting plastic in its uncured
form. The fibre ends are then inserted into the sleeve until
they touch each other; the plastic is subsequently cured.
The plastic matrix holds the fibres firmly in place and also
serves as an index-matching medium to reduce coupling losses.

8.3. OPTICAL TRANSITIONS

Optical waveguides which differ in their cross-sectional
forms, or in their sizes, or even in both, should not be
connected directly to each other by butt joints. In such
junctions, much of the incident power would be lost to radiation or converted into other, possibly unwanted, modes of
propagation. Instead we need transitions to join such different waveguides together. These transitions must be designed to match the field distributions of the waveguide on
the other side.

The transverse field distribution of the fundamental
mode in a number of optical waveguides resembles that of
Gaussian beam modes. We therefore first consider the transition from one Gaussian beam to another. To match different
Gaussian beams to each other, we only need to place a lens

of suitable focal length f at the proper distance between the incident and the transmitted beam. Fig.8.15 shows the arrangement of such a lens transformer between two different

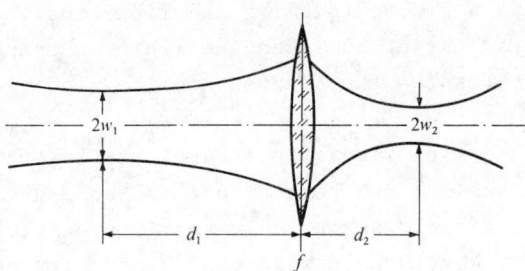

Fig.8.15. Transition between Gaussian beam modes with a matching lens.

Gaussian beams. If one beam has the minimum spot radius w_1 of its waist and the other w_2, any lens with a focal length

$$f > f_0 = \pi w_1 w_2 / \lambda \qquad (8.25)$$

will transform the field distribution of a beam mode in one beam to the corresponding field distribution in the other beam, if the beam waists are spaced by

$$d_1 = f \pm (w_1/w_2)(f^2 - f_0^2)^{1/2} \qquad (8.26)$$

and

$$d_2 = f \pm (w_2/w_1)(f^2 - f_0^2)^{1/2} \qquad (8.27)$$

from the centre plane of the matching lens (Kogelnik 1965). A lens which has as focal length the minimum length f_0 according to equation (8.25) must be placed right in the centre between both beam waists.

If a cladded-core or a graded-index fibre is to be excited in its fundamental mode by the fundamental Gaussian beam mode, a lens can match this beam mode to the fundamental fibre mode. The mismatch which occurs in such a transition and the radiation of power or its conversion into other modes depend on the deviations of the fundamental fibre mode from the Gaussian distribution of the beam mode. The graded-index fibre with a parabolic index profile has a fundamental mode with a nearly Gaussian field distribution. For all practical purposes, a suitable lens transformer represents a perfect transition from the fundamental mode of any Gaussian beam to the fundamental fibre mode. It must be designed to transform to the spot radius $w_2 = w_0$ of the parabolic index profile at the respective wavelength. For the fundamental mode excitation of cladded-core fibres with a homogeneous core, the matching lens must transform to the spot radius $w_2 = w_0$ according to Fig.8.5 which yields the maximum launching efficiency.

The transition between beam modes with lens transformers may be generalized and also used to match Gaussian beam waves of elliptical cross-section to each other. Such elliptical wave beams have different spot radii w_x and w_y in the direction of each of the principal axes of their cross-sectional ellipse. In the most general case, the beam waist occurs for each of the principal axes at a different position along the beam. Such a general Gaussian wave beam may be transformed to any other wave beam of the same general character with elliptical beam cross-sections when cylindrical lenses of suitable focal length are crossed with respect to each other and arranged according to equations (8.26) and (8.27) between the incident wave beam and the transmitted beam. The arrangement of Fig.8.16 deserves special attention. An elliptical wave beam with coplanar beam waists and minimum spot radii w_x and w_y is matched to a circular beam wave, which in its waist has the spot radius w_0. A typical application for such an arrangement is the transition from the fundamental HE_{00}- or EH_{00}-mode of a rectangular strip guide to the fundamental HE_{11}-mode of a round fibre. To obtain the highest possible coupling efficiency, the focal lengths f_x and f_y of the crossed cylindrical lenses and their spacing between the end faces of both guides must be chosen so as to transform to a circular beam of spot radius w_0 at the fibre front face. w_0 should be specified as the optimum spot radius from Fig. 8.5 for maximum launching efficiency of the fundamental mode

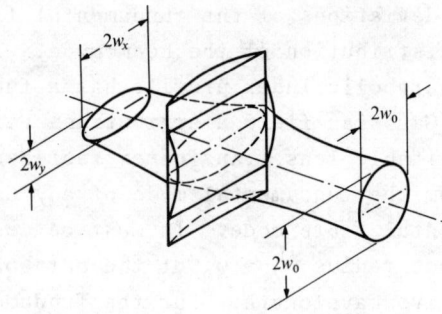

Fig.8.16. Transition from an elliptical to a circular wave beam with cylindrical lenses at right angles.

from a Gaussian beam. The equivalent spot radii w_x and w_y for the fundamental mode of the rectangular strip guide follow in a corresponding fashion as the spot radii of those Gaussian field distributions which best match the actual field distribution of the fundamental strip mode in the respective direction of the strip cross-section.

Alternative arrangements to Fig.8.16 utilize first a cylindrical lens to transform from the elliptical beam to a circular beam and then a spherical lens to match the spot radius of the circular beam to the optimum spot radius w_0 for maximum coupling efficiency into the fibre. These lenses for the transition from one waveguide to another do not need to be placed separately between the end faces of both guides; they can also be directly integrated with the waveguides. For such integration, the end face of one or both guides is formed to have the proper spherical shape either with or without astigmatism. An alternative for an integrated transition with lens transformers uses graded-index slabs or strips with parabolic index profiles. Depending on their index gradient and length, these graded-index guides perform as lenses of a certain focal length. If they connect directly to both end faces of the waveguides in the junction or if a matching medium fills the space between end faces of input and output waveguides and the transforming waveguide section in between, any reflection loss and interference is eliminated.

Optical waveguides with cross-sections of different size and shape may also be matched to each other by placing gradual transitions from one waveguide cross-section to the other in between. Such waveguide tapers are especially well suited when the transition to a fibre is to be integrated into a planar optical circuit. The waveguide in the tapered transition can be of the same material as the other guiding structures of the planar circuit and can have a similar form. It can therefore be fabricated together with the other components or with the whole circuit in the same process steps.

The performance of waveguide tapers is degraded by the interaction between guided modes and with the radiation field. In a uniform waveguide, all guided modes propagate independently from each other; in a tapered waveguide, however, the change of cross-section couples modes with each other and also to the radiation field. In single-mode structures, the fundamental mode couples to the radiation field and loses power to radiation. If the structure guides several modes, they exchange power with each other due to the interaction. The fundamental mode also loses power to other parasitic modes. Mode interaction is described by the coupled wave equations of section 6.1. Their solutions give power loss and conversion. To keep the loss of fundamental mode power below a certain specified limit, the cross-section should change its form and size only very gradually along the taper and the waveguide taper should connect to the uniform guides at both ends without any break in its contour. Slender and smooth tapers according to these specifications show hardly any radiation loss or mode conversion. They can conveniently be accommodated in optical circuits since, because of the microscopic dimensions of waveguide cross-sections, even very slender and relatively long tapers still remain quite short in absolute length.

For illustration, Fig.8.17 shows the radiation loss of single-mode fibre tapers from a core diameter $2a = 0 \cdot 8 \lambda$ to $2a = 0 \cdot 4 \lambda$ with a refractive index in the core of $n_1 = 1 \cdot 43$ and in the unlimited cladding of $n_2 = 1$. Starting at $1 \cdot 2$ dB for the radiation loss at the abrupt transition of a butt

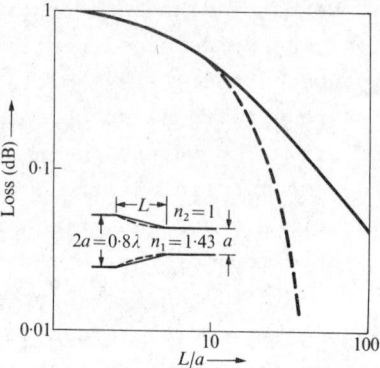

Fig.8.17. Contours and radiation losses of linearly (———) and exponentially (-----) tapered transitions between single-mode fibres (Marcuse 1970b).

joint, the loss decreases quite rapidly if the taper length extends beyond $L > 10a$. The radiation originates mainly at both ends of the taper where for the present example, the taper contour has a finite slope and therefore connects with a break in its contour to the uniform fibres. The taper end with the smaller size radiates more than the other end because the fundamental mode extends its field further into the cladding and the fibre guides it less effectively here. The exponential taper connects at this end with a smaller break in its contour to the uniform fibre. Therefore its radiation losses decrease much more rapidly for increasing taper length than in case of the linear taper.

A gradual transition that tapers from one waveguide to another waveguide of similar cross-sectional form but larger size will transform each guided mode of the smaller waveguide to the corresponding mode in the larger waveguide. If the taper is long and smooth enough, hardly any interaction occurs between modes so that the power which is incident in one particular mode of the smaller waveguide will all transfer to the corresponding mode of the larger waveguide. For illustration, we consider a multimode cladded-core fibre with a homogeneous core and unlimited cladding. Equation (4.20) relates the transverse phase parameter u of any particular mode to its propagation angle θ according to

$$u = n_1 k a \sin \theta .$$

For all modes in multimode fibres with large V-values, u remains nearly constant along the taper. Hence, as the radius a increases along the taper, θ must decrease. For a transition from a_1 to a_2, we obtain

$$a_1 \sin \theta_1 = a_2 \sin \theta_2 . \qquad (8.28)$$

θ_1 ranges up to the limiting angle θ_c of total reflection at the core-cladding boundary. According to this general law, all guided modes of the smaller fibre with propagation angles $\theta_1 < \theta_c$ transform along the fibre to the corresponding modes of the larger fibre with propagation angles

$$\theta_2 < \arcsin[(a_1/a_2)\sin \theta_c] . \qquad (8.29)$$

The larger fibre has higher-order modes with

$$\arcsin[(a_1/a_2)\sin \theta_c] < \theta_2 < \theta_c \qquad (8.30)$$

but through the long and smooth taper, these higher-order modes are not excited. If we let the light be incident in the large fibre, then only those modes with propagation angles θ_2 in the range of equation (8.29) will transfer their power through the taper to the smaller fibre. All higher-order modes with θ_2 according to equation (8.30) will radiate their power along the taper into the cladding. Fig.8.18 illustrates the radiation of power from a mode of high radial but zero circumferential order by tracing its meridional ray along a linear taper. As the ray proceeds on its zigzag path down the taper, its propagation angle θ increases from reflection to reflection until it eventually surpasses the limiting angle θ_c. A smoothly tapered transition from a larger to a smaller fibre transfers only the power in modes that are still guided by the core of the smaller fibre.

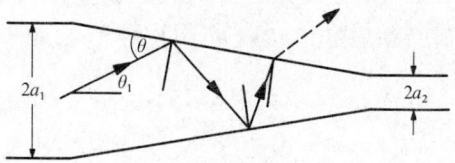

Fig.8.18. Linearly tapered transition between multimode waveguides and ray path for a transition into a radiating mode.

Equation (8.28) represents Abbe's sine theorem of optical imaging. According to this theorem light out of an aperture of radius a_1 can only be focused through an aperture of radius a_2 with propagation angles smaller than θ_2 when its propagation angle at a_1 is smaller than θ_1 according to equation (8.2). The imaging system must be free of any aberration in order to accomplish this transformation.

Just as the imaging system must be aberration free, the waveguide taper must be smooth and slender. To avoid any mode interaction and to transfer the full power of each mode from one end of the taper into the corresponding mode at the other end, the taper must be infinitely long. Tapers of finite length will always have some mode mixing and transmission loss. We estimate the minimum length of a linear taper which transfers at least low-order modes with little loss by tracing a ray that enters the wide end of the taper with zero propagation angle. With $\theta_1 = 0$, this ray represents modes which are very far above cut-off at the wide end of the taper. Fig. 8.19 traces a ray which is incident with $\theta_1 = 0$ at a_1, just inside the taper cross-section. The repeated reflections of this ray at the taper walls let the ray in the polygon of Fig.8.19 traverse successive sectors (Williamson 1952). It intersects the circle at the inner polygon of the narrow taper end with a propagation angle

$$\theta_2 = (a_1/L)(a_1/a_2 - 1) , \qquad (8.31)$$

where L designates the taper length. As long as θ_2 remains smaller than the limiting angle θ_c of total reflection, the waveguide which connects to the narrow taper end will accept

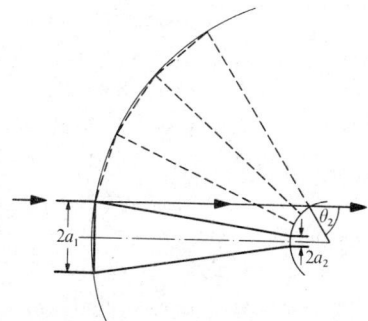

Fig.8.19. Ray trace in equivalent sectors of a linear taper.

this ray. Hence all rays which are incident at $\theta_1 = 0$ at the wide end of the taper will be accepted by the waveguide at the narrow end only if the taper is longer than

$$L = a_1(a_1/a_2 - 1)[1 - (n_2/n_1)^2]^{-1/2} . \qquad (8.32)$$

In this equation θ_c has been assumed small enough to be approximated by

$$\theta_c = [1 - (n_2/n_1)^2]^{\frac{1}{2}} . \qquad (8.33)$$

According to equation (8.32), the minimum length of a multimode waveguide taper depends on the waveguide size a_1, the taper ratio a_1/a_2, and the relative index difference between core and cladding. Usually the taper must be substantially longer than the waveguide is wide.

8.4. SOURCE TO FIBRE TRANSITIONS

Optical fibre communication systems with directly modulated lasers or luminescent diodes couple the fibre immediately to the source. Any transition between source and fibre serves to improve the launching efficiency. For launching power into the fundamental mode of a single-mode fibre, the source must generate a spatially coherent light wave which a suitable transition then transforms and matches to the fundamental fibre mode. Luminescent diodes emit light with no coherence in space; they are therefore not suited to excite single-mode fibres. For an acceptable launching efficiency into such fibres, even lasers must oscillate with a sufficient degree of spatial coherence. Preferably they should radiate only in their fundamental transverse order.

Gas and solid-state lasers have optical cavities with spherically curved mirrors. Their modes of oscillation have in the fundamental transverse order the field distribution of the fundamental Gaussian beam mode. Lenses or lens-like graded-index waveguide sections transform and match this distribution to the field distribution of the fundamental mode of single- and multimode fibres with stepped-index profiles as well as with graded-index cores. Launching efficiencies close to 100 per cent are possible for properly designed transformers between the spherical optical laser resonator and the fibre.

Semiconductor lasers in the form of the double heterojunction diode with stripe geometry oscillate under perfect conditions in the fundamental transverse mode of the strip-guide resonator. To transform this mode and match its field distribution to the fundamental mode field of a given fibre, we proceed as indicated in Fig.8.16. First we determine the equivalent spot radii w_x and w_y of an elliptical Gaussian beam wave which best matches the fundamental mode out of the strip-guide resonator of the laser. We then use cylindrical lenses at right angles to each other or other lens-like structures to transform the elliptical beam wave to a circular beam wave. The minimum spot radius w_0 in the waist of this circular beam wave is chosen for maximum launching efficiency of the fundamental fibre mode.

A tapered waveguide transition can also be used to transform the fundamental mode of the strip guide to the fundamental fibre mode. In the case of single-mode fibres with small index differences between core and cladding, the equivalent spot radius w_0 of the fundamental fibre mode is in the range between the corresponding spot radii w_x and w_y of the fundamental strip guide mode out of the laser. If under these conditions, the fibre is gradually flattened towards the end, its cross-section becomes elliptical there, with major and

minor axis of the core ellipse determined by the degree of
fibre end-flattening and taper stretching. In this process,
the elliptical core cross-section should be adjusted in its
size and ellipticity for its fundamental mode to match in
transverse field distribution the fundamental mode of the
laser strip.

In transitions between sources and multimode cladded-core
or graded-index fibres, excitation is usually not restricted
to the fundamental fibre mode. The transition should rather
be designed to transfer as much power as possible from the
light source into the guided modes of the fibre. The particular distribution of power between modes of the fibre is in
general not specified but a transition that launches more
power into the lower-order modes is to be preferred. To minimize the transmission loss and signal distortion in fibres
with random imperfections, the transition should preferably
excite the modes with the same power distribution as the
statistical mode of lowest loss exhibits for the respective
imperfection. But if such a special requirement need not be
met, multimode fibres can be coupled to lasers with less
spatial coherence or even to incoherent sources such as luminescent diode. The power which a properly designed transition
launches into all the guided modes of the fibre may then be
estimated from the radiation characteristic of the source
and its aperture.

Fig.8.20 shows schematically the radiating aperture of the
double heterojunction semiconductor laser in stripe geometry.

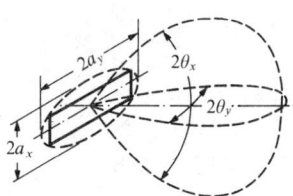

Fig.8.20. Radiating aperture and main lobe of radiation pattern of a double heterojunction laser in stripe geometry.

It has the effective width $2a_y$ and the effective height $2a_x$. The cross-section of the active strip guide in stripe-geometry lasers is up to ten times wider than its thickness. Nearly the same ratio also results for the effective width $2a_y$ to the effective height $2a_x$. The radiation pattern of the laser hence has a main lobe which usually is much wider perpendicular to the wide side of the strip than it is parallel to it. Cylindrical lenses perpendicular to each other (as in Fig.8.16) in a suitable position between the end faces of the laser and fibre allow the laser radiation to be focused on to the fibre front face, so that nearly all the radiation is incident within the numerical aperture

$$\text{N.A.} = \sin \theta_{0c} = (n_1^2 - n_2^2)^{\frac{1}{2}} \qquad (8.34)$$

of the fibre core.

According to Abbe's sine theorem in equation (8.28), an aberration-free imaging system focuses all rays out of the multimode laser on to the front face of the fibre within the core diameter $2a$ and within the acceptance angle θ_{0c} which radiate with

$$a_x \sin \theta_x \leq a \sin \theta_{0c} \simeq a(2n\, \Delta n)^{\frac{1}{2}}$$
$$a_y \sin \theta_y \leq a \sin \theta_{0c} \simeq a(2n\, \Delta n)^{\frac{1}{2}} \, . \qquad (8.35)$$

Single and double heterojunction lasers have effective radiation apertures and patterns with typically

$$a_x = 1 \ldots 3 \text{ μm} \qquad \theta_x = 10° \ldots 20°$$
$$a_y = 10 \ldots 50 \text{ μm} \qquad \theta_y = 5° \ldots 10° \, . \qquad (8.36)$$

Because these lasers radiate with some partial coherence in space, the smaller values for the lobe width θ go together with the larger aperture dimension. Under normal circumstances, both $a_x \sin \theta_x$ and $a_y \sin \theta_y$ remain smaller than 5 μm. As long as the fibre core has a product $a \sin \theta_{0c}$ which is larger than both $a_x \sin \theta_x$ and $a_y \sin \theta_y$, all the power which

FIBRE JUNCTIONS AND TRANSITIONS   731

radiates from the laser may, by proper imaging, be launched
into guided modes of the multimode fibre. In the typical case
of $n = 1 \cdot 5$ and $\Delta n = 0 \cdot 01$, the condition $a \sqrt{2n\Delta n} > 5$ μm requires
a fibre core with a diameter larger than $2a = 58$ μm.

If the conditions (8.35) are not met, part of the laser
power cannot be launched into fibre modes. To estimate the
launching efficiency, we assume a multimode laser which, over
its elliptical aperture in Fig.8.20, radiates uniformly into
an elliptical main lobe of widths θ_x and θ_y. A useful quantity to describe the radiation of an incoherent or partially
coherent source is its radiance or brightness B; it measures
the optical power which each infinitesimal part of the emitting
surface radiates into a specific direction, per unit area of
the emitting surface and per unit solid angle. Under most
general conditions, this brightness varies over the emitting
surface and also depends on the direction in space. For incoherent sources, however, it is constant over the emitting
surface and has a circularly symmetric radiation lobe that
depends only on the angle θ with respect to the normal to the
emitting surface. The total power is found when we integrate
the brightness over the full solid angle of the positive half
space and over the emitting surface. The above assumptions
of uniform radiation from the elliptical aperture of the multimode laser into an elliptical radiation lobe imply a brightness that is constant across the elliptical aperture and, in
the plane parallel to the strip width, depends on θ as

$$B(\theta) = \begin{cases} B & \theta \leq \theta_y \\ 0 & \theta > \theta_y \end{cases} \qquad (8.37)$$

while in the perpendicular plane, it depends on θ as

$$B(\theta) = \begin{cases} B & \theta \leq \theta_x \\ 0 & \theta > \theta_x \end{cases} \qquad (8.38)$$

Both angles θ_x and θ_y of the radiation lobe are assumed to
obey the condition

$$a_x \sin \theta_x = a_y \sin \theta_y . \qquad (8.39)$$

Actual radiation patterns of multimode lasers show a complicated fine structure which differs from laser to laser and also depends on operating conditions. The uniform brightness across the aperture and within the elliptical cone suffices as a model to estimate launching efficiency.

We now use two cylindrical lenses perpendicular to each other to image the elliptical laser aperture into a round aperture of radius a_s. Fig.8.21 shows source, lens, and the

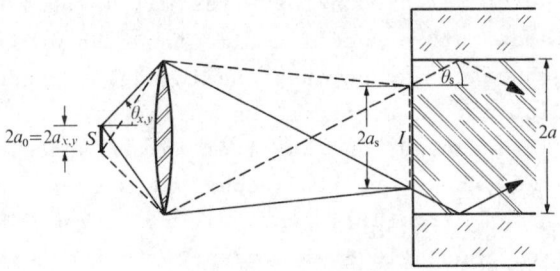

Fig.8.21. Imaging of source aperture on to fibre front face for the enhancement of launching efficiency.

image on the front face of the fibre in one longitudinal section. With equation (8.39) and according to Abbe's sine theorem (8.28), the elliptical radiation lobe of the laser aperture transforms into a round lobe of width $2\theta_s$. Both a_s and θ_s follow from

$$a_s \sin \theta_s = a_x \sin \theta_x = a_y \sin \theta_y . \qquad (8.40)$$

This round image has a constant brightness across its aperture that is circularly symmetric and depends on the propagation angle θ as

$$B_s(\theta) = \begin{cases} B & \theta \leq \theta_s \\ 0 & \theta > \theta_s . \end{cases} \qquad (8.41)$$

For the total power P_s which radiates from this round aperture, we integrate B_s over the aperture and the solid angle of the positive half-space:

$$P_s = 2\pi^2 a_s^2 B(1-\cos\theta_s) . \tag{8.42}$$

Of this power only that fraction is accepted and guided by the fibre which is incident within the core area $r < a$ at an angle θ smaller than the acceptance angle θ_{0c} of the fibre. For a graded-index fibre, θ_{0c} depends according to equation (5.47) on the radial distance r from the core centre. As long as this acceptance angle is smaller than the angle θ_s of illumination in the transformed laser aperture and as long as the core radius a is smaller than the radius a_s of this aperture, the fibre accepts the power

$$P_f = 4\pi^2 B \int_0^a \int_0^{\theta_{0c}} r \sin\theta \, d\theta \, dr . \tag{8.43}$$

Integration over the fibre acceptance angle leads to

$$P_f = 4\pi^2 B \int_0^a [1 - (1-n^2(r)+n_2^2)^{\frac{1}{2}}] \, r dr . \tag{8.44}$$

The launching efficiency is under these conditions given by

$$p = \frac{P_f}{P_s} = \frac{2}{a_s^2(1-\cos\theta_s)} \int_0^a [1 - (1-n^2(r)+n_2^2)^{\frac{1}{2}}] \, r dr \tag{8.45}$$

for $a \leq a_s$ and $\theta_{0c} \leq \theta_s$. For weakly guiding fibres with $(n^2(r)-n_2^2) \ll 1$, equation (8.45) reduces approximately to

$$p = \frac{1}{a_s^2(1-\cos\theta_s)} \int_0^a (n^2(r)-n_2^2) \, r dr . \tag{8.46}$$

For cladded-core fibres with a homogeneous core, the accurate and approximate expressions yield respectively the following launching efficiencies:

$$p = \left(\frac{a}{a_s}\right)^2 \frac{1-\cos\theta_{0c}}{1-\cos\theta_s} \approx \left(\frac{a}{a_s}\frac{\theta_{0c}}{\theta_s}\right)^2 . \tag{8.47}$$

In this case, the optical imaging of the laser aperture on to the fibre front face launches all laser output power into the fibre if $a_s \leq a$ and $\theta_s \leq \theta_{0c}$. Otherwise the launching efficiency decreases as the square of the ratios of core area to image area and of acceptance angle to illumination angle. For a graded-index fibre with a parabolic index profile, the accurate and approximate expressions yield respectively the launching efficiencies:

$$p = \frac{a^2}{a_s^2(1-\cos\theta_s)} \left\{ 1 - \frac{2}{3(n_1^2-n_2^2)} [1 - (1-n_1^2+n_2^2)^{3/2}] \right\} \simeq \frac{1}{2}(\frac{a}{a_s} \frac{\theta_0}{\theta_s})^2 ,$$

(8.48)

where $\theta_0 \simeq (n_1^2-n_2^2)^{\frac{1}{2}}$ designates the acceptance angle in the core centre. For the same difference $n_1^2-n_2^2$, the parabolic index profile guides only half the number of modes as the homogeneous core. Its launching efficiency is hence also only half that of the homogeneous core.

Within the limits $a \leq a_s$ and $\theta_{0c} \leq \theta_s$, for which equations (8.45) to (8.48) hold, we obtain the best possible launching efficiency when $a = a_s$ and $\theta_{0c} = \theta_s$. If we still maintain $\theta_{0c} \leq \theta_s$ but let $a \geq a_s$, the integration over r extends only to $r = a_s$ and the launching efficiency is

$$p = \frac{2}{a_s^2(1-\cos\theta_s)} \int_0^{a_s} [1 - (1-n^2(r)+n_2^2)^{\frac{1}{2}}]r dr \simeq \frac{1}{a_s^2(1-\cos\theta_s)} \int_0^{a_s} [n^2(r)-n_2^2]r dr.$$

(8.49)

For cladded-core fibres with a homogeneous core, these expressions reduce to

$$p = \frac{1-\cos\theta_{0c}}{1-\cos\theta_s} \simeq (\theta_{0c}/\theta_s)^2 , \qquad (8.50)$$

while for the graded-index fibre with a parabolic index profile, the launching efficiency is from equation (8.49)

$$p = \frac{1}{(1-\cos\theta_s)} \left[1 - \frac{2a^2}{3(n_1^2-n_2^2)a_s^2} \left\{ [1 - (n_1^2-n_2^2)(1-a_s^2/a^2)]^{3/2} - (1-n_1^2+n_2^2)^{3/2} \right\} \right]$$
$$\simeq (\theta_0/\theta_s)^2 [1 - \frac{1}{2}(a_s/a)^2] . \quad (8.51)$$

FIBRE JUNCTIONS AND TRANSITIONS

θ_s and a_s in these launching efficiencies depend on each other according to

$$a_s \sin \theta_s = C_s , \qquad (8.52)$$

where C_s is a characteristic quantity of the radiating source. If we let the source image on the fibre front face become smaller, the decrease in a_s will let the angle θ_s increase according to

$$\sin \theta_s = C_s/a_s . \qquad (8.53)$$

In the approximation of equation (8.51) for small angles θ_{0c} and $\theta_s \simeq C_s/a_s$, the launching efficiency is maximized by $a_s = a$. Imaging the source aperture with $a_s = a$ on to the fibre front face therefore gives the best launching efficiency for the parabolic index profile.

When power from a multimode laser is to be launched into a graded-index fibre, the image radius a_s of the source should, if at all possible, be made small enough so that the acceptance angle θ_{0c} at $r = a_s$ is still larger than the illuminating angle θ_s from equation (8.52) for the particular laser source. As long as $\theta_{0c}(a_s) \geq \theta_s$ the graded-index core will accept all the power. If, however, the source constant C_s is not small enough to meet this condition, a radius $a_c < a_s$ exists on the illuminated core face at which $\theta_{0c}(a_c) = \theta_s$. Within this radius the core accepts all the incident power, but in the annular core region $a_c < r < a_s$, the core accepts only that part of the power which is incident with an angle $\theta \leq \theta_{0c}$. The launching efficiency is under these conditions given by

$$p = \frac{a_c^2}{a_s^2} + \frac{2}{a_s^2(1-\cos\theta_s)} \int_{a_c}^{a_s} [1 - (1+n_2^2-n^2(r))^{\frac{1}{2}}] \, r dr$$

$$\simeq \frac{a_c^2}{a_s^2} + \frac{2}{a_s^2 \theta_s^2} \int_{a_c}^{a_s} [n^2(r)-n_2^2] \, r dr . \quad (8.54)$$

By evaluating the approximate expression for the parabolic index profile, we obtain

$$p \simeq \frac{\theta_0^2}{\theta_s^2}\left[1 - \frac{1}{2}\left(\frac{a_s^2}{a^2} + \frac{a^2}{a_s^2}\right)\right] + \frac{a^2}{a_s^2}\left(1 - \frac{\theta_s^2}{2\theta_0^2}\right) \quad (8.55)$$

for any source image that illuminates the core face with $a_s \leq a$ and $\theta_s \leq \theta_0$. For full illumination of the core face with $a_s = a$ but $\theta_s \leq \theta_0$, the launching efficiency into the parabolic index core reduces to

$$p \simeq 1 - \theta_s^2/(2\theta_0^2) \,. \quad (8.56)$$

It approaches 50 per cent in the limit of $\theta_s = \theta_0$.

Incoherent sources such as light-emitting diodes (LEDs) allow a more universal and accurate description of their radiation characteristic than the rather crude model that we have used here for multimode lasers. The emission from any point interior to the emitting layer is isotropic in all directions; however, due to refraction and internal reflection at the surface of the LED, the radiation pattern assumes an angular dependence.

Experimental observations indicate that LED surfaces radiate very nearly with Lambert's cosine law of diffuse emission from a plane surface. According to this law the radiance or brightness B at any point of the surface depends as $\cos \theta$ on the angle θ which the direction of radiation makes with the normal to the emitting surface. We have therefore the circularly symmetric radiation pattern

$$B(\theta) = B_0 \cos \theta \,, \quad (8.57)$$

where B_0 is the maximum radiance in the direction normal to the emitting surface. B is constant over the emitting surface. A circular plane emitter of radius a_0 and this brightness radiates the total power

$$P_0 = \pi^2 a_0^2 B_0 \,. \quad (8.58)$$

If this emitting surface is placed directly against a graded-index fibre, concentric with its core, the fibre core accepts only the power which is incident at $r < a$ and at angles $\theta < \theta_{0c}(r)$. For an emitting surface with $a_0 > a$, this power is given by

$$P_f = 4\pi^2 B_0 \int_0^a \int_0^{\theta_{0c}} r \cos\theta \sin\theta \, d\theta \, dr . \qquad (8.59)$$

Evaluating the integral over θ leads to

$$P_f = 2\pi^2 B_0 \int_0^a [n^2(r) - n_2^2] \, r \, dr . \qquad (8.60)$$

We relate this power to the total power emitted by the LED in equation (8.58) and obtain the following launching efficiency for $a_0 > a$:

$$p = \frac{2}{a_0^2} \int_0^a [n^2(r) - n_2^2] \, r \, dr . \qquad (8.61)$$

For a cladded-core fibre with a stepped index profile, this launching efficiency reduces to

$$p = (a/a_0)^2 \, (n_1^2 - n_2^2) , \qquad (8.62)$$

while for a graded-index fibre with a parabolic index profile, it evaluates to give

$$p = (a/a_0)^2 \, (n_1^2 - n_2^2)/2 . \qquad (8.63)$$

The maximum launching efficiency for the butt connection of LED and fibre results when the radiating surface just matches the core cross-section of the fibre and we have $a_0 = a$.

If with $a_0 < a$, the emitting surface is smaller than the core cross-section, and if it is again placed against the graded-index fibre concentric with its core, then the integration over r in equation (8.60) for the power into the fibre extends only to $r = a_0$, and the launching efficiency follows as

$$p = \frac{2}{a_0^2} \int_0^{a_0} [n^2(r) - n_2] \, r \, dr \, . \qquad (8.64)$$

This expression reduces for the stepped index profile to the same formula

$$p = (n_1^2 - n_2^2) \qquad (8.65)$$

as equation (8.62) for $a_0 = a$. For a butt joint of a stepped index fibre with $a_0 \le a$, the launching efficiency is hence equal to the square of the numerical aperture of the fibre core. In the case of a parabolic index profile, equation (8.64) yields

$$p = (n_1^2 - n_2^2)[1 - (a_0/a)^2/2] \, , \qquad (8.66)$$

which approaches the square of the maximum of the numerical aperture in the fibre centre when the emitting surface narrows down to a small spot at the centre of the core face.

Luminescent diodes can be driven up to a certain density of injection currents before they suffer any degradation and reduction in lifetime. This upper limit of injection current density can be higher for smaller diode areas. Luminescent diodes with a smaller emitting surface therefore radiate with more brightness. The launching efficiency of such small-area diodes into fibres with larger core cross-sections may be enhanced by imaging the diode area on to the core cross-section as in Fig.8.21. The image of the source on the fibre frontface should have a radius a_s that equals the core radius a. The brightness B_0 of the source in the direction normal to its plane surface remains the same in its image and only its radiation pattern contracts from the cosine characteristic in equation (8.57) to a characteristic which is by the inverse of the magnification a_s/a_0 narrower than the cosine characteristic. However, for the small acceptance angles θ_{0c} of weakly guiding fibres, this change in brightness remains insignificant. With proper imaging according to $a_s = a$, we therefore again obtain for the launching efficiencies equations (8.61), (8.62), and (8.63), in which we now have $a > a_0$. The improvement compared to equations (8.61), (8.65), and (8.66) for the butt connection is as large as the ratio $(a/a_0)^2$ of core to source area for the stepped index profile, but is still substantial for graded-index profiles.

It even seems from equations (8.62) and (8.63) that no limit is set to the launching efficiency, if only a/a_0 is large enough. Actually, however, since these equations as well as equation (8.61) neglect the change in radiation pattern due to the imaging, they apply only to moderate magnification factors a/a_0. For larger magnification, the launching efficiency will also be limited by aberrations in the imaging system. In addition, this system will introduce some loss which also increases with the magnification factor.

A simple system to image the LED-area on to the face of the fibre core consists of the lens in Fig.8.21 between the LED and the front face of the fibre. To contract all the useful LED radiation, the lens should be quite wide and close to the LED. Although this system is simple in principle, its adjustment for the maximum launching efficiency is difficult. Furthermore, the wide aperture which the lens must have and its short focal length for a location close to the LED require the lens to have such a large numerical aperture that its aberrations reduce the launching efficiency substantially.

Another arrangement that improves the launching efficiency when the source aperture is smaller than the acceptance area of the fibre front-face is shown in Fig.8.22. It uses a lens directly attached to the fibre front face. For a small

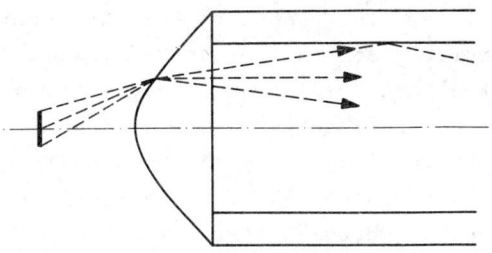

Fig.8.22. Hyperbolic lens on fibre front-face with representative rays from a source aperture in the focal plane.

source area placed at short distance from the lens on the fibre axis, this lens increases the acceptance angle of the fibre to a certain extent. However, it cannot provide the imaging which is required for the maximum launching efficiency.

In order that the lens on the fibre front-face accepts a substantial amount of the wide angle radiation from the small source, it must have a large numerical aperture. Since its effective diameter extends only to the diameter of the fibre core, its focal length must be quite short and hence its surface extremely convex. A spherical lens of these characteristics has large aberrations, which limit the launching efficiency. Better launching efficiencies are obtained with this arrangement when the lens on the fibre front-face presents a hyperbolically curved surface to the light source on the fibre axis (Kurokawa and Becker 1975). Any light ray from the focal point of this hyperbolic lens will be refracted into a direction parallel to the fibre axis. Light rays which originate off axis from a cross-section through the focal point will be refracted into directions with small propagation angles θ as long as their radial offset in the focal plane is not too large. Thus the fibre core will accept nearly all the power that is incident on the hyperbolic lens from small source apertures near the focal point.

Lenses on the front face of the fibre are, of course, not only suited for LED to fibre connections, but also for improving the coupling efficiency between lasers and fibres. When the radiation from multimode lasers is to be launched into multimode fibres, the situation is quite similar to the connection of a small-aperture LED with the multimode fibre. The numerical aperture of the lens should again be large enough to accept all the radiation from the laser. If the ellipticity of the laser aperture is too large for a hyperbolic lens of rotational symmetry to refract all rays within the acceptance angle of the fibre, a hyperbolic cylindrical lens might be considered, or better still a hyperbolic lens with sufficient astigmatism to compensate for the ellipticity of the source aperture.

REFERENCES

Gloge, D. and Marcatili, E.A.J. (1973). Multimode theory of graded-core fibres. *Bell Syst.tech. J.* 52, 1563-78.

———, Smith, P.W., Bisbee, D.L., and Chinnock, E.L.(1973). Optical fibre end preparation for low loss splices. *Bell Syst.tech. J.* 52, 1579-88.

Guttmann, J. and Krumpholz, O. (1973). Theoretische und experimentelle Untersuchungen zur Verkopplung zweier Glasfaser-Lichtwellenleiter. *Wiss.Ber.AEG-Telefunken* 46, 8-15.

———, ——— and Pfeiffer, E. (1975). A simple connector for glass fibre optical waveguides. *Archiv Elektron. & Übertragungstech.* 29, 50-52.

Heyke, H. (1970). Launching of fibre modes by Gaussian beams. *Archiv Elektron. & Übertragungstech* 24, 521-2.

Kogelnik, H. (1964). *Coupling and conversion coefficients for optical modes*. Proceedings of the Symposium on Quasi-Optics. Polytechnic Press, Brooklyn.

——— (1965). Imaging of optical mode-resonators with internal lenses. *Bell Syst. tech.J.* 44, 455-94.

Kurokawa, K. and Becker, E.E. (1975). Laser fibre coupling with a hyperbolic lens. *IEEE Trans. MTT* 23, 309-11.

Marcuse, D. (1970a). Radiation losses of the dominant mode in round dielectric waveguides. *Bell Syst.tech. J.* 49, 1665-93.

——— (1970b). Excitation of the dominant mode of a round fibre by a Gaussian beam. *Bell Syst.tech. J.* 49, 1695-1703.

——— (1974). *Theory of dielectric optical waveguides*. Academic Press, New York and London.

Miller, C.M. (1975). Loose tube splices for optical fibres. *Bell Syst.tech. J.* 54, 1215-25.

Neumann, E.G. and Weidhaas, W. (1976). Loss due to radial offsets in dielectric optical waveguides with arbitrary index profiles. *Archiv Elektron. & Übertragungstech.* 30,

Silver, S. (1949). *Microwave antenna theory and design*. McGraw Hill, New York.

Williamson, D.E. (1952). Cone channel condenser optics. *J.opt.Soc.Amer.* 42, 712-15.

INDEX

Abbe's sine theorem, 726
aberration, 726, 738
absorption, 23, 29, 402
  coefficient, 9-10, 11-12
  edge, 29, 36
  guided mode, 106
  intrinsic, 30
acceptance angle, 308, 733
acousto-optic interaction, 276
active component, 267, 277
Airy
  equation, 160
  function, 160, 480
alkali
  lead silicate, 650, 668
  oxide, 649
aluminosilicate, 668
aluminium arsenide, 277
amplification, 112
analytic continuation, 371
angular frequency, 9
annular fibre, 321
apparent reflector, 22, 105
argon
  laser, 274
  plasma, 275
aspect ratio, 211, 235
astigmatism, 722, 740
asymmetric mode, 316
asymmetry parameter, 81, 82
attenuation, 460
  constant, 26, 106, 461
    mean, 128
  guided mode, 401
  leaky-mode, 376, 462
  vector, 11
autocorrelation, 125, 129, 566
  distance, 129, 566
  function, 45, 564
Avogadro's number, 28
axially symmetric mode, 316, 333

bandwidth, 113
beam, 139
  Gaussian, 148
  mode, 192
  parasitic, 144
  shift, 22, 105
  substrate, 142, 145
  wave, 700
  waveguide, 191
beat wavelength, 345, 563

bending, micro, 569
bends, 157
Bessel function, 185, 211, 312
Boltzmann's constant, 26
bonds, dangling, 31
boric oxide, 650
borosilicate, 271, 651, 665
boundary,
  concave, 158
  condition, 14
  imperfect, 116
Bragg
  angle, 146
  effect, 146
brightness, 731
bulge
  guide, 238
  mode, 246
  parameter, 242
    modified, 247
bulk
  absorption, 112
  loss, 112
  scattering, 115
buried
  layer, 172, 183, 271
  strip, 197
burner,
  diffusion, 659
  ring, 659
Butterworth power spectrum, 585
butt joint, 699, 709, 211

cable
  connector, 695
  design, 693
  strength, 690
  cabling of fibres, 682
caustic, 299, 461
characteristic equation, 63, 177,
  202, 220, 300, 313, 315, 324, 391
chemical
  etching, 285
  polishing, 268
  vapour
    deposition, 654
    transport, 283
circuit,
  hybrid, 277
  monolithic, 277
  pattern, 272
  planar, 268

INDEX   743

circular
  harmonic expansion, 211
  harmonics, 210
circumferential order, 300, 317
cladded-core fibre, 291
cladding
  mode, 292, 418, 458
    absorption, 403
  parameter, 418
  unlimited, 321
coating bath, 683
coherence, 44
  time, 46
coherent light, 356
collapse of tube, 658
communication system, optical, 290
complete power transfer, 257
complex wave number, 402
compound
  glass, 668
  mode number, 573, 621
compressibility, isothermal, 26
computer-generated solution, 210
concave interface, 168
confinement, transverse, 189
confluent hypergeometric function, 180
confocal lens sequence, 196
conformable photo-mask, 273
connector, 709
  detachable, 699
converging film lens, 190
convex interface, 168
core mode, 292
correlation
  distance, 24, 127, 567
  exponential, 135
coupled
  line equation, 558
  power equations, 602
  strips, 264
  wave equations, 245, 553
  waveguides, 252
coupler
  length, 258
  directional, 252
  grating, 138
  prism, 148
  waveguide, 253
coupling
  coefficient, 254, 265, 558, 571, 580
  distributed, 149
  efficiency, 709
  factor, 568
  length, 610, 636, 645
    effective, 255
  loss, 258

coupling, cont.
  next-neighbour, 574
  parasitic, 253
crossed cylindrical lenses, 721
cross talk, 256, 388, 683
crystal lattice, 277
curing furnace, 680
current coefficient, 557
curvature,
  concave, 158
  coupling, 568
    coefficient, 573
  effect, 157
  loss, 157, 569, 575, 644
  mode, 579
  of fibre, 568
  power spectrum, 642
curved
  interface, 158
  slab, 157
cushion, 690
cut-off, 71
  approximation, 332
  condition, 231, 279, 336
  limit, 331
cylinder function, 312
cylindrical lens, 721

degeneracy, 334, 341, 351
degenerate mode, 456
delay, 361
  difference, 252, 433, 507
  dispersion, 426
  factor, 414
  group, 34
  phase, 34
  spread, 339, 366, 434, 442, 519, 528
  time, 235
depletion layer, 280
devitrification, 653
dielectric
  film, 58
  loss, 409
  rod waveguide, 320
  slab, symmetrical, 83
  tube waveguide, 320
  waveguide, round, 319
  wire, 319
differential equation of the parabolic cylinder, 180
diffraction, 190
  loss, 194
  order, higher, 144
diffused wave guide, 280
diffusion
  approximation, 628
  coefficient, 504

744 INDEX

diffusion, cont.
  depth, 280
  equation, 504, 625
  time, 504
Dirac's delta function, 140
direct semiconductor, 277
directional coupler, 252, 275
directivity, 258
dispersion
  characteristic, 303, 357
  coefficient, 53, 415, 618
  frequency, 36
  of LP-mode, 357
  plane-wave, 34
dominant
  mode, 214
  rib guide mode, 235
double
  crucible, 668
  hetero-junction diode, 728
  rib guide, 234
  step fibre, 420
drawing process, 660
dye laser, thin-film, 276

eccentric, 717
effective
  coupling length, 255
  electron mass, 277
  index, 314, 340, 516
  mode radius, 413
  refractive index, 114, 190, 414
  width, 103, 168
efficiency, launching, 146
EH-mode, 326
eigenvalue equation, 63, 108
eikonal equation, 174
elastic modulus, 641
electron concentration, 278
electronic states, 29
electro-optic material, 267
elliptical
  integral, 526
  polarization, 349
  wave beam, 721
embedded strip, 197
emission, 112
energy gap, effective, 30
ensemble average, 564
epitaxial
  growth, 279
    by melting, 284
    in flux, 285
  layer, 279
  regrowth, 286
epoxy
  film, 270
  resin, 270
Euler's constant, 325

evanescent,
  field region, 162
  wave, transversely, 11, 13, 143
E-wave, 13, 64, 94
excitation
  coefficient, 95, 315, 323
  fibre, 385
excitonic
  absorption, 37
  levels, 30
exponential profile, 184
extraordinary refractive index, 283
extruder, 679

fabrication, pattern, 268, 273
fibre
  annular, 321
  bundle, 681
  cable, 569, 681
  cabling, 682
  coating, 682
  connector, 699
  curvature, 571
  drawing, 659
  excitation, 385
  fabrication, 649
  graded-index, 445
  imperfection, 292, 388, 542
  joint, 700
  liquid-core, 444
  mode, 309
    excitation, 700
  model, 310
  packaging, 685
  parameter, 300, 314, 340, 453
  pulling, 670
  ribbons, 695
  splices, 694, 699
  taper, 723
field
  analysis, 179
  components in fibre, 322
  distribution, 346
    of LP-mode, 355
  solutions, 92
film-beam mode, 193
film
  deposition, 269
  dielectric, 58
  lens, 189
    converging, 190
    guide, 191
  modes, 61, 62
  parameter, 69, 185, 204
films
  sputtered, 116
  symmetrical, 58
film-wave beam, 191

INDEX

film waveguides, 58
flexural rigidity, 641
fluorine, 651
focal length, 192, 720
focusing, periodic, 189
Fourier
  integral, 40
  spectrum, 42
  transform, 39
four-layer
  film guide, 216
  structure, 153
fracture, 714
free carrier absorption, 279
  concentration, 277
frequency,
  fundamental, 145
  in space, 127
  parameter, 69
  spatial, 128
Fresnel coefficient, 164
frustration of total reflection,
    150, 167, 296, 305, 370
fundamental
  film mode, 279
  mode, 77, 90, 152, 214, 317,
    356, 412, 469, 471
    attenuation, 423
    of bulge, 245
furnace,
  electric, 656
  resistance-heated, 656
fused silica, 27, 30, 271, 338,
    649

gain, 112
  constant, 113
GaAlAs, 281
gallium arsenide, 277
gap, tapered, 155
garnet film, 285
Gaussian
  autocorrelation, 46
  beam, 148, 719
    wave, 703
  dip, 666
  distribution, 189, 194, 271
  field, 721
  function, 469
  intensity distribution, 189
  pulse, 42, 634
  response, 635
  time function, 614
geometric optics approximation, 173
germanium, 652
getter, 276
glass
  fibre, 290

glass cont.
  for fibres, 649
Goos-Haenchen shift, 22, 105, 168
graded index, 171
  fibre, 445
  film, 172
  slab, 722
grating
  coupler, 138, 275
  periodic, 139
Green's theorem, 555
groove alignment, 719
grooves,
  random, 136
  rectangular, 139, 144
  transverse, 136
group
  delay, 235, 251, 361, 432, 508
  index, 51, 362
  velocity, 34, 38
growth rate, 270
guided mode
  attenuation, 401
  power, 407

Hamiltonian, 193
Hankel function, 211, 312
HE-mode, 326
Hermite polynomial, 182, 194
heterojunction laser, 282
H-mode, 63
hollow-tube waveguide, 317
homojunction laser, 280
Huygens
  principle, 191
  source, 192
H-wave, 13, 94
hybrid
  character of strip mode, 200
  circuit, 277
  integration, 268
  mode, 316
  planar circuits, 268
hydroxyl, 662
hydroxyl-ion, 32
hyperbolic
  lens, 740
  -secant profile, squared, 183
hypergeometric
  function, 244
  series, 664

impedance of film mode, 217
imperfection, fibre, 552
implantation energy, 284
impulse
  function, 535
  response, 534, 617, 634, 645

impurity, 32
  absorption, 33
incoherent source, 431
index
  difference, relative, 178, 339
  dip, 658, 665
  gradient, 176
  profile, 172, 453
    fourth-order, 523
    optimum, 531
indiffusion, 283
induction heating, 657
inhomogeneous medium, 173
injection
  current, 113, 738
  laser, 277
integrated optics, 58
integration
  hybrid, 268
  monolithic, 268, 277
intensity distribution, 354
interaction,
  strong, 607
  weak, 606
intermodal pulse broadening, 545
intramodal pulse broadening, 545
intrinsic loss, 31
ion
  -beam etching, 285
    machine, 276, 285
  bombardment, 269, 276
  exchange, 271, 671
  exchange length, 673
  implantation, 271, 275, 284
  migration, 271, 275
iron-ion, 33

jacket, 292
  lossy, 292, 406
  metallic,

Kevlar, 684
Kirchhoff's diffraction theory, 702

Laguerre polynomial, 468
Lambert's cosine law, 736
Laplace transform, 629
Laplacian, 10, 311, 468
laser, 31, 44, 356, 549, 728
  active material, 267
  beam, 146, 189, 357
    exposure, 274
  CO_2, 659
  double hetero, 112
  hetero, 112
  output beam, 357
lattice
  constant, 280

lattice, cont.
  defect, 279
  displacement, 279
  match, 280
  vibrations, 37, 279
launching, 338
  efficiency, 146, 700, 703, 733
  layer, buried, 172, 183, 271
lead silicate film, 270
leaky
  mode, 142, 281, 296, 304, 369, 458, 625
    attenuation, 151, 155, 307, 370, 376, 462
    fibre, 420
  wave, 60, 151, 229, 246, 394, 420
lens
  film, 189
  guide, 189
  transformer, 720
lift-off, 276, 285
  metal, 274
light
  beam, 19
  emitting diode, 431, 549, 736
linear polarization, 352
linearly polarized mode, 352
line
  form, normalized, 48
  shape, 536
  source, effective, 132, 139
liquid
  -core fibre, 444
  film, 276
  guide, 276
  phase epitaxy, 284
lithium, 271
  niobate, 267, 283
  tantalate, 283
loading-strip pattern, 287
long-distance transmission, 290
Lorentz
  curve, 136
  reciprocity theorem, 260
loss
  factor, 435
  penalty, 634
  spectrum, 661
lossy jacket, 406, 423, 439
LP-mode, 352, 355, 388
luminescent diode, 31, 282, 738

magnetooptic material, 267
magnification, 739
manganese, 276
matching
  fluid, 700
  lens, 720

material
  dispersion, 46, 426, 511, 546
    inhomogeneous, 512,
  laser-active, 267
  electro-optic, 267
  magnetooptic, 267
  photoelectric, 267
matrix, statistical, 602
Maxwell's equations, 9, 173, 310
meriodional ray, 293
metal lift-off, 274
metallic
  ions, 29
  jacket, 319
  waveguide, 317
microbending, 569, 638, 685
  loss, 575
microcrack, 682
microwave plasma, 565
mode
  attenuation, 434
  conversion, 560
  count, 455, 514
  coupling, 553
  delay factor, 363
  dispersion, 362, 414, 426, 546
  expansion, 234
  family, 326
  field distribution of, 355
  index, 434
  leaky, 142
  mixing, 620, 716
  number, compound, 573
  radius, 469
    effective, 413, 577/578
  size, 215
  suppression, 440
  volume, 439
  width, effective, 105, 131
modified Hankel, 414
modulation,
  amplitude, 39
  phase, 39
modulator, 276
  optical, 267
modulus of elasticity, 687
molecular
  beam epitaxy, 286
  weight, 28
mole fraction, 28, 281, 510
moment of inertia, 642
monolithic
  circuit, 277
  integration, 268, 277
monomer, 270
multimode
  fibres, 430
  laser, 431, 730

multimode, cont.
  sources, 431

Navier-Stokes equation, 683
neckdown region, 659
neodymium, 267
Neumann number, 367
next-neighbour coupling, 574, 621
non-meridional ray, 295
non-uniform plane wave, 11, 18
number of leaky modes, 383
numerical aperture, 61, 293, 308,
                    340, 431, 463
nylon jacket, 684

offset, 703, 707, 711
optical
  communication,
    long-distance, 649
    system, 290
  isolation, 683
  modulator, 267
opto-electronic coefficient, 113
ordinary index, 284
organo-silicon film, 270
orthogonality, 555, 702
orthogonal polarization, 302, 353
orthonormality, 556, 604, 613, 627
out-diffusion, 283
overlap integral, 147, 156
oxide film, 270
oxygen-hydrogen burner, 656

packaging of fibres, 641
'parabolic'
  bulge, 243
  index profile, 454, 459, 465, 572
  profile, 178
    truncated, 472
parallel-plate waveguide, 209
parasitic coupling, 253
paraxial approximation, 176
passive
  circuit, planar, 268
  component, 267
  waveguide, 268
pattern,
  circuit, 272
  fabrication, 268, 273, 285
periodic focusing, 189
perturbation analysis, 110, 153
phase
  condition, 177, 298, 299
  constant, 65, 452
  front, 173
    transformer, 192
  integral, 177, 299
  parameter, 70, 185, 204, 314, 332,
             340, 454

phase, cont.
  separation, 32, 653
  synchronism, 256, 346
  vector, 11
phosphorus pentoxide, 652
phosphosilicate, 653, 665
photoconductor, 283
photodetector, 267
photoelastic coefficient, 26
photoelectric material, 267
photo-locking material, 275
photo-mask, 273
  conformable, 273
photon, 29
  energy, 30
photo-resist, 273, 285
planar
  circuit, 268, 723
    hybrid, 268
  passive circuit, 268
  waveguide technology, 267
Planck's constant, 29, 37
plane
  boundary, 13
  wave,
    non-uniform, 11, 18
    solution, 10
    uniform, 8
plasma, 277
  frequency, 278
  polymerization, 269, 270
plastic
  -clad fibres. 678
  cladding, 678
  coating, 684
  jacket, 682
platinum, 669
Pockel's electro-optic effect, 283
polarizability, electronic, 35
polarization,
  elliptical, 349
  linear, 352
  orthogonal, 302, 353
  uniform, 352
polishing,
  chemical, 268
  mechanical, 268
polymer
  clad fibres, 678
  cladding, 678
  film, radiation sensitive, 273
  pattern, 273
  radiation-sensitive, 272
polypropylene, 689
polyurethane, 270, 690
power, 100, 366
  coupling, 639
  diffusion, 620

power, cont.
  flow, 366
    density, 25
    in fibres, 598
  formula, 102
  guided by a mode, 367
  -law profile, 512
  -series analysis, 487
  spectral density, 643
  spectrum, 42, 46, 127, 564
    of curvature, 641
  transfer, complete, 257
  transients, 610
  transmission coefficient, 164
Poynting vector, 11, 366, 409
preform, 657
prism
  beam, 155
  coupler, 148
profile
  dispersion, 512, 516
  function, 453, 510
    parabolic, 454
  index, 172
  modes, 176
  symmetric, 172
proton
  bombardment, 271
  energy, 280
pulse
  broadening, intermodal, 545
    intramodal, 545
  spreading, 617
  tail, 588
  widening, 43
  width, 42
    root-mean-square, 541, 635

quantum efficiency, 113
quarter-wave layer, 701
quartz,
  fused, 37
  -glass, 654
  synthetic, 654
quasi mode, 575

radial
  attenuation parameter, 314
  order, 300, 317
  phase parameter, 313
radiance, 731
radiating
  aperture, 729
  mode, 291, 458
  wave, 161
radiation
  characteristic, 729
  lobe, 731

radiation, cont.
 mode, 59, 176
 pattern, 730
 sensitive polymer film, 273
 two-dimensional, 134
radiative transition, 112
raised strip, 197
random imperfection, 563
ray
 analysis, 302, 445
 equation, 176, 445
  paraxial, 178
 optics approach, 173
 picture of core mode, 292
Rayleigh
 law, 115, 131
 -Ritz method, 487
 scattering, 26, 402, 662
reciprocity, 118, 140, 146, 150
reconversion, 560
rectangular
 dielectric waveguide, 208
 strip guide, 198
 waveguide, 208
reflection, 8, 17
 coefficient, 16, 219
refocusing, 190
refraction, 17, 15
refractive index, 9
 effective, 114, 190, 414
repeater, 290
resonance frequency, 35
 effective, 36
retention length, 688
rib
 guide, 225
 parameter, 227
ridge guide, 225
rigidity, flexural, 687
rock crystal, 654
rod-in-tube method, 676
roughness, 130
 surface, 129

sapphire substrate, 282
scanning electron
  beam lithography, 274
  microscope, 272
scattering, 23
 guided mode, 114
 loss, 23, 214
 surface, 138
scoring blade, 717
semiconductor,
 direct, 277
 laser, 549, 728
separation
 condition, 11, 294, 450

separation, cont.
 constant, 311
shadow mask, 286
signal distortion, 39, 537, 586
silicate glass, 29, 338
silicone
 -clad fibre, 680
 resin, 679
single
 crystal component, 268
  semiconductor layer, 282
 material
  bulge, 243
  rib guide, 229
  waveguide, 229
 mode
  bulge, 252
  condition, 250
  fibre, 411
  guidance, 248
  guide, 412
  limit, 237
  operation, 500
  propagation, 233
  range, 338
skew ray, 295
slab
 coupling, 264
 curved, 157
 dielectric, 58
 guide, 83
 -supported rib, 230
sleeve, 718
Snell's law, 15, 176, 293, 447
soda
 aluminosilicate, 650
 borosilicate, 650, 653, 668, 671
 lime silicate, 29, 30, 650, 653, 668
softening
 point, 650
 temperature, 27
solution deposition, 275
space
 beam, 145
 harmonic, 145
 mode, 59/60, 292
spatial frequency, 562
splice, 709
splicing, 338
spontaneous emission, 44
spot
 radius, 469
 size, 148, 195 196
  effective, 577, 581
sputter etching, 276, 285
sputtering, 269
 tank, 269
 yield, 269

750 INDEX

squared hyperbolic-secant profile, 183
standing wave pattern, 298
stationary
  distribution, 628, 634
  form, 490
  response, 636
  state, 605
statistical
  matrix, 602
  mode, 603, 628
    attenuation, 606
steady-state power flow, 598
stiffness, 687
stimulated emission, 44, 112
strain relief, 690, 715/6
stranding, 690
strength member, 691
strip
  buried, 197
  embedded, 197
  guide, 197,
    coupler, 264
    rectangular, 198
  -loaded film, 214
  mode, 199, 200
  parameter, 203
  raised, 197
stripe geometry, 728
strips, coupled, 264
strong interaction, 607, 619
substrate
  beam, 145
  modes, 60
superluminescent diode, 431
surface
  scattering, 138
  roughness, 129
  variation, sinusoidal, 140
susceptibility, electric, 35

taper,
  exponential, 724
  waveguide, 723
Taylor expansion, 41
Teflon, 678
temperature,
  absolute, 26
  softening, 27
TE
  -mode, 63
  -wave, 13
tensile
  strength, 684
  stress, 683
thallium, 671
thin-film light deflector, 189
tight coupling, 257

tilt, 703, 708
time
  -harmonic field, 9
  moment of impulse response, 541
titanium, 276
  dioxide, 653
TM-wave, 13, 64
total
  number of modes, 306, 307
  reflection, 15, 18, 60, 199, 293, 452
    frustrated, 296
transient state, 610
transition
  between source and fibre, 727
  element, 284
  optical, 719
  tapered, 723
transmission
  coefficient, 16
  factor, 193
  medium, 290
transverse
  attenuation parameter, 67, 329
  confinement, 189
  electric mode, 333
  electromagnetic mode, 348
  magnetic mode, 334
  order, 302
  phase parameter, 67, 329
  wave
    equation, 241, 311
    number, 18
    parameter, 69
tunneling, 167
  of power, 401
turning point, 161, 176, 240, 304, 450

ultraphosphate, 267
ultraviolet light, 29
uniaxial crystal, 283
uniform
  plane wave, 8
  polarization, 352

vacuum evaporation, 269
vapour phase epitaxy, 286
variational analysis, 487
varnish, 270
vector wave equation, 476
velocity,
  group, 34, 38
  phase, 11, 34, 296
vibrational state, 29
voltage coefficient, 557

wave
  admittance, 17
  beam, elliptical, 721
  coupling, 560
  equation, 10, 179, 310
    scalar, 466
  -front, 174
  function, 175
  -guide coupler, 253
  junction, 700
  laser, 112
  parameter, 221
  pattern, 286
  technology, planar, 267
  impedance, 16, 39, 323
    intrinsic, 353
  number, 10
  vector, 10

weak
  coupling, 256
  interaction, 606
weakly
  guided mode, 340
  guiding fibre, 336
    with jacket, 388
  film, 67
  strip, 205, 213
    raised strip, 205
weight factor, 691
W-fibre, 420, 583
whispering-gallery mode, 169
wire, dielectric, 319

yttrium-aluminium garnet, 267

zinc oxide film, 282